T0192392

Aromatic $C(sp^2)$–H Dehydrogenative Coupling Reactions

Aromatic C(sp^2)–H Dehydrogenative Coupling Reactions

Heterocycles Synthesis

Bagher Eftekhari-Sis

CRC Press
Taylor & Francis Group
Boca Raton London New York

CRC Press is an imprint of the
Taylor & Francis Group, an **informa** business

CRC Press
Taylor & Francis Group
6000 Broken Sound Parkway NW, Suite 300
Boca Raton, FL 33487-2742

First issued in paperback 2021

ISBN-13: 978-0-367-34729-1 (hbk)
ISBN-13: 978-1-03-208599-9 (pbk)

Visit the Taylor & Francis Web site at
http://www.taylorandfrancis.com

and the CRC Press Web site at
http://www.crcpress.com

To my wife Maryam and my children Elyar and Elshan

Contents

Preface

Heterocycles are an important class of organic compounds, accounting for nearly one-third of the publications of the last decade in the field of heterocyclic chemistry. Actually, two-thirds of organic compounds are heterocyclic compounds. Heterocycles are found in great quantities in natural products, and they are present in the most important molecules required for life such as DNA, RNA, chlorophyll, and heme. Moreover, they are present as a key scaffold in a wide variety of drugs, vitamins, synthetic pharmaceuticals, and agrochemicals. Most heterocyclic compounds exhibit biological activity, including anti-HIV, antitumor, antibiotic, anti-inflammatory, antidepressant, antiviral, antimalarial, antimicrobial, antidiabetic, antibacterial, antifungal, fungicidal, herbicidal, and insecticidal properties. Also, they have important applications in organic synthesis as organocatalysts, synthetic intermediates, chiral auxiliaries, and metal ligands in asymmetric catalysis. Sensory applications of the heterocyclic compounds have also been strongly reported. Consequently, the development of new efficient methods to synthesize heterocycles is of considerable interest to synthetic and medicinal scientists.

Direct functionalization of the unreactive aromatic $C(sp^2)$–H bond to $C(sp^2)$–Y (Y = C, N, O, S, or P) via dehydrogenative or oxidative coupling reactions have recently gained an increased interest and have become one of the most valuable and straightforward approaches to construct different kinds of heterocyclic compounds. However, C–H bond activation/functionalization still presents many challenges to be overcome. The opening contributions for $C(sp^2)$–H bond conversion are conveyed by palladium and rhodium catalysts. Recently, different metal catalysts, including Cu(II), Pd(II), Rh(III), Co(II), Ru(III), and Ir(III), etc., have been used as catalysts for activation and direct functionalization of the $C(sp^2)$–H bond, in which metal C–H bond activation occurred by the formation of a metal–carbon bond, with subsequent generation of cyclometalated species via oxidative addition to electron-rich metal centers or σ-bond metathesis. The presence of a directing group possessing an N, O, or S heteroatom is necessary in both intramolecular and intermolecular dehydrogenative coupling reactions in order to construct heterocyclic systems. These kinds of directing groups led exclusively to C–H activation and functionalization at the *ortho* position.

This book has the aim of covering the literature, showing the distribution of publications involving metal-catalyzed dehydrogenative or oxidative coupling reactions of aromatic $C(sp^2)$–H bonds to construct heterocyclic compounds.

Dehydrogenative coupling reactions in heterocycle synthesis occurred in different pathways. Frequently, in the intramolecular dehydrogenative coupling reactions, either $C(sp^2)$–$C(sp^3)$, $C(sp^2)$–$C(sp^2)$, and $C(sp^2)$–$C(sp)$ bonds or $C(sp^2)$–Y bonds were formed, in which the Y heteroatom containing the substitution on the benzene ring acts as the directing group to functionalization at the *ortho* position. The intermolecular cross-dehydrogenative-coupling (CDC) process is assumed to be conducted via four different pathways. In a pathway, coupling of heteroatom-substituted benzene with an alkene or alkyne afforded five-membered heterocycles, in which two carbons of alkene or alkyne were incorporated in the formation of heterocycles. In other pathways, CDC of the heteroatom bearing substituted benzene with an alkene or alkyne could occur in two different approaches, in which the alkene or alkyne moieties provided one or two carbon atoms of the final obtained heterocycles. Carbene precursors, diazocompounds, were also used in CDC reactions, leading to heterocyclic compounds, providing one carbon atom of the heterocycle.

Intramolecular CDC reactions regularly took place by directing group-assisted cyclometalation at the *ortho* position, followed by reductive elimination, while in the case of intermolecular CDC reactions, after cyclometalation at the *ortho* position of a directing group, a migratory insertion of a bond of another molecule occurred with subsequent reductive elimination of the metal catalyst. However, a few of the CDC processes occurred via single electron transfer and finally hydrogen abstraction.

Author

Prof. Bagher Eftekhari-Sis obtained his PhD from Sharif University of Technology under the supervision of Prof. Mohammed M. Hashemi in 2009 and then joined the Chemistry Department of the University of Maragheh. His research field involves the synthesis of heterocyclic compounds, supported catalysts, and polymeric and nano-materials. He has some reviews in the fields of organic transformations, especially in the heterocycle synthesis trends in the leading chemistry journal *Chemical Reviews*.

1 Five-Membered *N*-Heterocycles

1.1 INTRODUCTION

Five-membered nitrogen-containing heterocycles are important classes of heterocycles found in a number of biologically important molecules, numerous natural products, pharmaceuticals, and bioactive molecules such as antibacterial, antifungal, and cytotoxic. They also serve as neuroexcitatory agents, antibiotics, anti-HIV, anticancer, antipsychotic, antiviral, antihypertensive, immunomodulator, and glycosidase inhibitors (Eftekhari-Sis, Zirak, and Akbari 2013, Eftekhari-Sis, and Zirak 2014). They have a wide range of uses as organocatalysts, sensors (Eftekhari-Sis, and Ghahramani 2015, Eftekhari-Sis, Malekan, and Younesi Araghi 2018), building blocks in organic synthesis (Eftekhari-Sis et al. 2010, Khalili et al. 2013), and chiral auxiliaries and ligands for asymmetric syntheses. However, various types of one-step or multi-step synthesis of these types of heterocycles have been reported in the literature; the development of new methodologies to construct five-membered *N*-heterocycles with different functional groups are of interest to synthetic and medicinal scientists (Eftekhari-Sis, Zirak, and Akbari 2013, Khalili et al. 2008, Eftekhari-Sis and Zirak 2014, Eftekhari-Sis and Vahdati-Khajeh 2013, Eftekhari-Sis and Zirak 2017, Zirak and Eftekhari-Sis 2015, Eftekhari-Sis, Akbari, and Amirabedi 2011, Saraei et al. 2014, Amini et al. 2017).

Cross-dehydrogenative-coupling (CDC) (Gandeepan et al. 2018, Shaikh and Hong 2016) processes are widely used in the synthesis of various types of heterocyclic compounds. In this chapter, intramolecular and intermolecular CDC of aromatic $C(sp^2)$–H bonds are discussed to construct five-membered *N*-heterocyclic compounds, including indoles, oxindoles, isoindoles, carbazoles, indazoles, benzimidazoles, and benzotriazoles, etc.

1.2 INDOLES

A picolinamide directing group-controlled, Cu-catalyzed approach to indoles was described by Yamamoto et al. Treatment of enamides **1** with $Cu(OPiv)_2$ (30 mol%) in the presence of MnO_2 (2 equiv) and PivOH (1.5 equiv) in *N,N*-dimethylformamide (DMF) under N_2 atmosphere and microwave irradiation resulted in the construction of 2-aryl (or alkyl) substituted indoles **2** in 51–91% yields. Enamides derived from thiophene and benzothiophene afforded corresponding thiophene-fused pyrroles **2a,b** in lower yields (5–16%). Reactions initiated by ligand exchange of $Cu(OPiv)_2$ with the picolinamide enamide moiety of **1** generated the *N,N*-bidentate-coordinated Cu complex **I**, which underwent C–H cleavage to give **II**, followed by disproportionation with additional $Cu(OPiv)_2$, affording a Cu(III) species **III**. Reductive elimination of **III** generated indole framework **IV**, which underwent spontaneous hydrolysis to furnish *NH*-indole **2**. CuOPiv formed in situ was reoxidized by MnO_2 into $Cu(OPiv)_2$ to complete the catalytic cycle (Scheme 1.1) (Yamamoto et al. 2017).

3*H*-indole derivatives **5** were synthesized by Cu(II)-mediated intramolecular aromatic C–H dehydrogenative coupling reaction of 3,3-disubstituted *N*-aryl-enamino esters **3**. Reactions were conducted in mesitylene under air atmosphere, catalyzed with an equivalent amount of $Cu(2\text{-ethylhexanoate})_2$, and corresponding 3*H*-indoles **5** were obtained in 42–80% yields. Reactions with 3-monosubstituted *N*-aryl-enamino esters **3** afforded 1*H*-indoles **4**, via intramolecular dehydrogenative coupling, followed by tautomerization in 34–69% yields (Drouhin and Taylor 2015). A similar approach for the synthesis of indoles was reported using *tert*-butyl hydroperoxide (TBHP) as the oxidant in the presence of a catalytic amount of $n\text{-}Bu_4NI$ in a mixture of AcOH/DMF under argon atmosphere. Corresponding indoles **4** were produced in 71–99% yields. The reaction of enamino amide afforded indole **4**, in low yield (27%) (Jia et al. 2013). Also, I_2 catalyzed construction of indole-3-carboxylates **4** was achieved by subjecting *N*-aryl-enamino esters to I_2 and *N*-bromosuccinimide (NBS) in the presence of a base, K_2CO_3. The reaction with a 3-alkyl substituted enamino ester led to the corresponding indole **4** in low yield (36%), while 3-aryl substituted enamino esters afforded corresponding indoles **4** in 54–95% yields. In the proposed reaction mechanism, the *N*-iodo intermediate was generated by oxidation

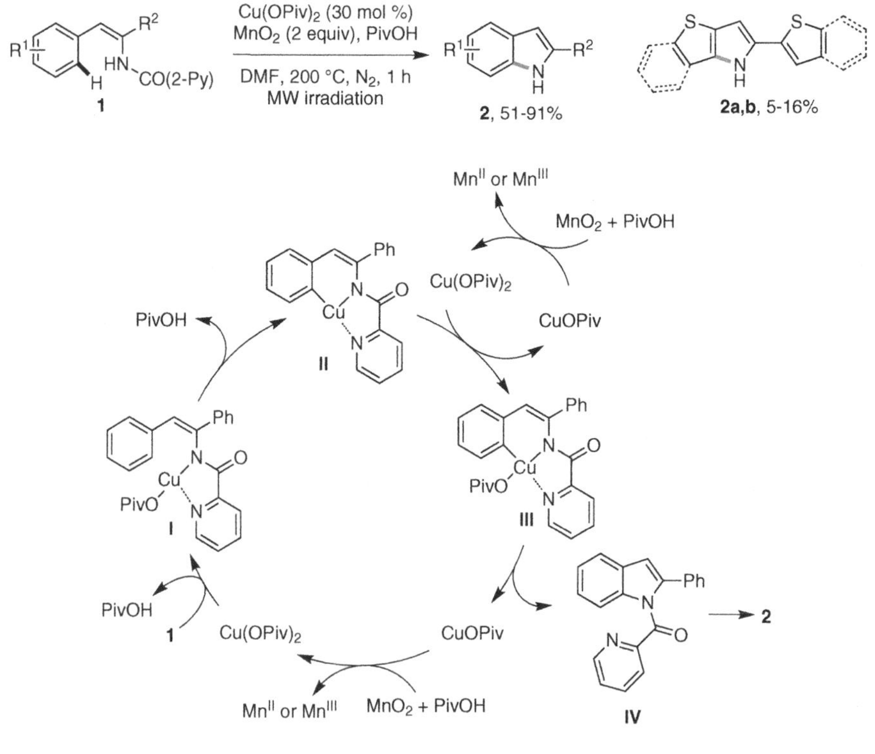

SCHEME 1.1

of an *N*-aryl-enamino ester **3** by hyperiodide generated in situ. The *N*-radical species generated by homolytic cleavage of the N–I bond underwent cyclization, followed by oxidation to form an indole. The hyperiodide species was regenerated by oxidation of atomic or molecular iodine with NBS, to further the catalytic cycle (He, Liu, and Li 2011). Ir(III)-catalyzed intramolecular dehydrogenative coupling to indole-3-carboxylates **4** was described in the presence of Co(II) (Wu et al. 2016) or O_2 (Liu et al. 2017). Reaction with $Co(dmgH)_2(4\text{-}CO_2Mepy)Cl$, and $Ir(ppy)_3$ under strictly deaerated conditions and blue LED ($\lambda = 450$ nm) irradiation afforded indole-3-carboxylate derivatives **4** in 55–97% yields. The reaction with O_2 was conducted in the presence of fac-$Ir(ppy)_3$ in dimethyl sulfoxide (DMSO) under irradiation with 3 W blue LEDs under air, leading to indole-3-carboxylates **4** in 71–99% yields (Scheme 1.2).

Pd(II)-catalyzed intramolecular dehydrogenative coupling of *N*-aryl enamino esters **3** was also developed to give indole derivatives **4** in 43–85% yields. Reactions were performed using $Pd(OAc)_2$ (2–10 mol%), $Cu(OAc)_2$ (3 equiv), and K_2CO_3 (3 equiv) in DMF under an atmosphere of argon. The plausible reaction mechanism is depicted in Scheme 1.3. Electrophilic palladation of the enamine, followed by deprotonation, generated palladium complex **I**, which underwent intramolecular C–H activation by σ-bond metathesis (**TS1**) or base-assisted deprotonation (**TS2**) to generate intermediate **II**. Subsequent reductive elimination produced the indole product **4** and a Pd(0) species, which was oxidized to a Pd(II) species by $Cu(OAc)_2$ (Scheme 1.3) (Neumann et al. 2011). Also, the synthesis of 2-substituted indoles via $Pd(OAc)_2$-catalyzed intramolecular cross-dehydrogenative coupling of aryl imines in the presence of molecular oxygen was reported by Shi and Glorius (2012), in which the Pd(II) catalytic active species was regenerated from Pd(0) by oxidation with O_2.

In 2009, Bernini et al. (2009) described the synthesis of 3-acyl indoles **7** via a Cu-catalyzed intramolecular dehydrogenative coupling reaction. By subjecting *N*-aryl enamino ketones **6** to CuI (5 mol%) and phen in the presence of Li_2CO_3 in DMF at 100°C under air, 3-acyl indoles **7** were obtained in 51–84% yields. Li et al. (2018) described the synthesis of multi-substituted indoles via a one-pot tandem copper-catalyzed Ullmann-type C–N bond formation/intramolecular cross-dehydrogenative coupling process of aryl iodides **8** and enamino esters **9**. Reactions were carried out using CuI (10 mol%) and a Johnphos ligand in DMSO at 130°C under air, in the presence of $KHCO_3$ as a base, leading to corresponding indole-3-carboxylates

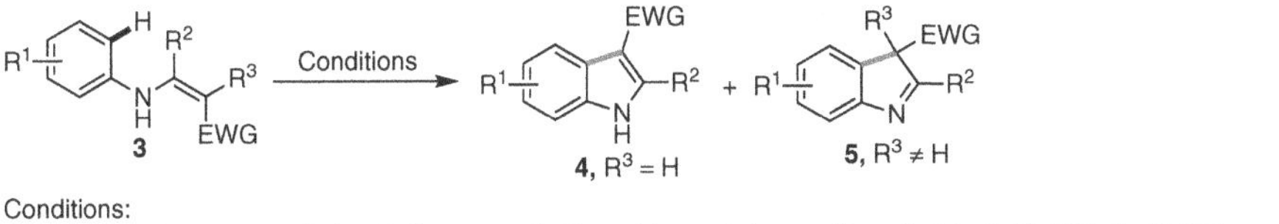

3 → Conditions → **4**, R^3 = H + **5**, $R^3 \neq$ H

Conditions:
Drouhin and Taylor 2015: Cu(2-ethylhexanoate)$_2$ (1 equiv), mesitylene, relfux, air, 2 h; **4**, 34-69%, **5**, 42-80%.
Jia et al. 2013: TBHP (2.5 equiv), *n*-Bu_4NI (30 mol %), AcOH/DMF, 100 °C, 24 h; **4**, 71-99%.
He, Liu, and Li 2011: I_2 (5 mol %), NBS (1.1 equiv), K_2CO_3 (1.2 equiv), DMF, N_2 (atm.), 100 °C, 1 h; **4**, 36-95%.
Wu et al. 2016: $Co(dmgH)_2(4\text{-}CO_2Mepy)Cl$ (7 mol %), $Ir(ppy)_3$ (3 mol %), DMF/*i*-PtOH, blue LEDs (λ = 450 nm), rt, 20 h; **4**, 55-97%.
Liu et al. 2017: fac-$Ir(ppy)_3$ (1 mol %), DMSO, 3 W blue LEDs, air, 75 °C, 5 h; **4**, 71-99%.

SCHEME 1.2

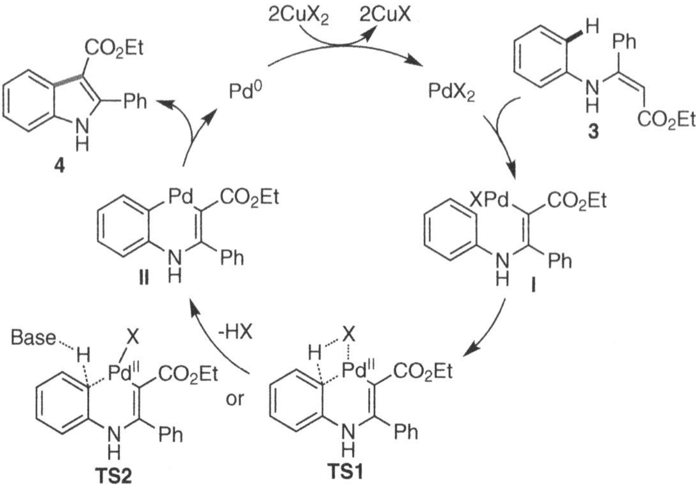

SCHEME 1.3

4 in 50–90% yields. By C–N bond formation via Ullmann-type reaction, *N*-aryl enamino esters **3** were obtained, which were converted to indole products **4** through intramolecular cross-dehydrogenative coupling (Scheme 1.4).

Synthesis of 2-perfluoroalkylated indoles **12** via a one-pot cascade Michael-type addition/palladium-catalyzed intramolecular CDC process, using molecular oxygen as the sole oxidant, was reported by Shen et al. in 2015. Treatment of aniline **10** with methyl perfluoroalk-2-ynoate **11** in the presence of $Pd(OAc)_2$ (10 mol%) and $NaHCO_3$ under an atmosphere of O_2 (1 atm) in a mixture of DMSO/PivOH at 100°C afforded indole-3-carboxylate derivatives **12** in 57–89% yields (Scheme 1.5) (Shen et al. 2015). Reactions proceeded by the formation of *N*-aryl enamino esters (like **3**) via Michael-type addition of aniline **10** into perfluoroalk-2-ynoate **11**, followed by intramolecular CDC reaction.

Zhang et al. (2016) developed Co(III)-catalyzed CDC of *N*-arylureas **13** and internal alkynes **14** to obtain indole derivatives **15**. Reactions were conducted using $[Cp^*Co(CO)I_2]$ complex (10 mol%) and $AgSbF_6$ (20 mol%) in the presence of NaH_2PO_4 and Ag_2CO_3 in dichloroethane (DCE) under O_2 atmosphere at 130°C,

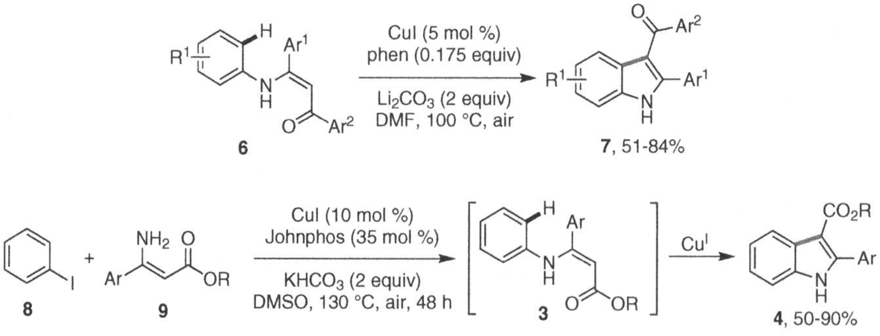

SCHEME 1.4

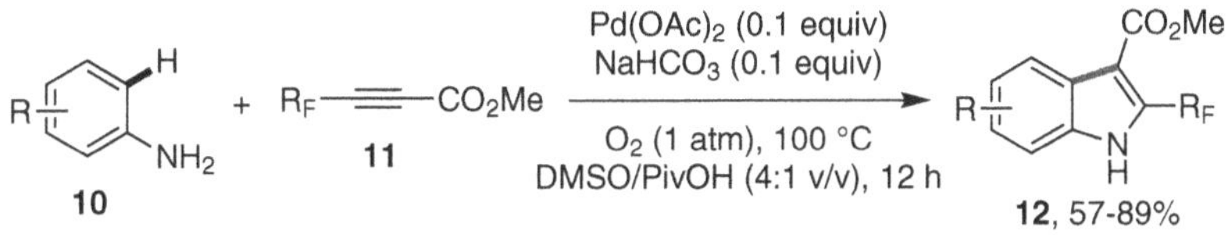

SCHEME 1.5

and the desired indoles **15** were produced in 39–94% yields. No reaction occurred in the case of a terminal alkyne. The plausible catalytic cycle involves the generation of active cationic Co(III) catalyst **I** by anion exchange between [Cp*Co(CO)I_2] and $AgSbF_6$, followed by the formation of six-membered metallacyclic intermediate **II**, via Co-mediated irreversible C–H bond activation of *N*-arylurea. Metallacyclic intermediate **II** underwent alkyne coordination and migratory insertion of the aryl group to a Co(III) alkenyl species **III**, which was transformed into intermediate **IV** through base-assisted elimination of HX. The C–N bond reductive elimination afforded indole **15** along with a Co(I) species **V**, which was oxidized with Ag_2CO_3 and O_2 to Co(III) active catalyst **I** for the subsequent catalytic cycle (Scheme 1.6).

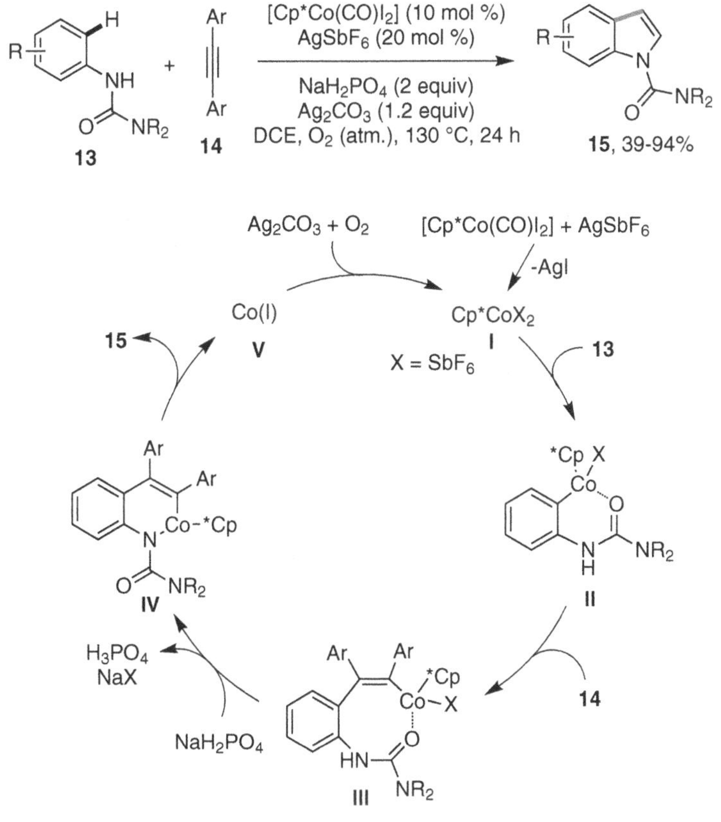

SCHEME 1.6

Co(III)-catalyzed synthesis of indole derivatives was reported by reaction between *N*-alkyl-*N*-arylhydrazines with an internal or terminal alkyne, via the *N*-amino directing group assisted C–H functionalization/cyclization reactions. With $[Cp^*CoI_2]_2$ (2 mol%) and $AgSbF_6$ (8 mol%) as the catalyst precursor and the assistance of $Zn(OTf)_2$, arylhydrazine **16** was reacted with the internal alkyne **17** at 50°C in DCE, yielding polysubstituted indoles **18**. Reaction with diarylalkynes and dialkylalkynes afforded corresponding indoles in 41–88% and 53–61% yields, respectively. Alkyl aryl alkynes resulted in the formation of 3-alkyl-2-arylindoles **18** in 65–90% yields. Reactions with terminal alkynes were conducted using $[Cp^*Co(MeCN)_3][SbF_6]_2$ (2 mol%) in the presence of $Zn(OTf)_2$ in DCE at 80°C. In the case of terminal alkynes with no H atom at the neighboring C atom of the C≡C bond **19**, such as *t*-BuC≡CH, TMSC≡CH, TESC≡CH, and TIPSC≡CH, corresponding 2-substituted indoles **20** were obtained in 39–83% yields. Reaction with linear-chain terminal alkynes **21** provided 3-alkyl-2-methylindoles **22** in 29–67% yields, by intriguing bond migration. The bond migration was proposed to proceed through Co(III)-promoted initial alkyne-allene isomerization (**21a** → **21b**), followed by carbocobaltation of the allene with the in situ generated five-membered cobaltacycle **I** via C–H cobaltation, to generate alkenyl cobalt intermediate **II**, which underwent cyclization to the final indole product **22** (Scheme 1.7) (Zhou et al. 2016).

Rhodium-catalyzed CDC reactions between *N'*-aryl acylhydrazides **23** and internal alkynes **17** have been reported by Li, Chen, and Liu (2014). The CDC process was catalyzed with $[Cp^*RhCl_2]_2$ (5 mol%) in the presence of NaOAc and 1,3-dinitrobenzene as the oxidant in DCE at 60°C under a nitrogen atmosphere, and 1-acylamidoindole derivatives **24** were obtained in 45–93% yields. In the proposed reaction mechanism, the active catalyst $[Cp^*Rh(OAc)_2]$ **I**, formed from the $[Cp^*RhCl_2]_2$/NaOAc catalytic system, directly promoted C–H activation of hydrazine **23**, giving the arene rhodation intermediate **II**, which underwent alkyne coordination and insertion, leading to the seven-membered rhodacycle **III**. By rearrangement of rhodacycle **III** to a more stable six-membered rhodium complex **IV**, followed by reductive elimination, the desired indole product **24** and [Cp*Rh(I)] **V** were formed. Finally, the active catalyst $[Cp^*Rh(OAc)_2]$ **I** was regenerated from **V** by oxidation with 1,3-dinitrobenzene (Scheme 1.8). Also, $[Cp^*RhCl_2]_2$ was used as the catalyst for the construction of *N*-acetyl indoles via CDC annulation of *N*-phenyl acetamide with alkynes. In the case of asymmetric alkylaryl-alkynes, corresponding 3-alkyl-2-aryl-*N*-acetyl indoles were obtained (Stuart et al. 2008).

Wang et al. (2013) described the triazene-directed C–H annulation of a triazenyl arene **25** with internal alkynes **17**, using $[Cp^*RhCl_2]_2$ (5 mol%) as a catalyst in the presence of $AgSbF_6$ and $Cu(OAc)_2{\cdot}H_2O$ as an oxidant in MeOH under an argon atmosphere. 2,3-Disubstituted-1*H*-indoles **26** were obtained in 42–94% yields. The catalytic cycle of the Rh(III)-catalyzed CDC indole synthesis is depicted in Scheme 1.9. 3-(1*H*-indol-1-yl)propanamide derivatives **28** were synthesized via Ru-catalyzed redox-neutral C–H activation reaction of pyrazolidin-3-one **27** with internal and terminal alkynes via N–N bond cleavage. Reactions were carried out using $[RuCl_2(p\text{-cymene})]_2$ (2.5 mol%) in the presence of NaOAc in chlorobenzene under an argon atmosphere at 110°C, and 3-(1*H*-indol-1-yl)propanamides **28** were obtained in 53–94% yields. Asymmetrical alkylaryl-alkynes afforded exclusively

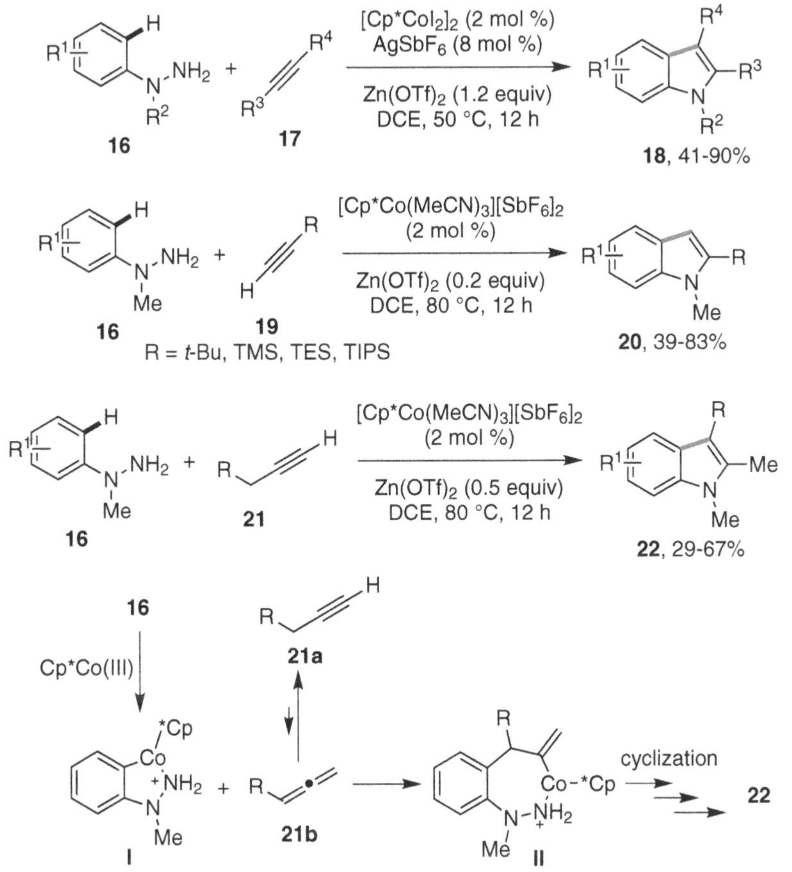

SCHEME 1.7

single regioisomers (2-aryl-3-alkylindoles). Also, terminal alkynes generated single regioisomeric 2-substituted indoles (Zhang, Jiang, and Huang 2014). Also, the synthesis of 1,2,3-trisubstituted indoles **30** was achieved by an Rh(III)-catalyzed CDC process of tertiary aniline *N*-oxides **29** with alkynes **17**. Reactions occurred by *ortho*-C–H functionalization of tertiary aniline *N*-oxides **29** via the N–O bond acting as a traceless directing group. Reactions were performed using $[Cp^*RhCl_2]_2$ (5 mol%) and $AgBF_4$ (2 equiv) in the presence of $MesCO_2H$ in *t*-BuOH at 115°C, affording corresponding indoles **30** in 33–98% yields. The plausible reaction mechanism involves two steps: first, Rh(III)-catalyzed CDC of aniline *N*-oxide **29** with alkyne **17**, and second, Ag(I)-assisted *N*-dealkylative cyclization (Scheme 1.9) (Li et al. 2016a).

Rh(III)-catalyzed dehydrogenative C–H annulation of nitrones **31** with symmetrical diaryl alkynes **14** was reported to prepare 2,3-diarylindoles **32**. One of the aryl substituents was derived from the N=C-aryl of the nitrone **31** and the other from the alkyne substrate. The synthesis procedure involves treatment of diaryl alkyne **14**

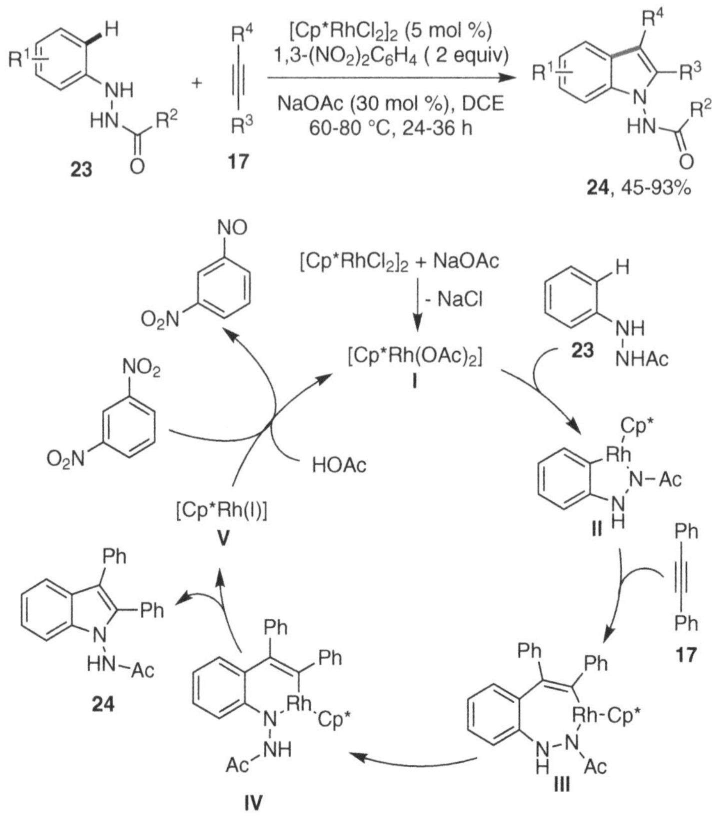

SCHEME 1.8

with 1.2 equiv of nitrone **31** in the presence of $[Cp^*RhCl_2]_2$ (2.5 mol%), $AgSbF_6$, and $Cu(OAc)_2$ in a mixture of DCE and 1,2-diethoxyethane, liberating 2,3-diarylindoles **32** in 52–88% yields. The tentative reaction mechanism involves the generation of seven-membered intermediate **II** via C–H bond activation of nitrone **31** to five-membered rhodacycle intermediate **I**, followed by alkyne coordination and insertion. Reductive elimination provides the cationic heterocyclic intermediate **III**, which tautomerizes to **IV** together with intramolecular O-atom transfer. Subsequent oxidative addition of **III** and intramolecular electrophilic addition of the imino moiety furnished intermediate **V**, which underwent transmetalation with a Cu salt through **VI** to form **VII** along with releasing the Rh catalyst. By subsequent hydride elimination to compound **IX**, followed by in situ generated CuH-assisted elimination of one molecule of PhCHO, the final indole product **32** was obtained (Yan et al. 2015). A similar methodology was developed in the absence of $Cu(OAc)_2$, by subjecting diaryl alkyne to 1.2 equiv of nitrone **31** in the presence of $[Cp^*RhCl_2]_2$ (4 mol%), $AgSbF_6$, PivOH, and 4 Å molecular sieve (MS) in dioxane at 50°C, yielding indolines **33** via a Rh(III)-catalyzed intermolecular dehydrogenative coupling of arylnitrones **31** with internal alkynes **14**, followed by O-atom transfer. 2,3-Diaryl-3-acylindolines **33**

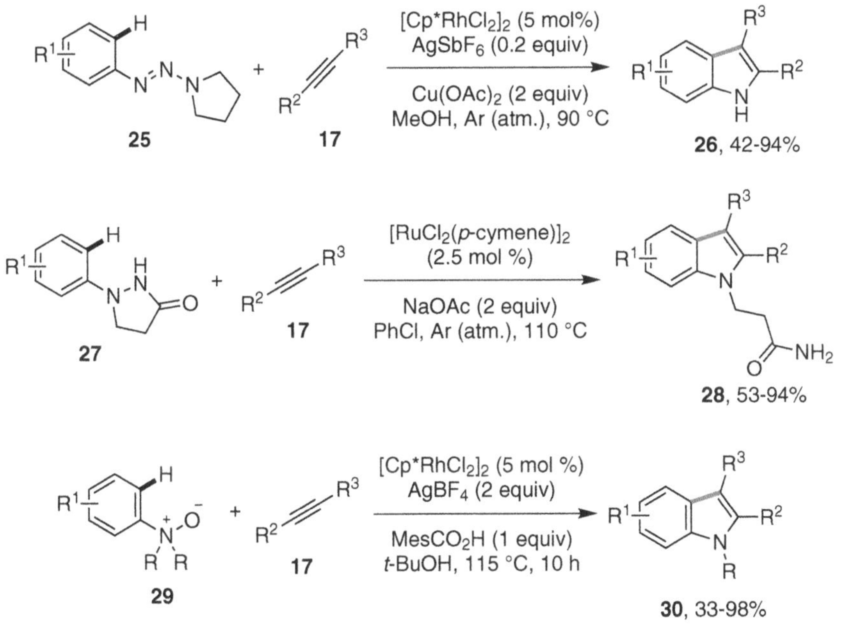

25 + 17 → 26, 42-94%
$[Cp^*RhCl_2]_2$ (5 mol%), $AgSbF_6$ (0.2 equiv), $Cu(OAc)_2$ (2 equiv), MeOH, Ar (atm.), 90 °C

27 + 17 → 28, 53-94%
$[RuCl_2(p\text{-cymene})]_2$ (2.5 mol %), NaOAc (2 equiv), PhCl, Ar (atm.), 110 °C

29 + 17 → 30, 33-98%
$[Cp^*RhCl_2]_2$ (5 mol %), $AgBF_4$ (2 equiv), $MesCO_2H$ (1 equiv), *t*-BuOH, 115 °C, 10 h

SCHEME 1.9

were obtained in 35–83% yields, with moderate to high diastereoselectivity (Scheme 1.10) (Dateer and Chang 2015b).

2,3-Disubstituted *NH*-indoles **4** were synthesized by Rh(III)-catalyzed intermolecular coupling between arylnitrones **31** and diazocompounds **34** by C–H activation/[4+1] annulation with a $C(N_2)$–C(acyl) bond cleavage. Reactions were conducted using $[Cp^*RhCl_2]_2$ (5 mol%), $AgSbF_6$, NaOAc, and $Cu(OAc)_2$ in DCE under air, and corresponding indoles **4** were obtained in 35–94% yields. The reaction was initiated by C–H bond activation of nitrone **31** with in situ generated catalytically active rhodium species to form five-membered rhodacycle intermediate **I**, which reacted with α-acyldiazoacetate **34** to generate an Rh(III) carbene intermediate **II**. By migratory insertion of the Rh–C bond into the activated carbene, six-membered rhodacyclic intermediate **III** was generated, which was transformed to intermediate **IV** by an intramolecular electrophilic attack of the imino moiety. The final desired indole product **4** was obtained by generation of intermediate **V** via transmetalation of the Rh catalyst with the Cu salt and subsequent β-hydride elimination to **VI**, followed by elimination of acetic acid through hydrolysis with water (Guo et al. 2017). When reactions were performed in the presence of $[Cp^*RhCl_2]_2$, $AgSbF_6$, PivOH, and 4 Å MS as the water adsorbent in 1,4-dioxane at room temperature, corresponding *N*-hydroxyindoline-3,3-dicarboxylates **35** were obtained in 61–91% yields. Cyclization is assumed to proceed through a redox-neutral pathway via **VII**, generated from intermediate **III**, followed by protonation with PivOH (Dateer and Chang 2015a). Also, Rh(III)-catalyzed coupling of *N*-Boc-hydrazines **36** with diazoketoesters **34** was reported, providing *N*-amino indoles **37**. Reactions were performed using

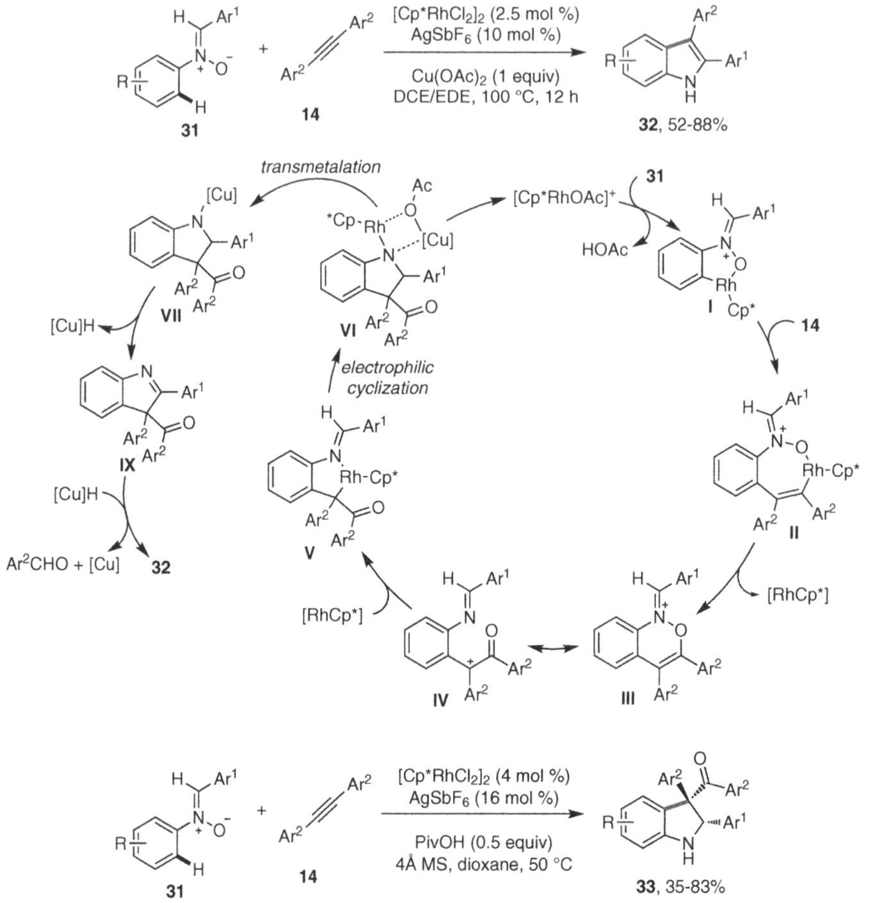

SCHEME 1.10

$[Cp^*RhCl_2]_2$ (2 mol%) in the presence of CsOAc in 2,2,2-trifluoroethanol (TFE), via Rh(III)-catalyzed C–C coupling, intramolecular C–N_α cyclization, and then *N*-Boc cleavage. Various electron-withdrawing and electron-donating substituted aryl hydrazines, and also different alkyl and aryl substituted diazoketoesters **34**, were tolerated under reaction conditions, affording *N*-aminoindole-3-carboxylates **37** in 51–93% yields (Scheme 1.11) (Shi et al. 2017).

Rh(III)-catalyzed C–H activation/annulation of imidamides **38** with α-diazo β-ketoesters **34** was developed for the construction of *N*-unprotected 2-alkylindole-3-carboxylates **4**. Subjecting imidamides **38** to α-diazo β-ketoesters **34** in the presence of $[Cp^*RhCl_2]_2$ (4 mol%), CsOAc, and HOAc in DCE at 80°C afforded 2-alkylindole-3-carboxylates **4** in 56–93% yields. Several substituted *N*-arylalkylimidamides **38** worked well in the reaction conditions, while *N*-phenylbenzimidamide reacted with poor efficiency. The proposed reaction mechanism involves the in situ formation of active catalyst $[Cp^*Rh(OAc)_2]$ through anion exchange, followed by coordination with imidamide **37** and cyclometalation to generate a rhodacyclic intermediate **I**,

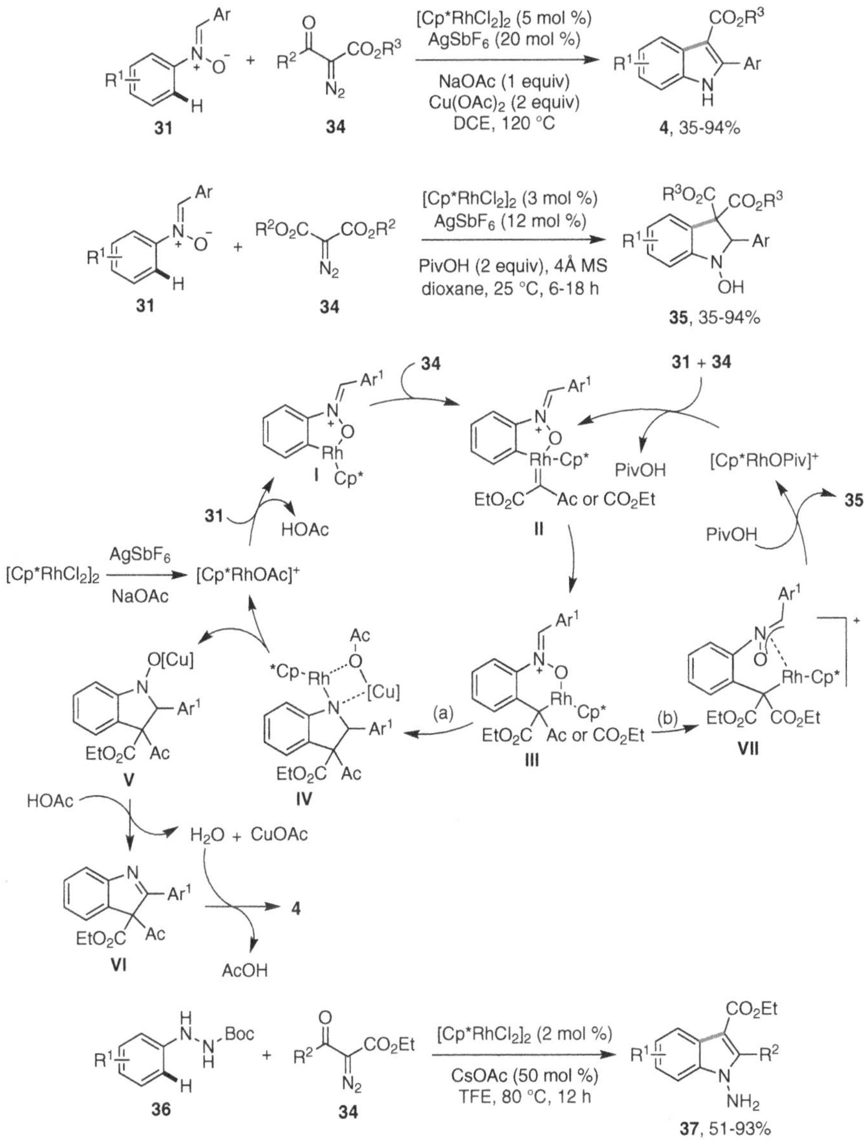

SCHEME 1.11

which gave rhodium carbene species **II** through elimination of N_2. By migratory insertion of the Rh–Ar bond into the carbene unit, intermediate **III** was generated, which furnished an amide species **IV** by the second migratory insertion of the Rh–C(alkyl) bond into the C=N bond. Upon protonolysis and intramolecular nucleophilic addition, intermediate **V** was formed, along with the regeneration of the active Rh(III) catalyst. The product **4** was formed by elimination of one molecule of acetamide (Qi, Yu, and Li 2016). Moreover, 3-acylindoles **40** were accessed through the Rh(III)-catalyzed C–H activation and annulation cascade of *N*-phenylamidines **38** with *α*-chloroketones **39**, in which *α*-chloroketones **39** served as unusual one-carbon (sp^3) synthons (Scheme 1.12) (Zhou et al. 2018b).

Cross-dehydrogenative-coupling annulation of *N*-hetaryl anilines with alkynes were investigated with different catalytic systems. In all cases the existence of a coordinative directing group such as 2-pyridyl or 2-pyrimidyl is essential (Ackermann and Lygin 2012, Xu et al. 2018a, Chen et al. 2010, 2011). An Ru-catalyzed electrochemical CDC process was performed using $[RuCl_2(p\text{-cymene})]_2$ (5 mol%) in the presence of KPF_6 and NaOAc (20 mol%) in H_2O/*i*-PrOH mixture at reflux, under electrochemical conditions of 10 mA ($J_{anode} \cong 0.13$ mA cm^{-2}) and 2.3–4.9 F mol^{-1}, in which *N*-2-pyrimidyl-indoles **42** were obtained in 39–99% yields. Asymmetrical alkylaryl-alkynes afforded 3-alkyl-2-arylindoles, exclusively (Scheme 1.13) (Xu et

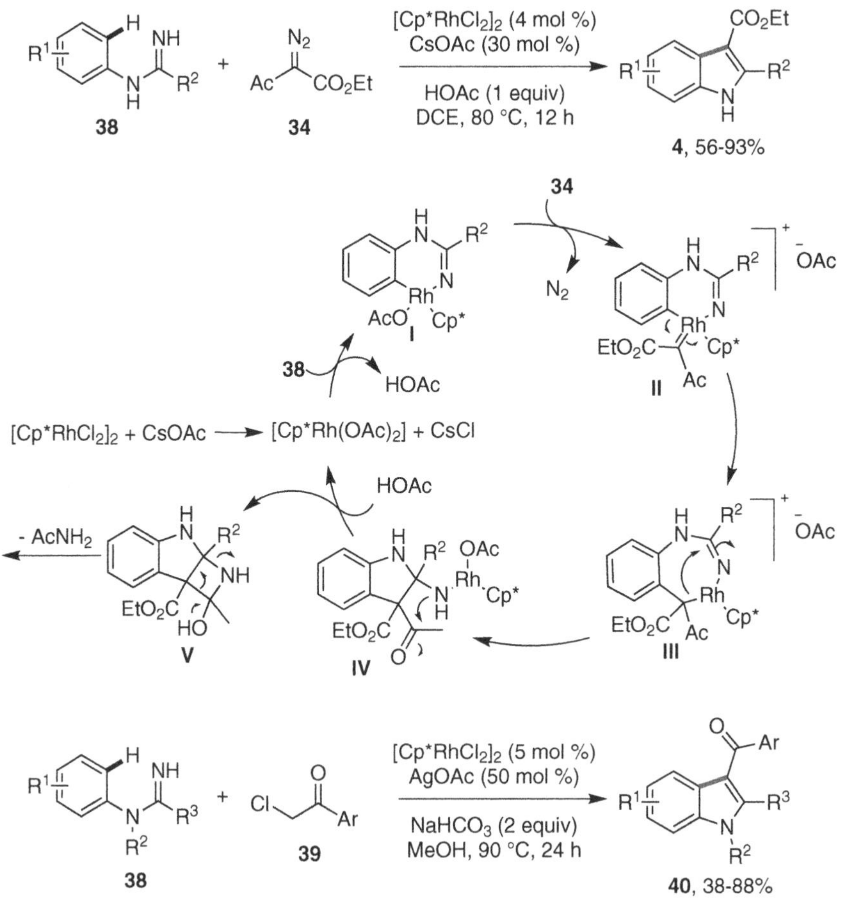

SCHEME 1.12

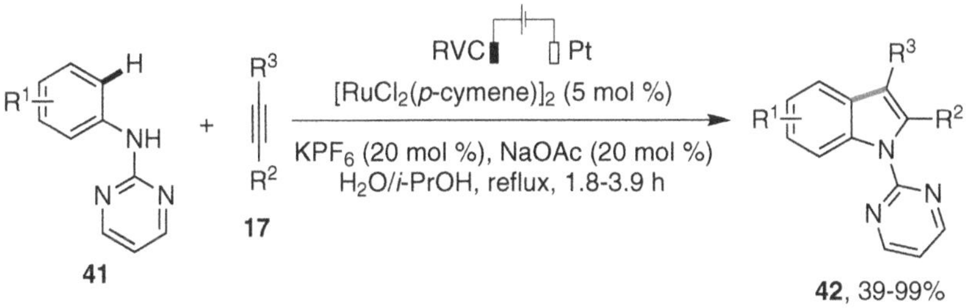

SCHEME 1.13

al. 2018a). In another work Ackermann and Lygin (2012) described a similar CDC process using $Cu(OAc)_2$ as the oxidant.

1,2,3-Trisubstituted indoles **45** were obtained by an Rh(III)-catalyzed carboamination of alkynylcycloalkanols **44** with arylamines **43**, via a C(sp^2)–H/C(sp^3)–C(sp^3) activation process. Reactions were performed using *N*-aryl-2-pyridinamines **43** and propargyl alcohols **44**, catalyzed with $[Cp^*Rh(CH_3CN)_3][SbF_6]_2$ (5 mol%) in the presence of AgOAc in *t*-AmOH at 50°C under argon atmosphere, liberating 2-acylindoles **45** in 40–82% yields. As the coordination of a 2-pyridyl directing group is necessary for starting the reaction, because of the strong coordination between Rh and electron-rich aniline nitrogen instead of pyridine nitrogen, *N*-(4-methoxyphenyl)-2-pyridinamine did not afford the desired indole. By initial coordination of pyridyl nitrogen of the substrate **43** to the active Rh(III) species with subsequent aryl C–H cleavage, six-membered rhodacycle **I** was obtained via a concerted metalation/deprotonation process. Then, Rh(III) species **II**, generated by migratory insertion of alkynyl alcohol **44** to Rh(III) complex **I**, underwent *β*-carbon elimination leading to another six-membered rhodacycle intermediate **III**. Finally, the 1,2,3-trisubstituted indole **45** was produced through an intramolecular nucleophilic attack of aniline nitrogen on the Rh(III) ion, followed by metal protonation to give complex **IV** and then reductive elimination, along with release of the Rh(I) catalysts. The Rh(III) active catalyst was regenerated by oxidation with Ag(I) (Scheme 1.14) (Hu et al. 2017).

Synthesis of 2-arylindoles **47** was achieved via a Pd(II)-catalyzed cyclization reaction of anilines with vinyl azides. Treatment of *N*-2-pyridylailine **43** with 1.5 equiv of vinyl azide **46** in the presence of $Pd(TFA)_2$ (10 mol%), 1,4-diazabicyclo[2.2.2]octane (DABCO), $K_2S_2O_8$, and trifluoroacetic acid (TFA) in toluene under air at 75°C afforded 2-arylindoles **47** in 40–85% yields. Reaction with 1,2-diphenylvinyl azide provided 2,3-diphenylindole in 52% yield (Scheme 1.15). However, another pathway including 2*H*-azirine, generated in situ by thermal decomposition of vinyl azide **46**, was also proposed (Jie et al. 2018).

Very recently, Zhou et al. (2018a) developed a three-component reaction of aryl iodide **8**, alkynes **17**, and diaziridinone **48**, giving indole derivatives **49**. Reactions were conducted using 1.1 equiv of aryl iodide and 1 equiv of diaziridinone in the presence of $Pd(OAc)_2$ (3 mol%), PPh_3, Cs_2CO_3, and KOPiv in DMF at 105°C under

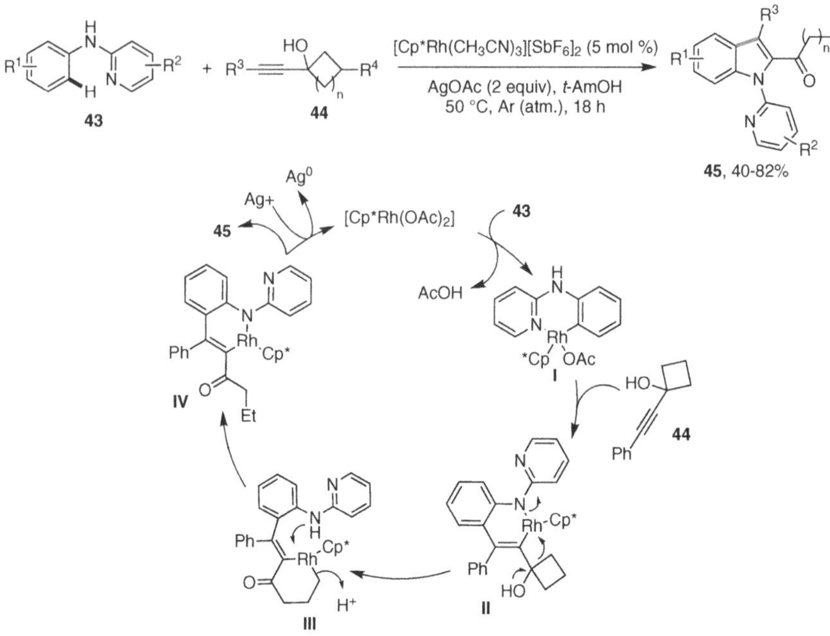

SCHEME 1.14

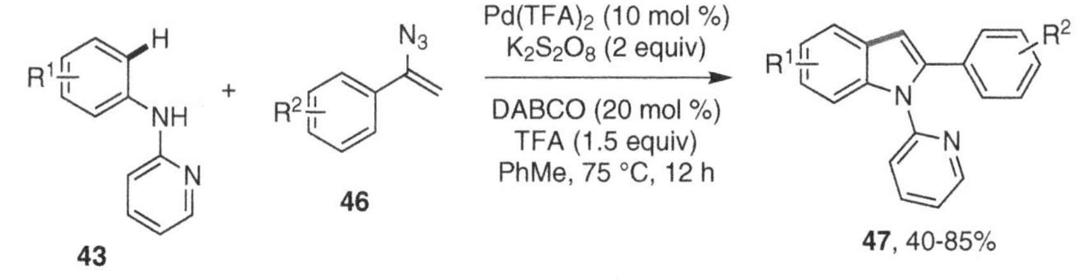

SCHEME 1.15

an atmosphere of N_2 for 12 h, and indoles **49** were obtained in 36–99% yields. Diaryl alkynes afforded corresponding indoles in high to excellent yields, while dialkyl alkynes gave corresponding indoles in moderate yields (54–61%). In the case of asymmetrical alkyl aryl alkynes, 3-alkyl-2-arylindoles were obtained. Reaction with 2-phenylacetylene carboxylate led to the formation of 3-phenylindole-2-carboxylate as a major product. The catalytic cycle involves the oxidative addition of iodobenzenes to Pd(0), giving aryl Pd(II) species **I**, which underwent migratory insertion to generate vinyl Pd(II) species **II**. By intramolecular C–H activation of palladacycle **III**, followed by oxidative addition to the N–N bond of diaziridinone, a pallada(IV) cycle **IV** was formed, which was transformed to the eight-membered intermediate **V** through reductive elimination. The final desired product **49** was obtained via *β*-*N* elimination (intermediate **VI**) with subsequent reductive elimination, along with regeneration of the Pd(0) catalyst (Scheme 1.16).

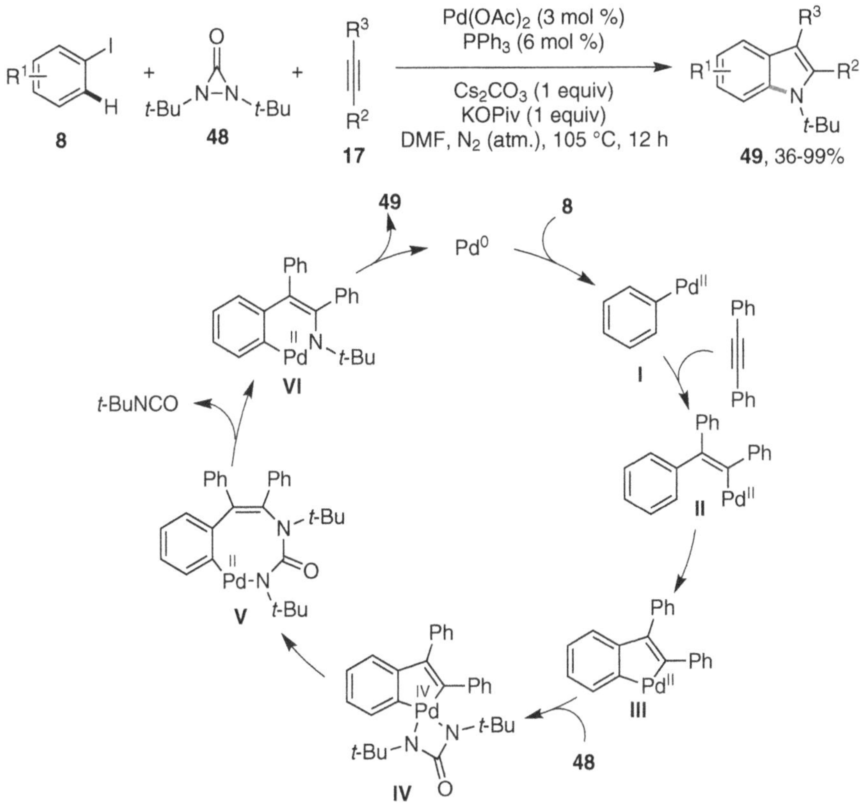

SCHEME 1.16

1.3 OXINDOLES

Jia and Kündig (2009) and Dey, Larionov, and Kuendig (2013) have reported intramolecular dehydrogenative coupling of C(sp^2)–H and C(sp^3)–H by treatment of amide **50** with $CuCl_2$ (2.2 equiv) and *t*-BuONa in DMF under N_2 atmosphere at 110°C, affording oxindole derivatives **51** in 38–97% yields. The plausible reaction mechanism involves the formation of amidyl radical **I** by oxidation of amide enolate. Intramolecular cyclization of **I** generated cyclohexadienyl radical **II**, which aromatized to the final oxindole **51** by dehydrogenation or one-electron oxidation followed by deprotonation (via intermediate **III**) (Scheme 1.17). In 2012, Dey and Kündig (2012) reported a similar methodology for the construction of aza-oxindoles.

A metal-free synthesis of oxindoles via intramolecular C(sp^2)–H and C(sp^3)–H bond activation was reported. Reactions were conducted using Di(*t*-butyl)peroxide (DTBP) (3 equiv) as radical initiator in DCE under air at 110°C, resulting in 3-nitrile substituted oxindoles **53** in moderate to high yields. In the case of substrates with NH functionality (**52b**) or the absence of an α-methyl substituent (**52a**), reactions failed. In the proposed reaction mechanism, a *tert*-butoxy radical, generated in situ by homolytic cleavage of DTBP on heating, initiated the reaction by abstraction of

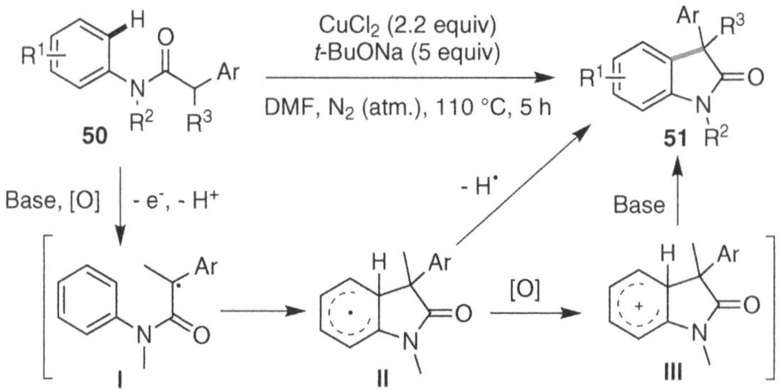

SCHEME 1.17

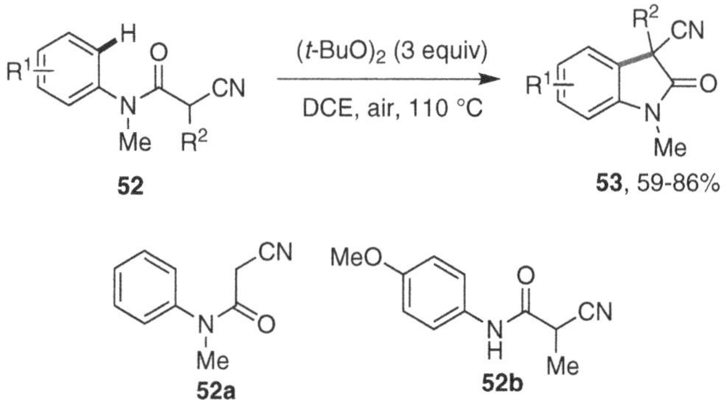

SCHEME 1.18

one hydrogen atom from the α-position of the nitrile group of substrate **52** (Scheme 1.18) (Mondal and Roy 2015). Donald, Taylor, and Petersen (2017) reported the transition metal-free CDC protocol for the synthesis of 3,3-disubstituted oxindoles by subjecting β-*N*-arylamido esters with potassium hexamethyldisilazide (KHMDS) in THF, followed by addition of iodine. In addition, C3-fluorinated oxindoles were synthesized in 38–91% yields through electrochemical cross-dehydrogenative coupling of C(sp^3)–H and C(sp^2)–H bonds from malonate amides (Zhang et al. 2017).

Synthesis of 3-acetyloxindoles **55** was achieved by intramolecular dehydrogenative C(sp^2)–C(sp^3) coupling of *N*-substituted acetoacetanilides **54** using Ag_2O as the oxidant. Reactions were performed in DMF using 1.1 equiv of Ag_2O and 2 equiv of Cs_2CO_3 at 70°C, affording 3-acetyloxindoles **55** in 25–71% yields. Reactions of *N*-substituted 2-methylacetoacetanilides **54** (R^3=Me) under the same reaction conditions resulted in the formation of 3-acetyl-3-methyloxindoles **56** in 46–74% yields (Yu, Ma, and Yu 2010). 2-Oxindoles **58** bearing a quaternary stereogenic center at the benzylic position were synthesized in 52–90% yields, via a one-pot *C*-alkylation of β-*N*-arylamido esters **57** with alkyl halides using *t*-BuOK, followed by metal-free I_2-promoted intramolecular dehydrogenative coupling (Scheme 1.19) (Kumar et al. 2016, Ghosh et al. 2012). A similar methodology was applied using 2,3-dichloro-5,6-dicyano-1,4-benzoquinone (DDQ) as the oxidant for the construction of 2-oxindole derivatives (Bhunia et al. 2013).

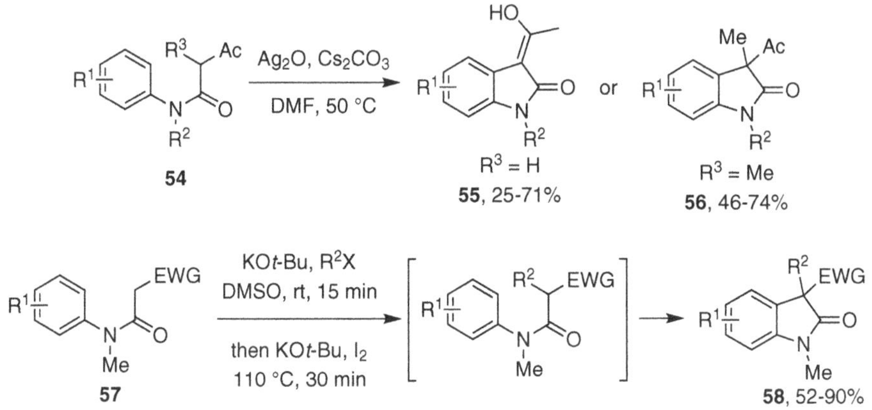

SCHEME 1.19

In 2015, Chan, Lee, and Yu (2015) developed Cu-catalyzed CDC of *N*-arylacrylamides **59** with chloroform using *tert*-butyl peroxybenzoate (TBPB) as oxidant for the synthesis of trichloromethylated 2-oxindoles **60**. Reactions were performed using 5 mol% $CuBr_2$ and 2 equiv of TBPB in $CHCl_3$ at 120°C to give 2-oxindoles **60** in 34–92% yields. Reaction with NH-unprotected substrates did not occur. Cinnamic acid-derived acrylamide underwent trichloromethyl radical-mediated cyclization to afford dihydroquinolinone **61** in 78% yield. In the proposed reaction mechanism, by Cu(II)-mediated decomposition of TBPB, Cu(III)(OBz) species and *tert*-butoxy radicals were generated. The *tert*-butoxy radicals abstracted the hydrogen atom of chloroform to give CCl_3 radicals, which reacted with the C=C bond of **59**. A similar approach for the CDC of *N*-arylacrylamides **59** with carbonyl C(sp^2)–H bonds was developed using TBHP, leading to 3-(2-oxoethyl)indolin-2-ones **62** in 42–83% yields (Zhou et al. 2013). Also, *N*-hydroxyphthalimide (NHPI) **63** was subjected to *N*-arylacrylamides **59** to produce aminooxylated oxindoles **64** via a radical C–H functionalization. A phthalimide *N*-oxyl radical, generated from NHPI by sulfate radical anions, acted as a radical initiator, and was added to the C=C double bond with subsequent dehydrogenative cyclization with the aryl moiety to form the final desired aminooxylated oxindoles **64** (Zhang et al. 2018b). Fe(III)-catalyzed oxidative 1,2-alkylarylation of the C=C double bond of *N*-arylacrylamides **59** with an aryl C(sp^2)–H bond and a C(sp^3)–H bond adjacent to a heteroatom **65** was reported for the selective synthesis of functionalized 3-(2-oxoethyl)indolin-2-ones **66**. Reactions were conducted using $FeCl_3$ (5 mol%), 1,8-diazabicyclo[5.4.0]undec-7-ene (DBU, 10 mol%) as a ligand, and TBHP (2 equiv) as an oxidant in benzene at 120°C, under argon atmosphere, to furnish 3-(2-oxoethyl)oxindoles **66** in 40–87% yields (Scheme 1.20) (Wei et al. 2013).

A series of 3,3-dialkyloxindoles **67** was synthesized by a redox-neutral Ni-catalyzed intramolecular hydroarylation of the C=C bond of *N*-arylacrylamides **59**. Reactions were carried out using $Ni(cod)_2$ (10 mol%), an IPr·HCl ligand (30 mol%), *t*-BuOK (30 mol%), and DMF (2 equiv) in toluene, in which DMF played a proton shuttle role, avoiding the use of additional reluctants and oxidants. Corresponding

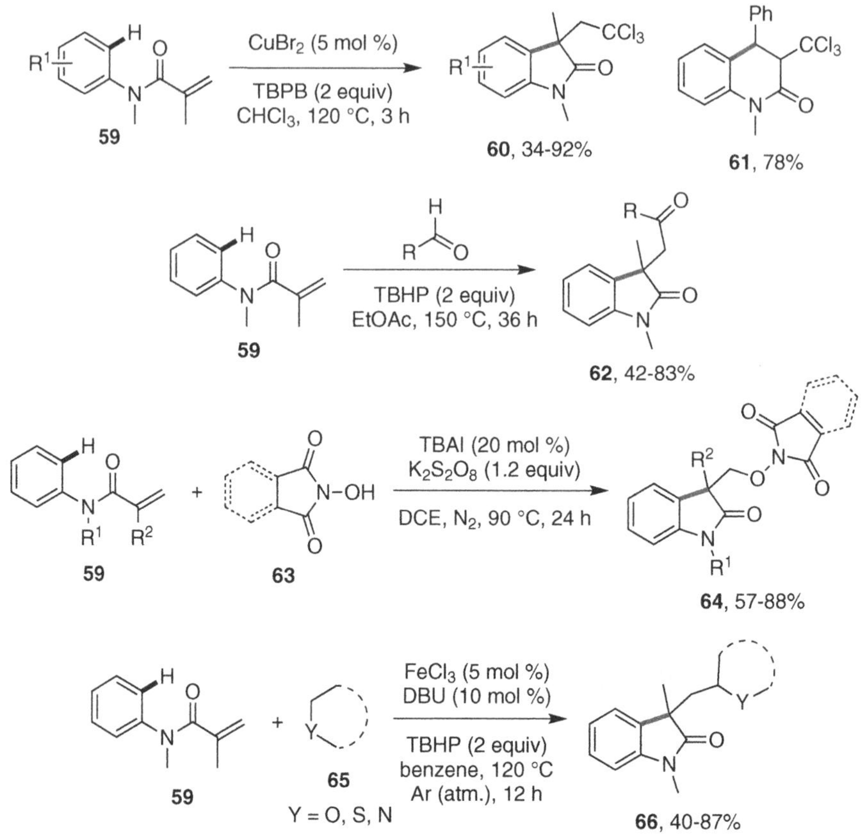

59 → 60, 34-92%; 61, 78%
$CuBr_2$ (5 mol %)
TBPB (2 equiv)
$CHCl_3$, 120 °C, 3 h

59 → 62, 42-83%
RCHO
TBHP (2 equiv)
EtOAc, 150 °C, 36 h

59 + 63 → 64, 57-88%
TBAI (20 mol %)
$K_2S_2O_8$ (1.2 equiv)
DCE, N_2, 90 °C, 24 h

59 + 65 (Y = O, S, N) → 66, 40-87%
$FeCl_3$ (5 mol %)
DBU (10 mol %)
TBHP (2 equiv)
benzene, 120 °C
Ar (atm.), 12 h

SCHEME 1.20

oxindoles **67** were obtained up to 99% yields. However, in the case of F and CF_3 substituents, corresponding tetrahydroquinolin-2-ones **68** were obtained via the 6-*endo* cyclization route. A typical radical pathway was proposed for 5-*exo* cyclization to oxindole **67**, in which intermediate **II**, generated in situ by hydrometallation of substrate **59** with Ni–H species **I**, was transformed to α-amido radical **III**. Subsequent cyclization with an aromatic ring, followed by a hydride transfer, furnished the final oxindole products **67** (Scheme 1.21) (Lu et al. 2018).

Synthesis of optically active 3-substituted 3-hydroxy-2-oxindoles **70** was achieved by the asymmetric intramolecular CDC process of α-ketoamides **69**. Reactions were carried out by heating a solution of α-ketoamides **69**, $[Ir(cod)_2]$ (BAr^F_4) (5 mol%), and an (*R*,*R*)-Me-BIPAM ligand (5.5 mol%) in 1,2-dimethoxyethane (DME) at 135°C, to give 3-hydroxy-2-oxindoles **70** in 73–99% yields with complete regioselectivity and high enantioselectivities (84–98% ee). In the proposed reaction mechanism, the [Ir((*R*,*R*)-Me-bipam)](BAr^F_4) complex ([Ir]), generated in situ from $[Ir(cod)_2](BAr^F_4)$ and (*R*,*R*)-Me-BIPAM, reacted with substrate α-ketoamides **69**, forming aryl iridium complex **I**, which was coordinated with the two carbonyl groups of the amide. Asymmetric hydroarylation of the ketone

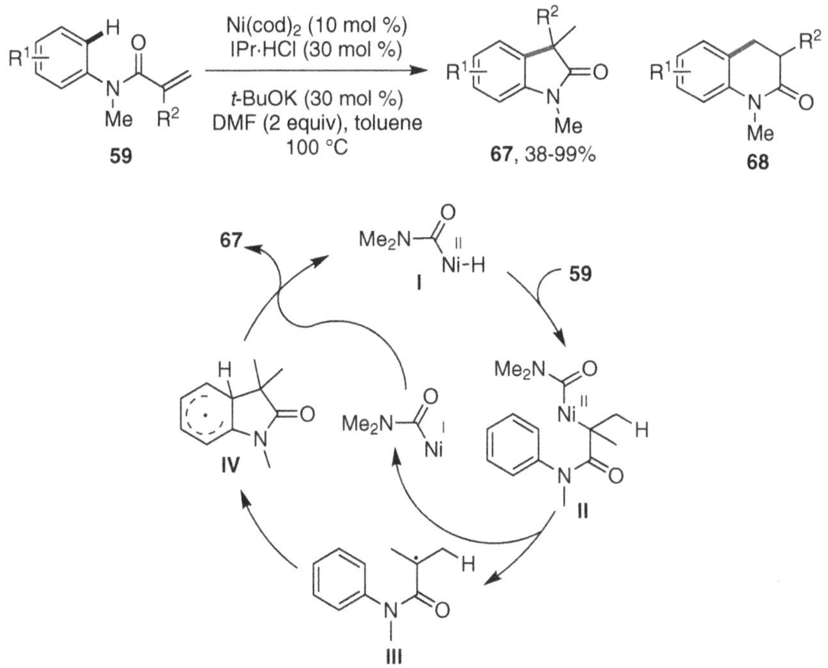

SCHEME 1.21

carbonyl group proceeded from **II**, to produce enantiomerically enriched iridium alkoxide species **III**. Finally, reductive elimination led to the product along with regeneration of the active iridium catalyst [Ir] (Scheme 1.22). The obtained high enantioselectivity could be rationalized by the *si*-face attack in intermediate **II**, which exhibits less steric effects (Shirai, Ito, and Yamamoto 2014, Shirai and Yamamoto 2015).

Hu et al. (2018) reported an unusual cobalt(III)-catalyzed CDC cyclization of aryl C–H bonds of *N*-nitrosoanilines **71** with α-diazo-β-ketoesters **34**, allowing for the assembly of quaternary 2-oxindoles **72**. Reactions were conducted using *N*-nitrosoanilines **71** (0.2 mmol) and α-diazo-β-ketones **34** (1.25 equiv) in the presence of $[Cp^*Co(CO)I_2]$ (10 mol%) and $Zn(OAc)_2$ (50 mol%) in DCE at 120°C for 24 h under an atmosphere of argon. Electron-donating and halide-substituted *N*-nitrosoanilines **71** afforded corresponding oxindoles **72** in good yields (55–81%), while electron-withdrawing (4-CN, 4-NO_2, and 4-CO_2Me) substituted *N*-nitrosoanilines **71** led to the corresponding oxindoles **72** in lower yields (56–63%). Besides the monosubstituted *N*-nitrosoanilines, 3,4-disubstituted *N*-nitrosoanilines underwent site-selective coupling cyclization to the multifunctional 2-oxindoles **72** in 75–82% yields. Also, fused tricyclic skeletons bearing 2-oxindole moiety **73** and **74** were prepared by this protocol. The proposed reaction mechanism involves the reversible C–H activation process, which reacted with diazoketoester **34** to form a Co(III)-carbenoid species with the extrusion of a N_2 molecule (Scheme 1.23).

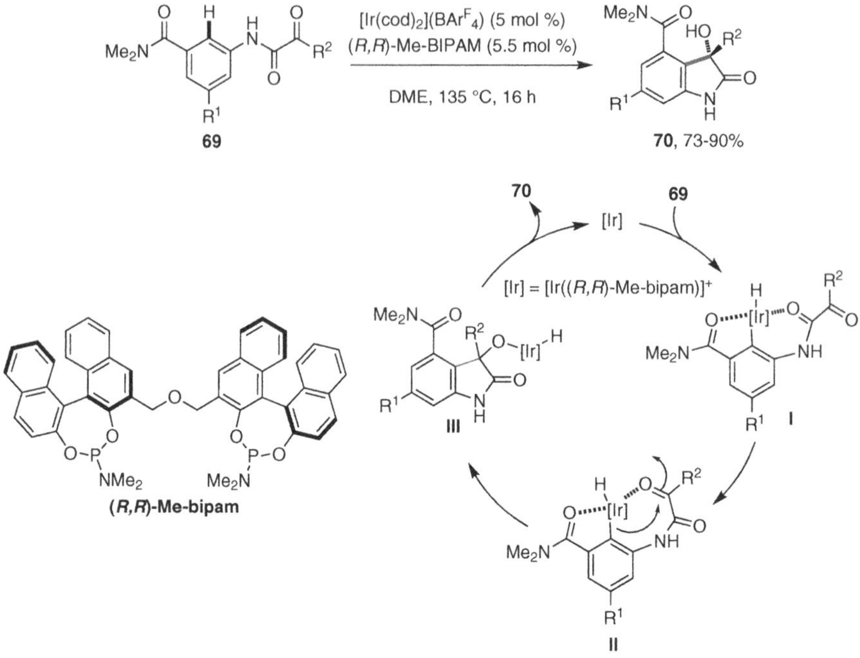

SCHEME 1.22

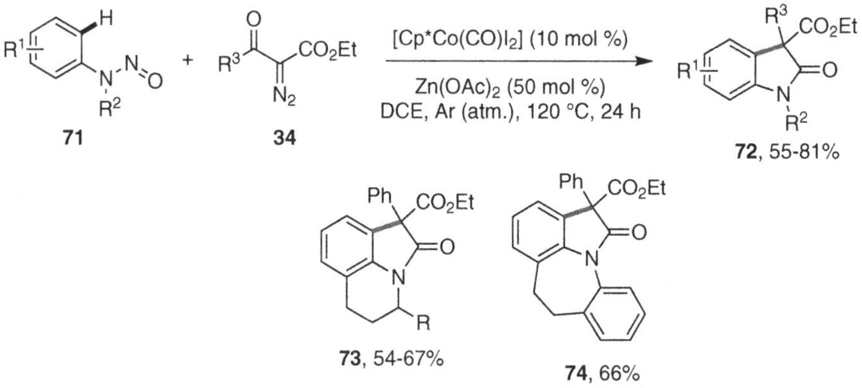

SCHEME 1.23

1.4 SPIRO-OXINDOLES

Cu(II)-mediated intramolecular radical CDC was reported by Hurst et al. (2018) to prepare spirocyclic oxindoles **76**. Treatment of amide **75** with $Cu(OAc)_2{\cdot}H_2O$ (10–100 mol%) under an atmosphere of O_2 in toluene at 110°C afforded spirocyclic oxindoles **76** in 55–68% yields. The oxindole core of Satavaptan, a vasopressin V_2 receptor antagonist, was synthesized using this protocol. Chiral iodine reagent-catalyzed intramolecular C(sp²)–H/C(sp³)–H CDC of N^1,N^3-diphenylmalonamides **77** was reported by Wu et al. (2014) Reactions were performed in CH_3NO_2 at room temperature for 16 h in the presence of organoiodine compound **I-Cat** as catalyst (15 mol%), $MeCO_3H$ (4.0 equiv), and CF_3CO_2H, giving spirocyclic bis-oxindoles **78** in 26–78% yields, with up to 90% ee. Sreenithya and Sunoj (2014) and Sreenithya et al. (2017) developed the synthesis of spirocyclic bis-oxindoles **78** via iodine(III)-promoted, intramolecular dual aryl C–H activation of dianilide **77**. In addition, synthesis of spirofurooxindoles **80** was reported by an enantioselective organocatalytic oxidative spirocyclization of alkyl 3-oxopentanedioate monoamides **79** via chiral aryliodine **I-Cat**-mediated cascade C–O and C–C bond formations. Treatment of alkyl 3-oxopentanedioate monoamides **79** with chiral catalyst **I-Cat** (20 mol%) in the presence of *meta*-chloroperoxybenzoic acid (*m*-CPBA) (2.5 equiv), and CF_3CO_2H (4 equiv) in CH_3CN/TFE mixture at room temperature gave spirofurooxindoles **80** in 46–77% yields, with 74–91% ee. The plausible reaction mechanism involves the formation of a C–I(Ph) bond intermediate **I** reacting **79** with the iodine(III) reagent, generated in situ from oxidation of chiral iodobenzene with *m*-CPBA. Intermediate **I** underwent intramolecular C–O bond formation to give intermediate **II**, which underwent further oxidation, resulting in another C–I bond to afford the chiral intermediates **III**, with subsequent intramolecular C–C bond formation/C–I bond cleavage to optically active spirofurooxindoles **80**. The enantioselectivity of the reaction could be rationalized by the aniline moiety in **III** facing away from the bulky substituent, making the *si*-face of the 2-furanone moiety open to nucleophilic attack by the benzene ring (Scheme 1.24) (Cao et al. 2016).

A Cu(II)-catalyzed intramolecular CDC spirocyclization of indole-2-carboxamides **81** using TBHP as an oxidant was developed, yielding C2-spiro-pseudoindoxyls **82**. Reactions were carried out using 3 equiv of TBHP and 5 mol% of $Cu(OTf)_2$ in DCE at 60°C under N_2 atmosphere, to give C2-spiro-pseudoindoxyls **82** in 46–89% yields (Scheme 1.25) (Kong et al. 2016).

1.5 ISOINDOLES

Pd-catalyzed $C(sp^2)$–H activation/cycloimidoylation of 2-isocyano-2,3-diarylpropanoates **83** to construct 1,1-disubstituted 1*H*-isoindoles **84** has been developed. Subjecting **83** to aryliodide (1.2 equiv) in the presence of $Pd(MeCN)_2Cl_2$ (10 mol%), bidentate ligand bis(diphenylphosphino)butane (DPPB) (10 mol%), Cs_2CO_3, and PivOH in $PhCF_3$ at 80°C under argon atmosphere led to the formation of 1*H*-isoindole derivatives **84** in 57–98% yields. The selectivity for the formation of 1*H*-isoindole **84** was enhanced by using steric-hindered aryl iodides. Density functional theory (DFT) calculations suggested the transition state **TS** for this conversion, in which the η^2-coordination of the DPPB to the palladium center **I** greatly lowers the ring strain of six-membered palladacycle **II** and makes phenylic C–H bond activation favorable due to the strong *trans* effect of the phosphine ligand (Scheme 1.26) (Tang et al. 2018).

Hyster, Ruhl, and Rovis (2013) developed an Rh(III)-catalyzed CDC of *O*-pivaloyl benzhydroxamic acids **85** with α-diazo esters **86a** to provide isoindolones **87**. Reactions were carried out using $[Cp^*RhCl_2]_2$ (1 mol%) and CsOAc (20 mol%) in CH_3CN at 23°C, and corresponding 1-aryl-3-oxoisoindoline-1-carboxylates **87** were

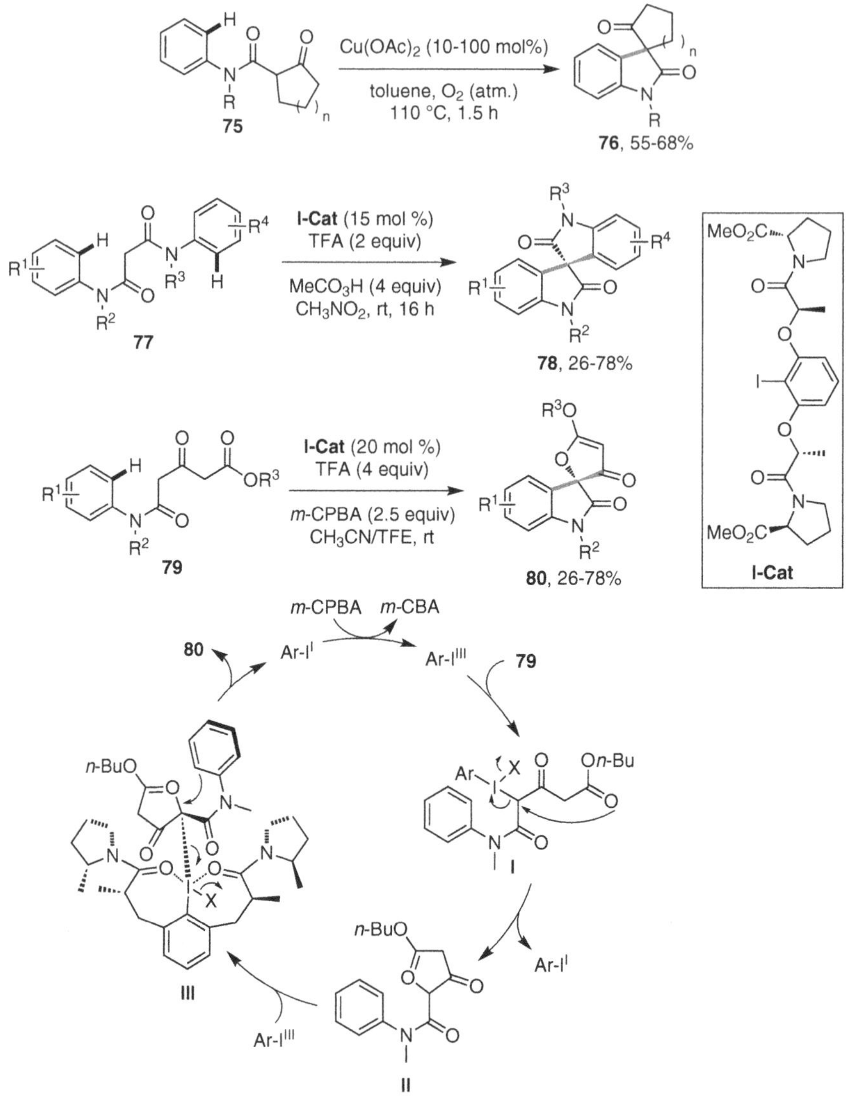

$Cu(OAc)_2$ (10-100 mol%)
toluene, O_2 (atm.)
110 °C, 1.5 h

75

76, 55-68%

I-Cat (15 mol %)
TFA (2 equiv)
$MeCO_3H$ (4 equiv)
CH_3NO_2, rt, 16 h

77

78, 26-78%

I-Cat (20 mol %)
TFA (4 equiv)
m-CPBA (2.5 equiv)
CH_3CN/TFE, rt

79

80, 26-78%

I-Cat

m-CPBA *m*-CBA

Ar-I[I] Ar-I[III]

I **II** **III**

SCHEME 1.24

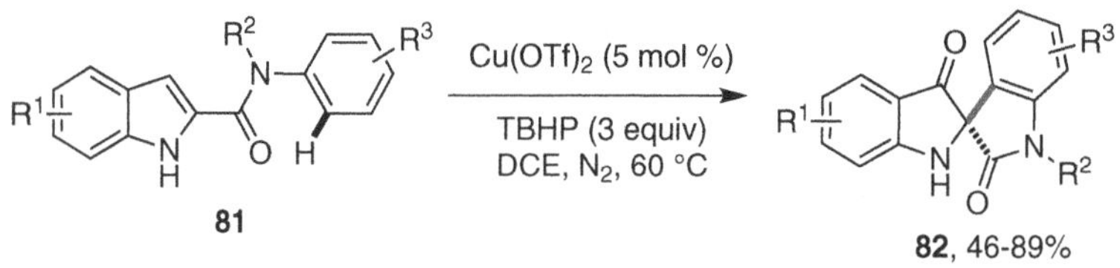

$Cu(OTf)_2$ (5 mol %)
TBHP (3 equiv)
DCE, N_2, 60 °C

81

82, 46-89%

SCHEME 1.25

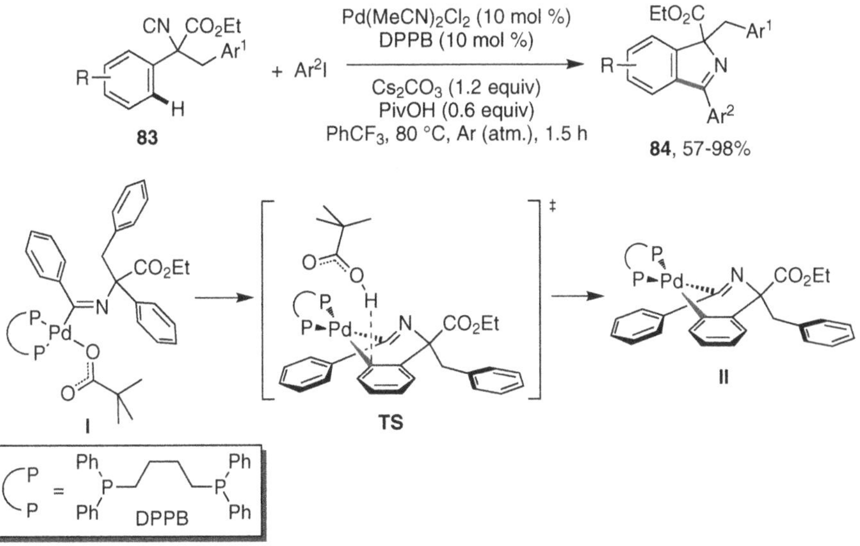

SCHEME 1.26

obtained in 71–99% yields. Reaction with heteroaryl diazo acetate resulted in the formation of the corresponding isoindolones in lower yields (27–37%). Alkyl diazo esters were also tolerated, in which the corresponding products were produced in 68–78% yields at elevated temperature. Cyclic diazo substrates afforded the desired spiro compounds **88** in 64–80% yields under the reaction conditions. In the screening electron-withdrawing groups on the diazocompound, it was found that the 1-phenyl-2,2,2-trifluoro diazoethane worked well under reaction conditions, leading to desired product **87** in 20–97% yields. Similarly, asymmetric synthesis of isoindolones **87** was reported by chiral cyclopentadienyl-Rh(III)-catalyzed C–H activation of *O*-pivaloyl benzhydroxamic acids **85** in 52–92% yields, with 92.5/7.5–96.5/3.5 ee, preferentially with (3*S*) stereoselectivity (Ye and Cramer 2014). Synthesis of dimethyl 3-oxoisoindoline-1,1-dicarboxylates **90** was achieved by a Rh(III)-catalyzed oxidative [4+1] cycloaddition of benzohydroxamic acids **89** and diazomalonate **86b**. Reactions proceeded through *N*-OAc-directing-group-assisted selective functionalization of *ortho*-C–H. Reactions were conducted using [Cp*Rh(OAc)$_2$] (5 mol%) in THF at 60°C for 4 h, to give dimethyl 3-oxoisoindoline-1,1-dicarboxylates **90** in 73–93% yields. Subjecting benzohydroxamic acid to α-diazo esters, bearing alkyl, aryl, or electron-withdrawing groups, such as phenylsulfone, cyano, and diethyl phosphonate groups, provided corresponding 3-oxoisoindolines in 30–90% yields. Spiroisoindolone **88a** was obtained from diazooxindole in 75% yield. The plausible reaction mechanism involves the formation of the five-membered rhodacycle via directing-group-assisted coordination of *O*-acetyl benzohydroxamic acid **89** to the [Cp*Rh(OAc)$_2$] by C–H/N–H deprotonation, together with elimination of acetic acid. Following coordination of the diazocompound, N_2 extrusion, migratory insertion, and then reductive eliminative formation of the C–N bond, the desired product

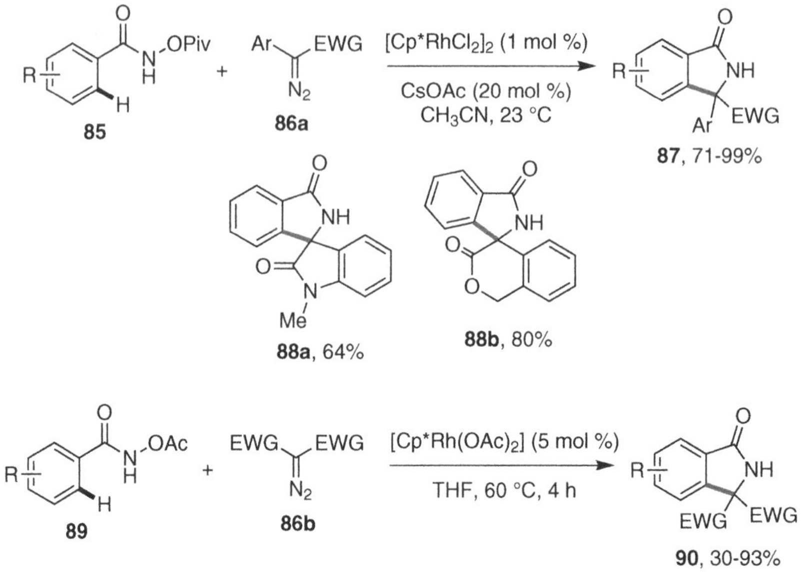

SCHEME 1.27

87 or **90** was obtained, in which the *N*-OAc moiety acted as an internal oxidant to regenerate the active Rh catalyst (Scheme 1.27) (Lam et al. 2014).

Rh(III)-catalyzed one-pot reaction of benzamides **85**, ketones, and hydrazine was developed for the construction of 3-alkyl-3-arylisoindolone. Reactions were performed by heating a solution of an alkyl aryl ketone (1.7 equiv) and hydrazine hydrate (2 equiv) in the presence of HOAc (20 mol%) in THF at 60°C, followed by addition of benzamides **85**, MnO_2 (14 equiv), CsOAc (1 equiv), and $[Cp^*RhCl_2]_2$ (2.5 mol%), and stirring at room temperature, affording 3-alkyl-3-arylisoindolones **91** in 45–80% yields. Diaryl ketones, including benzophenone, *p,p'*-dimethoxybenzophenone, and fluorenone, also furnished corresponding isoindolones **91** in 50–82% yields. Reaction with indoline-2,3-dione led to spiroisoindolone **88a**, in 70–76% yields. Additionally, under reaction conditions, *N*-(pivaloyloxy)-1*H*-pyrrole-1-carboxamide and *N*-(pivaloyloxy)-1*H*-indole-1-carboxamides delivered imidazolidinones **92** in 50–84% yields. Diazocompounds were generated in situ from the reaction of ketones with hydrazine in the presence of MnO_2, as the reaction intermediate (Scheme 1.28) (Zhang, Wang, and Cui 2015).

Wang et al. (2018) described Rh(III)-catalyzed CDC of *N*-methoxybenzamides **93** and ethenesulfonyl fluoride **94** to provide isoindolin-2-ones **95**. Reactions were conducted using 2 equiv of ethenesulfonyl fluoride **94** in the presence of $[Cp^*RhCl_2]_2$ (2.5 mol%), $AgSbF_6$ (1 equiv), and $Cu(OAc)_2$ (20 mol%) in dioxane under air at 80°C for 15 h, affording isoindolin-2-ones **95** bearing sulfonyl fluoride substituent in 31–91% yields. Reaction with thiophen-2-carboxamide led to the formation of corresponding (4*H*-thieno[2,3-*c*]pyrrol-4-yl)methanesulfonyl fluoride in 41% yield. The reactions proceeded by Rh-catalyzed dehydrogenative coupling between *N*-methoxybenzamides and ethenesulfonyl fluoride, followed by intramolecular

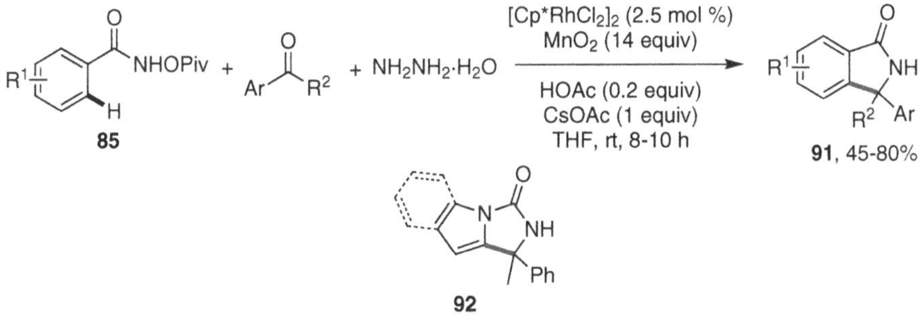

SCHEME 1.28

aza-Michael addition. Wang, Song, and Li (2010) reported an approach for the CDC olefination–Michael reactions between *N*-arylbenzamides **96** and acrylates **97** using $[Cp^*RhCl_2]_2$ (4 mol%) in the presence of Ag_2CO_3 as the oxidant in CH_3CN at 110°C, leading to corresponding isoindolin-2-ones **98** in 81–96% yields. Also, a similar method was developed using air as the oxidant (Lu et al. 2015). A tandem rhodium-catalyzed C–H olefination of *N*-benzoylsulfonamides **96a** with diethyl fumarate **99** followed by C–N bond formation via intramolecular aza-Michael type addition was described by Zhu and Falck (2012). Reactions were carried out by heating a solution of *N*-benzoylsulfonamides **96a** and diethyl fumarate **99** (1.2 equiv) in the presence of $[Cp^*RhCl_2]_2$ (4 mol%) and $Cu(OAc)_2{\cdot}H_2O$ (2.1 equiv) in toluene at 130°C to afford 3,3-disubstituted isoindolinones **100** in 46–85% yields. Other activated alkenes, including diethyl maleate, (4*E*)-octen-3,6-dione, and maleimide, were also investigated, in which corresponding 3,3-disubstituted isoindolinones were obtained in 51–80% yields (Scheme 1.29).

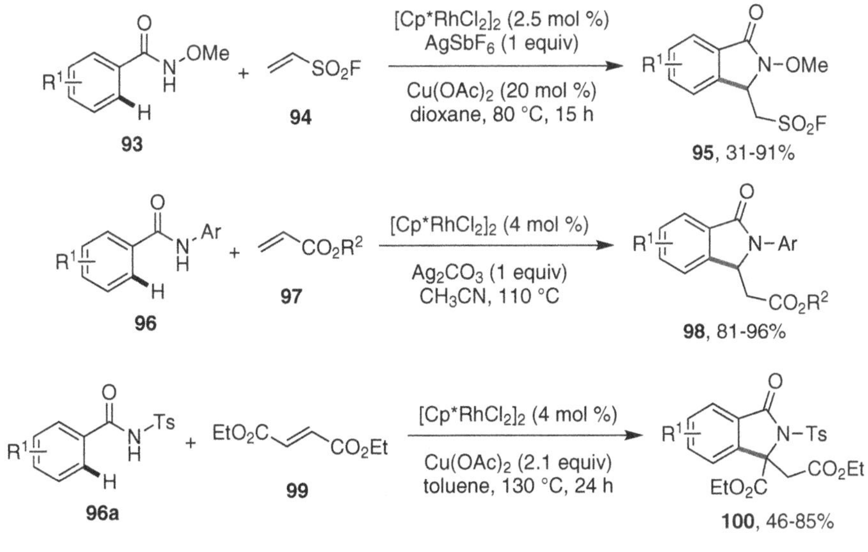

SCHEME 1.29

Isoindolone-incorporated spirosuccinimides **102** were synthesized via $Cu(OAc)_2$/Cy_2NMe-mediated dehydrogenative coupling of benzamides **96b** with maleimides **101**, with subsequent intramolecular aza-Michael-type addition. Treatment of benzamides **96b** with maleimides **101** (4 equiv) in the presence of $Cu(OAc)_2$ (4 equiv), PivOH, and Cy_2NMe in DMF under an atmosphere of N_2 at 80°C gave spiroisoindolones **102** in 19–99% yields (Miura, Hirano, and Miura 2015). A similar approach for the construction of spiroisoindolones **102** was developed using a catalytic amount of $Co(OAc)_2{\cdot}4H_2O$ (10 mol%) in the presence of pivalic acid and Ag_2CO_3 in DCE under N_2 at 100°C (Manoharan and Jeganmohan 2017). Cobalt-catalyzed synthesis of five-membered spirocycles **104** is reported from benzimidates **103** and maleimides **101** utilizing nitrobenzene as the promoter. Subjecting benzimidates **103** to maleimides **101** in the presence of $[Cp^*Co(CO)I_2]$ (10 mol%), AgOTf (20 mol%), and $PhNO_2$ (1 equiv) in DCE at 100°C liberated 3-alkoxyspiroisoindoles **104** in 35–95% yields (Lv et al. 2017). Also, a similar methodology was described by reacting benzimidates **103** with acrylates **97** catalyzed with $[RuCl_2(p\text{-cymene})]_2$ (5 mol%) in the presence of Adm-1-COOH in EtOH at room temperature, affording 3-ethoxy-1*H*-isoindole **105** in 45–91% yields (Scheme 1.30) (Manikandan, Tamizmani, and Jeganmohan 2017).

Stereoselective synthesis of (*E*)-3-methyleneisoindolin-1-ones **107** was achieved by palladium-catalyzed tandem dehydrogenative coupling of primary benzamides **106** with acrylates **97** via intermolecular *N*-alkenylation followed by intramolecular *C*-alkenylation. Reactions were performed by heating a mixture of benzamides **106** and an excess amount of alkene in the presence of $Pd(OAc)_2$ (10 mol%),

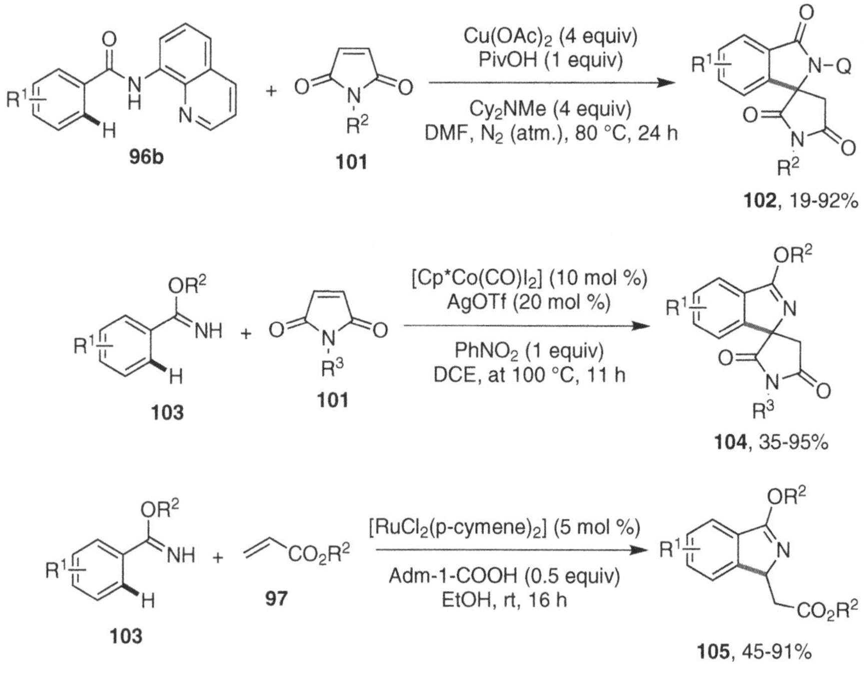

96b + 101 → 102, 19-92%
Cu(OAc)2 (4 equiv), PivOH (1 equiv) / Cy2NMe (4 equiv), DMF, N2 (atm.), 80 °C, 24 h

103 + 101 → 104, 35-95%
[Cp*Co(CO)I2] (10 mol %), AgOTf (20 mol %) / PhNO2 (1 equiv), DCE, at 100 °C, 11 h

103 + 97 → 105, 45-91%
[RuCl2(p-cymene)2] (5 mol %) / Adm-1-COOH (0.5 equiv), EtOH, rt, 16 h

SCHEME 1.30

$Cu(OAc)_2 \cdot H_2O$ (2 equiv), and pyridine in 1,4-dioxane at 110°C for 48 h, leading to (*E*)-3-methyleneisoindolin-1-ones **107** in 40–76% yields. The reaction mechanism was proposed according to DFT calculations, in which (*E*)-*N*-alkenylated benzamide **I** coordinated to Pd, forming complex **II**, which underwent rotation across the C–N bond to complex **III**. By intramolecular dehydrogenative cyclization, intermediate **IV** was formed, which was converted to the final desired product by releasing a Pd(0) species. Cu(II) oxidized the Pd(0) to the active catalytic species Pd(II) for the next catalyst cycle. As depicted in Scheme 1.31, due to the hindered rotation across the C–N bond (**V**), the *Z*-enamide did not undergo intramolecular oxidative cyclization (Laha et al. 2017).

Regio- and stereoselective synthesis of (*E*)-3-arylmethyleneisoindolin-1-ones **109** via Pd(II)/Cu(II)-catalyzed one-pot C–C/C–N bond-forming sequence between amides **96a** and styrenes **108** was reported. Reactions were conducted by heating a solution of *N*-tosylbenzamides **96a** with 2 equiv of styrene **108** in the presence of $Pd(TFA)_2$ (5 mol%) and $Cu(OAc)_2$ (40 mol%) in toluene at 120°C under air conditions, and (*E*)-3-arylmethyleneisoindolin-1-ones **109** were obtained in 30–92% yields (Scheme 1.32) (Youn et al. 2018).

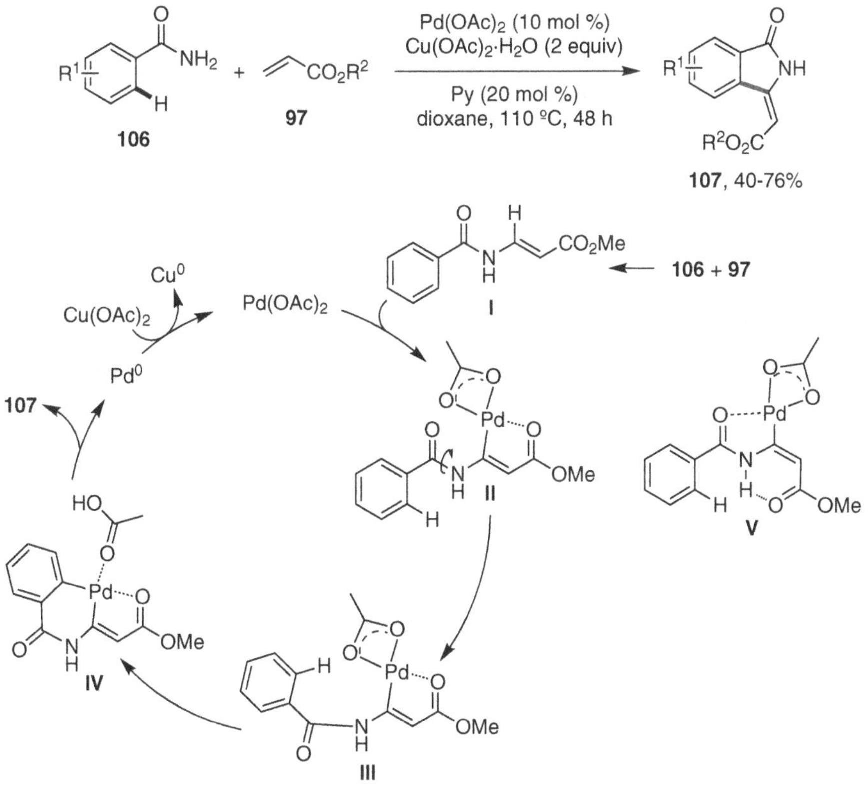

SCHEME 1.31

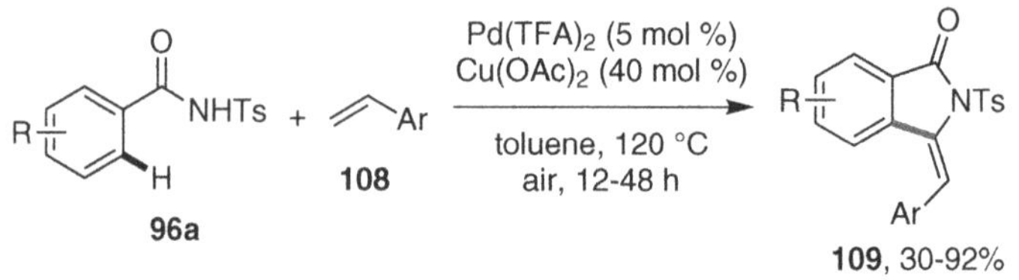

96a + 108 → 109, 30-92%

Pd(TFA)$_2$ (5 mol %)
Cu(OAc)$_2$ (40 mol %)
toluene, 120 °C
air, 12-48 h

SCHEME 1.32

The synthesis of 3-alkylidene isoindolinones **111** was achieved using a redox-neutral bimetallic Rh(III)/Ag(I) relay catalysis. Reactions were conducted using $[Cp^*Rh(CH_3CN)_3][SbF_6]_2$ (5 mol%), $AgSbF_6$ (20 mol%), and CsOPiv (1 equiv) in trifluoroethanol at 60°C, to furnish 3-alkylidene isoindolinones **111** in 28–81% yields. The *para*-substituted *N*-tosylbenzamides **96a** afforded corresponding 3-alkylidene isoindolinones **111** in good yields. The *meta*-substituted *N*-tosylbenzamides **96a** produced only one regioisomer, with the reaction occurring at the less sterically hindered positions, while the *ortho*-substituents *N*-tosylbenzamides **96a** led to corresponding 3-alkylidene isoindolinones **111** in low yields (28–32%). Reactions proceeded by the Rh(III)-catalyzed dehydrogenative C–H monofluoroalkenylation reaction, followed by the Ag(I)-salt-promoted cyclization. Initially, the *N*-Ts-directing-group-assisted Rh(III)-catalyzed C–H activation occurred to generate intermediate **I**, which underwent coordination with the alkene (intermediate **II**), followed by a regioselective alkene insertion to give seven-membered palladacycle **III**. A *syn*-coplanar β-F elimination provided the *Z*-type monofluoroalkenylation product **IV** with good stereoselectivity. The Ag(I)-induced cyclization took place by activation of olefin via formation of a π-complex **V**, which underwent intramolecular 5-*exo* cyclization to give **VI-a**, with, subsequently, an anti-coplanar β-F elimination, leading to the formation of 3-alkylidene isoindolinone **111**, in a stereospecific manner (Scheme 1.33) (Ji et al. 2017).

Synthesis of isoindolinones **112** was reported by Liang et al. (2015) via Pd-catalyzed C–H functionalization of carboxamides **96b** with carboxylic acids or anhydrides. Reactions with acetic anhydride (2 equiv) were performed using Pd(TFA)$_2$ (5 mol%) in toluene at 130°C for 12 h, giving corresponding 3-methyleneisoindolinones **112** in 60–99% yields. Subjecting thiophene-derived carboxamide to acetic anhydride under reaction conditions afforded 4-methylene-4*H*-thieno[2,3-*c*]pyrrol-6(5*H*)-one **112a** in 21% yield. Other anhydrides were also used, which resulted in the formation of isoindolinones **112** bearing alkyl or aryl groups at the terminal olefinic position in 50–93% yields. Reactions with carboxylic acids were carried out in the presence of $(t\text{-BuCO})_2O$ (2 equiv) and Pd(TFA)$_2$ (10 mol%) in toluene at 130°C under argon atmosphere for 12 h, to furnish corresponding isoindolinones **112** in 43–90% yields. $(t\text{-BuCO})_2O$ acts as the activator of carboxylic acids by converting them to the corresponding asymmetrical anhydride (Scheme 1.34).

Dong, Wang, and You (2014) developed copper-mediated tandem oxidative C(sp^2)–H/C(sp)–H cross-coupling, with subsequent intramolecular annulation of arenes with terminal alkynes. By subjecting *N*-(quinolin-8-yl)benzamides **96b** to terminal alkynes (2.5 equiv) in the presence of Cu(OAc)$_2$ (3 equiv) in *t*-AmylOH at

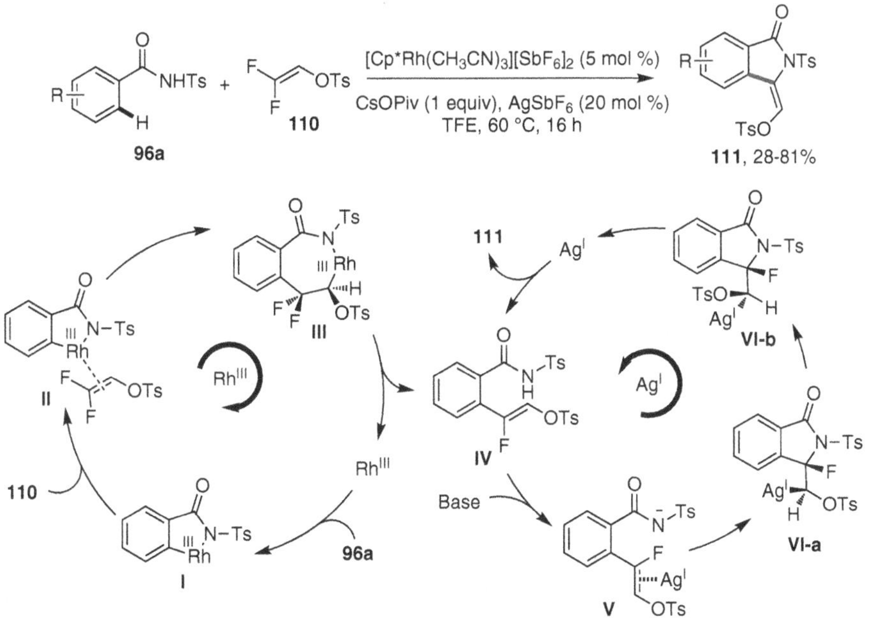

SCHEME 1.33

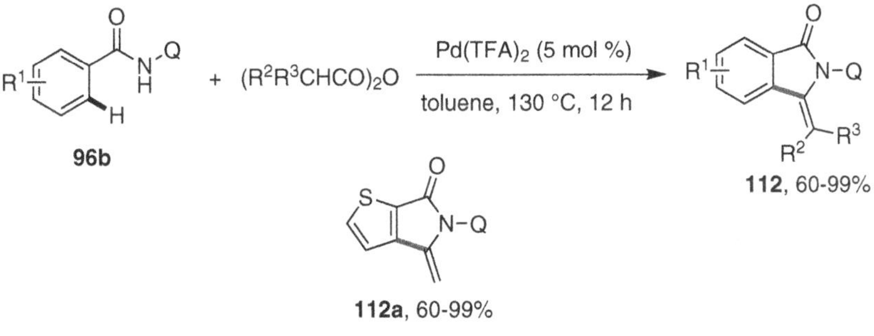

SCHEME 1.34

120°C, 3-methyleneisoindolin-1-one scaffolds **113** were obtained in 62–93% yields. The terminal alkyne possessing an alkyl substituent led to the corresponding 3-methyleneisoindolin-1-one in lower yield (42%), even at higher temperature (Scheme 1.35). A similar approach was developed using 2-aminophenyl-1*H*-pyrazole as a directing group (Lee, Wang, and Li 2018). Ni-catalyzed CDC of 2-benzamidopyridine 1-oxide with terminal alkynes was also reported, using pyridine *N*-oxide as a directing group (Zheng et al. 2016). Co-catalyzed cyclization of *N*-(quinolin-8-yl)benzamides **96b** with terminal alkynes using an Ag cocatalyst was reported by Zhang et al. (2015). Treating *N*-(quinolin-8-yl)benzamides **96b** with a terminal alkyne in the presence of $Co(OAc)_2{\cdot}4H_2O$ (20 mol%), Ag_2CO_3 (4 equiv), and tetrabutylammonium iodide

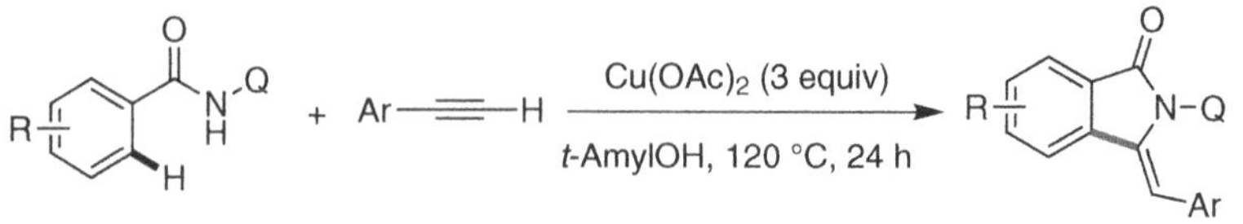

SCHEME 1.35

(TBAI) in $PhCF_3$ under N_2 atmosphere at 120°C produced 3-methyleneisoindolin-1-ones **113** in 41–80% yields.

Ruthenium-catalyzed redox-neutral [4+1] annulation of *N*-ethoxybenzamides **93** and propargyl alcohols **114** via C–H bond activation was developed by Wu et al. (2017a). Reactions were performed with 2 equiv of propargyl alcohol **114** using $[RuCl_2(p\text{-cymene})]_2$ (5 mol%) and CsOAc (1 equiv) in anhydrous DCE at 60°C to give quaternary isoindolinones **115** in 36–90% yields. *N*-ethoxybenzamides bearing electron-withdrawing substituent at the *para* position afforded corresponding isoindolinones in lower yields (33–42%). *Ortho*-substituted *N*-ethoxybenzamides provided the desired product in decreased yields. Coordination of benzamide **93** to the active Ru(II) species generated by anion exchange with cesium acetate and subsequent *ortho*-C–H bond activation formed a five-membered ruthenacycle **I**, which underwent regioselective coordination and migratory insertion with propargyl alcohol **114** to provide a seven-membered intermediate **II**. The abstraction of the allylic proton by the ruthenium complex or 1,2-hydride migration afforded a π-allylic ruthenacycle intermediate **III**, which transformed into intermediate **IV** by reductive elimination and enol–keto tautomerism. An oxidative addition to break the N–O bond generated intermediate **V**, which underwent reductive elimination to release the final desired product **115** and concomitantly regenerate the Ru(II) catalyst to the next catalytic cycle. In 2017, the same research group reported a similar approach for the construction of quaternary isoindolinones using an Rh(III) catalyst (Wu et al. 2017b). Also, reaction of *N*-methoxybenzamides with propargyl alcohols (1.5 equiv) in the presence of $[Cp^*RhCl_2]_2$ (2.5–4 mol%), AgOAc (10–16 mol%), and Ag_2CO_3 (1.5 equiv) in MeCN under air atmosphere at 30°C was reported to afford quaternary *N*-methoxyisoindolinones **115** in 25–83% yields (Xu et al. 2017). Also, the preparation of *N*-methoxyisoindolinones **115** was achieved by coupling of *N*-methoxybenzamides **93** with α-allenols **116** via an Rh(III)-catalyzed C–H activation. Reactions were performed by stirring a solution of **93** and α-allenols **116** (3 equiv) in the presence of a catalytic amount of $[Cp^*RhCl_2]_2$ (2.5 mol%) and AgOAc (2 equiv) in CH_3CN at room temperature under air for 24 h, resulting in the formation of the final isoindolinone products **115** in 51–74% yields (Scheme 1.36) (Zhou, Liu, and Lu 2016).

Isoindolin-1-ones **118** bearing 3-vinyl substituent were synthesized via an Rh-catalyzed C–H activation/allene formation/cyclization sequence. Reactions were carried out by refluxing a mixture of *N*-alkoxybenzamide **93** (2 equiv) and 2-alkynylic acetates **117** in the presence of $[Cp^*RhCl_2]_2$ (4 mol%), NaOAc (30 mol%), and HOAc (1 equiv) in water, to give 3,3-disubstituted isoindolin-1-ones **118**

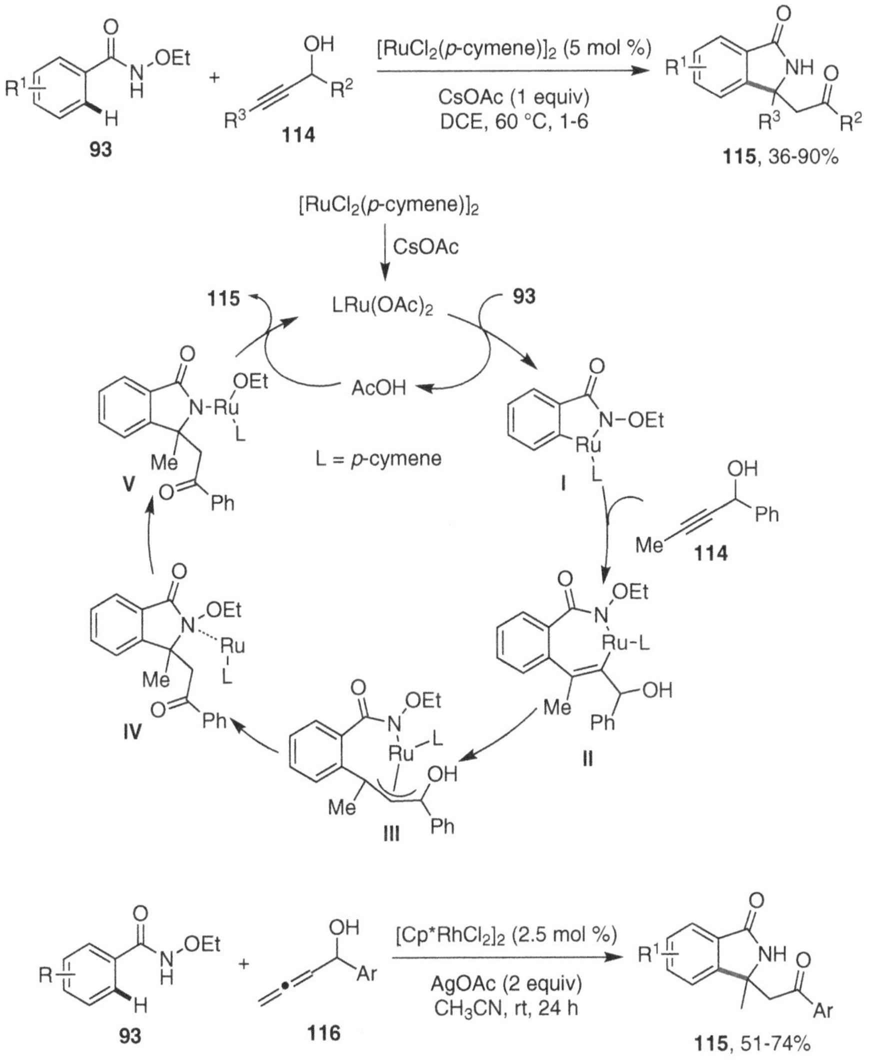

SCHEME 1.36

in 60–86% yields. No reaction occurred with aryl 2-alkynic acetate (R^2=Ph) and unsubstituted 2-alkynic acetate (R^3=R^4=H). In the proposed reaction mechanism, the cyclic intermediate **I** was formed via a dehydrogenative rhodation, which underwent propargylic-acetate-induced regiospecific insertion of the alkyne moiety by the coordination of the carbonyl group with Rh (intermediate **II**), leading to the carbonyl oxygen-coordinated rhoda-tricyclic **III**. By β-OAc elimination, and an allene moiety coordination with Rh(III), intermediate **IV** was formed, in which the regioselective azametalation of the allene unit (**V**) followed by protonolysis with AcOH afforded the 3,3-disubstituted isoindolin-1-one **118**, along with regeneration of the catalytically active Rh(III) species (Scheme 1.37) (Wu et al. 2018).

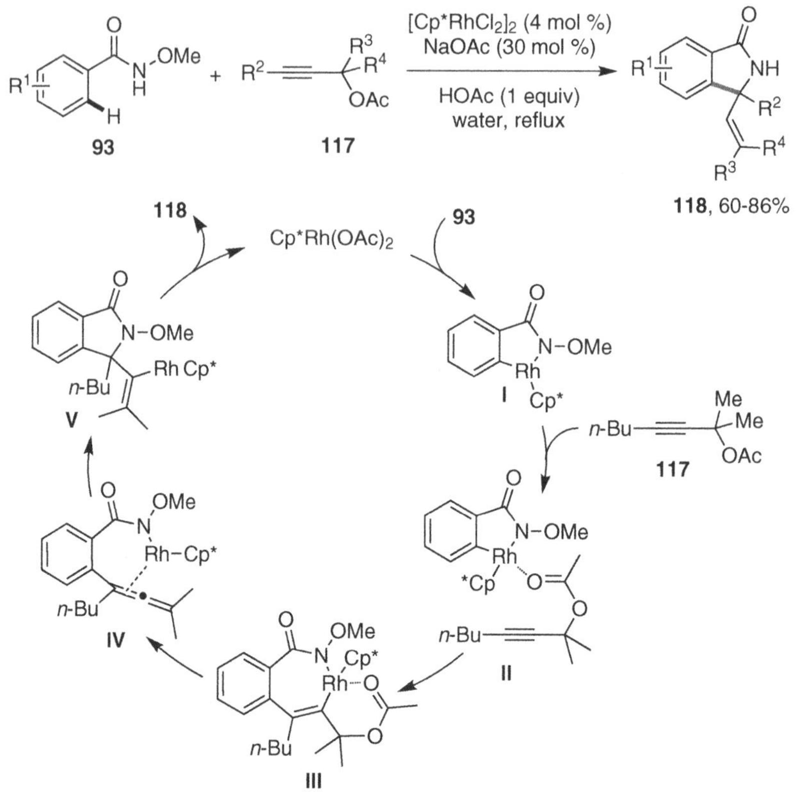

SCHEME 1.37

Also, 3-alkynyl substituted isoindolin-1-ones **120** were prepared through Rh-catalyzed redox-neutral [4+1] annulation via C–H bond activation. Treatment of *N*-methoxybenzamide **93** with 1.2 equiv of *gem*-difluoromethylene alkyne **119** in the presence of $[Cp^*RhCl_2]_2$ (2 mol%), KOAc (30 mol%), and 3 Å molecular sieve in MeOH at 40 or 80°C for 12–24 h resulted in the formation of corresponding 3-alkynyl substituted isoindolin-1-ones **120** in 42–85% yields. Reactions proceeded by the formation of five-membered rhodacycle **I**, followed by a regioselectively migratory insertion to the seven-membered rhodacycle **III** (via intermediate **II**). Cleavage of one of two C–F bonds occurred to selectively afford allene **IV**, which underwent intramolecular aminorhodation to generate alkenyl rhodium intermediate **V**. Finally, the second *β*-F elimination delivered the desired product **120** through a migratory reconstruction of the C≡C triple bond accompanied by catalyst regeneration. During the reaction, relocation of the C≡C triple bond took place (Scheme 1.38) (Wang et al. 2017).

Ruthenium-catalyzed intramolecular C(sp²)–H carbonylation of oxalyl amide (OA)-protected benzylamines **121** with isocyanate **122** as the carbonyl source was developed for the synthesis of isoindolin-1-ones **123**. By subjecting OA-protected benzylamines **121** to 1.2 equiv of isocyanate **122** in the presence of $[RuCl_2(p\text{-cymene})]_2$

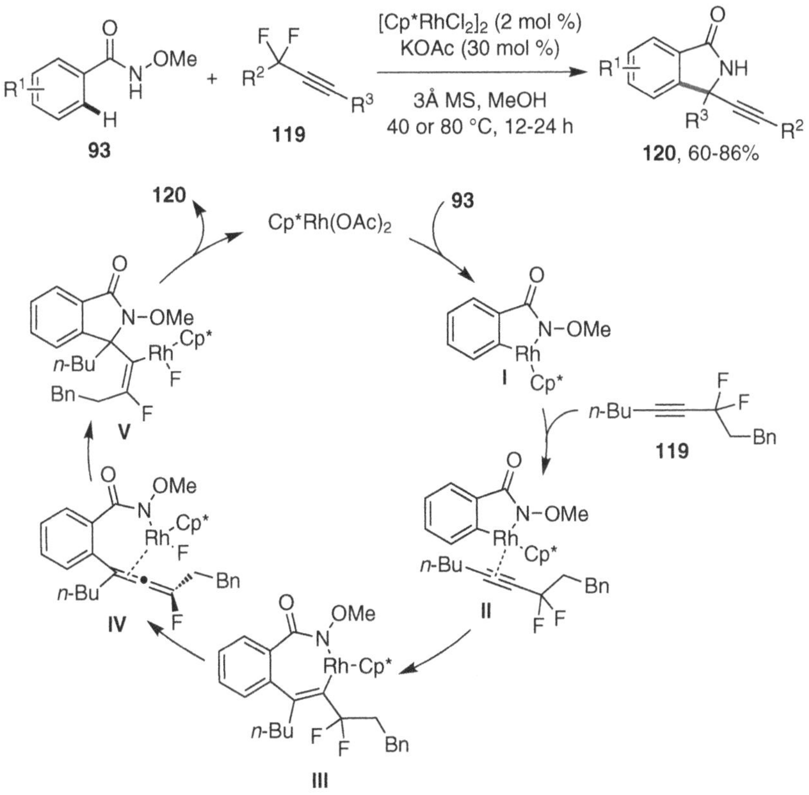

SCHEME 1.38

(5 mol%) and NaOAc (4 equiv) in DCE at 150°C, isoindolin-1-one derivatives **123** were obtained in 31–96% yields (Han et al. 2017). Also, palladium-catalyzed C–H carbonylation of primary benzylamines **124** using NH_2 as the directing group under an atmospheric pressure of CO has been achieved, leading to isoindolin-1-one derivatives **125** in 31–96% yields (Scheme 1.39). Substrates with unsubstituted and monosubstituted benzyl positions afforded corresponding isoindolin-1-ones in good to high yields. However, no reaction occurred in the case of a substrate with dimethyl substitutions at the benzyl position (Zhang et al. 2018a).

Co-catalyzed electrochemical intramolecular C–H/N–H carbonylation was developed for the construction of phthalimides **126**. Reactions were performed in an H-type divided cell under a constant current (15 mA), using *N*-quinolin-8-yl arylamides **96b**, $Co(OAc)_2{\cdot}4H_2O$ (15 mol%), NaOPiv (1 equiv), n-Bu_4NBF_4 (2 equiv), and 4 Å molecular sieves in DMF (anode), and NaOAc (4 equiv) in water/AcOH (7/1 V/V) mixture (cathode), under a CO balloon (1 atm) at 40°C for 2 h, providing phthalimides **121** in 40–85% yields. Also, heterocyclic oxamides such as thiophene-2-carboxamide, thiophene-3-carboxamide, benzothiophene-2-carboxamide, and *N*-methyl indole-2-carboxamide were investigated, in which corresponding imides were obtained in 52–73% yields (Zeng et al. 2018). Also, an efficient approach for the C(sp^2)–H bond

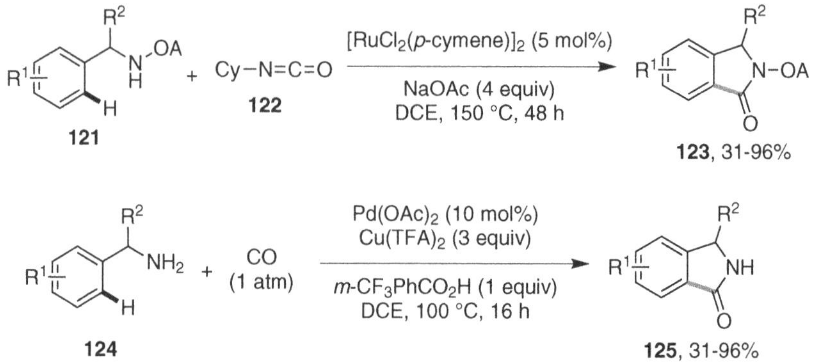

SCHEME 1.39

carbonylation of benzamides has been developed using $Co(OAc)_2{\cdot}4H_2O$ as the catalyst, Ag_2CO_3 as the oxidant, and azodicarboxylates as the carbonyl source (Ni et al. 2016). Wu, Zhao, and Ge (2015) reported the synthesis of phthalimides **126** by a direct aerobic carbonylation of aromatic $C(sp^2)$–H bonds of benzamides **96b** via Ni/Cu synergic catalysis with the assistance of a bidentate directing group using DMF as the CO source. By subjecting *N*-quinolin-8-yl benzamides **96b** to NiI_2 (10 mol%), $Cu(acac)_2$ (20 mol%), Li_2CO_3 (0.4 equiv), and THAB (1 equiv) under O_2 atmosphere (1 atm) in DMF at 160°C for 24 h, *N*-quinolin-8-yl phthalimides **126** were obtained in 51–90% yields. *N*-methyl-*N*-methylenemethanaminium species, generated in situ from DMF via a sequential decarbonylation, nucleophilic addition, and elimination process under copper catalysis with oxygen as the external oxidant, acted as the CO source. Additionally, the synthesis of 3-iminoisoindolinones **127** was reported by a Co-catalyzed annulation of amides **96b** with isocyanides via a $C(sp^2)$–H activation process. Reactions were performed using $Co(OAc)_2$ (20 mol%), TBPB (2 equiv), and Na_2CO_3 (2 equiv) in 1,4-dioxane at 110°C, under argon atmosphere and anhydrous conditions, which resulted in the creation of 3-iminoisoindolinones **127** in 35–86% yields (Scheme 1.40).

1.6 CARBAZOLES

Zirconium oxide-supported palladium hydroxide ($Pd(OH)_2/ZrO_2$) catalyzed the oxidative intramolecular couplings of diarylamines **128** to carbazoles **129** via dual aryl C–H bond functionalizations with molecular oxygen as the sole oxidant. Subjecting diarylamines **128** to 10 wt% $Pd(OH)_2/ZrO_2$ (50 mg) in a mixture of AcOH/1,4-dioxane at O_2 pressure afforded carbazoles **129** in 50–86% yields. Reaction with *o*-substituted diarylamines occurred in low yields (8–20%), due to the inhibition of the coordination of the nitrogen atom to Pd(II) species by steric hindrance (Ishida et al. 2014). Also, the oxidative cyclization of diarylamines to the corresponding carbazoles was performed by the $MoCl_5/TiCl_4$ mixture in CH_2Cl_2 at 0°C (Trosien, Böttger, and Waldvogel 2013). Palladium-catalyzed direct synthesis of carbazoles **129** was reported via one-pot *N*-arylation and dehydrogenative biaryl coupling. Treating aryltriflates **130** with aniline **10** (1.1 equiv), in the presence

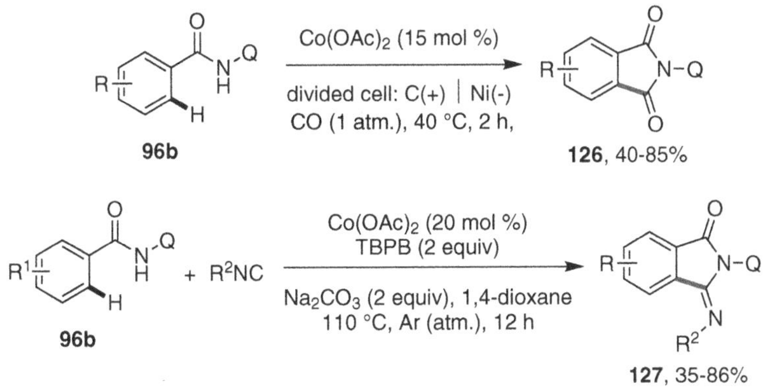

SCHEME 1.40

of $Pd(OAc)_2$ (10 mol%), ligand **L** (15 mol%), and Cs_2CO_3 (1.2 equiv) in toluene/AcOH under O_2 atmosphere at 100°C resulted in the formation of carbazoles **129** in 20–99% yields. No reaction occurred when both triflate and aniline possessed electron-withdrawing groups at the *para* position (Scheme 1.41) (Watanabe et al. 2009, Watanabe et al. 2007).

The synthesis of carbazole derivatives **129** was also achieved via a Pd-catalyzed dehydrogenative aromatization/C(sp^2)–C(sp^2) coupling sequence. Reactions were performed between cyclohexanone **131** and aniline **10** (1.4 equiv) in the presence of $Pd(OAc)_2$ (10 mol%) and $Cu(OAc)_2$ (6 equiv) in PivOH at 140°C under N_2 atmosphere for 24 h, to produce carbazoles **129** in 32–84% yields. The reaction was initiated by enamine formation between aniline and cyclohexanone (Wen et al. 2016). Intramolecular oxidative coupling of *N*-alkyl-3-(arylamino)cyclohex-2-enones **132** mediated by $Pd(OAc)_2$ under an O_2 atmosphere was developed to access *N*-alkylated carbazolones **133**. Oxidative coupling was conducted by heating a solution of *N*-substituted 3-(arylamino)cyclohex-2-enones **131** with 10 mol% of $Pd(OAc)_2$ in AcOH at 100°C under an air flow, providing carbazolones **133** in 25–85% yields. The six-membered palladacycle **I** was proposed as the key intermediate of this

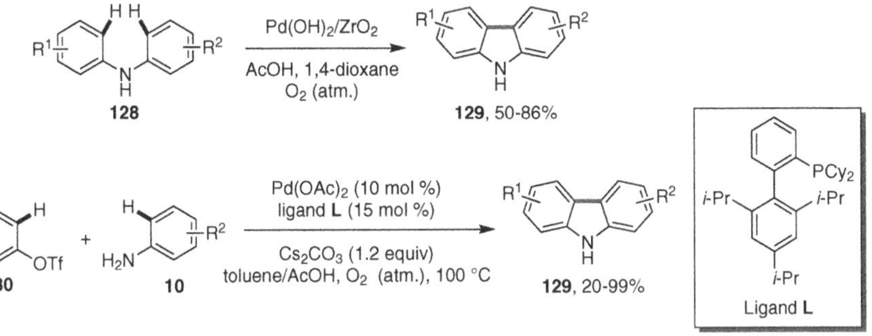

SCHEME 1.41

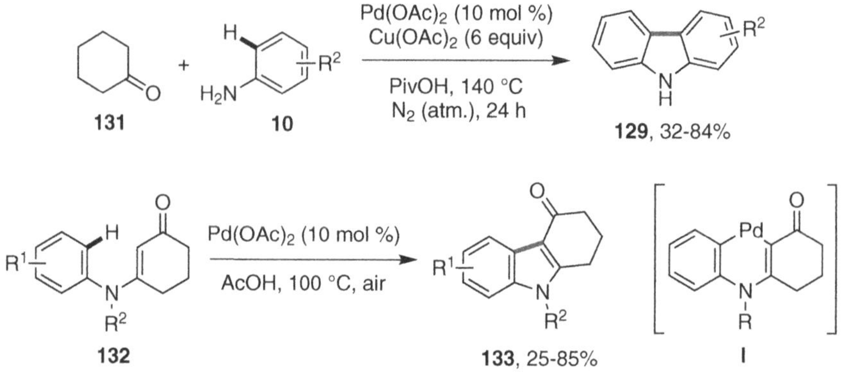

SCHEME 1.42

conversion, which underwent reductive elimination to achieve the desired product (Scheme 1.42) (Bi et al. 2010).

Cu-catalyzed intramolecular oxidative C–N bond formation has been developed for the construction of carbazoles **135** starting from *N*-substituted amidobiphenyls **134** using hypervalent iodine(III) as an oxidant. *N*-substituted amidobiphenyls **134** were subjected to $Cu(OTf)_2$ (5 mol%), and $PhI(OAc)_2$ (1.5 equiv) in 1,2-dichloroethane for 10 min at 50°C, and carbazole derivatives **135** were obtained in 47–98% yields. Also, reactions were conducted in the absence of $Cu(OTf)_2$ using $PhI(OAC)_2$ or $PhI(TFA)_2$, leading to carbazoles **135** in 8–78% or 9–98% yields, respectively (Cho, Yoon, and Chang 2011). Reddy, Kannaboina, and Das (2017) reported the synthesis of pyrano[2,3-*c*]carbazoles **137** via intramolecular oxidative C–N bond formation reactions. Reactions were conducted by heating a solution of *N*-Ts-2-aminobiaryls **136** in DMF, in the presence of $Pd(OAc)_2$ (10 mol%) and K_2CO_3 (3 equiv) under O_2 atmosphere at 100°C for 12 h, and pyrano[2,3-*c*]carbazoles **137** were obtained in 74–83% yields. Interestingly, treating *N*-Ts-2-aminobiaryls **136** with $Pd(OAc)_2$ (10 mol%), and oxone (1 equiv) in the presence of $PTSA{\cdot}H_2O$ (0.5 equiv) in a mixture of pivalic acid and DMF at room temperature for 12 h resulted in the formation of indeno-chromenes **138**. An Rh(III)-catalyzed reaction of biarylboronic acid **139** with aryl azide **140** to obtain unsymmetrical carbazoles **141** has been developed via construction of dual distinct C–N bonds via aromatic $C(sp^2)$–H activation and rhodium nitrene insertion. Reactions were performed using 1.2 equiv of biarylboronic acid **139** in the presence of $[Cp^*RhCl_2]_2$ (2.5 mol%), AgOAc (1.2 equiv), and Na_2CO_3 (1.5 equiv) in dioxane under air at 80°C for 12 h, furnishing carbazoles **141** in 50–91% yields. Reactions proceeded by transmetalation of biarylboronic acid **139** with Rh(III), followed by acetate-assisted intramolecular C–H bond activation to form a five-membered rhodacycle, which transformed into an Rh(V)-nitrenoid intermediate by coordination with azide, followed by release of N_2. By migratory insertion, with subsequent reductive elimination, carbazole **141** was obtained, along with formation of Rh(I) species. Oxidation of the Rh(I) species with an AgOAc-regenerated active Rh(III) catalyst in order to finish the catalytic cycle (Scheme 1.43) (Xu et al. 2018b).

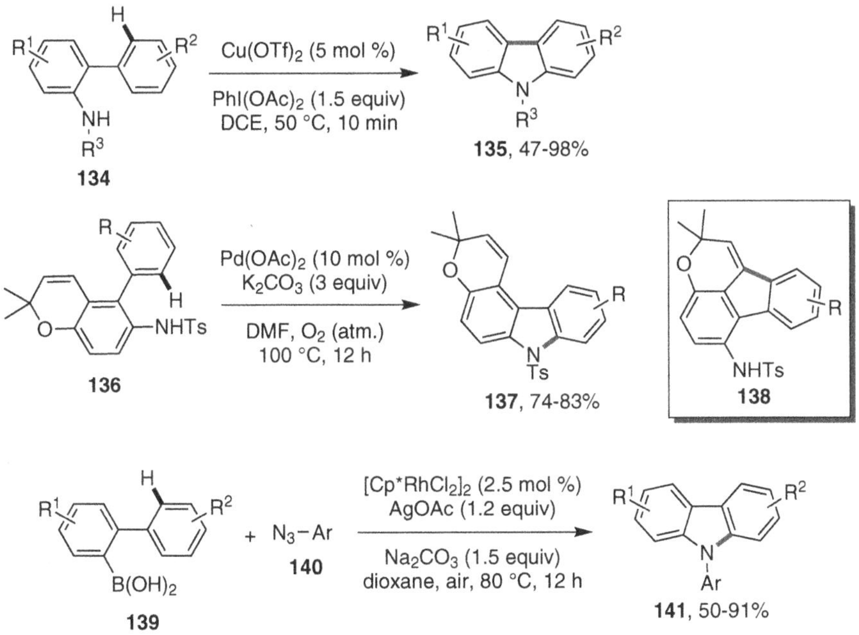

SCHEME 1.43

1.7 FUSED INDOLES

1.7.1 Pyrido-Indoles

The synthesis of pyrido[2,3-*b*]indoles **143** was achieved in 66–96% yields by subjecting 1,2,3,4-tetrazoles **142** bearing C8-substituted arenes to 5 mol% of $[Cp^*IrCl_2]_2$ in the presence of $AgSbF_6$ (20 mol%) in benzene at 130°C for 12–24 h. Reactions proceeded by Ir(III)-catalyzed intramolecular denitrogenative transannulation/C(sp^2)–H amination. In the plausible reaction pathway, the species **II** was generated by coordination of the N1 atom of the tetrazole **142b** to active catalytic species **I**, generated in situ by treating dimeric $[Cp^*IrCl_2]_2$ with $AgSbF_6$. Then, the metal nitrene **III**, formed by the loss of N_2, underwent rearrangement of its π-electron, and then 4-electron-5-atom electrocyclization to form the C–N bond in intermediate **IV**. A 1,5-H shift from the resonating structure **V** afforded intermediate **VI**, which liberated the product **143** with the regeneration of the active catalytic species **I** (Scheme 1.44) (Das et al. 2018).

1.7.2 Chromeno- and Quinolino-Indoles

A Pd-catalyzed intramolecular CDC process of 4-arylamino substituted coumarins, quinolinones, and pyrones **144** was reported to access indole-fused polyheterocycles. CDC of 4-arylamino coumarins **144** (X=O) was carried out either by heating a solution of **144** with $Pd(OAc)_2$ (15 mol%) in AcOH under air conditions

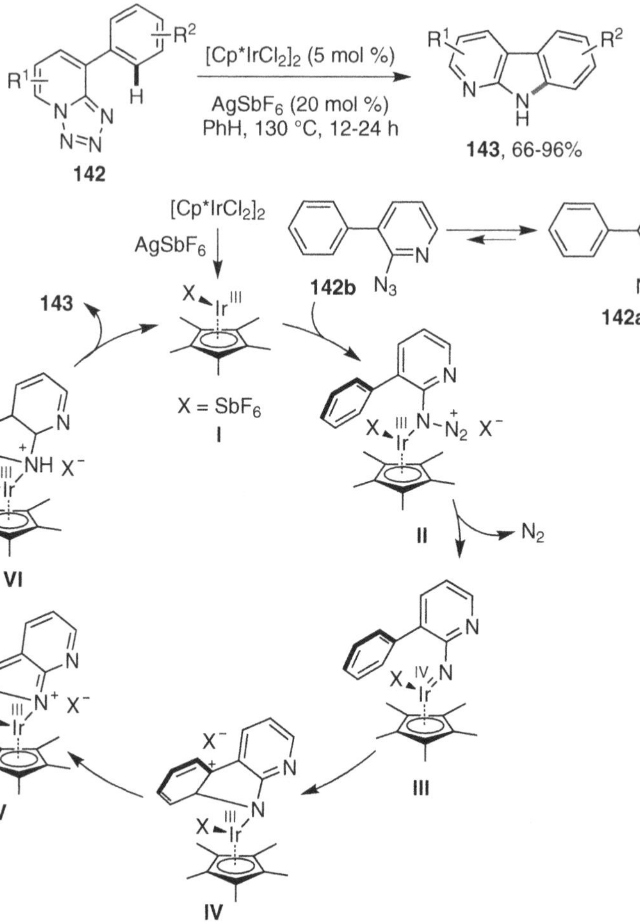

SCHEME 1.44

at 100°C, or by heating a solution of **144** with $Pd(OAc)_2$ (10 mol%), and AgOAc (2 equiv) in AcOH at 100°C, giving indolo[3,2-*c*]coumarins **145** in 76–99% or 71–95% yields, respectively. In the case of *meta*-substituted arylamino groups, a mixture of two regioisomers were obtained with a preference for the less hindered products. Also, indolo[3,2-*c*]quinolinones **146** and indolo[3,2-*c*]pyrones **147** were synthesized via the intramolecular CDC process of 4-arylamino quinolinones **144** (X=N), and 4-arylamino pyrone **144** (X=O), under the palladium/AgOAc catalytic system, in 87–97% and 76–94% yields, respectively (Cheng et al. 2016a). The same research group reported the construction of indolo[3,2-*c*]coumarins **145** in 63–92% yields, by subjecting 4-arylamino coumarins **144** (X=O) to $Pd(OAc)_2$ (10 mol%), AgOAc (2 equiv), and CsOAc (2 equiv) in PivOH at 100°C, for 5–12 h (Cheng et al. 2016b). One-pot CDC of 4-hydroxy coumarins **148** and anilines **10** was conducted in the presence of $Pd(OAc)_2$ (10 mol%) in DMF under an O_2 balloon at 140°C, affording indolo[3,2-*c*]coumarins **145** in 53–81% yields. The reaction was initiated by one-pot

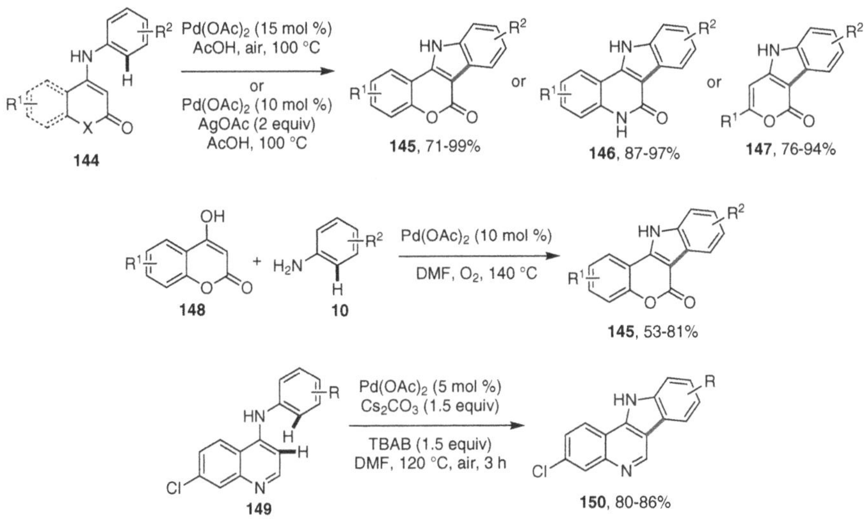

SCHEME 1.45

condensation of 4-hydroxy coumarin **148** with aniline **10**, followed by tautomerization to 4-arylamino coumarins **144** (Dey et al. 2017). Pd-catalyzed direct intramolecular aryl–heteroaryl C–C bond formation through C–H activation sequences was developed for the construction of quinolino-fused indole derivatives. Reactions were carried out by heating a solution of 4-amino-substituted quinolones in anhydrous DMF in the presence of 5 mol% of $Pd(OAc)_2$, Cs_2CO_3 (1.5 equiv), and TBAB (1.5 equiv) at 120°C under open air conditions for 3 h, giving quinolino-indoles in 80–86% yields (Scheme 1.45) (Kumar et al. 2015).

1.8 INDAZOLES

1.8.1 1H-INDAZOLES

Cyclization of benzophenone tosylhydrazone **151** to 3-arylindazoles **152** was reported by Inamoto et al. (2007). Reactions were catalyzed with $Pd(OAc)_2$ (10 mol%), $Cu(OAc)_2$ (1 equiv), and AgOTf (2 equiv) in DMSO at 50°C for 10–24 h, and corresponding indazoles **152** were obtained in 9–99% yields. The reaction of benzophenone tosylhydrazone bearing two *para*-MeO groups was sluggish, while that of benzophenone tosylhydrazone bearing two *meta*-MeO groups occurred in excellent yield (96%), with cyclization preferred at the less hindered position. Due to the steric and electronic factors, in the case of the substrate possessing two different substituents at the *meta* position, C–H activation occurred exclusively at the 6-position on the more electron-rich benzene ring. Cyclization of monosubstituted benzophenone tosylhydrazones occurred regioselectively on the benzene ring with high electron density. Reactions proceeded by *N*-atom-directed association of Pd(II) followed by C–H bond activation to form a six-membered palladacycle, which transformed to 3-arylindazole **152** by reductive elimination, along with generation of Pd(0). Catalytically active Pd(II) species were regenerated by oxidation of Pd(0) using $Cu(OAc)_2$ and AgOTf. A similar approach for the construction of 3-arylindazole scaffolds was developed using $Cu(OAc)_2$ (10 mol%) in the presence of DABCO (30 mol%) and K_2CO_3 (1 equiv) in DMSO under O_2 atmosphere at 120°C for 12 h. 3-Arylindazoles **152** were obtained in 62–86% yields (Li et al. 2013). Also, 3-aminoindazoles **154** were accessed by intramolecular ligand-free Pd-catalyzed C–H amination reaction of aminohydrazones **153**. Reactions were conducted by generation in situ of aminohydrazones **153**, followed by subjection to 10 mol% $Pd(OAc)_2$ and CsOPiv (2 equiv) in toluene under air conditions, affording 3-aminoindazoles **154** in 15–70% overall yields. Cyclic and acyclic amines can be incorporated at the C-3 position of the desired indazoles. No reaction occurred in the case of secondary hydrazonamide (Cyr et al. 2015). Also, $FeBr_3/O_2$-mediated C–H activation/C–N bond formation reactions have been reported for the formation of substituted 1*H*-indazoles **156**. Reactions were conducted by heating a solution of $FeBr_3$ (10 mol%) and arylhydrazone **155** in toluene at 110°C for 16–24 h, under an O_2 balloon, to afford corresponding indazoles **156** in 60–91% yields. For arylhydrazone of benzopheneones bearing two different substituents, cyclization occurred regioselectively at the electron-rich benzene ring (Scheme 1.46) (Zhang, and Bao 2013). Moreover, a TEMPO-induced C–H activation/C–N bond formation for the construction of 3-alkyl and 3-arylindazoles via a similar electron transfer process in the presence of O_2 was developed (Hu et al. 2014).

In 2013, Glorius et al. reported Rh(III)/Cu(II)-cocatalyzed synthesis of 3-alkoxy-1*H*-indazoles **158** through C–H amidation and N–N bond formation. Rh(III)

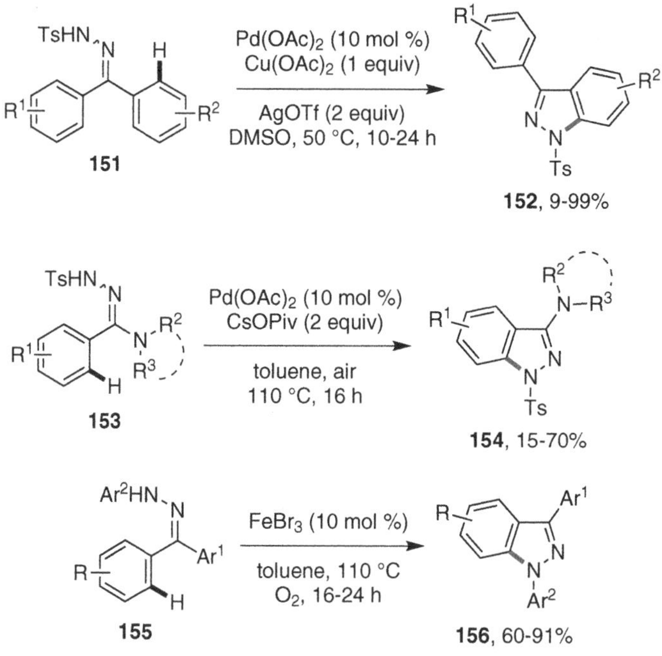

SCHEME 1.46

catalyzed the C–H activation and C–N bond formation stages, and Cu(II) catalyzed the N–N bond formation between arylimidates **157** and organoazides. By treatment of arylimidate **157** with 2.5 equiv of organic sulfonyl azides in the presence of $[Cp^*RhCl_2]_2$ (2.5 mol%), $AgSbF_6$ (10 mol%), $Cu(OAc)_2$ (25 mol%), and 4 Å MS under an atmosphere of O_2 (1 atm) in 1,2-dichloroethane at 110°C for 24 h, 3-alkoxy-1*H*-indazoles **158** were formed in 46–79% yields. Reactions were initiated by in situ generation of cationic [Cp*Rh(III)] **I** active catalyst in the presence of $AgSbF_6$, which coordinated with imidate **157** and underwent C–H activation to afford rhodacyclic complex **II**. The Rh(III) amido species **IV**, formed by coordination of azide with rhodium (intermediate **III**), followed by migratory insertion with release of N_2, underwent protonation to give the amidated product **V** along with regeneration of active [Cp*Rh(III)] catalyst **I**. Coordination of **V** with $Cu(OAc)_2$ gave complex **VI**, which was transformed to the final product **158** either by oxidation to Cu(III) complex **VII** by O_2, followed by N–N bond formation through reductive elimination (path a), or by double single-electron transfer pathway (path b), together with the generation of Cu(I) complex **VIII**, which was oxidized by O_2 in the presence of acid to regenerate $Cu(OAc)_2$ (Yu, Suri, and Glorius 2013). Tandem C–N/N–N bond formation to afford *N*-*tert*-butyl-3-aminoindazoles was developed by Cu(I)-catalyzed $C(sp^2)$–H amidation of *N*-*tert*-butylarylamidine with TsN_3. Reactions were conducted in 1,2-dichlorobenzane at 115°C under argon atmosphere using 30 mol% of copper(I) thiophene-2-carboxylate (CuTc) in the presence of hexafluoroisopropanol (HFIP), to furnish 3-amino-1*H*-indazoles in 34–61% yields (Scheme 1.47) (Peng et al. 2014).

Wang and Li (2016) reported the synergistic Rh(III)/Cu(II)-catalyzed synthesis of 1*H*-indazoles **162** from imidates **157** and nitrosobenzenes **161**, by subjection to $[Cp^*Rh(MeCN)_3](SbF_6)_2$ (5 mol%) and $CuCl_2$ (30 mol%) in the presence of 4 Å MS in $PhCF_3$ at 80°C under N_2 for 24 h. A variety of substituted imidates **157** and nitrosobenzenes **161** were tolerated in reaction conditions, affording corresponding 1*H*-indazoles **162** in 55–86% yields. The imidate bearing an *ortho*-fluoro group gave the corresponding indazole in lower yield (35%) (Scheme 1.48).

Cooperative Co(III)/Cu(II)-catalyzed C–N/N–N coupling of imidates **157** with anthranils **163** was developed as an approach to 3-alkoxy-1*H*-indazoles **164** via C–H activation. Treating imidates **157** with anthranil **163** (3 equiv) in the presence of $[Cp^*Co(MeCN)_3](SbF_6)_2$ (10 mol%), and $Cu(OAc)_2$ (2 equiv) in DCE at 100°C under N_2 atmosphere for 20 h afforded 3-alkoxy-1*H*-indazoles **164** in 37–93% yields. In the proposed reaction mechanism, intermediate **II** was generated via cyclometalation of the imidate **157** to give a five-membered metallacyclic intermediate **I**, followed by coordination of anthranil **163**. By intramolecular N–O bond cleavage, a nitrene intermediate **III** was formed, which underwent migratory insertion of the Co–aryl bond into the nitrene, generating a tripodal intermediate **IV**. Aminated intermediate **V** was formed through coordination of an imidate **163** to **IV** and subsequent C–H activation, along with the regeneration of Co(III) species **I**. Aminated intermediate **V** was transformed into the final 3-alkoxy-1*H*-indazole product **164** in a sequential copper-catalyzed cycle by coordination with $Cu(OAc)_2$, followed by extrusion of HOAc, leading to a Cu(II) species **VI**, which underwent double single-electron transfer to generate indazole **164** and Cu(I) intermediate **VIII**. The $Cu(OAc)_2$ active

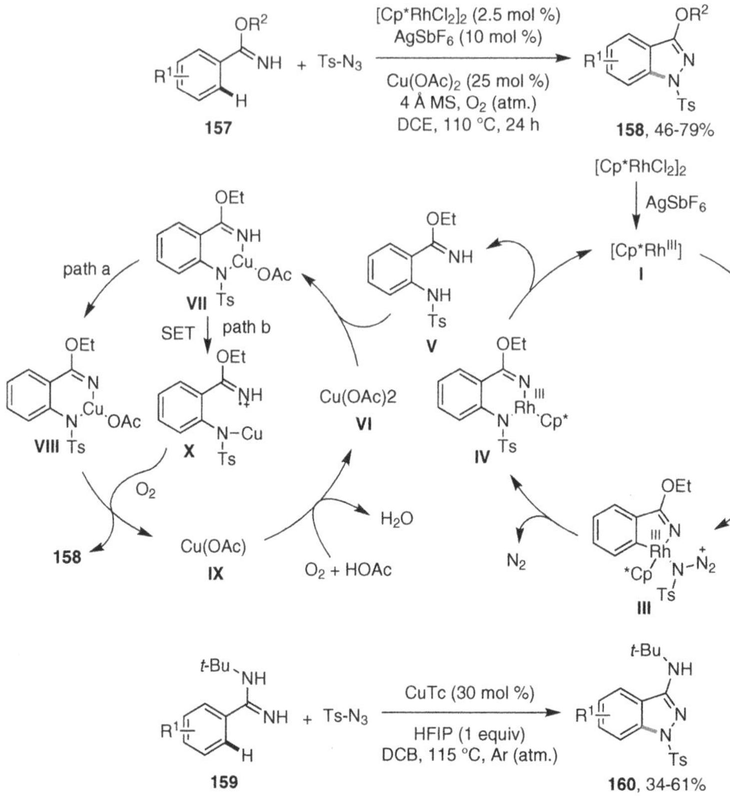

SCHEME 1.47

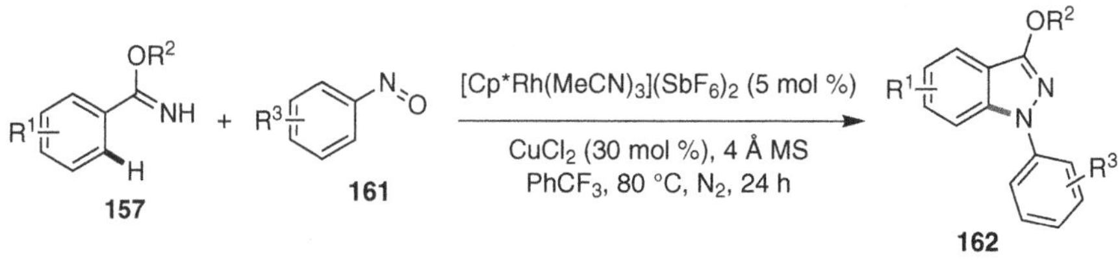

SCHEME 1.48

catalyst was regenerated by oxidation of Cu(I) intermediate **VIII**, with another molecule of anthranils **163** in the presence of AcOH (Scheme 1.49) (Li et al. 2016).

1.8.2 2H-Indazoles

An approach for the synthesis of 2,3-dihydro-1*H*-indazoles **166** has been reported by an Rh(III)-catalyzed CDC process of 1,2-disubstituted arylhydrazines with active alkenes via aromatic C(sp^2)–H bond activation followed by an intramolecular aza-Michael reaction. Treatment of arylhydrazines **165** with 2 equiv of alkenes in the

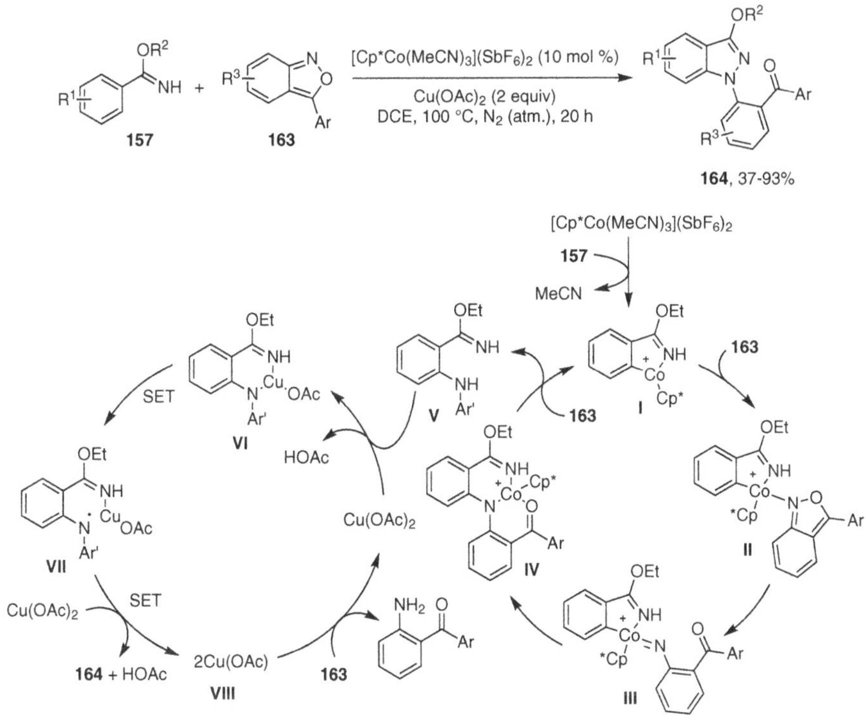

SCHEME 1.49

presence of $[RhCp^{*}Cl_2]_2$ (2.5 mol%) and $Cu(OAc)_2$ (50 mol%) in MeCN under air at 100°C for 20 h liberated 2,3-dihydro-1*H*-indazoles **166** in 23–82% yields. Reactions of both electron-donating and electron-withdrawing substituents at the *para* position of arylacetohydrazines occurred smoothly. Also, the reaction of *meta*-substituted arylacetohydrazides preferentially occurred at the less hindered position, furnishing the corresponding products as a single regioisomer. Products with decreased yields were obtained in the case of *ortho*-substituted arylacetohydrazides (Scheme 1.50) (Han et al. 2014).

Synthesis of 2*H*-indazole derivatives was achieved via Rh(III)-catalyzed C–H activation of azobenzenes. Treatment of azobenzenes **167** with acrylates **97** in the presence of $[Cp^{*}RhCl_2]_2$ (5 mol%) and $Cu(OAc)_2$ (2 equiv) in DCE under N_2 atmosphere furnished 2*H*-indazoles **168** in 52–86% yields. A variety of electron-donating

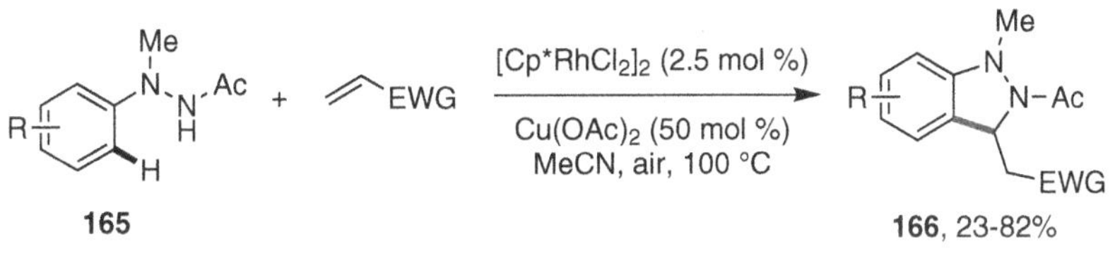

SCHEME 1.50

and electron-withdrawing substituents on azobenzene afforded the indazole products **168**, while the stronger electron-withdrawing group, CF_3, only gave trace product. When the reaction was conducted under an O_2 atmosphere, 3-acyl indazoles **169** were obtained in 63–81% yields, with the capture of molecular oxygen (Cai et al. 2016). Also, *N*-aryl-2*H*-indazoles **170** were synthesized by treatment of azobenzene **167** with aldehyde (2 equiv) in the presence of $[Cp^*RhCl_2]_2$ (5 mol%) and $AgSbF_6$ (20 mol%) in dioxane for 24 h, via rhodium(III)-catalyzed C–H bond addition of azobenzenes to aldehydes. Reactions proceeded by azo-group-directed *ortho*-C–H bond activation with Rh(III), followed by the addition to aldehyde to provide alcohol **171**, which was converted to the desired 2*H*-indazole **170** via cyclization by intramolecular nucleophilic substitution, followed by rapid aromatization (Scheme 1.51). In the case of unsymmetrical azobenzenes, C–H functionalization occurred regioselectivity on the more electron-rich phenyl ring (Lian et al. 2013).

The synthesis of 3-acyl (2*H*)-indazoles **173** was developed by Rh(III)-catalyzed annulation reaction of azobenzenes **167** with sulfoxonium ylides **172**. Reactions were catalyzed with $[Cp^*RhCl_2]_2$ (2.5 mol%) and $AgSbF_6$ (10 mol%) in the presence of $Cu(OAc)_2$ (50 mol%) and $CuCO_3{\cdot}Cu(OH)_2$ (1 equiv) in DCE at 110°C for 24 h under air, leading to the formation of 3-acyl (2*H*)-indazoles **173** in 29–95% yields. Various substituted azobenzenes **167** and arylacyl sulfoxonium ylides **172** were tolerated under reaction conditions. However, alkylacyl sulfoxonium ylides **172** resulted in formation of the corresponding 3-alkylacyl (2*H*)-indazoles **173** in lower yields

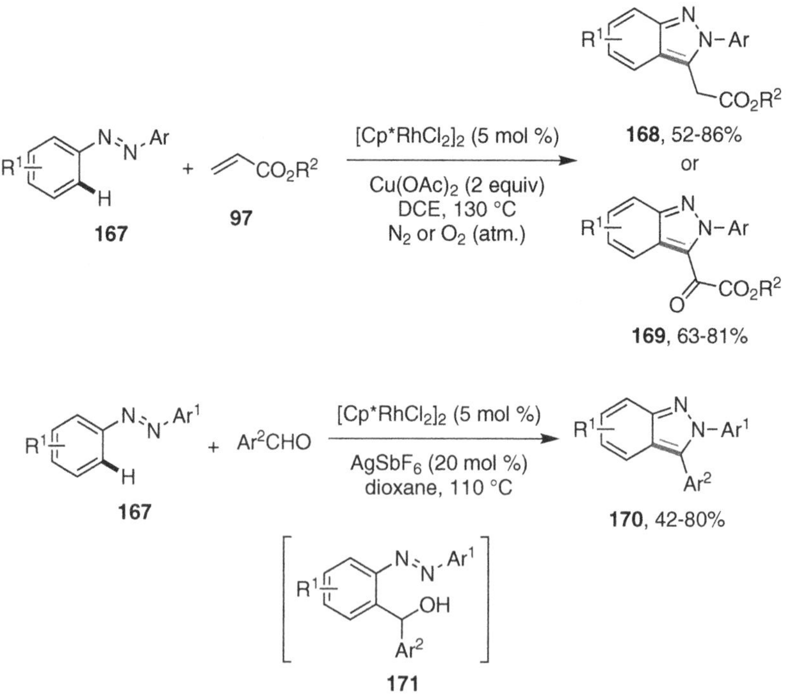

SCHEME 1.51

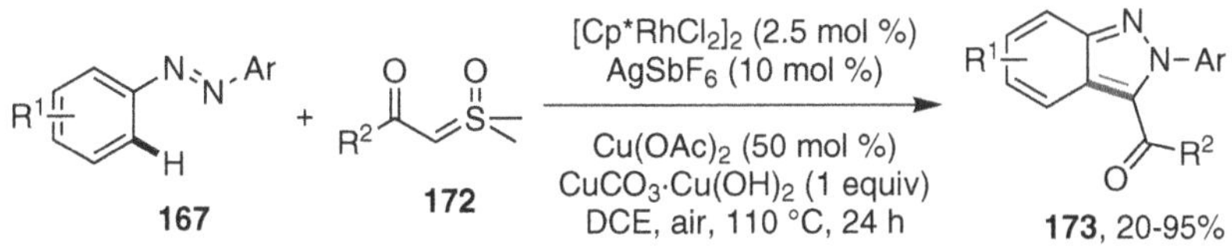

$[Cp^*RhCl_2]_2$ (2.5 mol %)
$AgSbF_6$ (10 mol %)
$Cu(OAc)_2$ (50 mol %)
$CuCO_3 \cdot Cu(OH)_2$ (1 equiv)
DCE, air, 110 °C, 24 h

167 + **172** → **173**, 20-95%

SCHEME 1.52

(20–44%). In the case of unsymmetrical azobenzenes, the reactions predominantly occurred at the electron-rich aromatic ring. This conversion was initiated by coordination of the azo group of **167** to a cationic Rh(III) catalyst, followed by C–H bond cleavage to deliver a rhodacycle, which was transformed to an α-oxo Rh-carbene species via coordination of sulfoxonium ylide **172**, followed by α-elimination of a DMSO molecule (Scheme 1.52) (Oh et al. 2018).

An Rh(III)-catalyzed tandem C–H alkylation/intramolecular decarboxylative cyclization of azoxy compounds **174** with α-diazocarbonyl compounds **34** for the synthesis of 3-acyl-2*H*-indazoles **173** was developed by Long et al. (2017). A main advantage of this reaction is a complete regioselectivity for unsymmetrical azoxybenzenes, in which *ortho*-C–H activation occurs at a benzene ring possessing N–O functionality. By heating a solution of azoxybenzenes **174** with 1.5 equiv of α-diazocarbonyls **34** in the presence of $[Cp^*RhCl_2]_2$ (2.5 mol%), $AgSbF_6$ (10 mol%), and PivOH (1 equiv) in DCE/dioxane under N_2 atmosphere at 130°C for 24 h, 3-acyl-2*H*-indazoles **173** were obtained in 38–96% yields. The proposed reaction mechanism involves the generation of active [Cp*Rh(III)] species **I** by the anion exchange of $AgSbF_6$ with $[Cp^*RhCl_2]_2$, followed by coordination with the azoxy group and then a directed reversible C–H bond cleavage to form the five-membered rhodacycle **II**. Coordination to a diazoester formed the diazonium intermediate **III**, which was transformed to the six-membered rhodacycle intermediate **V**, either by extrusion of N_2 to give a metal–carbene intermediate **IV**, followed by a migratory insertion of the carbene into the rhodium–carbon bond (path a), or by a direct intramolecular 1,2-migratory insertion of the aryl group (path b). By protonation of **V**, the alkylated intermediate **VI** was formed, along with regeneration of the active rhodium species **I**. Finally, 3-acyl-2*H*-indazole product **173** was liberated by an intramolecular nucleophilic addition of the malonate ester by the oxygen of the azoxy (intermediate **VII**), followed by CO_2 extrusion and intramolecular cyclization. Also, a rhodium-catalyzed regioselective C–H activation/cyclization of azoxy compounds **174** with alkynes **17** has been developed by the same research group to construct a variety of 2*H*-indazoles **170**. Reactions were carried out using $[Cp^*RhCl_2]_2$ (2.5 mol%), $AgSbF_6$ (10 mol%), $Cu(OAc)_2$ (1 equiv), and $Zn(OTf)_2$ (20 mol%) in $(CF_3)_2CHOH$ under N_2 at 80°C for 12 h, and the desired 2*H*-indazole derivatives **170** were obtained in 36–87% yields. Annulation of monoaryldiazene oxides with aliphatic alkynes occurred in low yields (25–28%). In the proposed reaction mechanism, an alkyne insertion into the intermediate **II** took place, giving a seven-membered rhodacyclic intermediate **III'**, which was transformed into a seven-membered rhodacycle **IV'** by an alteration of the coordination atom from nitrogen to oxygen. Reductive elimination afforded the intermediate **V'** and an Rh(I) species. The tautomerization of **V'** to **VI**, followed

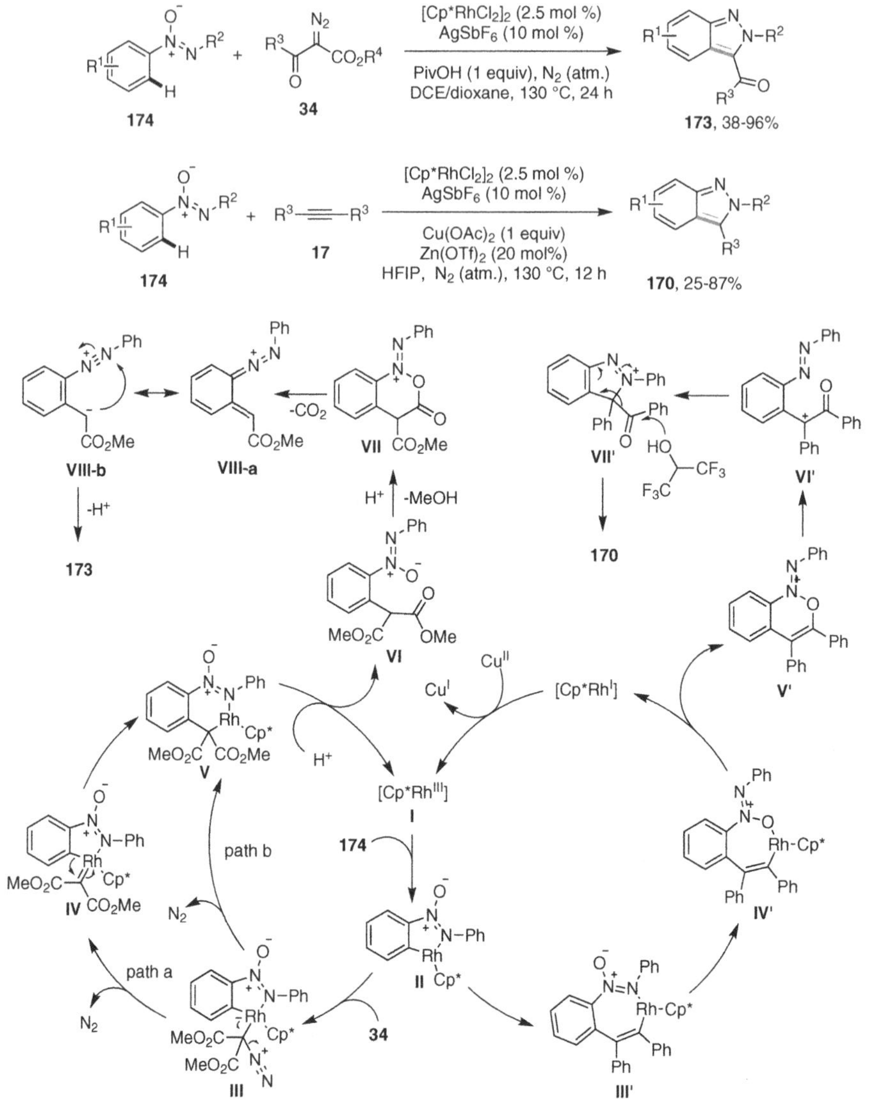

SCHEME 1.53

by intramolecular nucleophilic addition at the cationic carbon center, delivered a cationic heterocyclic intermediate **VII'**, which underwent nucleophilic attack of HFIP, leading to the desired product **170** (Scheme 1.53) (Long, Yang, and You 2017).

1.9 BENZIMIDAZOLES

Mahesh, Sadhu, and Punniyamurthy (2015) developed a Cu-catalyzed preparation of benzimidazoles **175** via an amination of *N*-aryl imines **176** in a one-pot multi-component reaction between anilines **10**, aromatic aldehydes, and $TMSN_3$. Reactions were conducted by treatment of a solution of aniline **10** and benzaldehyde (1.2 equiv) in DMSO, with CuI (10 mol%), $TMSN_3$ (2 equiv), and TBHP (1 equiv) at 90°C for 9–14 h, to give corresponding benzimidazoles **175a** in 49–84% yields. No reaction took place in aniline with a strong electron-withdrawing group, 4-nitro aniline, suggesting the essential role of the electronic nature of the aryl ring. Additionally, an aliphatic aldehyde, isobutyraldehyde, furnished 2-isopropylbenzimidazole in 64% yield under reaction conditions. Aryl imines **176**, generated in situ from anilines and aldehydes, acted as a directing group by chelating to the metal center. The same research group reported a similar approach, by in situ generation of aryl imines **176** from the reaction of alkyl- or benzyl amines with anilines **10** in the presence of $Cu(OAc)_2$ and TBHP, via domino C–H functionalization and transimination, followed by subjection to NaN_3 to afford benzimidazole derivatives **175b** through an *ortho* selective amination and cyclization sequence (Mahesh, Sadhu, and Punniyamurthy 2016). In another work, treating anilines **10** with methylarenes (10 equiv) and $TMSN_3$ (2 equiv) in the presence of $Cu(OAc)_2$ (20 mol%) and TBHP (3 equiv) in DMSO at 80°C gave benzimidazoles **175a** in 43–77% yields, in which aryl imine intermediates **176** were generated in situ by the reaction of anilines **10** and methylarenes in the presence of TBHP catalyzed with Cu(II) (Scheme 1.54) (Mahesh et al. 2017).

2-Arylbenzimidazole derivatives **175** were also synthesized by Cu-catalyzed C–H functionalization/C–N bond formation. Subjecting *N*-aryl amidines **38** to 15 mol% $Cu(OAc)_2$ in the presence of HOAc in DMSO under O_2 at 100°C for 18 h led to the formation of benzimidazoles **175** in 68–89% yields. In addition to *C*-aryl amidines, *t*-butyl amidines worked well in the reaction conditions, and corresponding 2-*t*-butylbenzimidazoles **175** were obtained in 83–89% yields (Brasche and Buchwald 2008). Mono-*N*-arylation of secondary benzamidines **38** with arylboronic

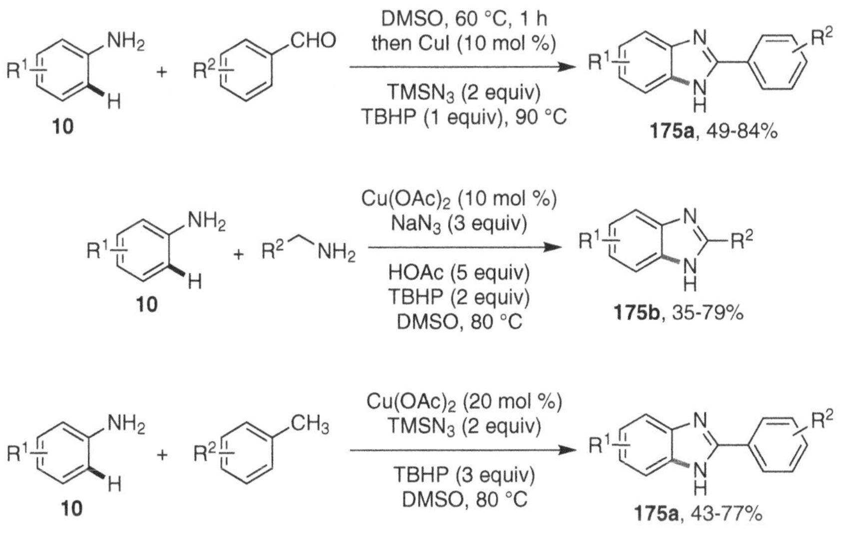

SCHEME 1.54

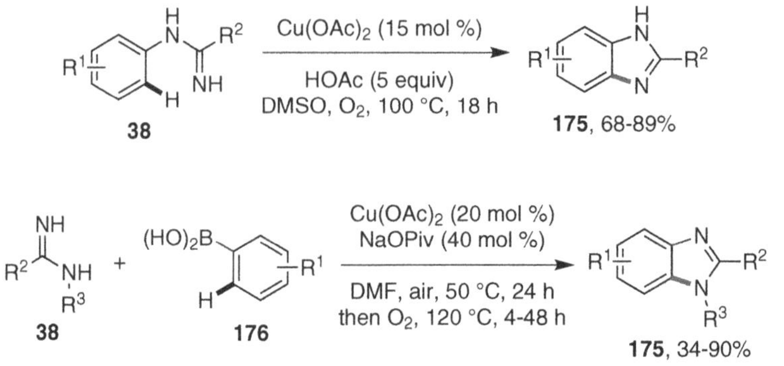

SCHEME 1.55

acids **176**, followed by an intramolecular direct C–H bond functionalization, was achieved in the presence of a catalytic amount of $Cu(OAc)_2$ (20 mol%) and NaOPiv (40 mol%) under O_2 atmosphere at 120°C, affording benzimidazoles **175** in 34–90% yields (Scheme 1.55) (Li et al. 2012).

N-arylsulfonyl-2-aryl benzimidazoles **178** were obtained through Ir-catalyzed annulation of imidamides **38** with sulfonyl azides **177**. Reactions were catalyzed with $[Cp^*IrCl_2]_2$ (4 mol%) in the presence of $AgNTf_2$ (16 mol%) and phenylacetic acid (1 equiv) in DCE at 80°C for 12 h to furnish benzimidazoles **178** in 50–99% yields. Alkylsulfonyl azides were also tolerated in this reaction, leading to corresponding *N*-alkylsulfonyl benzimidazoles in 80–99% yields. The reaction was initiated by generation of active catalyst species $[Cp^*Ir(NTf_2)_2]$ through anion exchange, followed by coordination with **38**, with subsequent cyclometalation to generate iridacyclic intermediate **I**, which was converted to iridium carbene species **III** by coordination of TsN_3 (intermediate **II**) followed by elimination of nitrogen. By migratory insertion of the Ir–Ar bond into the carbene unit, intermediate **IV** was formed, which underwent a second migratory insertion of the Ir–N bond into the C=N bond to afford amide species **V**. The product **178** was obtained by elimination of the active Ir(III) catalyst upon protonolysis (Scheme 1.56) (Xu et al. 2017).

Moreover, Rh(III)-catalyzed synthesis of 2-alkylbenzimidazoles **180** from imidamides **38** and *N*-hydroxycarbamates **179** was developed by Li et al. (2018). Reactions were conducted using $[Cp^*RhCl_2]_2$ (2.5 mol%), Ag_2CO_3 (1 equiv), PivOH (2 equiv), and *t*-BuOK (50 mol%) in DCE under air at 100°C for 36 h, to construct 2-alkylbenzimidazoles **180** in 56–97% yields. A variety of *ortho*-, *meta*-, and *para*-substituted *N*-arylimidamides were tolerated under reaction conditions. In the case of the *meta*-substituent, C–H functionalization occurred exclusively at the less hindered *ortho* site. In the plausible reaction mechanism, a six-membered rhodacycle **I** was formed through the coordination of *N*-phenylalkylimidamides **38** with the active rhodium catalyst $[Cp^*Rh(OPiv)_2]$, with subsequent *ortho* C–H bond activation. Then, by coordination of **V**, generated in situ by oxidation of **179**, intermediate **II** was formed, which underwent migratory insertion to give the intermediate **III**. Subsequently, intramolecular nucleophilic attack at the N atom with the concomitant

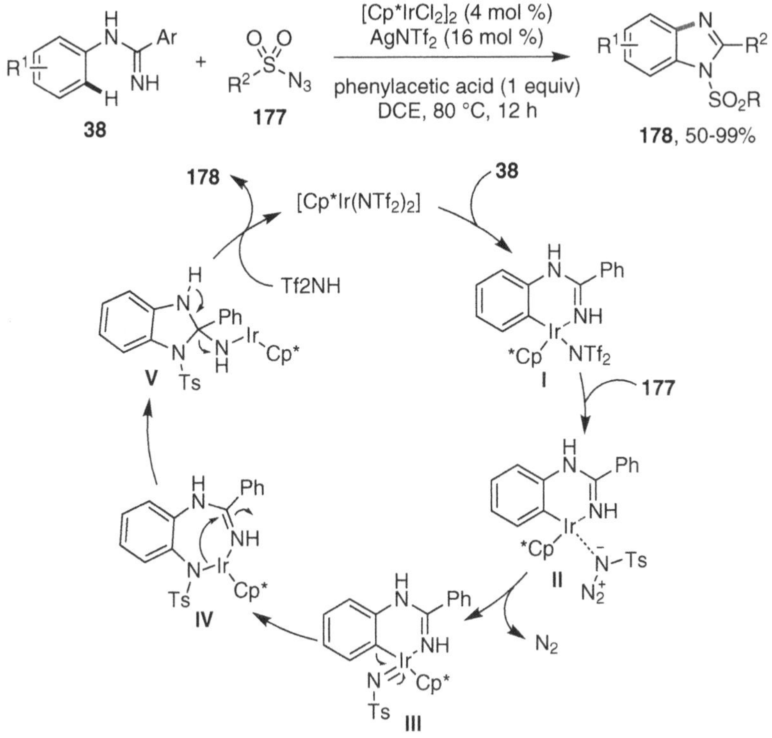

SCHEME 1.56

N–O cleavage afforded amide species **IV**, which was transformed to the final desired product **180** through β-hydride elimination, together with regeneration of the catalytically active rhodium(III) species (Scheme 1.57).

1.10 BENZOTRIAZOLES

Pd/TBHP-catalyzed synthesis of 2-aryl-2*H*-benzotriazoles **181** was described by *ortho*-C–H amination of azoarenes **167** using $TMSN_3$ as the source of nitrogen via an intermolecular *o*-azidation (C–N bond formation) followed by an intramolecular N–N bond formation through nucleophilic attack by one of the azo nitrogens on the *o*-azide nitrogen. Reactions were performed using 2 equiv of $TMSN_3$ in the presence of $Pd(OAc)_2$ (20 mol%) and TBHP (2 equiv) in DMSO at 100°C under argon atmosphere, leading to 2-aryl-2*H*-benzotriazoles **181** in 35–87% yields. In the case of strongly electron-withdrawing groups (CF_3), the corresponding 2-aryl-2*H*-benzotriazole was produced only in low yield (8%). In the case of unsymmetrical azobenzenes, reaction occurred in the electron-rich ring, with moderate selectivity. By cyclopalladation between the "azo moiety" of azobenzene **167** and the Pd(II) catalyst, intermediate complex **I** was generated, which reacted with the azide radical, generated in situ by the reaction of $TMSN_3$ and TBHP, resulting in the formation of

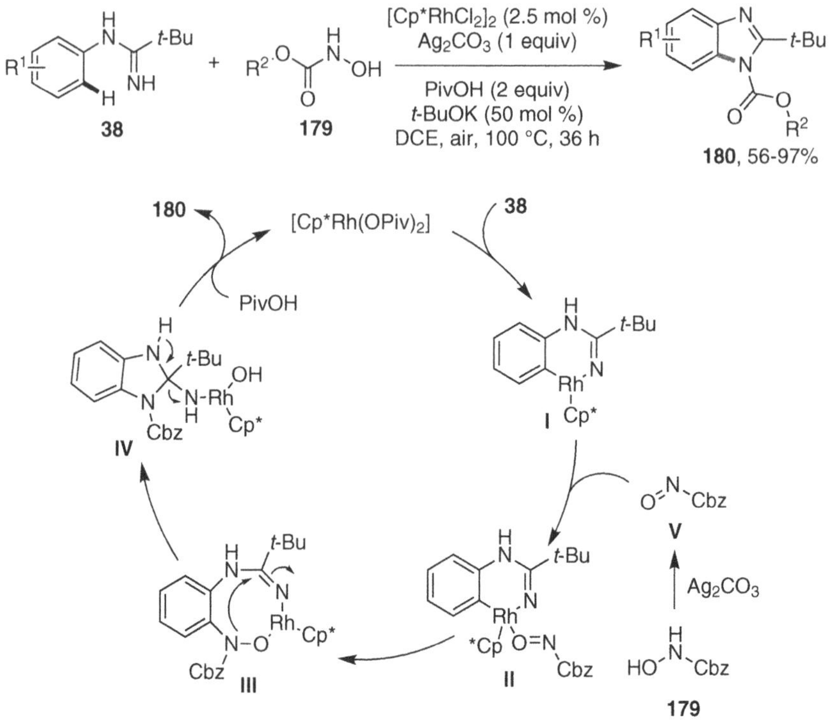

SCHEME 1.57

Pd(III) intermediate **II**. A further oxidation of intermediate **II** by TBHP led to the formation of a Pd(IV) intermediate **III**, which underwent reductive elimination to an *ortho* azido azobenzene **IV** and regenerated the Pd(II) catalyst for the next cycle. In the final stage, an attack by one of the azo nitrogens on the *o*-azide nitrogen of **IV** gave cyclization product **181**, with the expulsion of a molecule of N_2 (Scheme 1.58) (Khatun et al. 2015).

1.11 CONCLUSION

In summary, aromatic C(sp²)–H dehydrogenative coupling processes are widely applied in the construction of five-membered nitrogen-containing heterocyclic compounds, including indoles, oxindoles, isoindoles, carbazoles, some heterocycle-fused indoles, indazoles, benzimidazoles, and benzotriazoles. Dehydrogenative coupling could also be developed for the synthesis of various substituted five-membered benzoid heterocycles that were previously thought to be inaccessible. However, many different methods were developed for the construction of these types of heterocycles, due to the quantitative and one-step synthesis and a broad spectrum of substitution on the synthesized heterocycles, the aromatic C–H bond direct functionalization approach could be of interest in the synthesis of pharmaceutical, medicinal, and natural products.

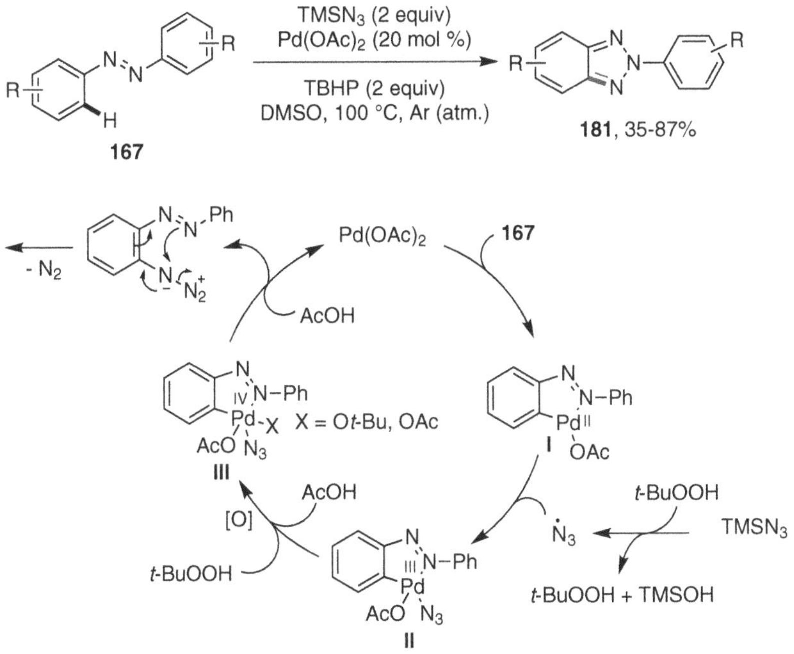

SCHEME 1.58

REFERENCES

Ackermann, Lutz, and Alexander V Lygin. 2012. “Cationic ruthenium (II) catalysts for oxidative C–H/N–H bond functionalizations of anilines with removable directing group: Synthesis of indoles in water.” *Organic Letters* no. 14(3):764–767.

Amini, Mojtaba, Sajjad Bahadori Tekantappeh, Bagher Eftekhari-Sis, Parviz Gohari Derakhshandeh, and Kristof Van Hecke. 2017. “Synthesis, characterization and catalytic properties of a copper-containing polyoxovanadate nanocluster in azide–alkyne cycloaddition.” *Journal of Coordination Chemistry* no. 70(9):1564–1572.

Bernini, Roberta, Giancarlo Fabrizi, Alessio Sferrazza, and Sandro Cacchi. 2009. “Copper-catalyzed C-C bond formation through C-H functionalization: Synthesis of multisubstituted indoles from N-aryl enaminones.” *Angewandte Chemie International Edition* no. 48(43):8078–8081.

Bhunia, Subhajit, Santanu Ghosh, Dhananjay Dey, and Alakesh Bisai. 2013. “DDQ-mediated direct intramolecular-dehydrogenative-coupling (IDC): Expeditious approach to the tetracyclic core of ergot alkaloids.” *Organic Letters* no. 15(10):2426–2429.

Bi, Wenying, Xiliu Yun, Yanfeng Fan, Xiuxiang Qi, Yunfei Du, and Jianhui Huang. 2010. “Syntheses of N-alkylated carbazolones via Pd (OAc) 2-mediated intramolecular coupling of N-substituted 3-(arylamino) cyclohex-2-enones.” *Synlett* no. 2010(19):2899–2904.

Brasche, Gordon, and Stephen L Buchwald. 2008. “C-H functionalization/C-N bond formation: Copper-catalyzed synthesis of benzimidazoles from amidines.” *Angewandte Chemie International Edition* no. 47(10):1932–1934.

Cai, Shangjun, Songyun Lin, Xiangli Yi, and Chanjuan Xi. 2016. "Substrate-controlled transformation of azobenzenes to indazoles and indoles via Rh (III)-catalysis." *The Journal of Organic Chemistry* no. 82(1):512–520.

Cao, Yang, Xiang Zhang, Guangyu Lin, Daisy Zhang-Negrerie, and Yunfei Du. 2016. "Chiral aryliodine-mediated enantioselective organocatalytic spirocyclization: Synthesis of spirofurooxindoles via cascade oxidative C–O and C–C Bond formation." *Organic Letters* no. 18(21):5580–5583.

Chan, Chun-Wo, Pui-Yiu Lee, and Wing-Yiu Yu. 2015. "Copper-catalyzed cross-dehydrogenative coupling of N-arylacrylamides with chloroform using tert-butyl peroxybenzoate as oxidant for the synthesis of trichloromethylated 2-oxindoles." *Tetrahedron Letters* no. 56(20):2559–2563.

Chen, Jinlei, Qingyu Pang, Yanbo Sun, and Xingwei Li. 2011. "Synthesis of N-(2-pyridyl) indoles via Pd (II)-catalyzed oxidative coupling." *The Journal of Organic Chemistry* no. 76(9):3523–3526.

Chen, Jinlei, Guoyong Song, Cheng-Ling Pan, and Xingwei Li. 2010. "Rh (III)-catalyzed oxidative coupling of N-Aryl-2-aminopyridine with alkynes and alkenes." *Organic Letters* no. 12(23):5426–5429.

Cheng, Chao, Wen-Wen Chen, Bin Xu, and Ming-Hua Xu. 2016a. "Access to indole-fused polyheterocycles via Pd-catalyzed base-free intramolecular cross dehydrogenative coupling." *The Journal of Organic Chemistry* no. 81(22):11501–11507.

Cheng, Chao, Wen-Wen Chen, Bin Xu, and Ming-Hua Xu. 2016b. "Intramolecular cross dehydrogenative coupling of 4-substituted coumarins: Rapid and efficient access to coumestans and indole [3, 2-c] coumarins." *Organic Chemistry Frontiers* no. 3(9):1111–1115.

Cho, Seung Hwan, Jungho Yoon, and Sukbok Chang. 2011. "Intramolecular oxidative C–N bond formation for the synthesis of carbazoles: Comparison of reactivity between the copper-catalyzed and metal-free conditions." *Journal of the American Chemical Society* no. 133(15):5996–6005.

Cyr, Patrick, Sophie Regnier, William S Bechara, and Andre B Charette. 2015. "Rapid access to 3-aminoindazoles from tertiary amides." *Organic Letters* no. 17(14):3386–3389.

Das, Sandip Kumar, Satyajit Roy, Hillol Khatua, and Buddhadeb Chattopadhyay. 2018. "Ir-catalyzed intramolecular transannulation/C (sp^2)–H amination of 1, 2, 3, 4-tetrazoles by electrocyclization." *Journal of the American Chemical Society* no. 140(27):8429–8433.

Dateer, Ramesh B, and Sukbok Chang. 2015a. "Rh (III)-catalyzed C–H cyclization of arylnitrones with diazo compounds: Access to N-hydroxyindolines." *Organic Letters* no. 18(1):68–71.

Dateer, Ramesh B, and Sukbok Chang. 2015b. "Selective cyclization of arylnitrones to indolines under external oxidant-free conditions: Dual role of Rh (III) catalyst in the C–H activation and oxygen atom transfer." *Journal of the American Chemical Society* no. 137(15):4908–4911.

Dey, Amrita, Md Ashif Ali, Sourav Jana, Sadhanendu Samanta, and Alakananda Hajra. 2017. "Palladium-catalyzed synthesis of indole fused coumarins via cross-dehydrogenative coupling." *Tetrahedron Letters* no. 58(4):313–316.

Dey, Chandan, Evgeny Larionov, and E Peter Kuendig. 2013. "Copper (ii) chloride mediated (aza) oxindole synthesis by oxidative coupling of C sp^2–H and C sp^3–H centers: Substrate scope and DFT study." *Organic and Biomolecular Chemistry* no. 11(39):6734–6743.

Dey, Chandan, and E Peter Kündig. 2012. "Aza-oxindole synthesis by oxidative coupling of C sp^2–H and C sp^3–H centers." *Chemical Communications* no. 48(25):3064–3066.

Donald, James R, Richard JK Taylor, and Wade F Petersen. 2017. "Low-temperature, transition-metal-free cross-dehydrogenative coupling protocol for the synthesis of 3, 3-disubstituted oxindoles." *The Journal of Organic Chemistry* no. 82(20):11288–11294.

Dong, Jiaxing, Fei Wang, and Jingsong You. 2014. "Copper-mediated tandem oxidative C(sp^2)–H/C(sp)–H alkynylation and annulation of arenes with terminal alkynes." *Organic Letters* no. 16(11):2884–2887.

Drouhin, Pauline, and Richard JK Taylor. 2015. "A copper-mediated oxidative coupling route to 3H-and 1H-indoles from N-Aryl-enamines." *European Journal of Organic Chemistry* no. 2015(11):2333–2336.

Eftekhari-Sis, B, A Akbari, and M Amirabedi. 2011. "Synthesis of new N-alkyl (aryl)-2, 4-diaryl-1H-pyrrol-3-ols via aldol Paal–Knorr reactions." *Chemistry of Heterocyclic Compounds* no. 46(11):1330.

Eftekhari-Sis, Bagher, and Fatemeh Ghahramani. 2015. "Synthesis of 2-{5-[4-((4-nitrophenyl) diazenyl) phenyl]-1, 3, 4-oxadiazol-2-ylthio} ethyl acrylate monomer and its application in a dual pH and temperature responsive soluble polymeric sensor." *Designed Monomers and Polymers* no. 18(5):460–469.

Eftekhari-Sis, Bagher, Fatemeh Malekan, and Hessamaddin Younesi Araghi. 2018. "CdSe quantum dots capped with p-nitrophenyldiazenylphenyloxadiazole: A nanosensor for Cd2+ ions in aqueous media." *Canadian Journal of Chemistry* no. 96(4):371–376.

Eftekhari-Sis, Bagher, and Saleh Vahdati-Khajeh. 2013. "Ultrasound-assisted green synthesis of pyrroles and pyridazines in water via three-component condensation reactions of arylglyoxals." *Current Chemistry Letters* no. 2(2):85–92.

Eftekhari-Sis, Bagher, and Maryam Zirak. 2014. "Chemistry of α-oxoesters: A powerful tool for the synthesis of heterocycles." *Chemical Reviews* no. 115(1):151–264.

Eftekhari-Sis, Bagher, and Maryam Zirak. 2017. "α-Imino esters in organic synthesis: Recent advances." *Chemical Reviews* no. 117(12):8326–8419.

Eftekhari-Sis, Bagher, Maryam Zirak, and Ali Akbari. 2013. "Arylglyoxals in synthesis of heterocyclic compounds." *Chemical Reviews* no. 113(5):2958–3043.

Eftekhari-Sis, Bagher, Maryam Zirak, Ali Akbari, and Mohammed M Hashemi. 2010. "Synthesis of new 2-aryl-4-chloro-3-hydroxy-1H-indole-5, 7-dicarbaldehydes via Vilsmeier-Haack reaction." *Journal of Heterocyclic Chemistry* no. 47(2):463–467.

Gandeepan, Parthasarathy, Thomas Müller, Daniel Zell, Gianpiero Cera, Svenja Warratz, and Lutz Ackermann. 2018. "3d transition metals for C–H activation." *Chemical Reviews* no. 119(4):2192–2452.

Ghosh, Santanu, Subhadip De, Badrinath N Kakde, Subhajit Bhunia, Amit Adhikary, and Alakesh Bisai. 2012. "Intramolecular dehydrogenative coupling of sp^2 C–H and sp^3 C–H bonds: An expeditious route to 2-oxindoles." *Organic Letters* no. 14(23): 5864–5867.

Guo, Xin, Jianwei Han, Yafeng Liu, Mingda Qin, Xueguo Zhang, and Baohua Chen. 2017. "Synthesis of 2, 3-disubstituted NH indoles via rhodium (III)-catalyzed C–H activation of arylnitrones and coupling with diazo compounds." *The Journal of Organic Chemistry* no. 82(21):11505–11511.

Han, Jian, Ning Wang, Zhi-Bin Huang, Yingsheng Zhao, and Da-Qing Shi. 2017. "Ruthenium-catalyzed carbonylation of oxalyl amide-protected benzylamines with isocyanate as the carbonyl source." *The Journal of Organic Chemistry* no. 82(13):6831–6839.

Han, Sangil, Youngmi Shin, Satyasheel Sharma, Neeraj Kumar Mishra, Jihye Park, Mirim Kim, Minyoung Kim, Jinbong Jang, and In Su Kim. 2014. "Rh (III)-catalyzed oxidative coupling of 1, 2-disubstituted arylhydrazines and olefins: A new strategy for 2, 3-dihydro-1 H-indazoles." *Organic Letters* no. 16(9):2494–2497.

He, Zhiheng, Weiping Liu, and Zhiping Li. 2011. "I2-catalyzed indole formation via oxidative cyclization of N-aryl enamines." *Chemistry–An Asian Journal* no. 6(6):1340–1343.

Hu, Jiantao, Huacheng Xu, Pengju Nie, Xiaobo Xie, Zongxiu Nie, and Yu Rao. 2014. "Synthesis of indazoles and azaindazoles by intramolecular aerobic oxidative C-N coupling under transition-metal-free conditions C-N coupling under transition-metal-free conditions." *Chemistry–A European Journal* no. 20(14):3932–3938.

Hu, Xinwei, Xun Chen, Youxiang Shao, Haisheng Xie, Yuanfu Deng, Zhuofeng Ke, Huanfeng Jiang, and Wei Zeng. 2018. "Co (III)-catalyzed coupling-cyclization of aryl C–H bonds with α-Diazoketones involving wolff rearrangement." *ACS Catalysis* no. 8(2):1308–1312.

Hu, Xinwei, Xun Chen, Yong Zhu, Yuanfu Deng, Huaqiang Zeng, Huanfeng Jiang, and Wei Zeng. 2017. "Rh (III)-catalyzed carboamination of propargyl cycloalkanols with arylamines via Csp^2–H/Csp^3–Csp^3 activation." *Organic Letters* no. 19(13):3474–3477.

Hurst, Timothy E, Ryan Gorman, Pauline Drouhin, and Richard JK Taylor. 2018. "Application of copper (II)-mediated radical cross-dehydrogenative coupling to prepare spirocyclic oxindoles and to a formal total synthesis of satavaptan." *Tetrahedron* no. 74(45):6485–6496.

Hyster, Todd K, Kyle E Ruhl, and Tomislav Rovis. 2013. "A coupling of benzamides and donor/acceptor diazo compounds to form γ-lactams via Rh (III)-catalyzed C–H activation." *Journal of the American Chemical Society* no. 135(14):5364–5367.

Inamoto, Kiyofumi, Tadataka Saito, Mika Katsuno, Takao Sakamoto, and Kou Hiroya. 2007. "Palladium-catalyzed C–H activation/intramolecular amination reaction: A new route to 3-aryl/alkylindazoles." *Organic Letters* no. 9(15):2931–2934.

Ishida, Tamao, Ryosuke Tsunoda, Zhenzhong Zhang, Akiyuki Hamasaki, Tetsuo Honma, Hironori Ohashi, Takushi Yokoyama, and Makoto Tokunaga. 2014. "Supported palladium hydroxide-catalyzed intramolecular double CH bond functionalization for synthesis of carbazoles and dibenzofurans." *Applied Catalysis B: Environmental* no. 150:523–531.

Ji, Wei-Wei, E Lin, Qingjiang Li, and Honggen Wang. 2017. "Heteroannulation enabled by a bimetallic Rh (iii)/Ag (i) relay catalysis: Application in the total synthesis of aristolactam BII." *Chemical Communications* no. 53(41):5665–5668.

Jia, Yi-Xia, and E Peter Kündig. 2009. "Oxindole synthesis by direct coupling of $C(sp^2)$-H and $C(sp^3)$-H centers." *Angewandte Chemie* no. 121(9):1664–1667.

Jia, Zhenhua, Takashi Nagano, Xingshu Li, and Albert SC Chan. 2013. "Iodide-ion-catalyzed carbon–carbon bond-forming cross-dehydrogenative coupling for the synthesis of indole derivatives." *European Journal of Organic Chemistry* no. 2013(5):858–861.

Jie, Lianghua, Lianhui Wang, Dan Xiong, Zi Yang, Di Zhao, and Xiuling Cui. 2018. "Synthesis of 2-arylindoles through Pd (II)-catalyzed cyclization of anilines with vinyl azides." *The Journal of Organic Chemistry* no. 83(18):10974–10984.

Khalili, Behzad, Pedram Jajarmi, Bagher Eftekhari-Sis, and Mohammed M Hashemi. 2008. "Novel one-pot, three-component synthesis of new 2-alkyl-5-aryl-(1 H)-pyrrole-4-ol in water." *The Journal of Organic Chemistry* no. 73(6):2090–2095.

Khalili, Behzad, Faramarz Sadeghzadeh Darabi, Bagher Eftekhari-Sis, and Mehdi Rimaz. 2013. "Green chemistry: ZrOCl2·8H2O catalyzed regioselective synthesis of 5-amino-1-aryl-1H-tetrazoles from secondary arylcyanamides in water." *Monatshefte für Chemie-Chemical Monthly* no. 144(10):1569–1572.

Khatun, Nilufa, Anju Modi, Wajid Ali, and Bhisma K Patel. 2015. "Palladium-catalyzed synthesis of 2-aryl-2H-benzotriazoles from azoarenes and TMSN3." *The Journal of Organic Chemistry* no. 80(19):9662–9670.

Kong, Lingkai, Mengdan Wang, Fangfang Zhang, Murong Xu, and Yanzhong Li. 2016. "Copper-catalyzed oxidative dearomatization/spirocyclization of indole-2-carboxamides: Synthesis of 2-spiro-pseudoindoxyls." *Organic Letters* no. 18(23):6124–6127.

Kumar, Gopal Senthil, Mohamed Ashraf Ali, Tan Soo Choon, and Karnam Jayarampillai Rajendra Prasad. 2015. "Palladium-catalyzed regioselective aerobic oxidative cyclization via C–H activation in chloroquine analogues: Synthesis and cytotoxic study." *Monatshefte für Chemie-Chemical Monthly* no. 146(12):2127–2134.

Kumar, Nivesh, Santanu Ghosh, Subhajit Bhunia, and Alakesh Bisai. 2016. "Synthesis of 2-oxindoles via 'transition-metal-free' intramolecular dehydrogenative coupling (IDC) of sp^2 C–H and sp^3 C–H bonds." *Beilstein Journal of Organic Chemistry* no. 12:1153.

Laha, Joydev K, Mandeep Kaur Hunjan, Rohan A Bhimpuria, Deepika Kathuria, and Prasad V Bharatam. 2017. "Geometry driven intramolecular oxidative cyclization of enamides: An umpolung annulation of primary benzamides with acrylates for the synthesis of 3-methyleneisoindolin-1-ones." *The Journal of Organic Chemistry* no. 82(14):7346–7352.

Lam, Hon-Wah, Ka-Yi Man, Wai-Wing Chan, Zhongyuan Zhou, and Wing-Yiu Yu. 2014. "Rhodium (III)-catalyzed formal oxidative [4 + 1] cycloaddition of benzohydroxamic acids and α-diazoesters. A facile synthesis of functionalized benzolactams." *Organic and Biomolecular Chemistry* no. 12(24):4112–4116.

Lee, Wan-Chen Cindy, Wei Wang, and Jie Jack Li. 2018. "Copper (II)-mediated ortho-selective C (sp^2)–H tandem alkynylation/annulation and ortho-hydroxylation of anilides with 2-aminophenyl-1 H-pyrazole as a directing group." *The Journal of Organic Chemistry* no. 83(4):2382–2388.

Li, Bin, Hong Xu, Huanan Wang, and Baiquan Wang. 2016a. "Rhodium-catalyzed annulation of tertiary aniline N-oxides to N-alkylindoles: Regioselective C–H activation, oxygen-atom transfer, and N-dealkylative cyclization." *ACS Catalysis* no. 6(6): 3856–3862.

Li, Deng Yuan, Hao Jie Chen, and Pei Nian Liu. 2014. "Rhodium-catalyzed oxidative annulation of hydrazines with alkynes using a nitrobenzene oxidant." *Organic Letters* no. 16(23):6176–6179.

Li, Jihui, Sébastien Bénard, Luc Neuville, and Jieping Zhu. 2012. "Copper catalyzed N-arylation of amidines with aryl boronic acids and one-pot synthesis of benzimidazoles by a Chan–Lam–Evans N-arylation and C–H activation/C–N bond forming process." *Organic Letters* no. 14(23):5980–5983.

Li, Lei, He Wang, Songjie Yu, Xifa Yang, and Xingwei Li. 2016b. "Cooperative Co (III)/Cu (II)-catalyzed C–N/N–N coupling of imidates with anthranils: Access to 1 H-indazoles via C–H activation." *Organic Letters* no. 18(15):3662–3665.

Li, Xianwei, Li He, Huoji Chen, Wanqing Wu, and Huanfeng Jiang. 2013. "Copper-catalyzed aerobic C (sp^2)–H functionalization for C–N bond formation: Synthesis of pyrazoles and indazoles." *The Journal of Organic Chemistry* no. 78(8):3636–3646.

Li, Yanlin, Chunqi Jia, Huan Li, Linhua Xu, Lianhui Wang, and Xiuling Cui. 2018. "Rh (III)-catalyzed synthesis of 2-alkylbenzimidazoles from imidamides and N-hydroxycarbamates." *Organic Letters* no. 20(16):4930–4933.

Li, Yue, Jinsong Peng, Xin Chen, Baichuan Mo, Xue Li, Peng Sun, and Chunxia Chen. 2018. "Copper-catalyzed synthesis of multisubstituted indoles through tandem ullmann-type C–N formation and cross-dehydrogenative coupling reactions." *The Journal of Organic Chemistry* no. 83(9):5288–5294.

Lian, Yajing, Robert G Bergman, Luke D Lavis, and Jonathan A Ellman. 2013. "Rhodium (III)-catalyzed indazole synthesis by C–H bond functionalization and cyclative capture." *Journal of the American Chemical Society* no. 135(19):7122–7125.

Liang, Hong-Wen, Wei Ding, Kun Jiang, Li Shuai, Yi Yuan, Ye Wei, and Ying-Chun Chen. 2015. "Redox-neutral palladium-catalyzed C–H functionalization to form isoindolinones with carboxylic acids or anhydrides as readily available starting materials." *Organic Letters* no. 17(11):2764–2767.

Liu, Wen-Qiang, Tao Lei, Zi-Qi Song, Xiu-Long Yang, Cheng-Juan Wu, Xin Jiang, Bin Chen, Chen-Ho Tung, and Li-Zhu Wu. 2017. "Visible light promoted synthesis of indoles by single photosensitizer under aerobic conditions." *Organic Letters* no. 19(12): 3251–3254.

Long, Zhen, Zhigang Wang, Danni Zhou, Danyang Wan, and Jingsong You. 2017. "Rh (III)-catalyzed regio-and chemoselective [4 + 1]-annulation of azoxy compounds with diazoesters for the synthesis of 2 H-indazoles: Roles of the azoxy oxygen atom." *Organic Letters* no. 19(11):2777–2780.

Long, Zhen, Yudong Yang, and Jingsong You. 2017. "Rh (III)-catalyzed [4 + 1]-annulation of azoxy compounds with alkynes: A regioselective approach to 2 H-indazoles." *Organic Letters* no. 19(11):2781–2784.

Lu, Ke, Xing-Wang Han, Wei-Wei Yao, Yu-Xin Luan, Yin-Xia Wang, Hao Chen, Xue-Tao Xu, Kun Zhang, and Mengchun Ye. 2018. "DMF-promoted redox-neutral Ni-catalyzed intramolecular hydroarylation of alkene with simple arene." *ACS Catalysis* no. 8(5):3913–3917.

Lu, Yi, Huai-Wei Wang, Jillian E Spangler, Kai Chen, Pei-Pei Cui, Yue Zhao, Wei-Yin Sun, and Jin-Quan Yu. 2015. "Rh (III)-catalyzed C–H olefination of N-pentafluoroaryl benzamides using air as the sole oxidant." *Chemical Science* no. 6(3):1923–1927.

Lv, Ningning, Yue Liu, Chunhua Xiong, Zhanxiang Liu, and Yuhong Zhang. 2017. "Cobalt-catalyzed oxidant-free spirocycle synthesis by liberation of hydrogen." *Organic Letters* no. 19(17):4640–4643.

Mahesh, Devulapally, Pradeep Sadhu, and Tharmalingam Punniyamurthy. 2015. "Copper (I)-catalyzed regioselective amination of N-aryl imines using TMSN3 and TBHP: A route to substituted benzimidazoles." *The Journal of Organic Chemistry* no. 80(3):1644–1650.

Mahesh, Devulapally, Pradeep Sadhu, and Tharmalingam Punniyamurthy. 2016. "Copper (ii)-catalyzed oxidative cross-coupling of anilines, primary alkyl amines, and sodium azide using TBHP: A route to 2-substituted benzimidazoles." *The Journal of Organic Chemistry* no. 81(8):3227–3234.

Mahesh, Devulapally, Vanaparthi Satheesh, Sundaravel Vivek Kumar, and Tharmalingam Punniyamurthy. 2017. "Copper (II)-catalyzed oxidative coupling of anilines, methyl arenes, and TMSN3 via C (sp^3/sp^2)–H functionalization and C–N bond formation." *Organic Letters* no. 19(24):6554–6557.

Manikandan, Rajendran, Masilamani Tamizmani, and Masilamani Jeganmohan. 2017. "Ruthenium (II)-catalyzed redox-neutral oxidative cyclization of benzimidates with alkenes with hydrogen evolution." *Organic Letters* no. 19(24):6678–6681.

Manoharan, Ramasamy, and Masilamani Jeganmohan. 2017. "Cobalt-catalyzed oxidative cyclization of benzamides with maleimides: Synthesis of isoindolone spirosuccinimides." *Organic Letters* no. 19(21):5884–5887.

Miura, Wataru, Koji Hirano, and Masahiro Miura. 2015. "Copper-mediated oxidative coupling of benzamides with maleimides via directed C–H cleavage." *Organic Letters* no. 17(16):4034–4037.

Mondal, Biplab, and Brindaban Roy. 2015. "Di-tert-butyl peroxide (DTBP) promoted dehydrogenative coupling: An expedient and metal-free synthesis of oxindoles via intramolecular C (sp 2)–H and C (sp 3)–H bond activation." *RSC Advances* no. 5(85): 69119–69123.

Neumann, Julia J, Souvik Rakshit, Thomas Droege, Sebastian Würtz, and Frank Glorius. 2011. "Exploring the oxidative cyclization of substituted N-aryl enamines: Pd-catalyzed formation of indoles from anilines." *Chemistry–A European Journal* no. 17(26):7298–7303.

Ni, Jiabin, Jie Li, Zhoulong Fan, and Ao Zhang. 2016. "Cobalt-catalyzed carbonylation of C (sp^2)–H bonds with azodicarboxylate as the carbonyl source." *Organic Letters* no. 18(22):5960–5963.

Oh, Hyunjung, Sangil Han, Ashok Kumar Pandey, Sang Hoon Han, Neeraj Kumar Mishra, Saegun Kim, Rina Chun, Hyung Sik Kim, Jihye Park, and In Su Kim. 2018. "Synthesis of (2 H)-indazoles through Rh (III)-catalyzed annulation reaction of azobenzenes with sulfoxonium ylides." *The Journal of Organic Chemistry* no. 83(7):4070–4077.

Peng, Jiangling, Zeqiang Xie, Ming Chen, Jian Wang, and Qiang Zhu. 2014. "Copper-catalyzed C (sp^2)–H amidation with azides as amino sources." *Organic Letters* no. 16(18):4702–4705.

Qi, Zisong, Songjie Yu, and Xingwei Li. 2016. "Rh (III)-catalyzed synthesis of N-unprotected indoles from imidamides and diazo ketoesters via C–H activation and C–C/C–N bond cleavage." *Organic Letters* no. 18(4):700–703.

Reddy, K Ranjith, Prakash Kannaboina, and Parthasarathi Das. 2017. "Palladium-Catalyzed chemoselective switch: Synthesis of a new class of indenochromenes and pyrano [2, 3-c] carbazoles." *Asian Journal of Organic Chemistry* no. 6(5):534–543.

Saraei, Mahnaz, Bagher Eftekhari-Sis, Massomeh Faramarzi, and Roshanak Hossienzadeh. 2014. "Synthesis of new 1, 2, 3-triazole derivatives possessing 4H-Pyran-4-one moiety by 1, 3-dipolar cycloaddition reaction of azidomethyl phenylpyrone with various alkynes." *Journal of Heterocyclic Chemistry* no. 51(5):1500–1503.

Shaikh, Tanveer Mahamadali, and Fung-E Hong. 2016. "Recent developments in the preparation of N-heterocycles using Pd-catalysed C–H activation." *Journal of Organometallic Chemistry* no. 801:139–156.

Shen, Dandan, Jing Han, Jie Chen, Hongmei Deng, Min Shao, Hui Zhang, and Weiguo Cao. 2015. "Mild and efficient one-pot synthesis of 2-(perfluoroalkyl) indoles by means of sequential michael-type addition and Pd (II)-catalyzed cross-dehydrogenative coupling (CDC) reaction." *Organic Letters* no. 17(13):3283–3285.

Shi, Pengfei, Lili Wang, Shan Guo, Kehao Chen, Jie Wang, and Jin Zhu. 2017. "AC–H activation-based strategy for N-amino azaheterocycle synthesis." *Organic Letters* no. 19(16):4359–4362.

Shi, Zhuangzhi, and Frank Glorius. 2012. "Efficient and versatile synthesis of indoles from enamines and imines by cross-dehydrogenative coupling." *Angewandte Chemie International Edition* no. 51(37):9220–9222.

Shirai, Tomohiko, Hajime Ito, and Yasunori Yamamoto. 2014. "Cationic Ir/Me-BIPAM-catalyzed asymmetric intramolecular direct hydroarylation of α-ketoamides." *Angewandte Chemie International Edition* no. 53(10):2658–2661.

Shirai, Tomohiko, and Yasunori Yamamoto. 2015. "Scope and mechanistic studies of the cationic Ir/Me-BIPAM-catalyzed asymmetric intramolecular direct hydroarylation reaction." *Organometallics* no. 34(14):3459–3463.

Sreenithya, A, Chandan Patel, Christopher M Hadad, and Raghavan B Sunoj. 2017. "Hypercoordinate iodine catalysts in enantioselective transformation: The role of catalyst folding in stereoselectivity." *ACS Catalysis* no. 7(6):4189–4196.

Sreenithya, A, and Raghavan B Sunoj. 2014. "Mechanistic insights on iodine (III) promoted metal-free dual C–H activation involved in the formation of a spirocyclic bis-oxindole." *Organic Letters* no. 16(23):6224–6227.

Stuart, David R, Mégan Bertrand-Laperle, Kevin MN Burgess, and Keith Fagnou. 2008. "Indole synthesis via rhodium catalyzed oxidative coupling of acetanilides and internal alkynes." *Journal of the American Chemical Society* no. 130(49):16474–16475.

Tang, Shi, Sheng-Wen Yang, Hongwei Sun, Yali Zhou, Juan Li, and Qiang Zhu. 2018. "Pd-Catalyzed divergent C (sp^2)–H activation/cycloimidoylation of 2-isocyano-2, 3-diarylpropanoates." *Organic Letters* no. 20(7):1832–1836.

Trosien, Simon, Philipp Böttger, and Siegfried R Waldvogel. 2013. "Versatile oxidative approach to carbazoles and related compounds using MoCl5." *Organic Letters* no. 16(2):402–405.

Wang, Chengming, Huan Sun, Yan Fang, and Yong Huang. 2013. "General and efficient synthesis of indoles through triazene-directed C–H annulation." *Angewandte Chemie International Edition* no. 52(22):5795–5798.

Wang, Cheng-Qiang, Lu Ye, Chao Feng, and Teck-Peng Loh. 2017. "C–F bond cleavage enabled redox-neutral [4+ 1] annulation via C–H bond activation." *Journal of the American Chemical Society* no. 139(5):1762–1765.

Wang, Fen, Guoyong Song, and Xingwei Li. 2010. "Rh (III)-catalyzed tandem oxidative olefination– Michael reactions between aryl carboxamides and alkenes." *Organic Letters* no. 12(23):5430–5433.

Wang, Qiang, and Xingwei Li. 2016. "Synthesis of 1 H-Indazoles from imidates and nitrosobenzenes via synergistic rhodium/copper catalysis." *Organic Letters* no. 18(9):2102–2105.

Wang, Shi-Meng, Chen Li, Jing Leng, Syed Nasir Abbas Bukhari, and Hua-Li Qin. 2018. "Rhodium (iii)-catalyzed oxidative coupling of N-methoxybenzamides and ethenesulfonyl fluoride: A C–H bond activation strategy for the preparation of 2-aryl ethenesulfonyl fluorides and sulfonyl fluoride substituted γ-lactams." *Organic Chemistry Frontiers* no. 5(9):1411–1415.

Watanabe, Toshiaki, Shinya Oishi, Nobutaka Fujii, and Hiroaki Ohno. 2009. "Palladium-catalyzed direct synthesis of carbazoles via one-pot N-arylation and oxidative biaryl coupling: Synthesis and mechanistic study." *The Journal of Organic Chemistry* no. 74(13):4720–4726.

Watanabe, Toshiaki, Satoshi Ueda, Shinsuke Inuki, Shinya Oishi, Nobutaka Fujii, and Hiroaki Ohno. 2007. "One-pot synthesis of carbazoles by palladium-catalyzed N-arylation and oxidative coupling." *Chemical Communications* no. 21(43):4516–4518.

Wei, Wen-Ting, Ming-Bo Zhou, Jian-Hong Fan, Wei Liu, Ren-Jie Song, Yu Liu, Ming Hu, Peng Xie, and Jin-Heng Li. 2013. "Synthesis of oxindoles by iron-catalyzed oxidative 1, 2-Alkylarylation of activated alkenes with an aryl C(sp^2)-H Bond and a C(sp^3)-H Bond adjacent to a heteroatom." *Angewandte Chemie* no. 125(13):3726–3729.

Wen, Lixian, Lin Tang, Yu Yang, Zhenggen Zha, and Zhiyong Wang. 2016. "Ligand-free Pd-catalyzed domino synthesis of carbazoles via dehydrogenative aromatization/C (sp^2)–C (sp^2) coupling sequence." *Organic Letters* no. 18(6):1278–1281.

Wu, Cheng-Juan, Qing-Yuan Meng, Tao Lei, Jian-Ji Zhong, Wen-Qiang Liu, Lei-Min Zhao, Zhi-Jun Li, Bin Chen, Chen-Ho Tung, and Li-Zhu Wu. 2016. "An oxidant-free strategy for indole synthesis via intramolecular C–C bond construction under visible light irradiation: Cross-coupling hydrogen evolution reaction." *ACS Catalysis* no. 6(7):4635–4639.

Wu, Hua, Yu-Ping He, Lue Xu, Dong-Yang Zhang, and Liu-Zhu Gong. 2014. "Asymmetric organocatalytic direct C (sp^2)-H/C (sp^3)-H oxidative cross-coupling by chiral iodine reagents." *Angewandte Chemie* no. 126(13):3534–3537.

Wu, Shangze, Xiaoyan Wu, Chunling Fu, and Shengming Ma. 2018. "Rhodium (III)-catalyzed C–H functionalization in water for isoindolin-1-one synthesis." *Organic Letters* no. 20(10):2831–2834.

Wu, Xiaowei, Bao Wang, Yu Zhou, and Hong Liu. 2017a. "Propargyl alcohols as one-carbon synthons: Redox-neutral rhodium (III)-catalyzed C–H bond activation for the synthesis of isoindolinones bearing a quaternary carbon." *Organic Letters* no. 19(6):1294–1297.

Wu, Xiaowei, Bao Wang, Shengbin Zhou, Yu Zhou, and Hong Liu. 2017b. "Ruthenium-catalyzed redox-neutral [4+ 1] annulation of benzamides and propargyl alcohols via C–H bond activation." *ACS Catalysis* no. 7(4):2494–2499.

Wu, Xuesong, Yan Zhao, and Haibo Ge. 2015. "Direct aerobic carbonylation of C (sp^2)–H and C (sp^3)–H bonds through Ni/Cu synergistic catalysis with DMF as the carbonyl source." *Journal of the American Chemical Society* no. 137(15):4924–4927.

Xu, Fan, Yan-Jie Li, Chong Huang, and Hai-Chao Xu. 2018a. "Ruthenium-catalyzed electrochemical dehydrogenative alkyne annulation." *ACS Catalysis* no. 8(5):3820–3824.

Xu, Linhua, Lianhui Wang, Yadong Feng, Yudong Li, Lei Yang, and Xiuling Cui. 2017. "Iridium (III)-catalyzed one-pot access to 1, 2-disubstituted benzimidazoles starting from imidamides and sulfonyl azides." *Organic Letters* no. 19(16):4343–4346.

Xu, Shiyang, Baoliang Huang, Guanyu Qiao, Ziyue Huang, Zhen Zhang, Zongyang Li, Peng Wang, and Zhenhua Zhang. 2018b. "Rh (III)-catalyzed C–H activation of boronic acid with aryl azide." *Organic Letters* no. 20(18):5578–5582.

Xu, Youwei, Fen Wang, Songjie Yu, and Xingwei Li. 2017. "Rhodium (III)-catalyzed selective access to isoindolinones via formal [4+ 1] annulation of arylamides and propargyl alcohols." *Chinese Journal of Catalysis* no. 38(8):1390–1398.

Yamamoto, Chiaki, Kazutaka Takamatsu, Koji Hirano, and Masahiro Miura. 2017. "A divergent approach to indoles and oxazoles from enamides by directing-group-controlled Cu-catalyzed intramolecular C–H amination and alkoxylation." *The Journal of Organic Chemistry* no. 82(17):9112–9118.

Yan, Hao, Haolong Wang, Xincheng Li, Xiaoyi Xin, Chunxiang Wang, and Boshun Wan. 2015. "Rhodium-catalyzed C-H annulation of nitrones with alkynes: A regiospecific route to unsymmetrical 2, 3-diaryl-substituted indoles." *Angewandte Chemie International Edition* no. 54(36):10613–10617.

Ye, Baihua, and Nicolai Cramer. 2014. "Asymmetric synthesis of isoindolones by chiral cyclopentadienyl-rhodium (III)-catalyzed C-H functionalizations." *Angewandte Chemie International Edition* no. 53(30):7896–7899.

Youn, So Won, Tae Yun Ko, Young Ho Kim, and Yun Ah Kim. 2018. "Pd (II)/Cu (II)-catalyzed regio-and stereoselective synthesis of (E)-3-Arylmethyleneisoindolin-1-ones using air as the terminal oxidant." *Organic Letters* no. 20(24):7869–7874.

Yu, Da-Gang, Mamta Suri, and Frank Glorius. 2013. "RhIII/CuII-cocatalyzed synthesis of 1 H-indazoles through C–H amidation and N–N bond formation." *Journal of the American Chemical Society* no. 135(24):8802–8805.

Yu, Zhengsen, Lijuan Ma, and Wei Yu. 2010. "Ag2O-mediated intramolecular oxidative coupling of acetoacetanilides for the synthesis of 3-acetyloxindoles." *Synlett* no. 2010(17):2607–2610.

Zeng, Li, Haoran Li, Shan Tang, Xinlong Gao, Yi Deng, Guoting Zhang, Chih-Wen Pao, Jeng-Lung Chen, Jyh-Fu Lee, and Aiwen Lei. 2018. "Cobalt-catalyzed electrochemical oxidative C–H/N–H carbonylation with hydrogen evolution." *ACS Catalysis* no. 8(6):5448–5453.

Zhang, Chunhui, Yongzheng Ding, Yuzhen Gao, Shangda Li, and Gang Li. 2018a. "Palladium-catalyzed direct C–H carbonylation of free primary benzylamines: A synthesis of benzolactams." *Organic Letters* no. 20(9):2595–2598.

Zhang, Jitan, Hui Chen, Cong Lin, Zhanxiang Liu, Chen Wang, and Yuhong Zhang. 2015. "Cobalt-catalyzed cyclization of aliphatic amides and terminal alkynes with silver-cocatalyst." *Journal of the American Chemical Society* no. 137(40):12990–12996.

Zhang, Ming-Zhong, Na Luo, Rui-Yang Long, Xian-Tao Gou, Wen-Bing Shi, Shu-Hua He, Yong Jiang, Jin-Yang Chen, and Tieqiao Chen. 2018b. "Transition-metal-free oxidative aminooxyarylation of alkenes: Annulations toward aminooxylated oxindoles." *The Journal of Organic Chemistry* no. 83(4):2369–2375.

Zhang, Sheng, Fei Lian, Mengyu Xue, Tengteng Qin, Lijun Li, Xu Zhang, and Kun Xu. 2017. "Electrocatalytic dehydrogenative esterification of aliphatic carboxylic acids: Access to bioactive lactones." *Organic Letters* no. 19(24):6622–6625.

Zhang, Tianshui, and Weiliang Bao. 2013. "Synthesis of 1 H-Indazoles and 1 H-pyrazoles via FeBr3/O2 mediated intramolecular C–H amination." *The Journal of Organic Chemistry* no. 78(3):1317–1322.

Zhang, Yan, Dahai Wang, and Sunliang Cui. 2015. "Facile synthesis of isoindolinones via Rh (III)-catalyzed one-pot reaction of benzamides, ketones, and hydrazines." *Organic Letters* no. 17(10):2494–2497.

Zhang, Zhenxing, Hao Jiang, and Yong Huang. 2014. "Ruthenium-catalyzed redox-neutral C–H activation via N–N cleavage: Synthesis of N-substituted indoles." *Organic Letters* no. 16(22):5976–5979.

Zhang, Zhuo-Zhuo, Bin Liu, Jing-Wen Xu, Sheng-Yi Yan, and Bing-Feng Shi. 2016. "Indole synthesis via cobalt (iii)-catalyzed oxidative coupling of N-arylureas and internal alkynes." *Organic Letters* no. 18(8):1776–1779.

Zheng, Xin-Xiang, Cong Du, Xue-Mei Zhao, Xinju Zhu, Jian-Feng Suo, Xin-Qi Hao, Jun-Long Niu, and Mao-Ping Song. 2016. "Ni (II)-catalyzed C (sp^2)–H alkynylation/annulation with terminal alkynes under an oxygen atmosphere: A one-pot approach to 3-methyleneisoindolin-1-one." *The Journal of Organic Chemistry* no. 81(10):4002–4011.

Zhou, Bo, Zhuo Wu, Ding Ma, Xiaoming Ji, and Yanghui Zhang. 2018a. "Synthesis of indoles through palladium-catalyzed three-component reaction of aryl iodides, alkynes, and diaziridinone." *Organic Letters* no. 20(20):6440–6443.

Zhou, Jianhui, Jian Li, Yazhou Li, Chenglin Wu, Guoxue He, Qiaolan Yang, Yu Zhou, and Hong Liu. 2018b. "Direct synthesis of 3-acylindoles through rhodium (III)-catalyzed annulation of N-phenylamidines with α-Cl ketones." *Organic Letters* no. 20(23):7645–7649.

Zhou, Ming-Bo, Ren-Jie Song, Xuan-Hui Ouyang, Yu Liu, Wen-Ting Wei, Guo-Bo Deng, and Jin-Heng Li. 2013. "Metal-free oxidative tandem coupling of activated alkenes with carbonyl C (sp 2)–H bonds and aryl C (sp 2)–H bonds using TBHP." *Chemical Science* no. 4(6):2690–2694.

Zhou, Shuguang, Jinhu Wang, Lili Wang, Kehao Chen, Chao Song, and Jin Zhu. 2016. "Co (III)-catalyzed, internal and terminal alkyne-compatible synthesis of indoles." *Organic Letters* no. 18(15):3806–3809.

Zhou, Zhi, Guixia Liu, and Xiyan Lu. 2016. "Regiocontrolled coupling of aromatic and vinylic amides with α-Allenols to form γ-lactams via rhodium (III)-catalyzed C–H activation." *Organic Letters* no. 18(21):5668–5671.

Zhu, Chen, and John R Falck. 2012. "Rhodium catalyzed C–H olefination of N-benzoylsulfonamides with internal alkenes." *Chemical Communications* no. 48(11):1674–1676.

Zirak, Maryam, and Bagher Eftekhari-Sis. 2015. "Kojic acid in organic synthesis." *Turkish Journal of Chemistry* no. 39(3):439–496.

2 Six- and Higher-Membered *N*-Heterocycles

2.1 INTRODUCTION

Six-membered nitrogen-containing heterocycles are vastly found in a number of biologically active compounds, natural products and pharmaceuticals, such as antibacterial, antifungal, antibiotics, anti-HIV, anticancer, antipsychotic, antiviral, antihypertensive, immunomodulator, and glycosidase inhibitors (Eftekhari-Sis, Zirak, and Akbari 2013, Eftekhari-Sis and Zirak 2014). They are also widely used as organocatalysts (Eftekhari-Sis, Sarvari Karajabad, and Haqverdi 2017, Malkov et al. 2003) and sensors (Eftekhari-Sis and Mirdoraghi 2016, Eftekhari-Sis, Samadneshan, and Vahdati-Khajeh 2018, Eftekhari-Sis et al. 2018). However, various types of one-step or multi-step synthesis of these types of heterocycles have been reported in the literature; the development of new approaches to the synthesis of six-membered *N*-heterocycles with various functional groups are of interest to synthetic and medicinal scientists (Eftekhari-Sis, Zirak, and Akbari 2013, Eftekhari-Sis and Zirak 2014, Zirak and Eftekhari-Sis 2015).

Cross-dehydrogenative-coupling (CDC) (Gandeepan et al. 2018, Shaikh and Hong 2016) processes are widely used in the synthesis of a variety of types of heterocyclic compounds. In this chapter, intramolecular and intermolecular CDC of aromatic $C(sp^2)$–H bonds are discussed to construct six- and seven-membered *N*-heterocycles, including quinoline, isoquinoline, acridine, phenanthridine, cinnoline, quinazoline, benzacepine, and benzazocine, etc.

2.2 QUINOLINE

Intramolecular dehydrogenative coupling of *N*-substituted but-3-en-1-ylaniline derivatives **1a** were reported for the construction of dihydroquinoline **2** scaffolds by treating **1a** with $PdCl_2(CH_3CN)_2$ (5 mol%) in the presence of TsOH and *p*-benzoquinone in dioxane at room temperature or 70°C. Corresponding dihydroquinolines **2** were obtained in 11–89% yields. Reaction with conjugated alkenes **1b** in the presence of an oxidant in AcOH at 70°C afforded quinolines **3** in 32–62% yields. Reactions proceeded by electrophilic palladation to give intermediate **I**, which underwent migratory insertion, generating intermediate **II**, and then *β*-hydride elimination to generate quinoline intermediate **III** with an exocyclic double bond, which isomerized to an endocyclic double bond, forming the final desired quinoline product **2**. The generated Pd(0) species was oxidized by *p*-benzoquinone (BQ) to a Pd(II) active species for the next catalytic cycle (Scheme 2.1) (Carral-Menoyo et al. 2017).

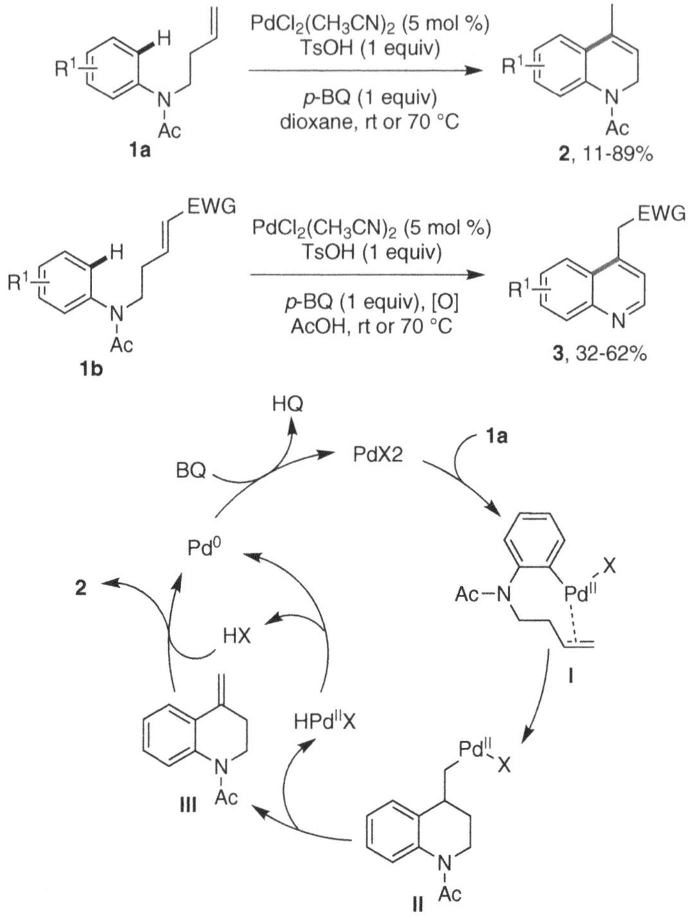

SCHEME 2.1

Rh-catalyzed synthesis of 1,2-dihydroquinolines **6** from the reaction of terminal alkynes **5** and aniline **4** was reported by heating a solution of 2.1 equiv of terminal alkynes **5** and anilines **6** in the presence of $[Cp^*RhCl_2]_2$ (2 mol%) and $HBF_4{\cdot}OEt_2$ (5 mol%) in toluene, in 34–89% yields. Reaction with alkyl alkynes gave corresponding 1,2-dihydroquinolines **6** in lower yields than aromatic alkynes. The proposed reaction mechanism involves the Rh-catalyzed hydroamination of alkyne **5** with aniline **4**, generating enamine intermediate **I**, which underwent *ortho*-C–H activation by electrophilic addition of Rh active species **II**, to give enaminyl alkene coordinated species **III**. The intermediate **V**, formed by coordinated alkene displacement by an alkyne (**IV**) followed by alkyne insertion, was converted to dihydroquinoline via 1,2-alkene insertion of the enamine to give intermediate **VI**, followed by protonolysis along with regeneration of the active Rh catalyst (Scheme 2.2) (Kumaran and Leong 2015).

1,2-Dihydroquinolines **9** were synthesized by Co-catalyzed C–H/N–H annulation of anilides **7** with allenes **8**. Reactions were performed using 1.5 equiv of allene

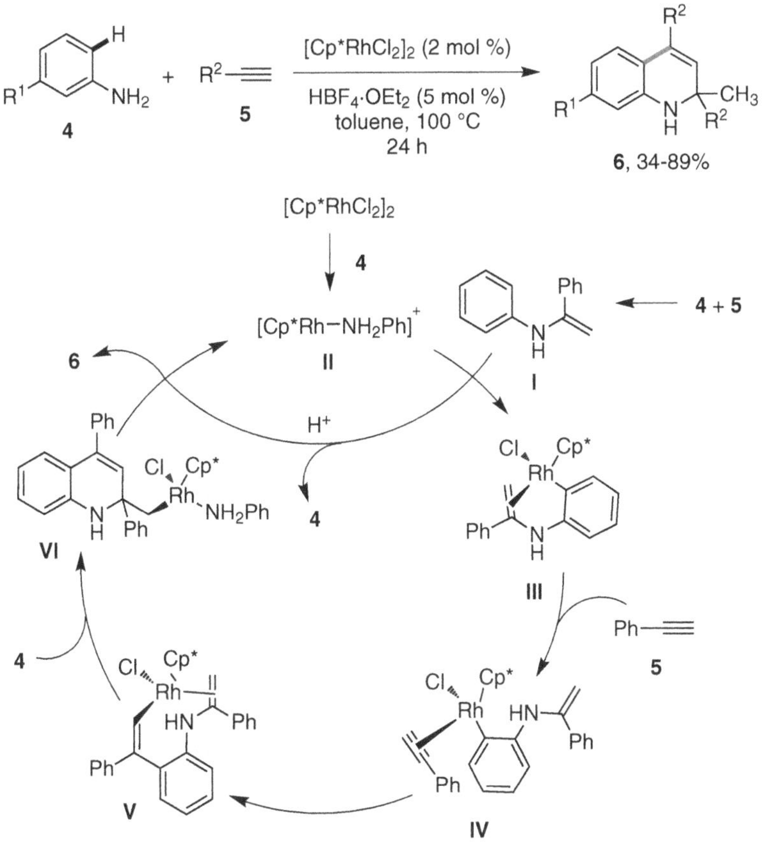

SCHEME 2.2

8 in nitromethane in the presence of [Cp*Co(CO)I_2] (9 mol%), $AgSbF_6$ (20 mol%), $Cu(OAc)_2{\cdot}H_2O$ (20 mol%), and Ag_2CO_3 (1 equiv) at 80°C for 15 h under N_2, affording 1,2-dihydroquinolines **9** in 61–91% yields. The reaction was initiated by the generation of active Co(III) complex [Cp*Co(OAc)]$^+$ from [Cp*Co(CO)I_2], $Cu(OAc)_2{\cdot}H_2O$ and $AgSbF_6$. By *ortho*-C–H metalation of anilide, followed by allene insertion, β-hydride elimination, and then reductive elimination, dihydroquinolines were obtained with generation of a Co(I) species, which was oxidized to an active Co(III) species by the silver ion (Scheme 2.3) (Kuppusamy et al. 2018).

Co-catalyzed construction of quinoline derivatives **11** was achieved by dehydrogenative coupling of anilides **7** with alkynes **10**. Reactions were conducted using 1.2 equiv of alkynes **10** in the presence of [Cp*CoCl$_2$]$_2$ (6 mol%) and $AgNTf_2$ (1 equiv) in 1,2-dichloroethane (DCE) under air atmosphere at 130°C for 16 h, leading to quinolones **11** in 54–96% yields. Reaction with unsymmetrical alkyl phenyl alkynes afforded corresponding 3-phenyl-4-alkylquinolines **11**, regioselectively. In the plausible reaction mechanism, by C–H bond activation of acetanilide **7** with an active cationic Co(III) catalyst, generated in situ from [Cp*CoCl$_2$]$_2$ with $AgNTf_2$,

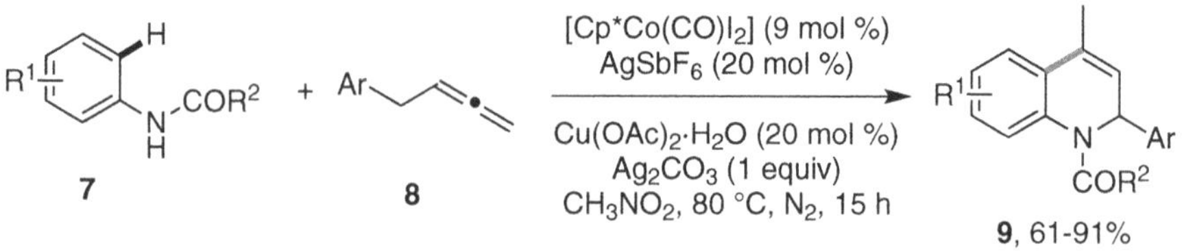

SCHEME 2.3

a six-membered metallacyclic intermediate **I** was formed, which underwent coordination and migratory insertion of alkyne **10** to give a Co(III) alkenyl species **II**. Protonolysis of the Co–O bond of Co(III) alkoxide species **III**, formed by the migratory insertion of the Co–C bond into the carbonyl group, produced alcohol **IV** with regeneration of the active catalyst. Upon dehydration of the alcohol **IV**, the final desired quinoline **11** was obtained (Scheme 2.4) (Kong et al. 2016, Lu et al. 2016).

Rh-catalyzed synthesis of benzoquinoline derivatives **13** was developed through *peri*-C–H functionalization. Treatment of naphthylcarbamates **12** with alkynes **10** (2 equiv) in the presence of $[Cp^*RhCl_2]_2$ (2.5 mol%) and Ag_2CO_3 (2 equiv) in *N,N*-dimethylformamide (DMF) under an atmosphere of argon at 70°C resulted in the formation of benzoquinolines **13** in 64–99% yields. Various polycyclic aromatic

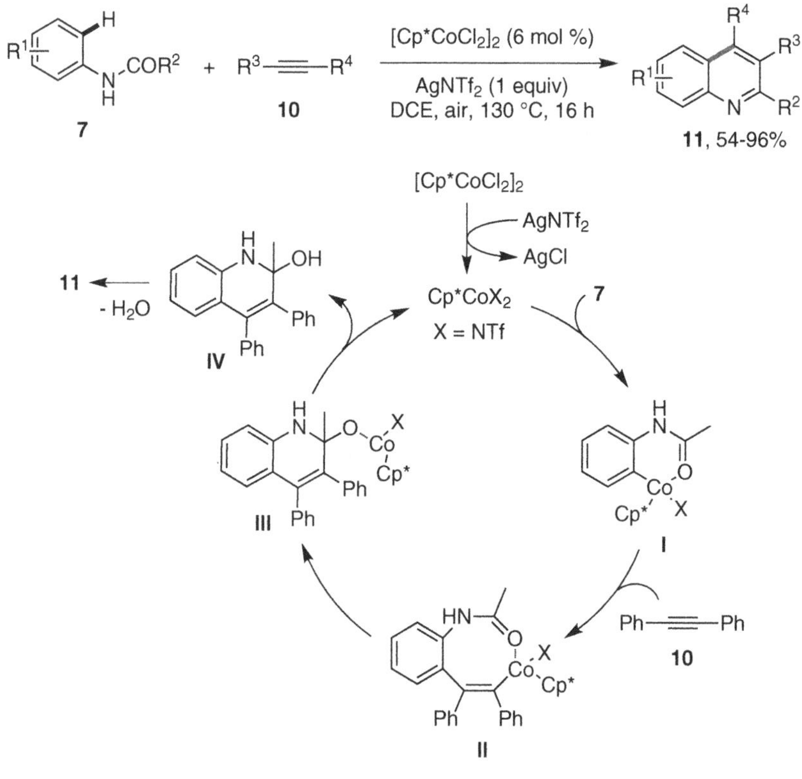

SCHEME 2.4

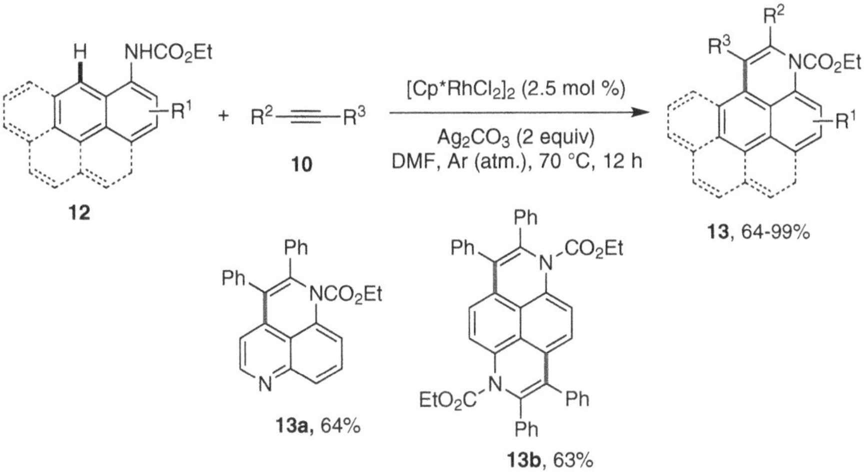

SCHEME 2.5

carbamates were also tolerated under reaction conditions, leading to the expected benzoquinolines. 1*H*-benzo[*ij*][2,7]naphthyridine carbamate **13a** was obtained from quinolin-5-ylcarbamate substrate in 64% yield. Double annulation of diethyl naphthalene-1,5-diyldicarbamate with diphenyl acetylene under reaction conditions afforded 1,6-dihydrobenzo[*lmn*][2,7]phenanthroline product **13b** in 63% yield. The proposed reaction mechanism involves the coordination of the neutral Rh(III) complex preferentially to the *N* atom of carbamate **12**, followed by cyclorhodation at the *peri*-C–H bond. By coordination and insertion of an alkyne followed by reductive elimination, benzoquinoline product **13** was obtained along with an Rh(I) complex, which was oxidized by Ag_2CO_3 to regenerate the active Rh(III) complex (Scheme 2.5) (Zhang et al. 2014).

2.3 QUINOLINONE

The Ag-catalyzed cascade cyclization of cinnamamides **14** with diphenylphosphine oxide was developed for the construction of 3,4-disubstituted dihydroquinolin-2(1*H*)-ones **15** through activation of the P–H bond and functionalization of the C(sp^2)–H bond. Reactions were performed using 2 equiv of diphenylphosphine oxide in the presence of $AgNO_3$ (15 mol%), $Mg(NO_3)_2 \cdot 6H_2O$, and 4 Å molecular sieve (MS) in dry MeCN at 100°C under N_2 atmosphere for 12 h, to produce dihydroquinolin-2(1*H*)-ones **15** in 52–78% yields. A similar approach was reported for the synthesis of cyanomethylated 3,4-dihydroquinolin-2(1*H*)-one derivatives **16** by heating a solution of *N*-methyl-*N*-arylcinnamamide and AgOAc (4 equiv) in CH_3CN at 130°C under N_2 for 36 h, in 58–62% yields. Also, Cu-catalyzed tandem oxidative cyclization of cinnamamides **14** with benzyl hydrocarbons has been developed in order to prepare 3-benzyl dihydroquinolin-2(1*H*)-ones **17** through the direct cross-dehydrogenative coupling of sp^3 and sp^2 C–H bonds. Reactions were carried out using 5 mol% of Cu_2O in the presence of *tert*-butyl peroxybenzoate (TBPB) (2 equiv) at

120°C for 28 h, leading to corresponding dihydroquinolin-2(1*H*)-ones **17** in 37–88% yields. The proposed reaction mechanism involves metal-induced hydrogen abstraction from diphenylphosphine oxide, CH_3CN, or toluene to generate radical species **I**, which was subsequently added to the double bond of **14** to give radical **II**. By cyclization of radical **II**, radical intermediate **III** was generated, which was converted to the final product via single-electron oxidation by Cu(II) or Ag(I) followed by the loss of H^+ (Scheme 2.6) (Zhang et al. 2016, Wang et al. 2018, Zhou et al. 2014).

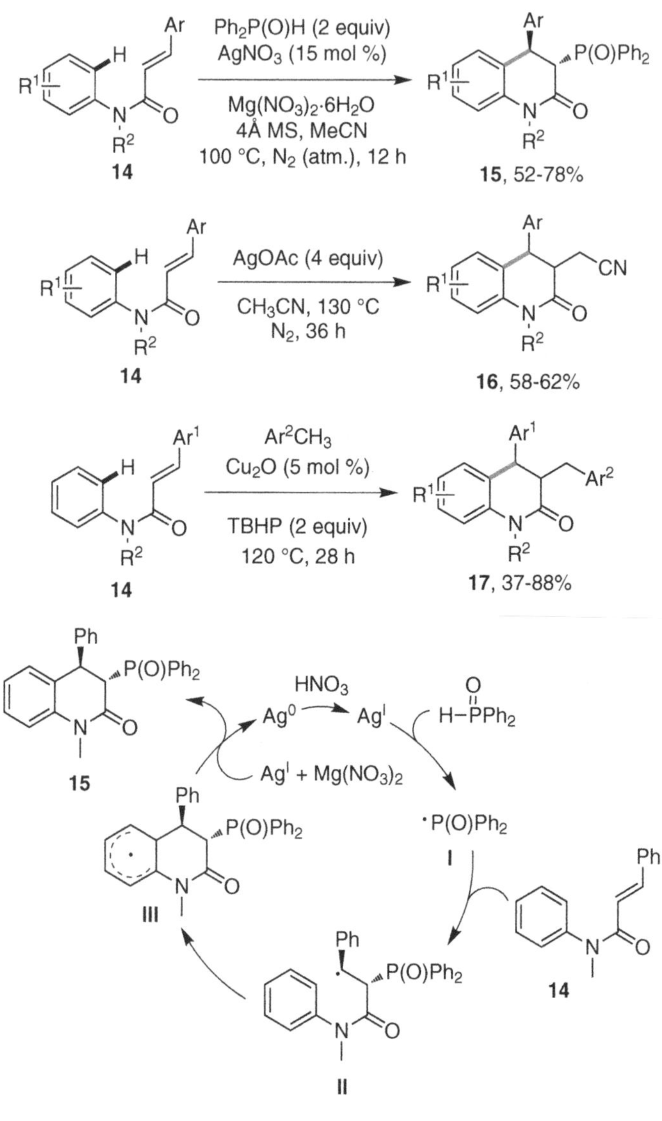

SCHEME 2.6

Quinoline-2-one derivatives **19** were synthesized by a Pd-catalyzed dehydrogenative intramolecular amidation process. Treating **18** with $PdCl_2$ (10 mol%) and $Cu(OAc)_2$ (50 mol%) in dimethyl sulfoxide (DMSO) under an O_2 atmosphere at 120°C for 14 h afforded quinoline-2-ones **19** in 38–98% yields. The reaction of the unsymmetrical substrate occurred at the aryl group with a *Z*-position to the amide moiety, specifically (Scheme 2.7) (Inamoto et al. 2010).

Ru-catalyzed dehydrogenative coupling cyclization of acetanilides **7** with propiolates **19** or acrylates **18** was developed to afford 2-quinolinones having diverse functional groups. Reactions with propiolates **19** were performed using 1.5 equiv of propiolate **19** in the presence of $[RuCl_2(p\text{-cymene})]_2$ (5 mol%) and $AgSbF_6$ (20 mol%), either in PivOH/*i*-PrOH or AcOH at 130°C for 24 h, to give 2-quinolinones **21** in 62–83% yields. Reactions of acrylates **18** (1.5 equiv) were conducted using $[RuCl_2(p\text{-cymene})]_2$ (5 mol%), $AgSbF_6$ (20 mol%), and $Cu(OAc)_2 \cdot H_2O$ (1.5 equiv) in DCE at 110°C for 12 h, leading to 2-quinolinones **20** in 46–64% yields. Fused 7-methylthieno[3,2-*b*]pyridin-5(4*H*)-one **21a** was also obtained in 83% yield from the reaction of *N*-(thiophen-3-yl)acetamide with ethyl 2-butynoate under the reaction conditions. In the tentative reaction pathway, coordination with subsequent *ortho*-metalation of acetanilide **7** with cationic Ru species, generated in situ by reaction of $AgSbF_6$ with $[RuCl_2(p\text{-cymene})]_2$, provided ruthenacycle **I**, which underwent alkyne insertion into the Ru–C bond to give intermediate **II**. Protonation at the Ru–C bond of intermediate **II** by RCO_2H afforded *ortho*-alkenylated anilide ***E*-III** with regeneration of the Ru species. By *trans–cis* isomerization of the double bond of **III**, followed by intramolecular nucleophilic addition of NH to the ester moiety and then by loss of the acetyl group, 2-quinolinone **21** was produced (Scheme 2.8) (Manikandan and Jeganmohan 2014). A similar approach for the construction of 2-quinolionones was described by Chen et al. (2010) by subjecting *N*-aryl-2-aminopyridines to benzyl acylate (2 equiv) in the presence of $[Cp^*RhCl_2]_2$ (1 mol%) and $Cu(OAc)_2$ (2.2 equiv) in DMF at 100°C under N_2, in 50–91% yields.

Rh-catalyzed dehydrogenative carbonylation and annulation of *N*-alkylanilines **4** with CO and alkynes **10** through N–H and C–H bond activation was developed for the synthesis of 2-quinolinones **22**. By treatment of an aniline with an alkyne (1.2 equiv) in the presence of $Rh(PPh_3)_3Cl$ (2 mol%), $Cu(OAc)_2$ (2 equiv), and Li_2CO_3 (0.5 equiv) in xylene under CO atmosphere at 130°C, 2-quinolinone derivatives **22** were obtained in 35–95% yields. Diaryl acetylenes as well as dialkyl acetylenes work well under reaction conditions, affording the expected 3,4-disubstituted 2-quinolinones. Unsymmetrical alkyl aryl alkynes led to corresponding 2-quinolinones **22** in 86–92%

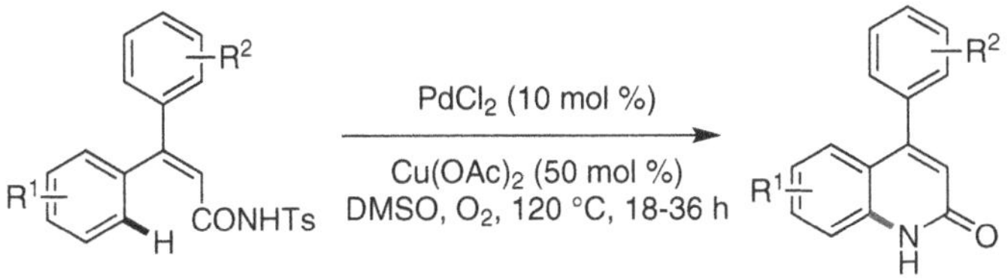

R¹ R² H CONHTs
$PdCl_2$ (10 mol %)
$Cu(OAc)_2$ (50 mol %)
DMSO, O_2, 120 °C, 18-36 h
R¹ R² N H O
18
19, 38-98%

SCHEME 2.7

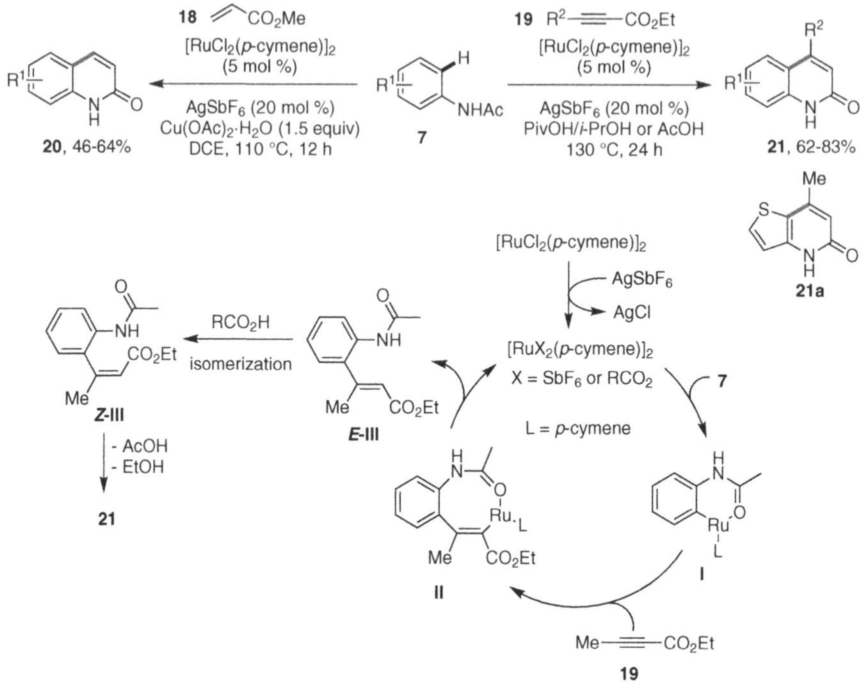

SCHEME 2.8

yields, with 2.6:1–2.7:1 regioselectivity, favoring 4-alkyl-3-aryl 2-quinolinone isomers. Also, fused 2-quinolinones, including pyrido[3,2,1-*ij*]quinolin-3(5*H*)-one **22a**, 3*H*-azepino[3,2,1-*ij*]quinolin-3-one **22b**, and benzo[6,7][1,4]oxazepino[2,3,4-*ij*]quinolin-1(12*H*)-one **22c**, were obtained starting from the corresponding substrates in 60%, 84%, and 35–41% yields, respectively. Macrocyclic alkyne, cyclododecyne, was also compatible with the reaction, affording the polymacrocyclic product **22d** in 55% yield. The reaction was initiated with the generation of Rh(III) via oxidation of Rh(I) with $Cu(OAc)_2$, followed by ligand exchange with CO to afford Rh(III)-CO species **I**. Subsequent coordination of aniline **4** to **I** followed by CO insertion gave Rh(III) complex **II**, which underwent the concerted metalation/deprotonation process to give rhodacycle **III**. The final 2-quinolinone product **22** was obtained by ligand exchange of **III** with alkyne **10** (Rh(III)-alkyne complex **IV**) and then insertion to generate seven-membered rhodacycle **V**, followed by reductive elimination, along with regeneration of the Rh(I) species, which reoxidized to a Rh(III) species in the presence of $Cu(OAc)_2$ (Scheme 2.9) (Li, Li, and Jiao 2015).

Synthesis of dihydro-1,10-phenanthrolin-2(1*H*)-one derivatives **24** was achieved by intramolecular benzylic $C(sp^3)$–H/$C(sp^2)$–H cross-dehydrogenative coupling. Reactions were catalyzed with $Pd(OAc)_2$ (10 mol%), in the presence of 2,2-bipyridine (15 mol%), and AgOAc (2 equiv) in HFIP/HOAc at 160°C under N_2 atmosphere, and dihydro-1,10-phenanthrolin-2(1*H*)-ones **24** were obtained in 48–82% yields. Reactions failed to produce expected products in the case of 3-alkyl, alkenyl, and

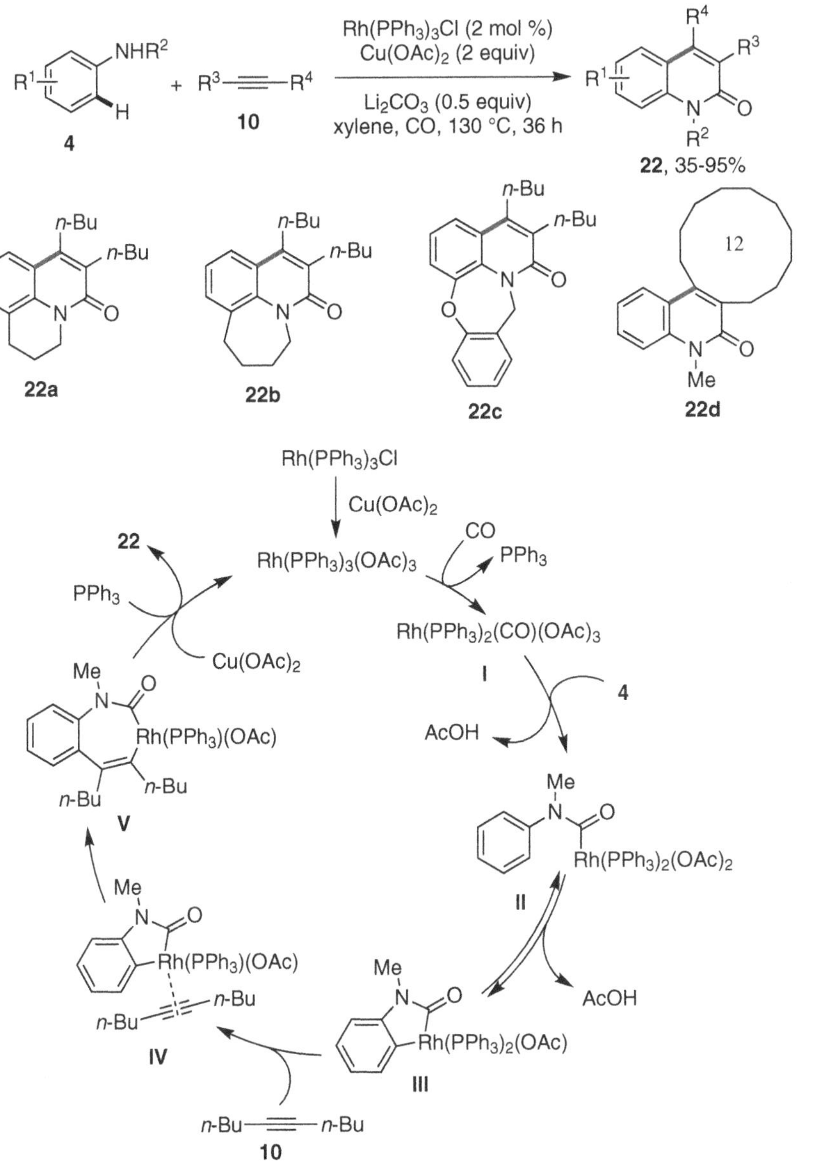

SCHEME 2.9

alkynyl substituted propionamides. Oxidation of the obtained dihydro-1,10-phenanthrolin-2(1*H*)-ones **24** with 2,3-dichloro-5,6-dicyano-1,4-benzoquinone (DDQ) gave rise to the formation of 2-hydroxy-1,10-phenanthrolines **25** in 82–96% yields. The proposed reaction mechanism involves 8-amino-quinoline-directed benzylic C(sp^3)–H bond activation, with subsequent C–N bond rotation and activation of the C(sp^2)–H bond of the 8-amino quinoline to generate palladacycle **I**. Reductive

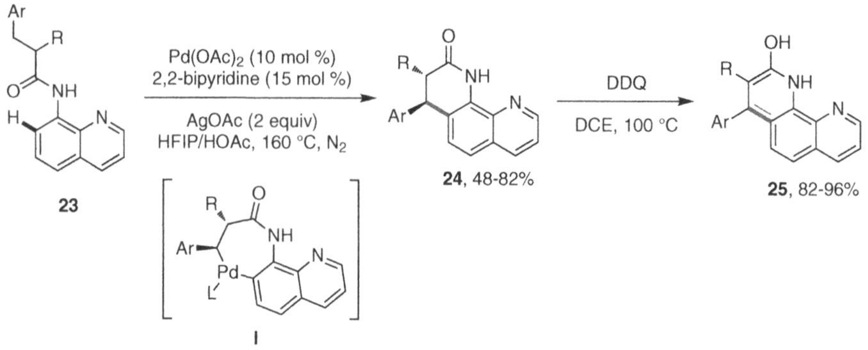

SCHEME 2.10

elimination afforded the final 4-phenyl-3,4-1,10-phenanthroline-2(1*H*)-one **24** (Scheme 2.10) (Jiang et al. 2018).

Quinolin-4(1*H*)-ones **27** were synthesized through palladium-catalyzed oxidative carbonylation of ketones **26**, amines, and carbon monoxide. Subjecting 1.2 equiv of an enolizable ketone **26** to aniline in the presence of 4 Å MS in $PhCH_3$ at 110°C for 6 h, followed by addition of Pd(dba)2 (20 mol%), Xantphos (20 mol%), KI (1 equiv), $CuBr(Me_2S)$ (1 equiv), and $PhCO_2Na$ (1.5 equiv) in $PhCH_3$/DMSO at 110°C under an atmosphere of $CO:O_2=3:1$ mixture for 24 h afforded 4-quinolinones **27** in 44–76% yields. In the proposed reaction mechanism, electrophilic addition of Pd(II) to enamine **II**, derived from the reaction of ketone with aniline, generated intermediate **III**, which underwent intramolecular C–H activation and CO insertion to give intermediate **IV**. Subsequently, reductive elimination of **IV** afforded the final product **27** and released a Pd(0) species, which was oxidized by copper and O_2 to regenerate the Pd(II) species for the next catalytic cycle (Scheme 2.11) (Wu et al. 2017).

2.4 ISOQUINOLINE

Pd-catalyzed site-selective C(sp²)–H activation/cycloimidoylation of 2-isocyano-2,3-diarylpropanoates **28** was developed for the construction of 3,4-dihydroisoquinolines **30** containing a C-3 quaternary carbon center. Reactions were performed using 1.2 equiv of aryl iodide **29** in the presence of $Pd(OAc)_2$ (10 mol%), Ad_2Pn-Bu (20 mol%), Cs_2CO_3 (1.2 equiv), and PivOH in 1,4-dioxane at 90°C under Ar atmosphere, to furnish 3,4-dihydroisoquinolines **30** in 60–85% yields. Reaction with *meta*-substituted 2-isocyano-2,3-diarylpropanoate **28** gave two regioisomers, favoring functionalization at the less hindered *para* position of the substituent (Tang et al. 2018). A similar approach was described by Luo et al. (2018) for the synthesis of pyrido[3,4-*b*]ferrocenes **32** by a Pd-catalyzed enantioselective isocyanide insertion/desymmetric C(sp²)–H bond activation reaction. Reactions were catalyzed with $Pd(OAc)_2$ (5 mol%) in the presence of a phosphoramidite ligand, Cs_2CO_3 (1.2 equiv), and PivOH in toluene at 75°C, leading to pyrido[3,4-*b*]ferrocenes **32** in 61–99% yields (Scheme 2.12).

Guimond and Fagnou (2009) reported Rh(III)-catalyzed synthesis of isoquinoline derivatives via dehydrogenative coupling of *N*-*tert*-butylbenzaldimines **33** with

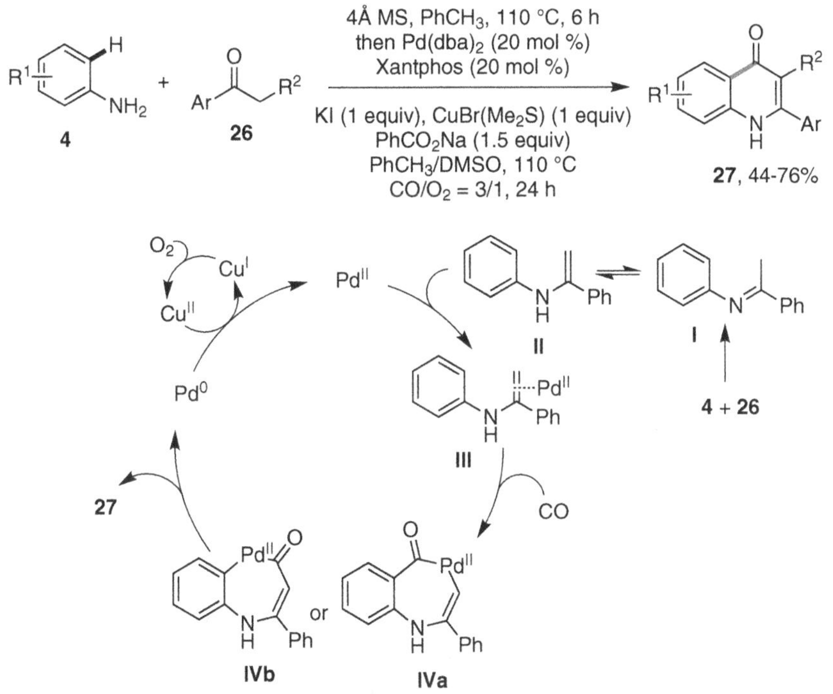

SCHEME 2.11

alkynes **10**. Reactions were conducted using $[Cp^*Rh(MeCN)_3][SbF_6]_2$ (2.5 mol%) in the presence of 2.1 equiv of $Cu(OAc)_2$ in DCE at reflux, to afford corresponding isoquinolones **34** in 30–81% yields. Reactions occurred by imine-induced *ortho*-C–H activation by Rh(III), followed by alkyne insertion, with subsequent reductive elimination to form the final isoquinoline core with loss of *iso*-butene and Rh(I) species. Synthesis of 1-unsubstituted isoquinoline derivatives **34** was also achieved by treatment of *N*-benzyl picolinamides **35** with alkyne **10** (1.2 equiv) in the presence of $Co(OAc)_2{\cdot}4H_2O$ (50 mol%) and KPF_6 (50 mol%) in PEG-400 at 140°C under O_2 for 20 h, in 42–94% yields. Terminal alkynes afforded corresponding 3-substituted isoquinolines **34** in 48–76% yields, regioselectively. Unsymmetrical alkyl aryl alkynes gave 4-alkyl-3-arylisoquinolines **34** in 79–91% yields with excellent regioselectivity. Reactions were initiated by Co(III) species generated in situ by the aid of oxygen from Co(II) species (Kuai et al. 2017). Reacting *N*-(*p*-methoxyphenyl) imine **36**, derived from benzophenone, with diphenyl acetylene **10** using $[Cp^*Co(CO)I_2]$ (10 mol%), AgOAc (10 mol%), and $AgBF_4$ (2 equiv) catalytic systems in DCE at 130°C for 24 h produced isoquinolin-2-ium salt **37a** in 82% yield (Prakash, Muralirajan, and Cheng 2016). Also, isoquinolin-2-ium salts **37b** were prepared by Rh-catalyzed dehydrogenative coupling of in situ generated aldimines with internal alkynes. Reactions were carried out by subjecting alkynes **10** with 1.2 equiv of aromatic aldehydes and 1.5 equiv of primary amines in the presence of $[Cp^*RhCl_2]_2$ (2 mol%), $AgBF_4$ (1 equiv), and $Cu(OAc)_2$ (1 equiv) in *t*-amyl alcohol at 110°C for 3

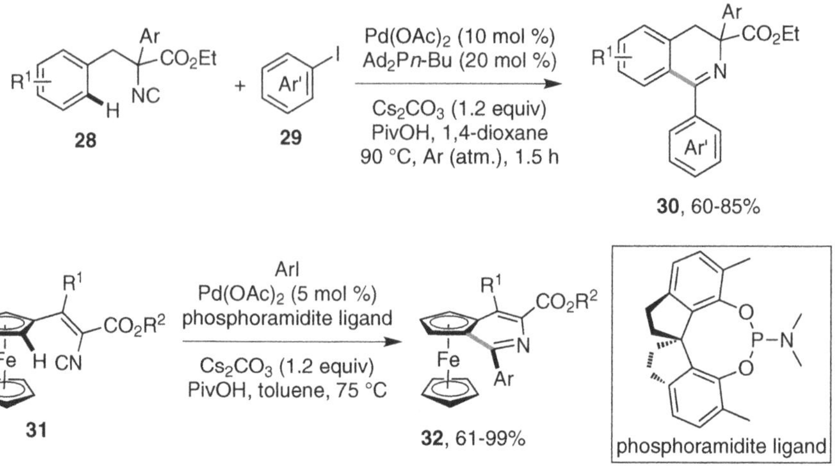

SCHEME 2.12

h, affording corresponding isoquinolin-2-ium salts **37b** in 58–96% yields (Scheme 2.13) (Jayakumar, Parthasarathy, and Cheng 2012).

Trifluoroethyl isoquinolines **40** were obtained by an Rh(III)–Cu(II) bimetallic catalytic system. Treatment of an alkyne **10** with 2 equiv of vinyl azide **38** and 1.8 equiv of Togni's reagent **39** in the presence of $[Cp^*RhCl_2]_2$ (9.5 mol%) and $Cu(OAc)_2$ (1 equiv) in CH_3CN at 90°C for 3 h under N_2 atmosphere resulted in the formation of isoquinolines **40** in 40–85% yields. Diaryl acetylenes as well as dialkyl acetylenes were compatible with the reaction, delivering corresponding isoquinoline derivatives **40**. The reaction of unsymmetrical alkynes, alkyl aryl alkynes, resulted in the formation of 4-alkyl-3-aryl isoquinolines **40** in 70–74% yields, regioselectively. 3-Phenylisoquinoline-4-carboxylate derivative **40a** was obtained in 40% yield from

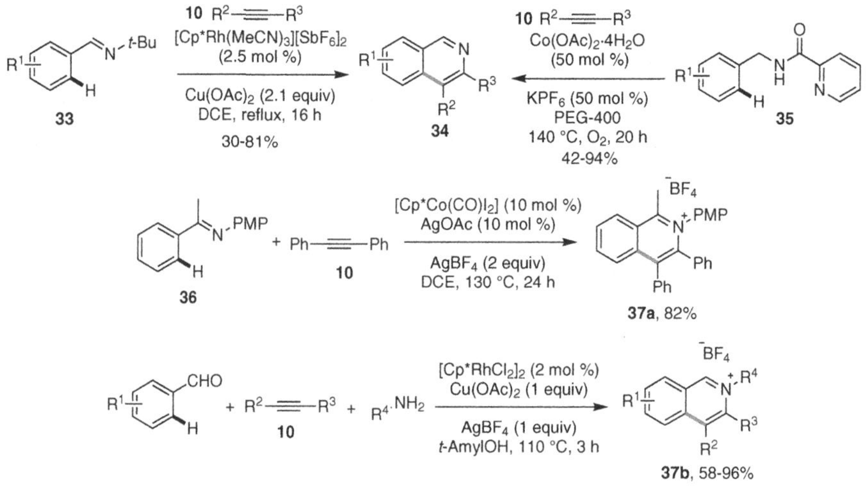

SCHEME 2.13

the reaction of ethyl phenylacetylene carboxylate. Fused thieno[2,3-*c*]pyridine **40b** was also prepared in 45% yield by reaction of diphenyl acetylene with 2-thienyl vinyl azide. In the proposed reaction mechanism, subjecting Togni's reagent to Cu(II) produced a CF_3 radical, which attacked the vinyl azide **38** to give the radical intermediate **I**, which could then be trapped by Cu(II) to give intermediate **II**. Rhodacyclic intermediate **IV**, generated from the reaction of intermediate **II** with Rh(III) via an iminyl rhodium intermediate **III**, underwent alkyne insertion to afford intermediate **V**, which then underwent reductive elimination to produce **40**, along with forming the Rh(I) species. A redox reaction between Rh(I) and Cu(III) regenerated the Rh(III) species for the next catalytic cycle (Scheme 2.14) (Liu et al. 2015).

Synthesis of 1-alkoxyisoquinoline derivatives **42** was achieved by C–H alkenylation/annulation of benzimidates **41** with alkynes **10**. In 2017, Gong et al. (2017) performed the reaction using [Cp*Co(CO)I_2] (10 mol%) in the presence of $AgNTf_2$ (20 mol%) and AcOH in DCE under air at 80°C, affording 1-alkoxyisoquinolins **42** in 29–85% yields. No regioselectivity was shown in the reactions with asymmetric alkynes. The terminal alkyne, phenyl acetylene, led to the corresponding 3-phenyl-1-alkoxyisoquinoline **42** in 29% yield. Terephthalimidate was also compatible with the reaction, giving 1,6-diethoxy pyrido[3,4-*g*]isoquinoline **42a** in 41% yield. The

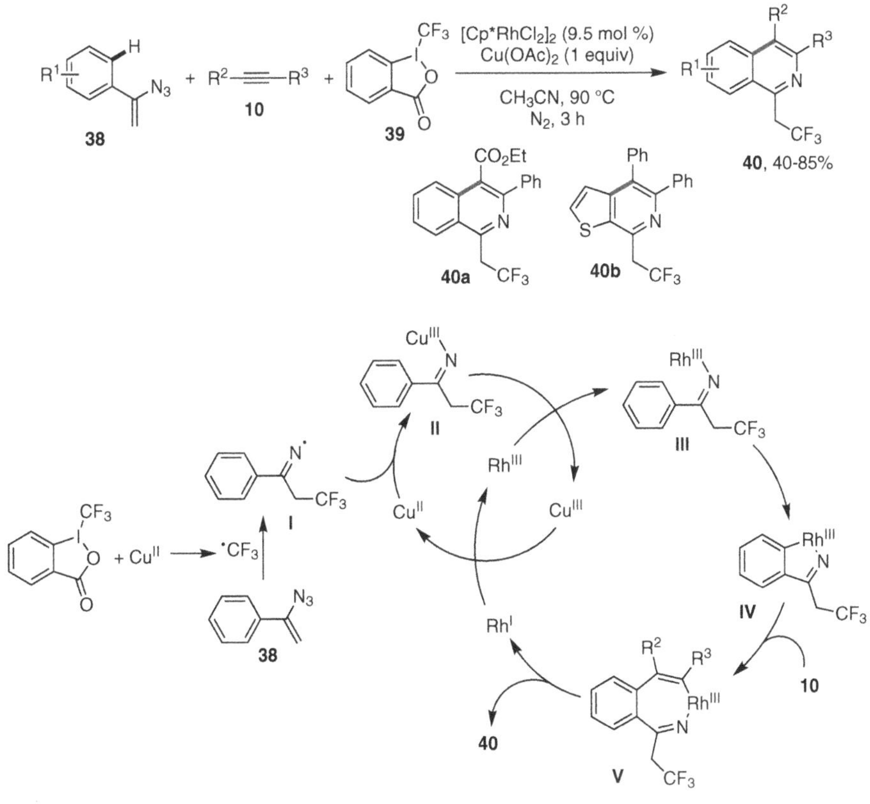

SCHEME 2.14

proposed reaction mechanism involves acetate-assisted C–H cobaltation with subsequent migratory insertion of alkyne, and then reductive elimination to form the isoquinoline product along with a Co(I) species. Oxidation of the Co(I) species with air generated the active Co(III) species. Anukumar, Tamizmani, and Jeganmohan (2018) reported an approach for the construction of 1-alkoxyisoquinolines **44** by dehydrogenative coupling of benzimidates **41** with substituted propargyl alcohols **43**. Reactions were carried out using [Ru(CH_3CN)$_3$(*p*-cymene)][SbF_6]$_2$ (10 mol%) in the presence of Na_2HPO_4 (1 equiv) in DCE at 100°C to furnish 1-alkoxyisoquinoline derivatives **44** in 45–85% yields. Two isomeric thiophene-fused isoquinolines, thieno[2,3-*c*]pyridine **44a** and thieno[3,2-*c*]pyridine **44b**, were prepared from 2-thienyl and 3-thienyl imidate substrates in 44% and 31% yields, respectively. Reactions occurred in a highly selective manner via *ortho*-C–H allenylation by generation of five-membered ruthenocycle **I**, followed by regioselective migratory insertion of an alkyne to generate seven-membered ruthenocycle **II** and then *β*-OH elimination. *Ortho*-allenylated benzimidate **III** was converted into the final isoquinoline product **44** in the presence of a base (Scheme 2.15).

1-Aminoisoquinolines **46** were also prepared via Rh(III)-catalyzed dehydrogenative coupling of benzamidines **45** with alkynes **10**. Subjecting benzamidine to alkyne (1.05 equiv) in the presence of [Cp*$RhCl_2$]$_2$ (4 mol%) and Cu(OAc)$_2$ (2.1 equiv) as the

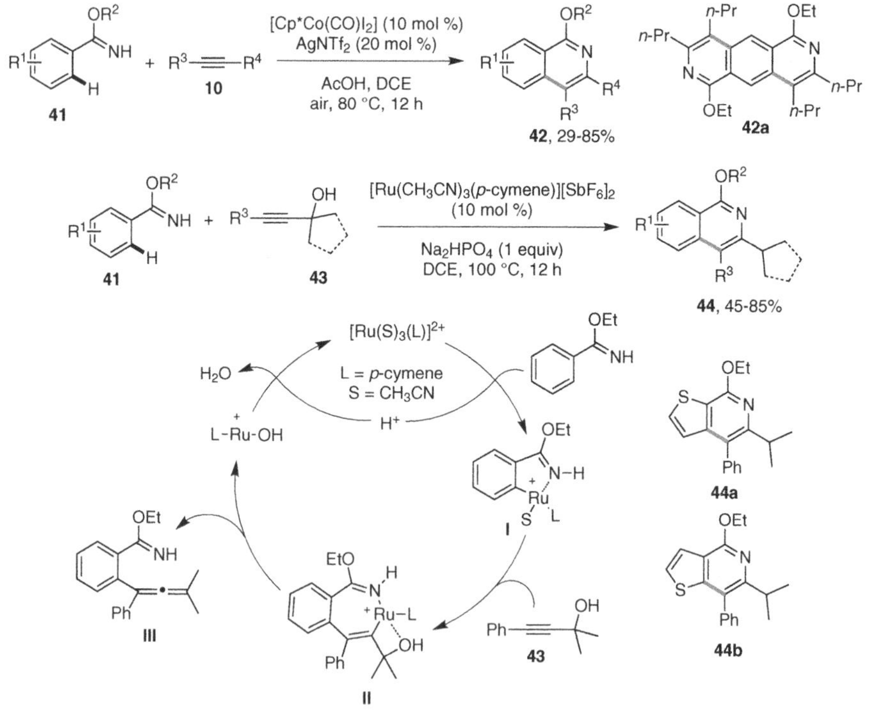

SCHEME 2.15

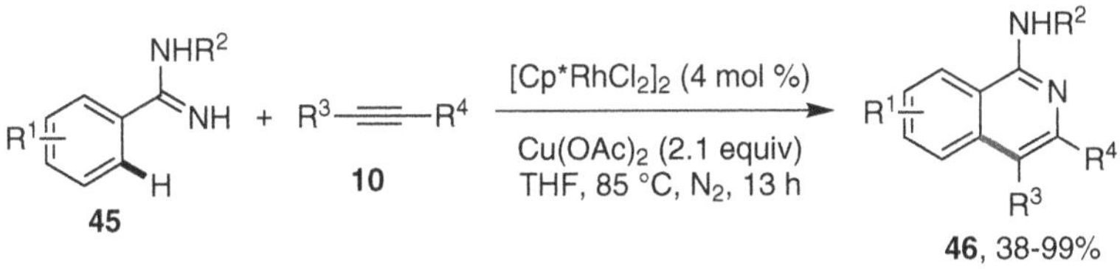

SCHEME 2.16

oxidant in THF at 85°C under N_2 resulted in the creation of 1-aminoisoquinolines **46** in 38–99% yields (Scheme 2.16) (Wei et al. 2011).

Yang et al. (2017a) and Yu et al. (2016) reported the construction of 1-aminoisoquinolines **46** via a Co(III)-catalyzed aromatic C–H dehydrogenative coupling process of 5-methyl-3-aryl-1,2,4-oxadiazole **47** or 3-aryl-1,2,4-oxadiazolone **48** with alkynes **10**, in which nitrogen atoms of the oxadiazole moiety acted as the directing group for activation of the *ortho*-C–H bond. Reactions of 5-methyl-3-aryl-1,2,4-oxadiazole **47** or 3-aryl-1,2,4-oxadiazolone **48** with 1.2 equiv of alkyne **10** were conducted using 10 mol% of [Cp*Co(CO)I_2] in the presence of $AgSbF_6$ (20 mol%) and LiOAc (20 mol%) or KOAc (30 mol%) and $AdCO_2H$ (20 mol%) in 2,2,2-trifluoroethanol (TFE) at 90°C or 120°C, leading to 1-aminoisoquinolines **46** in 40–98% or 40–89% yields, respectively. The proposed reaction mechanism involves the generation of Co(III) complex **I** by coordination of **47** to $[Cp^*Co(OAc)]^+$, generated in situ from the reaction of [Cp*Co(CO)I_2] and $AgSbF_6$ with the release of CO and iodide ligands, followed by coordination of AcO^-. C–H activation led to cobaltacycle **II**, which underwent coordination and then migratory insertion of alkyne **10** to furnish alkenyl cobalt intermediate **III**. By C–N cyclization and ring opening of the oxadiazole, intermediate **IV** was formed, along with the regeneration of the Co(III) complex. Deacetylation of **IV** provided **46** as the final product. In the case of 3-aryl-1,2,4-oxadiazolone, the final 1-aminoisoquinolines **46** were obtained by release of CO_2 (Scheme 2.17).

Similarly, an Ru(II)-catalyzed tandem C–H activation/cyclization/hydrolysis cascade process of 2*H*-imidazoles **49** and alkynes **10** for regioselective access to 1-acylisoquinolines **50** has been developed. Reactions were performed using $[RuCl_2(p\text{-cymene})]_2$ (5 mol%), $Cu(OTf)_2$ (1 equiv), CF_3CO_2Ag (1 equiv), and 1-$AdCO_2H$ (2 equiv) in TFE at 100°C for 6–18 h under air resulted in the formation of 1-acylisoquinoline derivatives **50** in 41–85% yields. Unsymmetrical alkynes 1-phenylhexyne and ethyl phenylacetylene carboxylate gave rise to produce the expected 4-butyl-3-phenyl-1-acetylisoquinoline and 3-phenyl-1-acetylisoquinoline-4-carboxylate in 64% and 46% yields, respectively. No reaction occurred to afford a 1-formylisoquinoline derivative when 5-unsubstituted 4-phenyl-2*H*-imidazole was used. Imine intermediate **I** was proposed as the product of Ru-catalyzed imidazole induced *ortho*-C–H activation, followed by migratory alkyne insertion and then reductive elimination, which was hydrolyzed to the corresponding 1-acylisoquinolines **50** (Scheme 2.18) (Wu and Dong 2018).

Rh(III)-catalyzed C–H/N–H bond functionalization for the synthesis of 1-aminoisoquinolines **52** from aryl amidines **45** and α-MsO/TsO/Cl arylketones **51** was

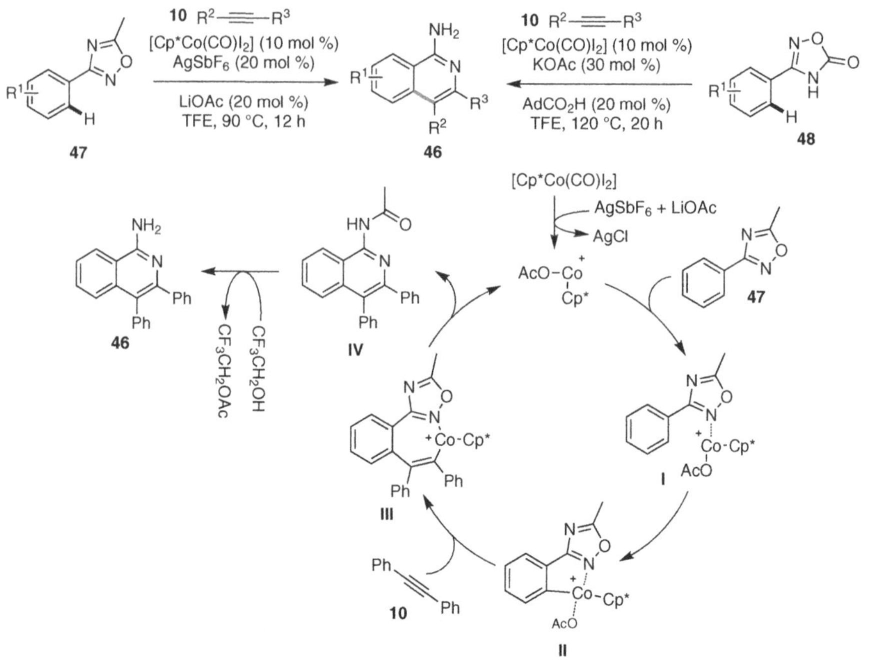

SCHEME 2.17

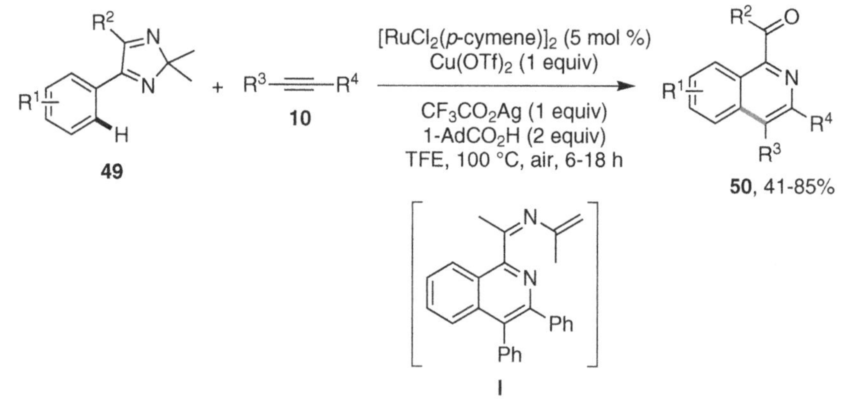

SCHEME 2.18

achieved by heating a solution of **45** and **51** (2 equiv) in the presence of $[Cp^*RhCl_2]_2$ (5 mol%), $AgBF_4$ (20 mol%), $Cu(OAc)_2$ (20 mol%), and KOAc (2 equiv) in TFE under Ar atmosphere at 80°C for 24 h. 1-Amino-3-arylisoquinolines **52** were obtained in 21–93% yields. α-TsO alkylketones **51** were also tolerated under reaction conditions, leading to expected 1-amino-3-alkylisoquinolines **52** in 40–84% yields. α-chlorocyclohexanone led to corresponding tetrahydrophenanthridin-6-amine **52a**

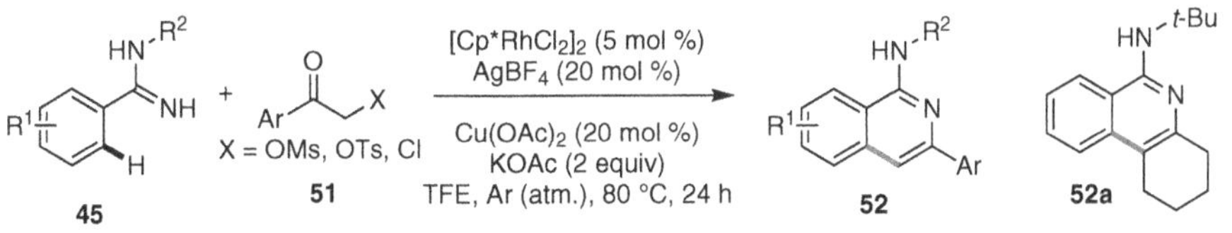

SCHEME 2.19

in 21% yield. Other α-halo-α-substituted ketones did not afford expected products. Reactions occurred via *ortho*-C–H activation to form a five-membered rhodacycle, followed by migratory insertion of the C–X bond, with subsequent intramolecular cyclization to the isoquinoline core (Scheme 2.19) (Li et al. 2016b).

Synthesis of isoquinoline-4-carboxylate derivatives **55** were reported by dehydrogenative coupling of primary benzyl amines **53** with 2-diazoacetoacetates **54**. Reactions were catalyzed with $[Cp^*RhCl_2]_2$ (5 mol%) in the presence of $AgSbF_6$ (20 mol%) in dry acetone for 12 h at 100°C, producing isoquinoline-4-carboxylates **55** in 55–80% yields. Also, 2-diazo-1,3-diketones were tolerated under reaction conditions to give corresponding 4-acylisoquinolines in 40–58% yields. The proposed reaction mechanism involves the generation of imine **I** from a reaction between benzyl amine **53** and acetone, which underwent coordination and *ortho*-C–H activation by $[Cp^*Rh(III)]^{2+}$ species, generated in situ from $[Cp^*RhCl_2]_2$ and $AgSbF_6$, giving rhodacycle intermediate **II**. Then, the insertion of diazo **54** to the C–Rh bond of **II** afforded rhodium carbene intermediate **III**, along with the loss of N_2. By migratory insertion of rhodium carbene **III**, intermediate **IV** was formed, which was transformed to intermediate **V** through protonation, where the cation Rh(III) species was regenerated. Then, intermediate **V** was converted into the final isoquinoline product **55** by hydrolysis to intermediate **VI**, followed by the intramolecular annulation of the amine and carbonyl groups, with subsequent aromatization of the generated 1,2-dihydroisoquinoline by the extrusion of H_2 (Chu et al. 2016). A similar approach was developed for the creation of 1-alkoxyisoquinoline-4-carboxylates **56** via Rh(III)-catalyzed reaction of arylimidates **41** with 2-diazoacetoacetates **54** in 40–95% yields. Reaction with diazotized Meldrum's acid **54a** under the same reaction conditions led to the formation of 1-alkoxy-3-hydroxyisoquinolines **57** in 51–90% yields, by an addition/elimination/decarboxylation sequence (Li et al. 2016c). Moreover, the synthesis of 1-alkoxy-3-hydroxyisoquinoline-4-carboxy lates **58** was successful through the Rh(III)-catalyzed dehydrogenative coupling of benzimidates **41** with diazomalonates **54b**. Reactions were performed by heating a solution of benzimidates **41** with diazomalonates **54b** (1.5 equiv) in the presence of $[Cp^*RhCl_2]_2$ (2 mol%), $AgSbF_6$ (8 mol%), and CsOAc (10 mol%) in DCE at 50°C for 5 h, to furnish 1-alkoxy-3-hydroxyisoquinoline-4-carboxylates **58** in 58–95% yields (Scheme 2.20) (Wang et al. 2016b).

Co(III)-catalyzed C–H/N–H bond functionalization for the synthesis of 1-aminoisoquinoline-4-carboxylates **59** from aryl amidines **45a** and 2-diazoacetoacetates **54** was reported by Li et al. (2016a). Treating **45a** with **54** (2 equiv) in the presence of $[Cp^*Co(CO)I_2]$ (10 mol%), $AgSbF_6$ (20 mol%), and KOPiv (20 mol%) in TFE under an atmosphere of argon at 110°C for 16 h afforded 1-aminoisoquinoline-4-carboxylates

SCHEME 2.20

59 in 42–83% yields. Reaction with 2-diazo-1,3-diphenylpropane-1,3-dione **54c** occurred through a two-step reaction sequence consisting of the C–H/N–H functionalization and a decarbonylation, to produce an unexpected product, 4-unsubstituted 1-aminoisoquinolines **60**, in 51–60% yields. A similar methodology was applied for the creation of 6-amino-3,4-dihydrophenanthridin-1(2*H*)-one **61** via Rh(III)-catalyzed dehydrogenative coupling of aryl amidines **45b** with cyclic 2-diazo-1,3-diketones **54d** in 65–86% yields (Scheme 2.21) (Zuo et al. 2018).

Isoquinolinium salts **63** were synthesized via Rh(III)-catalyzed C–H activation of presynthesized or in situ formed imines and coupling with α-diazo ketoesters **54e**. Reactions with imines **64** were conducted using Cp*Rh(OAc)$_2$ (8 mol%), Zn(NTf$_2$)$_2$ (50 mol%), NaNTf$_2$ (1 equiv), and HOAc (1 equiv) in TFE at 40°C for 12 h, leading to isoquinolinium salts **63** in 58–82% yields. Three-component reactions between an aldehyde or methyl ketone **62**, a primary amine, and an α-diazo ketoester **54e** were performed with [Cp*RhCl$_2$]$_2$ (4 mol%), Zn(OTf)$_2$ (1 equiv), and NaOTf (1 equiv) in TFE at 110°C, to furnish corresponding isoquinolinium salts **63** in 25–95% yields. Reaction of imines **64** with diazomalonate **54b** resulted in the formation of corresponding isoquinolin-3-one-4-carboxylates **66** in 48–86% yields. The reaction was initiated by cyclometalation of the imine, followed by coordination and N$_2$ extrusion of the diazocompound to give a rhodium carbenoid, with subsequent migratory insertion of the Rh–C bond to the carbene to form alkylated intermediate via protonolysis. Subsequent Zn(II)-mediated cyclization furnished the isoquinolinium salt **65–66** (Tian et al. 2018). Also, substituted isoquinoline *N*-oxides **68** were

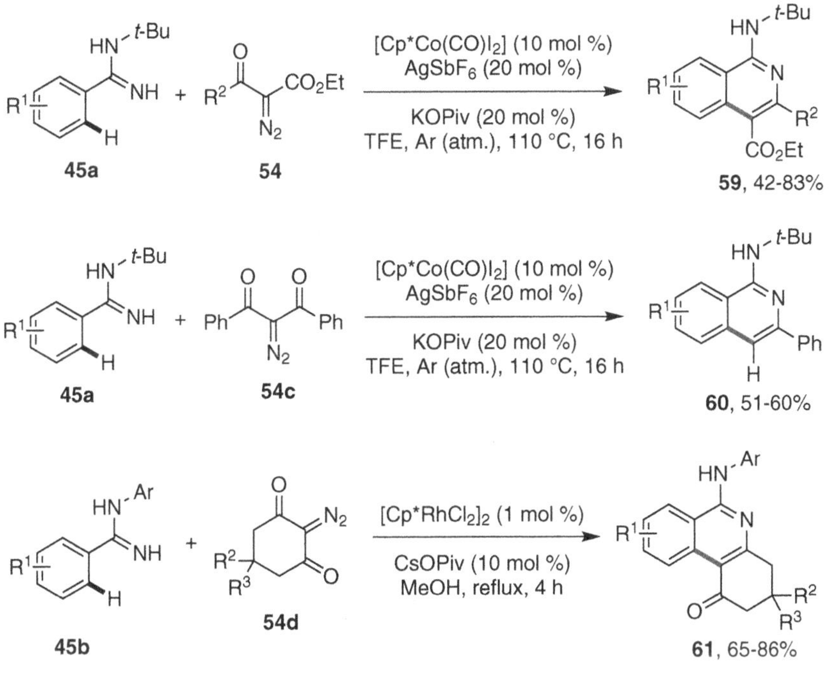

SCHEME 2.21

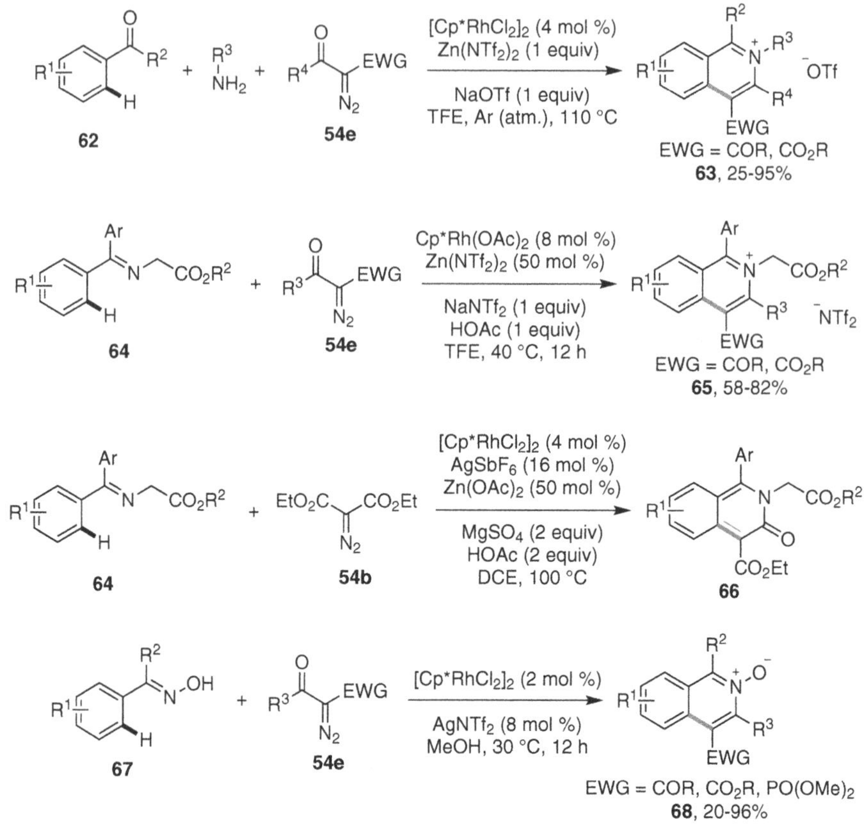

SCHEME 2.22

accessed by Ir(III)-catalyzed C–H activation and annulations of aryloxime **67** with α-diazocarbonyl compounds **54e** (Scheme 2.22) (Phatake, Patel, and Ramana 2015).

2.5 ISOQUINOLINONE

Pd-catalyzed aromatic C–H aminocarbonylation of phenethylamines **69** was described in order to synthesize 3,4-dihydroisoquinolinones **70**. Subjecting a solution of phenethylamines **69** with CO (1 atm) in the presence of $Pd(TFA)_2$ (10 mol%), 1,1′-bi-2-naphthol (BINOL) (40 mol%), and Ag_3PO_4 (5 equiv) in CH_3CN/AcOH at 100°C for 24 h led to the construction of 3,4-dihydroisoquinolinone derivatives **70** in 29–89% yields. Interestingly, the C–X bond was left unreacted during the reaction conditions (Scheme 2.23) (Taneda, Inamoto, and Kondo 2016).

Construction of 3,4-dihydroisoquinolinones **72** was reported by di-*tert*-butyl peroxide (DTBP)-promoted dehydrogenative coupling of *N*-allylbenzamide **71** with aryl aldehydes. Reactions were carried out by treating *N*-allyl-*N*-methylbenzamide **71** with aldehyde (3 equiv) in the presence of DTBP (4 equiv) in toluene at 120°C for 24 h under argon atmosphere to furnish 3,4-dihydroisoquinolinones **72** in 36–89%

yields. The reactions occurred by the addition of an acyl radical, generated by abstraction of hydrogen from the aldehyde by a *tert*-butoxy radical generated in situ, to the double bond of the alkene, followed by intramolecular radical cyclization and then dehydrogenation (Scheme 2.24) (Xu, Wang et al. 2018).

Ru-catalyzed synthesis of 3,4-dihydroisoquinolinones **76–77** via dehydrogenative C–H bond olefination of *N*-methoxybenzamides **73** with styrene **74** or 2,5-norbornadiene **75** was described by Li et al. (2012). Reactions were catalyzed with $[RuCl_2(p\text{-cymene})]_2$ (10 mol%) in the presence of NaOAc (2 equiv) in TFE at 50°C, and corresponding 3,4-dihydroisoquinolinones **76–77** were obtained in 47–87% yields. The proposed reaction mechanism involves the formation of a five-membered ruthenocycle by coordination of amide and *ortho*-C–H activation of benzamide, which was converted into the final desired product by alkene insertion, with a subsequent AcO-assisted reductive elimination (Scheme 2.25).

Rh(III)-catalyzed C–H activation/cycloaddition of benzamides **73b** and arylidene cyclopropanes **78** for the synthesis of spiro-dihydroisoquinolinones **79** was developed by Cui, Zhang, and Wu (2013). Subjecting benzamides **73b** to 2 equiv of arylidene cyclopropanes **78** in the presence of $[Cp^*RhCl_2]_2$ (2 mol%) and CsOAc (1 equiv)

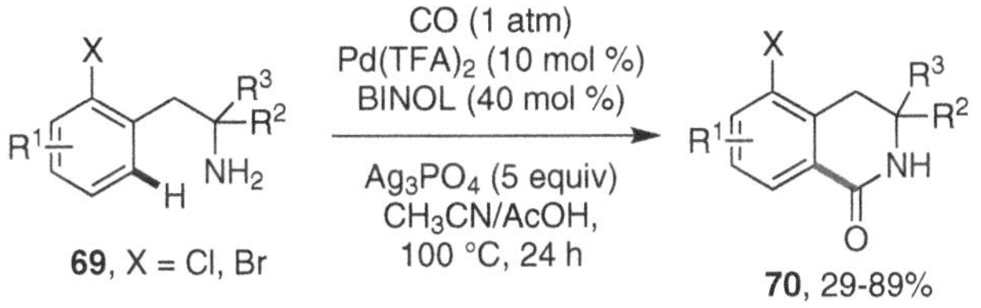

SCHEME 2.23

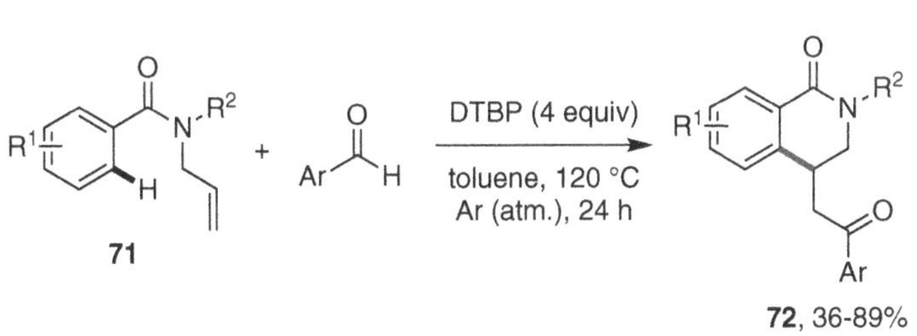

SCHEME 2.24

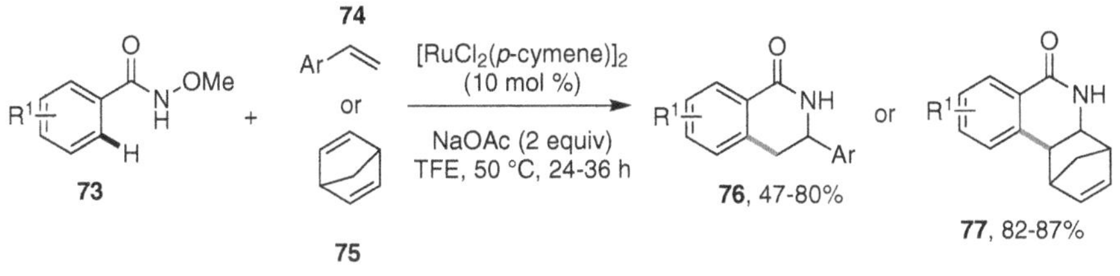

SCHEME 2.25

in MeOH at 30°C afforded 3'-aryl-2',3'-dihydro-1'*H*-spiro[cyclopropane-1,4'-isoquinolin]-1'-ones **79** in 55–95% yields. Alkylidene cyclopropane was also compatible with the reaction to give corresponding spiro-dihydroisoquinolinones **79** in 81–84% yields. The proposed reaction mechanism involves carboxylate-assisted C–H activation via a concerted metalation/deprotonation (CMD) pathway to form intermediate **I**, which underwent coordination and regioselective insertion of arylidene cyclopropane into the C–Rh bond to afford rhodacycle **II**. Then, C–N bond formation led to **III** along with N–O bond cleavage. Finally, **III** was protonated to furnish the product **79** with concomitant Rh(III) catalyst release (Scheme 2.26).

3-Vinyl substituted 3,4-dihydroisoquinolinones **81** were prepared by tandem Rh(III)-catalyzed C–H allylation/Pd(II)-catalyzed *N*-allylation of benzamide **73a** with 4-vinyl-1,3-dioxolan-2-one **80**. Reactions were performed using $[Cp^*Rh(CH_3CN)_3](SbF_6)_2$ (5 mol%), $Pd_2(dba)_3$ (5 mol%), CsOAc (1 equiv), and Cs_2CO_3 (0.2 equiv), in CH_3CN at 50°C for 3–48 h, delivering 3-vinyl-3,4-dihydroisoquinolinones **81** in 31–91% yields. In the proposed reaction mechanism, rhodacycle **I** was generated via benzamide C–H activation, followed by alkene coordination and insertion (**II**), which underwent β-oxygen elimination to form allyl alcohol **IV** via intermediate **III**. The resulting allyl alcohol **IV** was transformed to the final 3-vinyl substituted 3,4-dihydroisoquinolinone **81** via a Pd-catalytic cycle, in which a palladium π-complex **V** similar to intermediate **III** was initially formed, which underwent *syn*-azapalladation to give a σ-palladium complex **VI**, and then subsequent β-oxygen elimination (Zhang et al. 2014a). A similar approach was described by Wang, Lorion, and Ackermann (2017) to construct 1-alkoxy-3-vinylisoquinolines

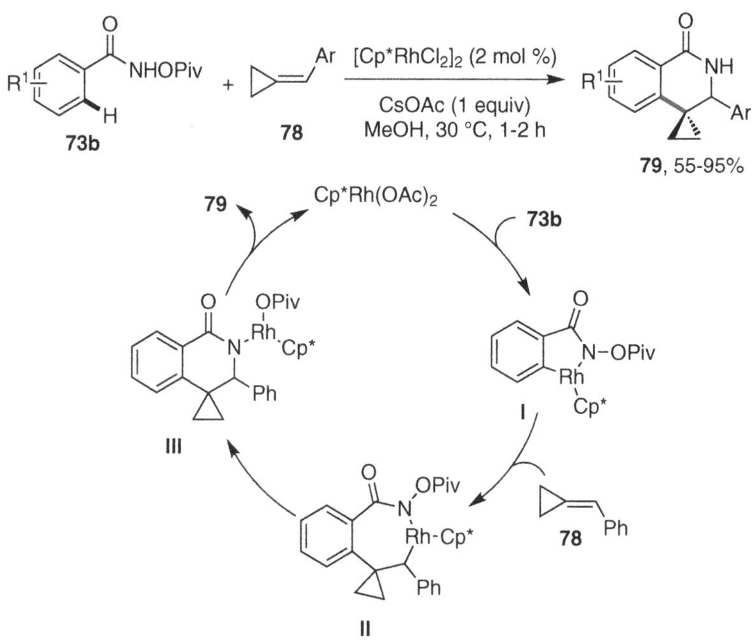

SCHEME 2.26

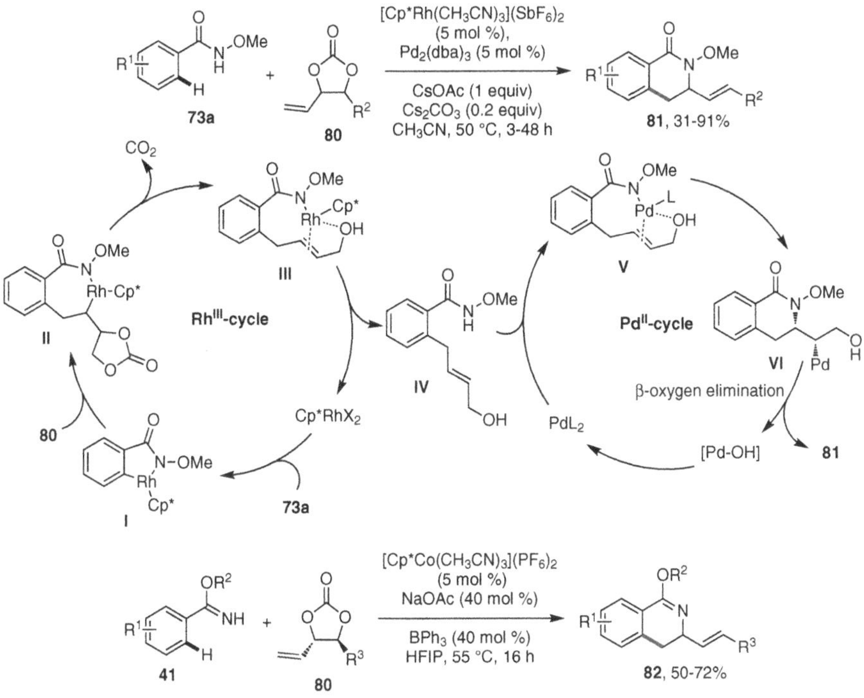

SCHEME 2.27

82 via domino C–H/N–H allylation of aryl imidates **41** with 4-vinyl-1,3-dioxolan-2-one **80** catalyzed with [Cp*Co(CH_3CN)$_3$](PF_6)$_2$ (5 mol%) in 50–72% yields, which could be converted to 3-vinylisoquinolinone by treating with aq. HCl in THF at 80°C (Scheme 2.27).

Kalsi et al. (2018) reported a merging Co- and photoredox catalysis for the synthesis of isoquinolin-2-ones **84** by reacting *N*-(quinolin-8-yl)benzamides **83** with alkynes **10**. Reactions were performed using Co(acac)$_2$ (10 mol%), NaOPiv (20 mol%), and Na$_2$[Eosin Y] (10 mol%) in the presence of O_2 under visible light irradiation in TFE at room temperature for 24 h, affording isoquinolin-2-ones **84** in 26–99% yields. Reactions with unsymmetrical alkyl aryl alkynes led to the formation of corresponding 4-alkyl-3-arylisoquinolin-2-ones **84** with high regioselectivity. Reaction with terminal alkynes produced 3-substituted isoquinolin-2-ones **84**, regioselectively. Also, diynes **85** were investigated under reaction conditions, affording 3-alkynylisoquinolin-2-ones **86** in 50–93% yields. The reaction was initiated by ligand exchange of Co(II) with *N*-(quinolin-8-yl)benzamide **83**, followed by 1e$^-$ oxidation via electron transfer to Co(III), assisted by reduction of Na$_2$Eosin Y*. *Ortho*-C–H activation with subsequent alkyne coordination and regioselective insertion between Co–C followed by reductive elimination liberated the final product, with generation of a Co(I) species, which was oxidized to a Co(II) species by photoexcited Na$_2$Eosin Y* (Scheme 2.28). Also, Rh(III)-catalyzed dehydrogenative coupling between secondary benzamides and alkynes was developed for the synthesis of isoquinolin-2-ones, in which

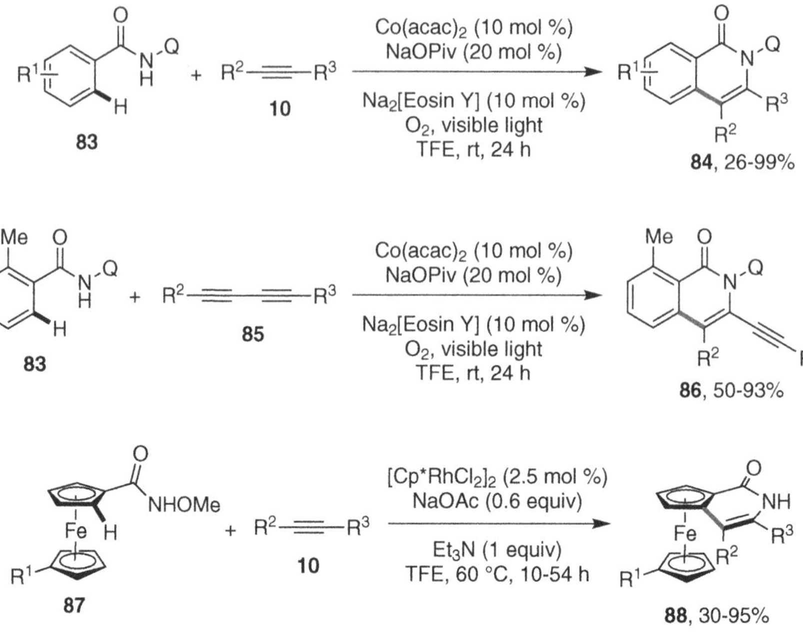

SCHEME 2.28

$[Cp^*RhCl_2]_2$ (4 mol%) and Ag_2CO_3 (1.3 equiv) were used as catalyst and oxidant, respectively, affording isoquinolin-2-ones in 36–98% yields. *N*-alkyl benzamides led to corresponding isoquinolin-2-ones in lower yields than *N*-aryl substituted benzamides (Song et al. 2010). A similar approach was developed for Pd(II)- or Rh(III)-catalyzed dehydrogenative C–H activation/annulation of ferrocenecarboxamides **87** with internal alkynes **10**, to give ferrocene[1,2-*c*]pyridine-3(4*H*)-ones **88** (Scheme 2.28) (Xie et al. 2014, Wang, Zheng, and You 2016). Moreover, electrochemical dehydrogenative C–H/N–H activation for the annulation of benzoyl haydrazides with alkyne was developed in accompaniment with a cobalt catalyst. Reactions were conducted in an undivided cell using $Co(OAc)_2$ (10 mol%) and PivOH (2 equiv) in TFE at 5.0 mA under N_2 for 16 h and corresponding isoquinoline-2-ones were obtained in 37–96% yields. Unsymmetrical alkyl aryl alkynes furnished 4-alkyl-3-arylisoquinoline-2-ones with high regioselectivity. Terminal alkynes led to the corresponding 3-substituted isoquinoline-2-ones. The proposed reaction mechanism involves the electro-oxidative generation of catalytically active species (Co(III)) by means of anodic oxidation, which underwent C–H activation, followed by coordination and migratory insertion of an alkyne. Then, reductive elimination produced the desired product along with generation of a Co(I) species (Mei et al. 2018).

Synthesis of 3-acylisoquinoline-2-ones **91** was achieved by Co(II)-catalyzed C–H bond functionalization of benzamides **89** with allenes **90**. Reactions were carried out by heating a solution of **89** and allene **90** (4 equiv), in the presence of $Co(OAc)_2{\cdot}4H_2O$ (20 mol%) and Ag_2CO_3 (2 equiv) in EtOH at 60°C under an atmosphere of O_2 for 16 h, followed by treating with Dess–Martin periodinane (DMP) (1 equiv) in dichloromethane (DCM) at room temperature for 1 h, to give 3-acylisoquinoline-2-ones **91** in 52–88% yields. In the proposed reaction mechanism, active catalytic species Co(III), generated in situ through oxidation of Co(II) by Ag_2CO_3, underwent coordination with hydrazide **89** followed by subsequent C–H activation to form **I**. By coordination and migratory insertion of allene **90** to the Co(III) center of **I**, seven-membered cobaltacycle **II** was generated, which was converted into compound **III** by reductive elimination, along with release of a Co(I) species. The action of a peroxy radical species, derived from coordination of Co(I) with molecular oxygen, on the intermediate **III**, provided intermediate **IVa**, which tautomerized to intermediate **IVb**. Oxidation of radical intermediate **IVb** afforded iminium **V**, which gave peroxide **VI** upon deprotonation and proton demetalation. Peroxide **VI** was transformed into 3-acylisoquinoline-2-one **91** through an elimination reaction. The active Co(III) species was finally regenerated through oxidation of Co(II) with Ag_2CO_3 for the next catalytic cycle (Scheme 2.29) (Zhai et al. 2018).

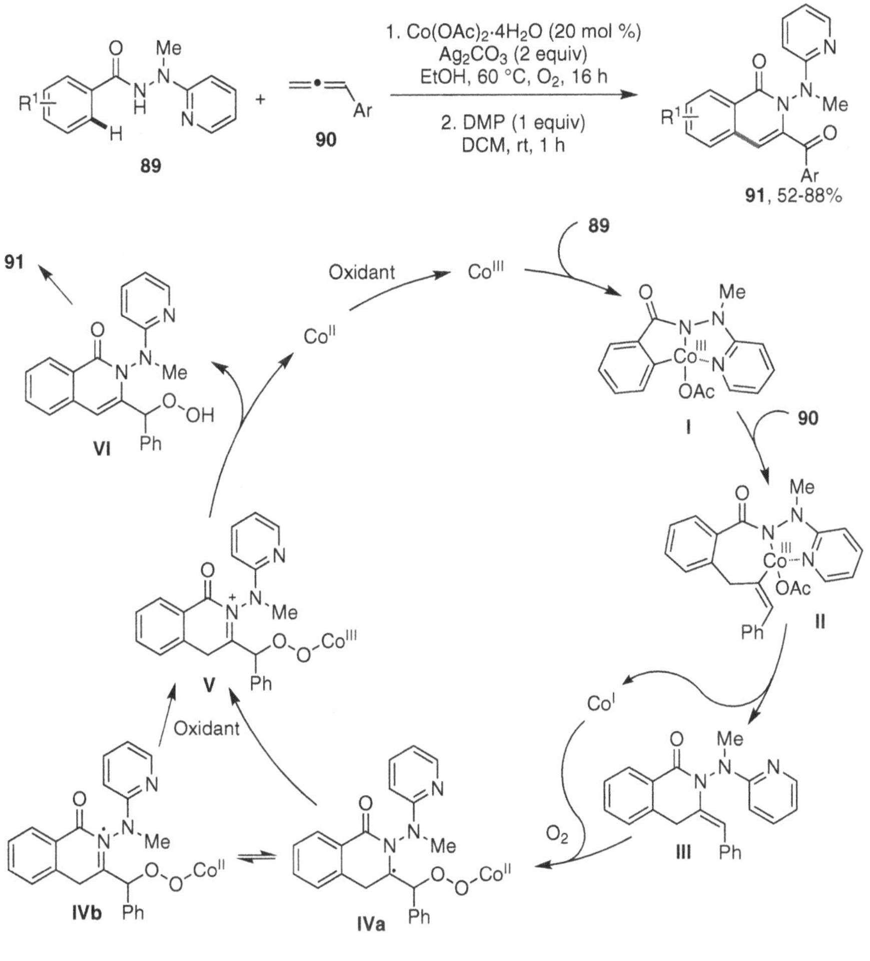

SCHEME 2.29

Also, electrooxidative allene **93** annulations by Co(II)-catalyzed benzamide **92** C–H activation were described in order to produce isoquinoline-2-one derivatives **94**. Reactions were conducted in an undivided cell under constant current at 2 mA, in the presence of $Co(OAc)_2{\cdot}4H_2O$ (10 mol%) and NaOPiv in MeOH at 40°C for 15 h, producing isoquinoline-2-ones **94a** in 45–97% yields. *Exo*-methylene isoquinolones **94b** were obtained in the reaction of internal allenes **93b** under similar reaction conditions. The proposed reaction mechanism involves anodic Co(II) oxidation, generating Co(III) species, which underwent coordination and C–H activation, followed by migratory insertion of an allene with subsequent reductive elimination to form product **94a**, along with the generation of a Co(I) species, which was electrochemically oxidized to a Co(III) active species. The *exo*-methylene isoquinolones **94b** were converted to isoquinoline-2-one upon isomerization (Scheme 2.30) (Meyer et al. 2018).

Bisannulation of *N*-(pivaloyloxy)benzamides **73b** and cyclohexadienone-tethered allenes **95** was accomplished through Rh(III)-catalyzed C–H activation to provide a 3-isoquinolonyl *cis*-hydrobenzofuran framework **96**. Reactions were catalyzed with $[Cp^*RhCl_2]_2$ (1 mol%) in the presence of CsOAc (10 mol%) in acetone at 70°C for 10 h, under an atmosphere of argon, and corresponding 3-isoquinolonyl *cis*-hydrobenzofurans **96** were obtained in 31–81% yields (Scheme 2.31) (Kong et al. 2018).

Pd-catalyzed *ortho*-C–H bond activation and intramolecular N–C annulation of benzamides **83** and α-bromo ketones **51** produced isoquinolin-1(2*H*)-ones

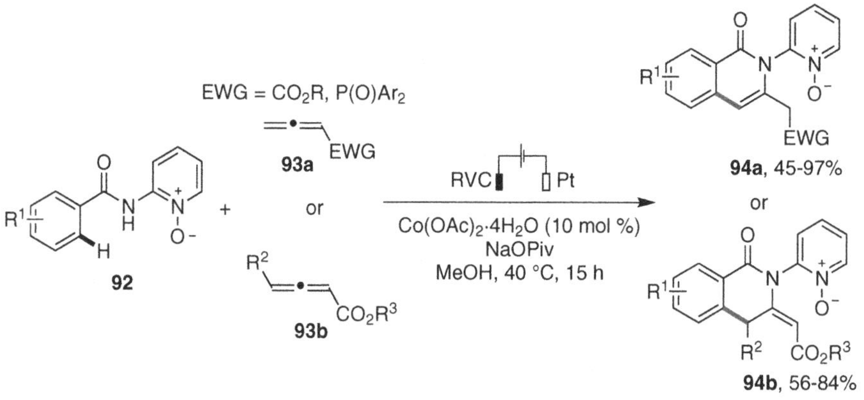

EWG = CO_2R, $P(O)Ar_2$

92 + **93a** or **93b** → RVC | Pt; $Co(OAc)_2{\cdot}4H_2O$ (10 mol %), NaOPiv, MeOH, 40 °C, 15 h → **94a**, 45-97% or **94b**, 56-84%

SCHEME 2.30

73b + **95** → $[Cp^*RhCl_2]_2$ (1 mol %); CsOAc (10 mol %), acetone, 70 °C, Ar, 10 h → **96**, 31-81%

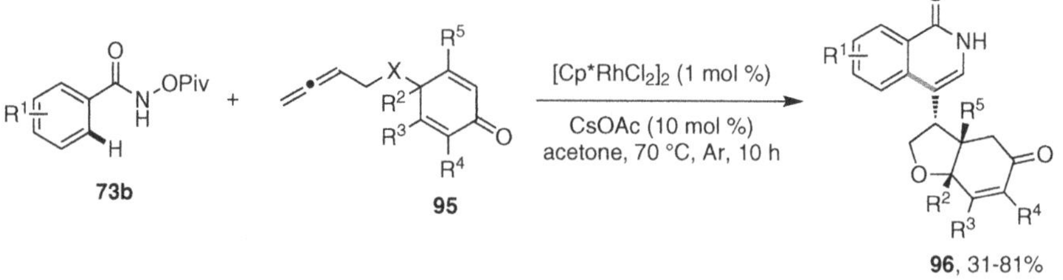

SCHEME 2.31

97. Reactions were performed using 4 equiv of α-bromo ketones **51**, $Pd(OAc)_2$ (10 mol%), *L*-alanine (20 mol%), and K_2CO_3 (2.5 equiv) in *t*-AmOH at 130°C, leading to corresponding isoquinolin-1(2*H*)-ones **97** in 51–84% yields (Xie et al. 2018). A Pd-catalyzed dehydrogenative cross-coupling/annulation reaction for the synthesis of isoquinolinones **99** was developed by treatment of benzamides **73** with β-keto esters **98** (2 equiv) in the presence of $Pd(TFA)_2$ (5 mol%) and $K_2S_2O_8$ (2 equiv) in AcOH at 60°C for 24 h. Isoquinolinones **99** were obtained in 50–87% yields. The reaction was initiated by formation of a five-membered palladacycle intermediate **I** by activation of the *ortho*-C(sp^2)–H bond in benzamide **73**, followed by oxidation of the Pd(II) center to Pd(IV) (**II**) by $K_2S_2O_8$. Then, the acetoacetate anion generated in situ coordinated with Pd(IV) via ligand exchange to form Pd(IV) intermediate **III**, which underwent reductive elimination to C–H functionalized intermediate **IV** and regenerated the Pd(II) catalyst. Finally, intramolecular condensation of **IV** via dehydration afforded the desired isoquinolinone derivative **99** (Scheme 2.32) (Xu and Huang 2017).

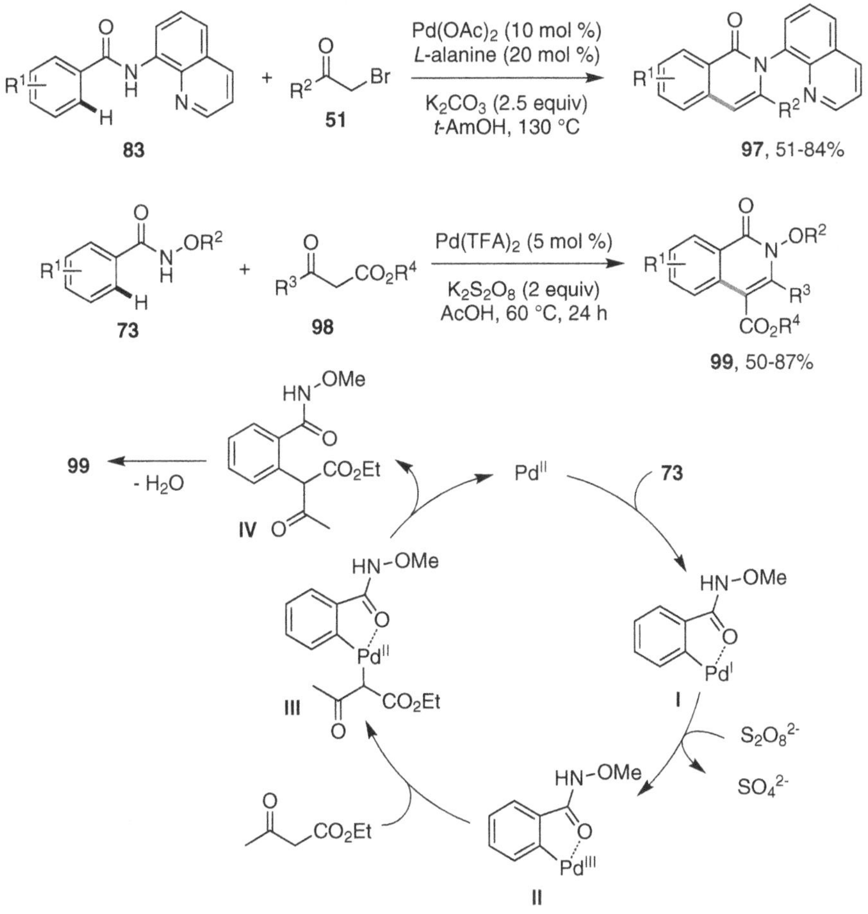

SCHEME 2.32

Ir(III)-catalyzed C–H carbenoid functionalization was developed to access isoquinolinedione and isoquinolinone derivatives. Reactions of *N*-methoxybenzamides **73a** with α-diazocarbonyl compounds **54f** were carried out in DCE using $[Cp^*IrCl_2]_2$ (2 mol%) as the catalyst in the presence of $AgNTf_2$ (8 mol%) at 35°C to afford isoquinolin-1-ones **100** in 58–88% yields. α-diazo ketoesters and α-diazo diketones were also tolerated under the reaction conditions, giving expected isoquinolinone derivatives in 58–64% yields. Reaction with α-diazotized Meldrum's acid **54d** under similar reaction conditions resulted in the formation of *N*-methoxyisoquinoline-1,3-dione derivatives **101** in 28–92% yields. (Benzofuran-2-yl) and 2-thienyl *N*-methoxycarboxamides led to corresponding fused heterocyclic systems in 62–80% yields. The plausible reaction mechanism involves the coordination and *ortho*-C–H activation of benzamide **73a** with cationic [Cp*Ir(III)] species to form the five-membered iridacyclic intermediate **I**, which underwent coordination of the diazocompound to give the diazonium species **II**, and then released the N_2, generating Ir-carbene species **III**. Subsequently, migratory insertion of carbene into the Ir–C bond furnished the six-membered iridacyclic intermediate **IV**, which was transformed into the alkylated intermediate **V** upon protonation, along with regeneration of the active [Cp*Ir(III)] species. Annulations via tandem addition, elimination, decarboxylation, and protonation of intermediate **V** delivered the desired product **101** (Scheme 2.33) (Phatake, Patel, and Ramana 2016). Also, $[Cp^*RhCl_2]_2$ (2.5 mol%) in the presence of CsOAc (0.5 equiv) in DCE at 100°C catalyzed dehydrogenative coupling of benzamides with α-diazo ketoesters to give isoquinolinone-4-carboxylate derivatives in 53–98% yields. No reaction occurred in the case of diazo malonate (Wu et al. 2016).

Synthesis of isoquinolin-3-ones **102** was accomplished through a cooperative $B(C_6F_5)_3$ and Cp*Co(III)-catalyzed C–H bond activation of imines **33** with diazo malonates **54**. Reactions were catalyzed with 5 mol% of $[Cp^*Co(CO)I_2]$ in the presence of CsOAc (10 mol%) and $B(C_6F_5)_3$ (20 mol%) in TFE at 120°C for 12 h to construct isoquinolin-3-one derivatives **102** in 56–99% yields. Secondary ketimines worked as well as primary ketimines under the reaction conditions. In the proposed reaction mechanism, *ortho*-alkylation of the imine occurred via Co-carbenoid species, which underwent intramolecular lactamization through an attack of the imine on the ester moiety of diazocompounds (Scheme 2.34) (Kim, Gressies, and Glorius 2016).

Rh(III)-catalyzed C–H conjugate addition/cyclization reaction between benzamides **7a** and ethyl 3-acyl acrylates **103** was developed to access isoquinoline-1,3-dione derivatives **104**. By heating a solution of benzamides **7a** (1.2 equiv) and ethyl 3-acyl acrylates **103** in DCE in the presence of $[Cp^*RhCl_2]_2$ (2.5 mol%) and $AgSbF_6$ (10 mol%) at 90°C for 18 h, isoquinoline-1,3-diones **104** were obtained in 66–93% yields. Also, reaction with benzyl amines **53b** was studied, in which corresponding isoquinoline-3-one derivatives **105** were obtained in 57–82% yields (Scheme 2.35) (Weinstein and Ellman 2016).

Also, Rh(III)-catalyzed annulations of *O*-pivaloyl oximes **67b** with ketenes **106** was reported to construct isoquinolin-4(3*H*)-ones **107**. Reactions were performed using $[Cp^*Rh(MeCN)_3]_2(SbF_6)_2$ (8 mol%), NaOAc (2 equiv) in the presence of 4 Å MS in DCM at 60°C, providing isoquinolin-4(3*H*)-ones **107** in 34–98% yields. *O*-pivaloyl oximes derived from 2-acetylpyrrole and 3-acetylthiophene were also tolerated under reaction conditions, delivering the expected fused pyrrolo[2,3-*c*]

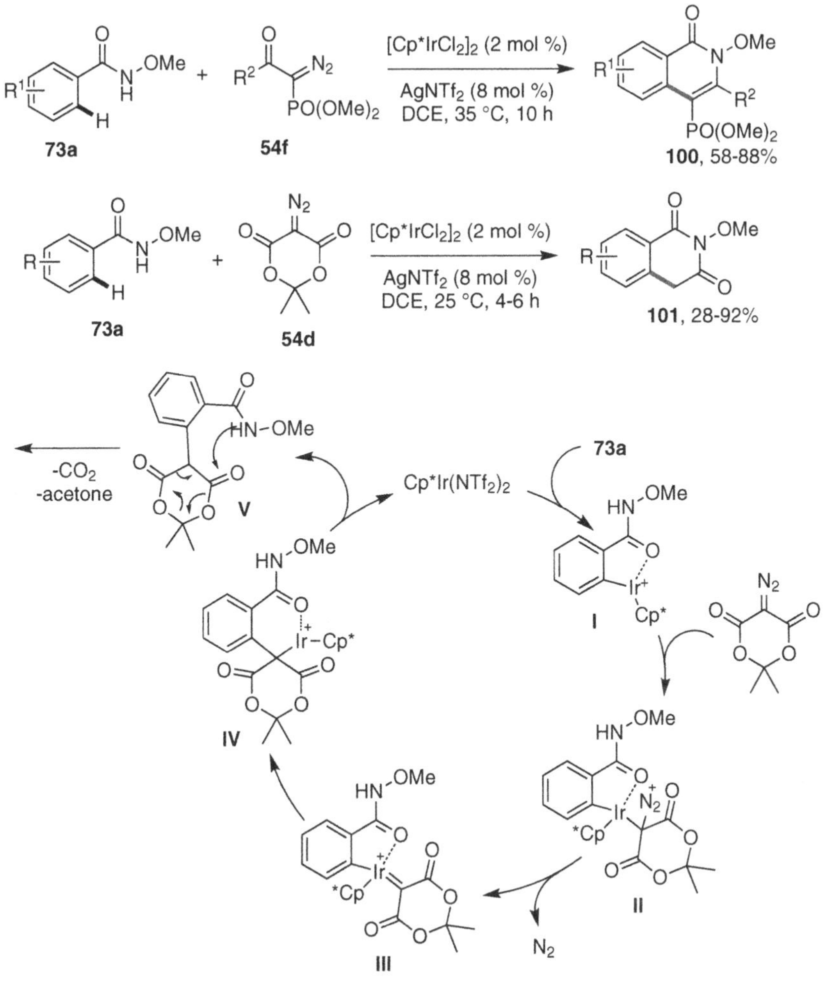

SCHEME 2.33

pyridin-4-one and thieno[3,2-*c*]pyridin-7-one in 34% and 98% yields, respectively. First, a cationic rhodium complex **I** was generated by the ligand exchange with oxime ester **67b**, which underwent *ortho*-C–H cleavage via concerted metalation/deprotonation to give a rhodacyclic intermediate **II**. By coordination of a ketene

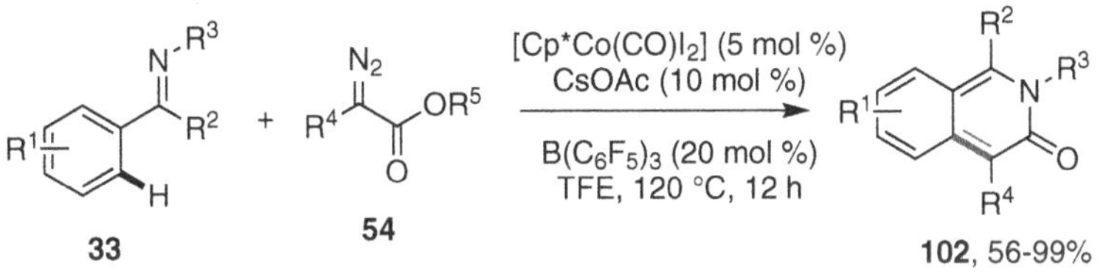

SCHEME 2.34

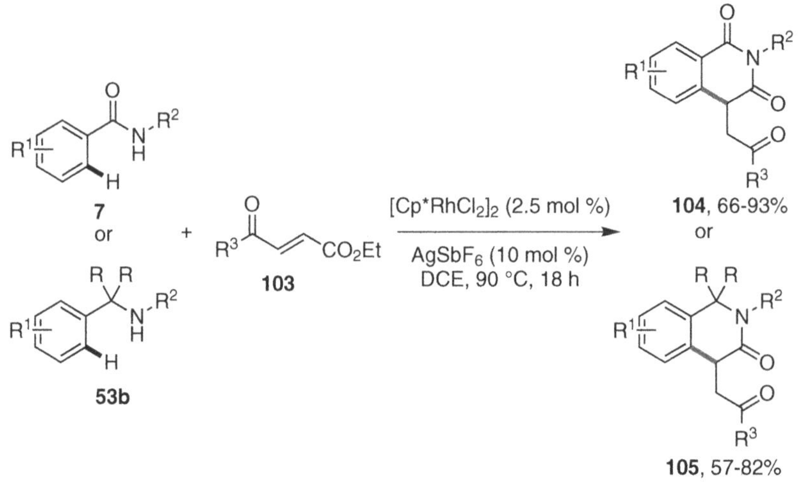

SCHEME 2.35

106, followed by migratory insertion of the Rh–C(alkyl) bond into the C=C bond of ketene, a fused rhodacycle **IV** was formed. Meanwhile, N–O bond cleavage and corresponding N–C bond formation occurred in a concerted fashion to furnish the coupled product **107**, together with regeneration of the active Rh(III) intermediate **I** by ligand exchange of oxime ester **67b** (Scheme 2.36) (Yang et al. 2018).

2.6 ACRIDINE

The synthesis of acridanes **109** through a Cu-catalyzed intramolecular radical cross-dehydrogenative coupling of 2-[2-(arylamino)aryl]malonates **108** was reported by Hurst and Taylor (2017). Treatment of **108** with Cu(2-ethylhexanoate)$_2$ (10 mol%) in toluene at reflux in an open flask afforded acridine-9,9(10*H*)-dicarboxylates **109** in 64–88% yields. In the proposed reaction mechanism, proton abstraction and single-electron oxidation with Cu(II) gave malonyl radical **I**, which underwent intramolecular homolytic aromatic substitution to give **II**, followed by further oxidation to give the cyclohexadienyl cation **III**, which delivered the final acridine product **109** upon aromatization (Scheme 2.37).

Wen et al. (2016) developed a Pd/Cu cocatalyzed dehydrogenative double C(sp^2)–H functionalization/carbonylation of diphenylamines **110** for the synthesis of acridones **111**. Reactions were conducted using $PdCl_2$ (10 mol%), Cu(OPiv)$_2$ (20 mol%), and di-*tert*-butyl peroxide (3 equiv) under CO atmosphere (1 atm) in DMSO at 100°C for 24 h, and corresponding acridin-9(10*H*)-ones **111** were obtained in 42–85% yields. The reaction was initiated by the electrophilic palladation of **110** with Pd(II) to generate arylpalladium species **I**, which reacted with CO to produce the intermediate **II**. By subsequent intramolecular C–H functionalization and reductive elimination of **III**, the acridone **111** was produced along with generation of a Pd(0) species, which oxidized to an active Pd(II) species with DTBP in the presence of Cu(II) (Scheme 2.38).

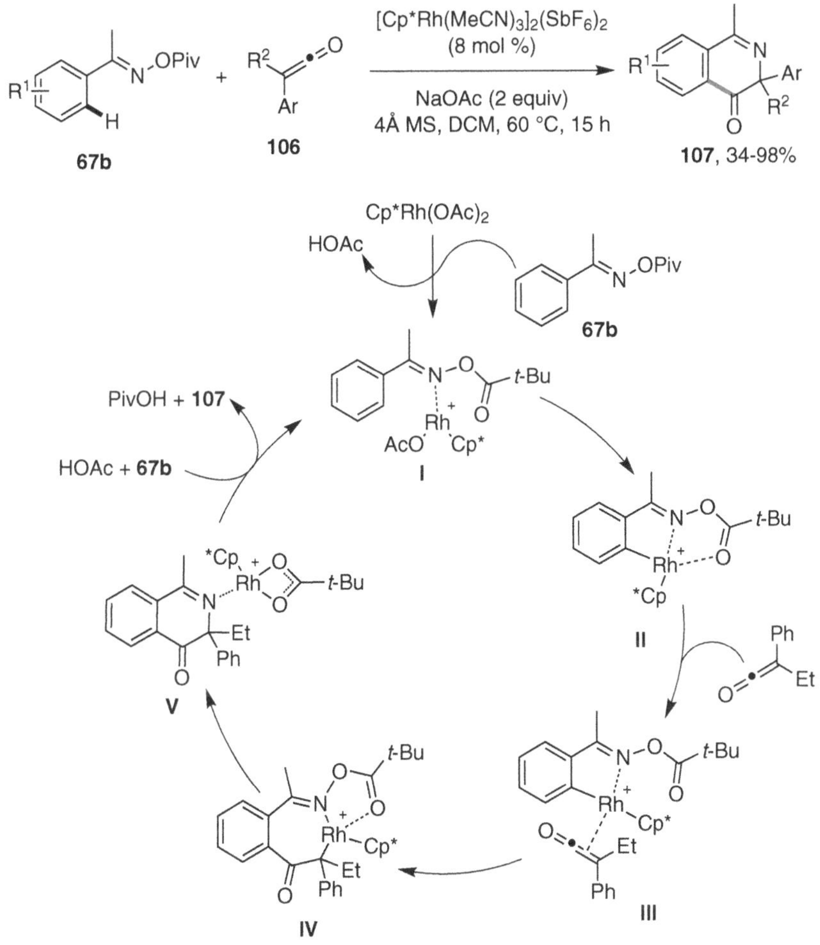

SCHEME 2.36

Cationic Rh(III)-catalyzed (sp^2)–H functionalization of benzaldehydes **38** or *N*-tosyl benzaldimines **33b** with anthranils **112** was developed by Kim et al. (2018a, 2018b) using $[Cp^*RhCl_2]_2$ (2.5 mol%) and $AgSbF_6$ (10 mol%) in the presence of AcOH (1 equiv) in DCE at 110°C, or PivOH (0.5 equiv) in DMF at 120°C, producing corresponding 4-acylacridines **113** in 31–90% or 21–83% yields, respectively. The reaction was preceded by formation of aldimine **II** from the reaction of benzaldehyde with **I**, derived from anthranil **112**, under acidic conditions. The C–H activation of aldimine **II** with a cationic Rh(III) led to a rhodacycle intermediate **III**, which underwent coordination with **112**, with subsequent migratory insertion to afford a rhodacycle intermediate **V**. Intermediate **VI** was generated upon protonolysis of intermediate **V** with regeneration of active Rh(III) catalyst. Next, intramolecular electrophilic cyclization followed by aromatization occurred to give 4-acyl acridine **113**, with release of 2-benzoyl aniline **I** (Scheme 2.39).

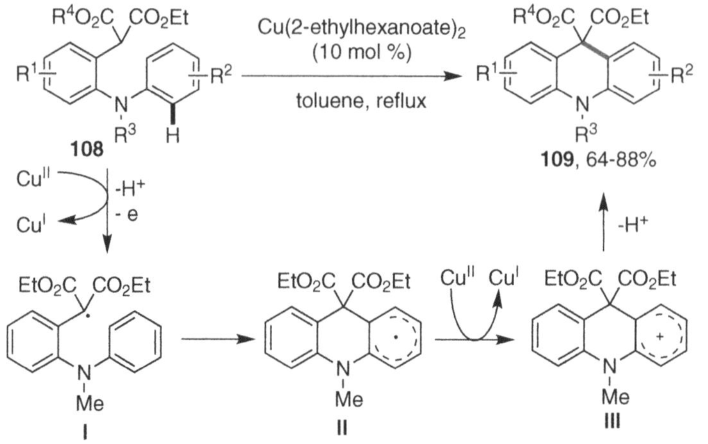

SCHEME 2.37

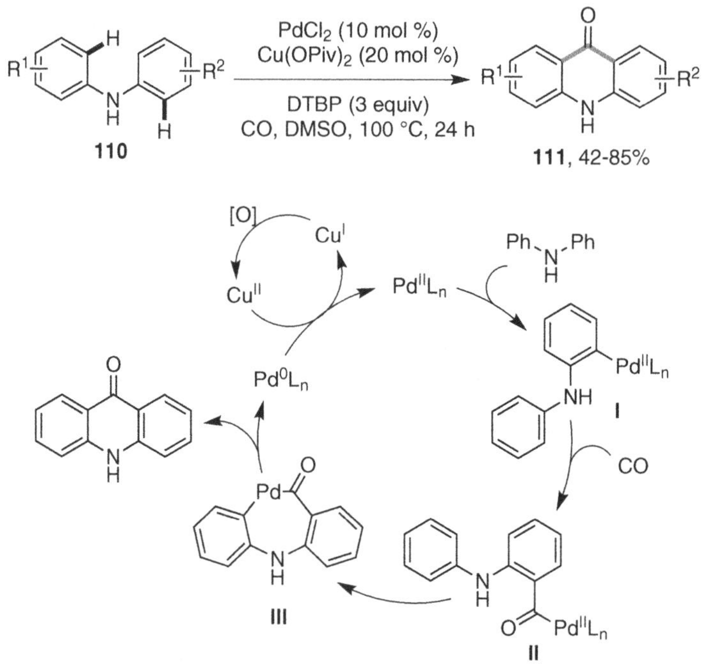

SCHEME 2.38

Also, the construction of acridines was reported by Rh(III)-catalyzed dehydrogenative coupling amination followed by intramolecular electrophilic aromatic substitution and aromatization. Reactions of aromatic aldehydes **38** and aryl azides **114** were performed in DCE, using $[Cp^*Rh(CH_3CN)_3](SbF_6)_2$ (10 mol%) and $BnNH_2$ (40 mol%) in the presence of $MgSO_4$ for 20 h, leading to acridine derivatives **115** in

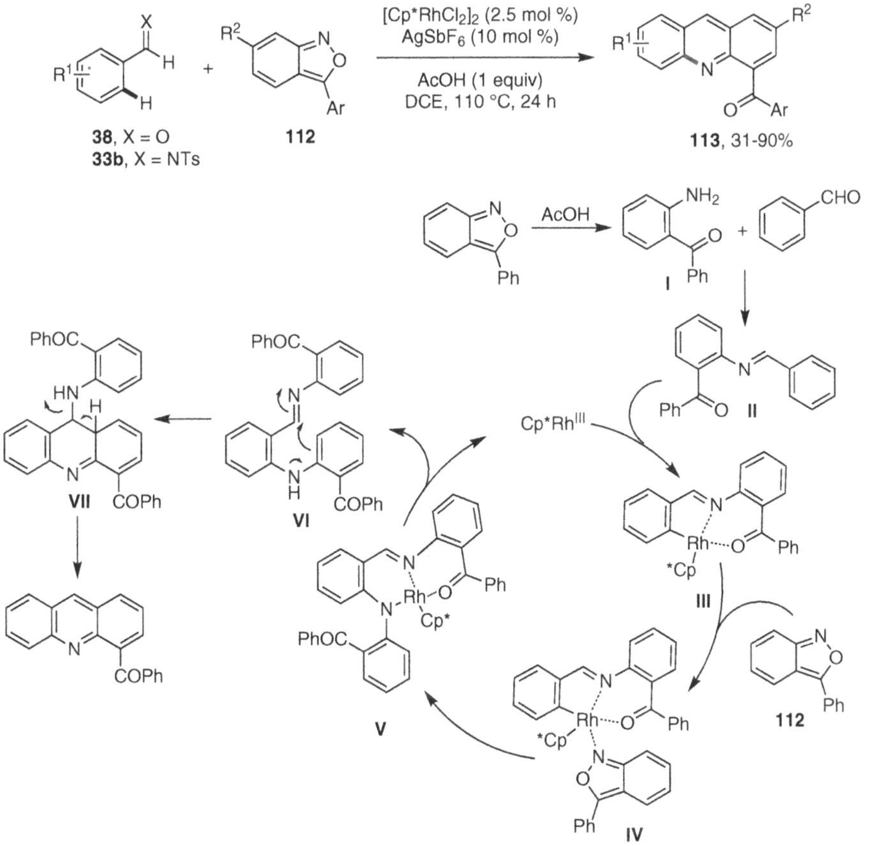

SCHEME 2.39

32–91% yields. Imines **33** derived from aromatic ketones were also compatible with the reaction, catalyzed with $[Cp^*Rh(CH_3CN)_3](SbF_6)_2$ (10 mol%) in the presence of Ac_2O (2 equiv) in DCE at 110°C for 20 h, affording 10-substituted acridines **116** in 51–91% yields. The tentative reaction pathway is illustrated in Scheme 2.40. The imine **33**, derived from aldehyde **38** and $BnNH_2$, underwent *ortho*-directed C–H bond activation to give metallacycle **I**, followed by coordination and migratory insertion of the azide to afford metallacycle **III**, which upon protonation released diarylamine **IV** and the Rh(III) active catalytic species. By intramolecular electrophilic aromatic substitution and subsequent aromatization (intermediate **V**), diarylamine **IV** was converted to the desired acridine **115** (Lian et al. 2013).

2.7 PHENANTHRIDINE

Pd(II)-catalyzed picolinamide-directed intramolecular dehydrogenative C–H amination of *N*-(*ortho*-arylbenzyl)picolinamides **117** followed by oxidation was described in order to prepare phenanthridines **118**. By treatment of picolinamide **117**

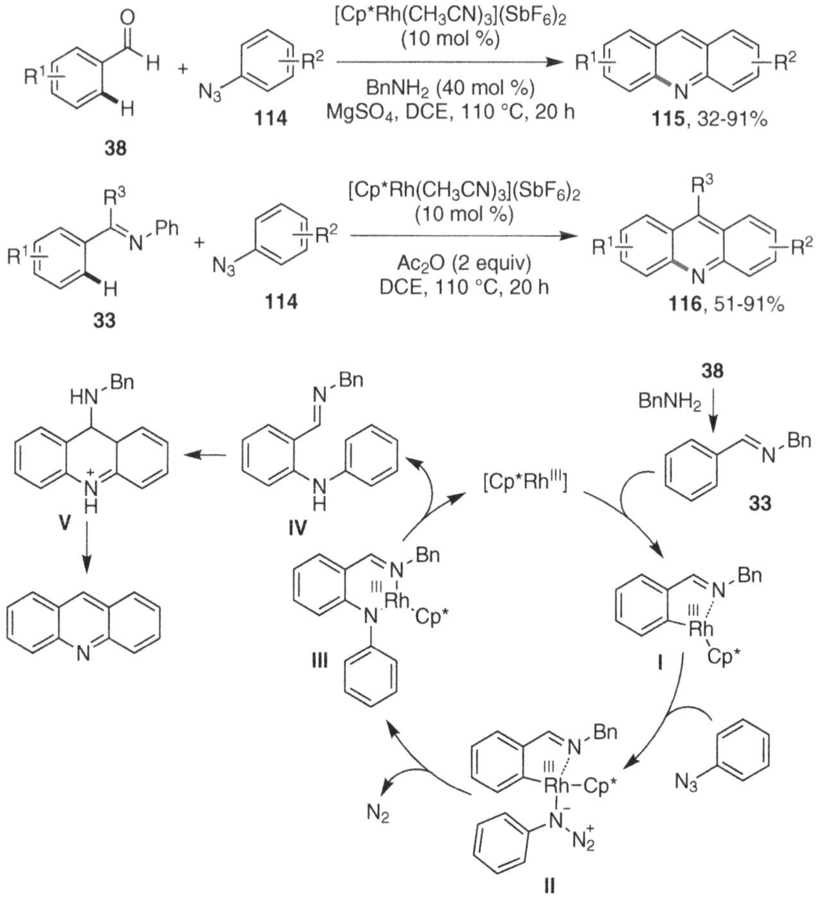

SCHEME 2.40

with $Pd(OAc)_2$ (10 mol%), $PhI(OAc)_2$ (2 equiv), and $Cu(OAc)_2$ (2 equiv) in anhydrous toluene under N_2 atmosphere at 120°C for 24 h, phenanthridines **118** were obtained in 38–58% yields (Scheme 2.41) (Pearson et al. 2013).

Domino *N*-benzylation/intramolecular direct arylation involving sulfonanilides **119** and 2-bromobenzyl bromides **120** was catalyzed by Pd(II) to provide *N*-sulfonyl 5,6-dihydrophenanthridines **121**. Subjecting a solution of an equimolar amount of sulfonanilides **119** and 2-bromobenzyl bromide **120** to $Pd(OAc)_2$ (10 mol%), PPh_3 (20 mol%), and Cs_2CO_3 (2.5 equiv) in dioxane at 110°C under N_2 atmosphere for 16 h furnished 5,6-dihydrophenanthridines **121** in 71–95% yields (Laha et al. 2014). A similar Pd-catalyzed nucleophilic substitution/C–H activation/aromatization cascade reaction of *N*-Ms-arylamines and 2-bromobenzyl bromide derivatives was developed to construct 6-unsubstituted phenanthridines in 31–85% yields. The reaction took place by the base-promoted nucleophilic substitution of 2-bromobenzyl bromide **120** with *N*-Ms-arylamines, producing *N*-benzylaniline, followed by

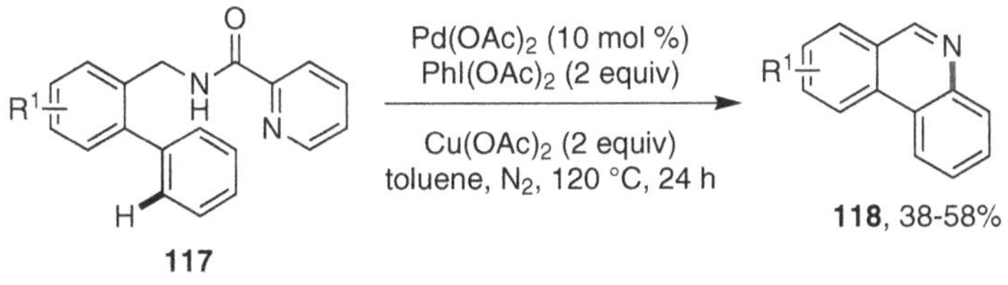

SCHEME 2.41

oxidative palladation of the C–Br bond, with subsequent C–H activation and then reductive elimination (Scheme 2.42) (Han et al. 2015).

Phenanthridines **125** were prepared through Pd/Cu-catalyzed reaction between potassium 2-aminobenzoates **122** and 2-haloarylaldehydes **123**, via cascade imination/C–H functionalization reactions. Reactions were conducted using $Pd(PPh_3)_2Cl_2$ (3 mol%), Cu_2O (1.5 mol%), 1,10-phenanthroline (4 mol%), tetra-*n*-butylammonium bromide (TBAB) (10 mol%), and 4 Å MS in dry DMF at 160°C for 15 min under microwave irradiation to afford phenanthridines **125** in 52–82% yields. The reaction of potassium 2-aminobenzoates **122** with 4-iodo-3-pyrazole carbaldehydes **124** was also investigated under the same reaction conditions, leading to corresponding pyrazolo[4,3-*c*]quinolone derivatives **126** in 67–71% yields. The plausible reaction mechanism involves the generation of imine intermediate **III** by cation exchange of potassium 2-aminobenzoate **122** with Cu to afford intermediate **I**, followed by condensation with intermediate **II**, generated in situ by oxidative addition of 2-bromobenzaldehyde **123** to a Pd(0) species. The formation of a six-membered transition state between the imine nitrogen and copper carboxylate rendered the decarboxylation. Simultaneously, the C–H bond activation led to the formation of **IV**, which underwent reductive elimination to give product **125** (Scheme 2.43) (Bhowmik, Pandey, and Batra 2013).

Kim, Lee, and Youn (2011) developed a Pd-catalyzed domino olefination–conjugate addition reaction for the synthesis of dihydrophenanthridines **128**, by treatment of *N*-tosyl 2-phenylanilines **127** with ethyl acrylate **18** (2 equiv), in the presence of $Pd(OAc)_2$ (3–10 mol%), $K_2S_2O_8$ (3 equiv), and *p*-TsOH (1 equiv) in TFA/CH_2Cl_2 at 25°C. Dihydrophenanthridines **128** were obtained in 46–90% yields. Quinolino[4,3-*j*] phenanthridine **128a** and thieno[3,2-*c*]quinoline **128b** were obtained starting from 2,2''-diamino-[1,1';4',1'']terphenyl derivative and 2-(2-thienyl)sulfanilide under the reaction conditions, in 47% and 68% yields, respectively. The reaction occurred by *N*-atom-directed C–H activation of *N*-tosyl 2-phenylanilines **127** with Pd(II) species,

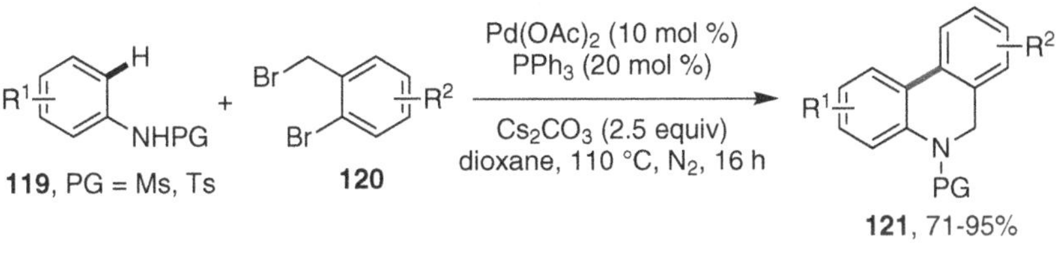

SCHEME 2.42

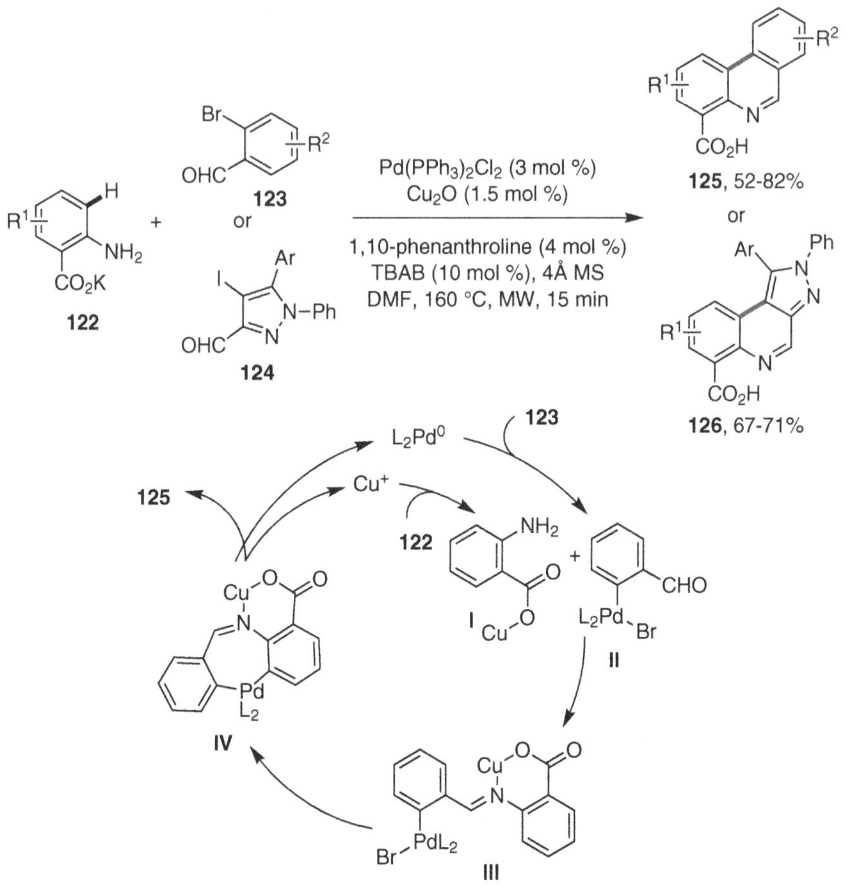

SCHEME 2.43

followed by carbopalladation of olefin, with subsequent hydride elimination to afford cross-coupling product **I**, which underwent an intramolecular conjugate addition to give the cyclized product **128** (Scheme 2.44). A similar approach was developed using $Pd(OAc)_2$ (5 mol%) as the catalyst and $Cu(OAc)_2{\cdot}H_2O$ (5 mol%) under air as the oxidant and NaOAc as a base in DMF at 100–120°C, to deliver corresponding *N*-arylsulfonyl dihydrophenanthridines in up to 99% yields (Miura et al. 1998).

Also, *N*-(*para*-hydroxyphenyl)dihydrophenanthridines **130** and **131** were prepared via dehydrogenative coupling of biaryl 2-iminoquinones **129** with activated alkenes **18**. Subjecting biaryl 2-iminoquinones **129** to acrylates **18** (3 equiv) in the presence of $Pd(OAc)_2$ (10 mol%) in PivOH under N_2 at 120°C for 24 h gave dihydrophenanthridines **131** in 63–94% yields. *N,N*-dimethyl acrylamide, (vinylsulfonyl)benzene, and diethyl vinyl phosphonates were also tolerated in reaction with biaryl 2-iminoquinones **129**, producing corresponding dihydrophenanthridines **131** in 88%, 77%, and 64% yields, respectively. When reactions were carried out using $Pd(OAc)_2$ (10 mol%) and $Cu(OAc)_2$ (0.5 equiv) in TFE at 120°C for 24 h,

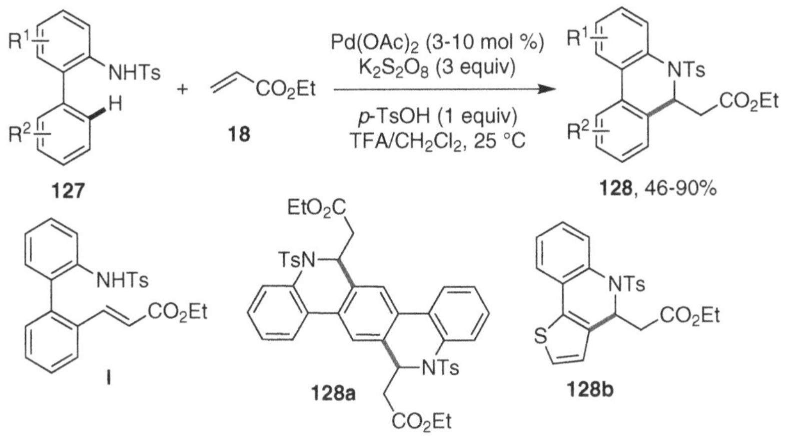

SCHEME 2.44

corresponding phenanthridin-6(5*H*)-ylidenes **130** were obtained in 40–94% yields. The reactions proceeded by an iminoquinone-directed *ortho*-C–H bond activation with Pd(II), generating the *ortho*-metalated complex **I**, which underwent migratory insertion of olefin to form an eight-membered palladacycle **III**. Subsequently, *β*-hydride elimination occurred to give rise an *ortho*-alkenylated hydride complex **IV**, which transformed to complex **V** through reducing the iminoquinone moiety by an internal hydride shift. Michael-type nucleophilic addition gave the intermediate **VI**. By protonation of **VI** under acidic solvent PivOH, dihydrophenanthridine **131** was produced, along with regenerating the Pd(II) active catalyst. Reductive elimination of **VI** led to the formation of phenanthridin-6(5*H*)-ylidene **130** through a redox-neutral pathway, with generation of a Pd(0) species, which was oxidized with Cu(II) under air conditions to Pd(II) for the next catalytic cycle (Scheme 2.45) (Raju et al. 2017).

A redox-neutral Rh(III)-catalyzed C–H bond annulation of *N*-methoxybenzamides **73a** with quinones **132** was developed for the construction of hydrophenanthridinones **133** and phenanthridinones **134**. Reactions were carried out using 2 equiv of quinones **132** in the presence of $[Cp^*RhCl_2]_2$ (2.5 mol%), CsOAc (30 mol%), and HOAc (50 mol%) in DCE/acetone at room temperature, which gave rise to hydrophenanthridinones **133** in 40–86% yields. Subjecting the obtained products **133** to Et_3N (1.5 equiv) and Tf_2O (1.2 equiv) at room temperature in a one-pot manner led to the formation of phenanthridinones **134** in 45–77% overall yields. The reaction was initiated by activation of the *ortho*-C–H bond via formation of a five-membered rhodacycle, which underwent migratory insertion of a quinone C=C double bond with subsequent protonolysis to an *ortho*-alkylates product, followed by intramolecular cyclocondensation of nitrogen with the carbonyl moiety of quinone (Scheme 2.46) (Yang et al. 2016).

Synthesis of phenanthridinone derivatives **137** was also achieved by Pd(II)-catalyzed C–H/N–H activation of *N*-methoxybenzamides **73a** with benzyne, generated in situ from reaction of the Kobayashi precursor **135** with CsF. Reactions were catalyzed with $Pd(OAc)_2$ (5 mol%) in the presence of $Cu(OAc)_2$ (2 equiv) and

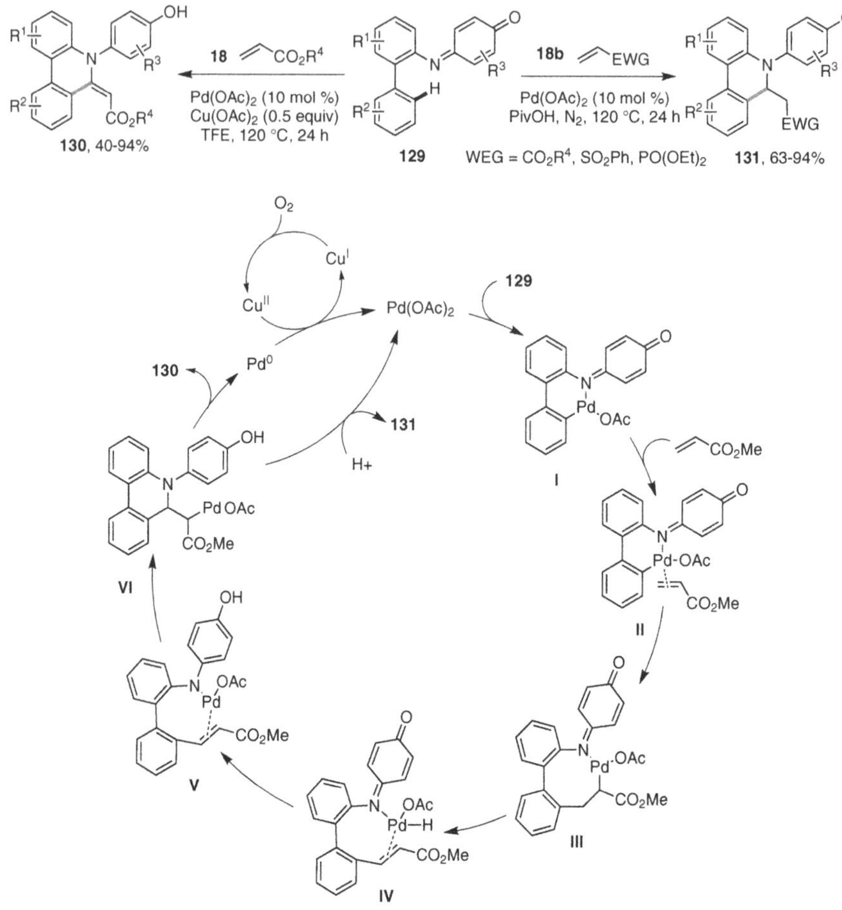

SCHEME 2.45

CsF (2.4 equiv) in DMSO/dioxane at 80°C, and the corresponding phenanthridinones **137** were obtained in 43–92% yields. Moreover, annulation of benzamides **73a** with cyclooctyne **136** under the same reaction conditions furnished the corresponding isoquinolinone derivatives **138** in 52–69% yields (Peng et al. 2014). *N*-(pyrazol-1-ylphenyl)benzamides **139** were annulated with polyfluoroarenes **140** in a dehydrogenative coupling manner to construct phenanthridinones **141**. Reactions were performed using $Cu(OAc)_2{\cdot}H_2O$ (1 equiv), K_2CO_3 (1.5 equiv), Ag_2CO_3 (1 equiv), and $Zn(OAc)_2{\cdot}H_2O$ (0.5 equiv) in DMSO at 70°C for 20 h, producing polyfluorophenanthridinones **141** in 57–92% yields. Benzo[*c*][2,6]naphthyridin-5(6*H*)-one **141a** and thieno[2,3-*c*]quinolin-4(5*H*)-one **141b** were also synthesized starting from pyridine-4- and thiophene-2-carboxamide derivatives, under reaction conditions. By complexation of benzamide **139**, with subsequent *ortho*-C–H bond activation with a Cu salt, cyclometalated species **I** was formed, which gave intermediate **IV** by transmetalation with in situ generated arylzinc reagent **II** followed by oxidation. Reductive elimination delivered the intermediate **V**, which underwent aromatic

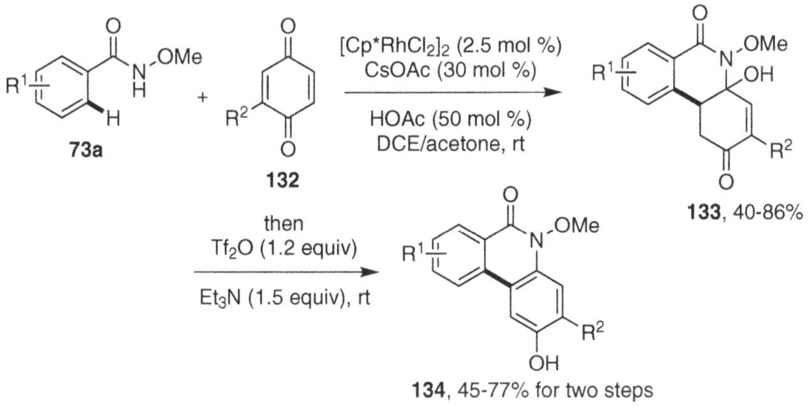

SCHEME 2.46

nucleophilic substitution reaction to the desired phenanthridinone **141** (Scheme 2.47) (Mandal et al. 2018).

Phenanthridinone derivatives were prepared by Rajeshkumar, Lee, and Chuang (2013), Liang et al. (2013a), and Liang et al. (2013b) independently via Pd-catalyzed C–H aminocarbonylation of *ortho*-biarylamines using CO as the carbonyl source in the presence of copper or silver salts. Rajeshkumar, Lee, and Chuang (2013) used $Pd(OAc)_2$ (10 mol%) and AgOAc (5 equiv) in anhydrous CH_3CN at 80°C for 24 h under a CO balloon, giving *N*-tosyl phenanthridinones **142** in 56–94% yields. In Liang et al. (2013a), reactions were conducted using $Pd(MeCN)_2Cl_2$ (5 mol%), $Cu(TFA)_2$ (1 equiv), and TFA (1 equiv) under a CO balloon (1 atm) in dioxane at 110°C, leading to *N*-unsubstituted phenanthridinones in 38–91% yields. The proposed reaction mechanism involves the coordination and C–H activation of *ortho*-biarylamine with Pd(II), followed by migratory insertion of coordinated CO into the aryl–Pd bond, with subsequent reductive elimination to give the final desired product with generation of a Pd(0) species, which was oxidized to an active Pd(II) species with Cu(II) or Ag(I) salts. Nageswar Rao, Rasheed, and Das (2016) developed a similar approach, utilizing DMF as the carbon source under oxygen to construct phenanthridinones **144**. Heating a solution of *ortho*-arylanilines **143** in DMF in the presence of $Pd(OAc)_2$ (10 mol%) and AgOTf (0.4 equiv) at 140°C under O_2 atmosphere for 24 h gave rise to phenanthridinones **144** in 70–90% yields. In the tentative reaction mechanism, the amine-directed *ortho*-C–H activation of *ortho*-arylanilines **143** with Pd(II) gave six-membered cyclopalladated intermediate **I**, which attacked iminium species **II**, generated in situ from DMF via a sequential decarbonylation, nucleophilic addition, and elimination process under silver catalysis with oxygen as the external oxidant, to provide the intermediate **III**. Reductive elimination afforded intermediate **IV** with generation of Pd(0), which was oxidized by Ag in the presence of terminal oxidant O_2 to complete the catalytic cycle. The final desired phenanthridinones **144** were produced from intermediate **IV** via oxidation and hydrolysis (Scheme 2.48). Also, Co-catalyzed $C(sp^2)$–H carbonylation of *ortho*-arylanilines for the synthesis of phenanthridinone scaffolds was accomplished using diisopropyl

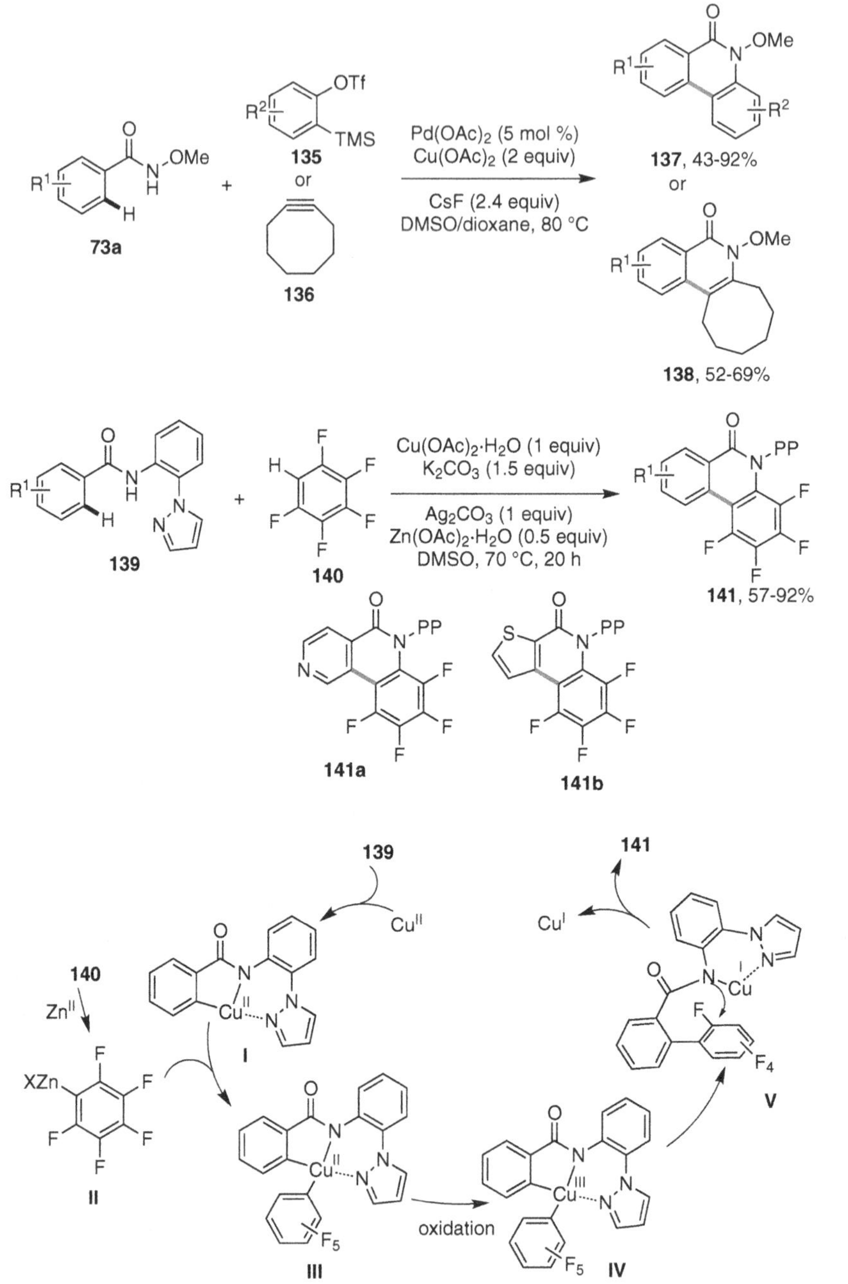

R1
O
N-OMe
H
H
73a
+
OTf
R2
TMS
135
or
136
Pd(OAc)2 (5 mol %)
Cu(OAc)2 (2 equiv)
CsF (2.4 equiv)
DMSO/dioxane, 80 °C
137, 43-92%
or
138, 52-69%
139
140
Cu(OAc)2·H2O (1 equiv)
K2CO3 (1.5 equiv)
Ag2CO3 (1 equiv)
Zn(OAc)2·H2O (0.5 equiv)
DMSO, 70 °C, 20 h
PP
141, 57-92%
141a
141b
CuII
CuI
141
I
ZnII
XZn
II
III
oxidation
IV
V
F5
F4

SCHEME 2.47

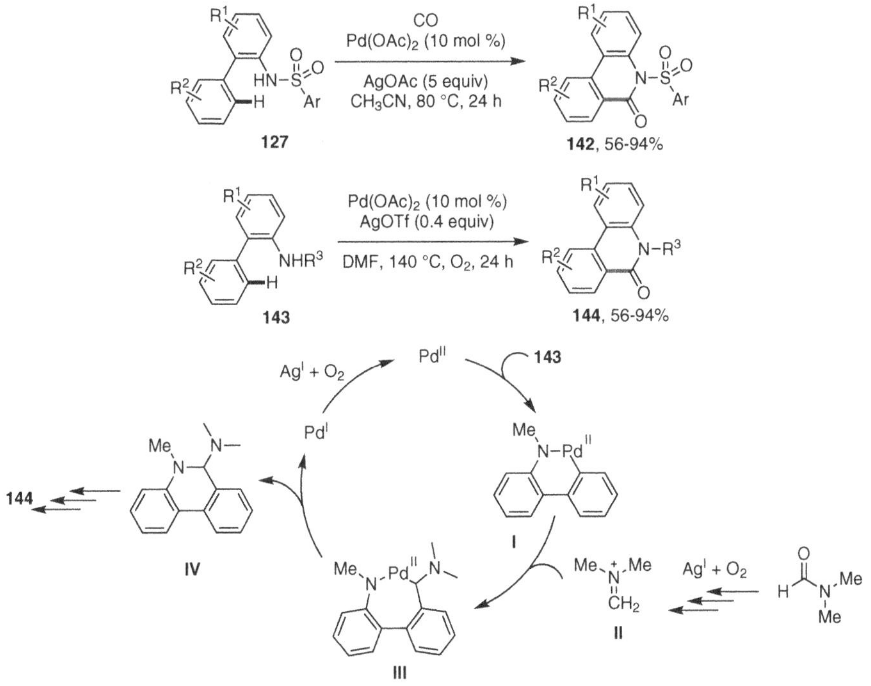

SCHEME 2.48

azodicarboxylate as the CO source and oxygen as the sole oxidant, in 54–97% yields (Ling et al. 2018).

2.8 CINNOLINES AND PHTHALAZINE

Rh-catalyzed construction of cinnolinium salts **146** was described by dehydrogenative coupling of azobenzenes **145a** with internal alkynes **10**, using $[Cp^*RhCl_2]_2$ (2 mol%), $Cu(OAc)_2$ (2 equiv), Ag_2CO_3 (10 mol%), and $NaBF_4$ (2 equiv) in *t*-amylOH at 110°C for 16 h, delivering the expected cinnolinium salts **146** in 71–94% yields. Reaction with unsymmetrical alkyl aryl acetylene and phenylacetylene carboxylate led to corresponding 4-alkyl-3-arylcinnoliniums and 4-phenyl cinnolinium-3-carboxylate in 72–78% and 82% yields, respectively. Terminal alkynes **5** were also compatible with the reaction, affording corresponding 4-unsubstituted cinnolinium salts **147** in 75–83% yields. When the reaction was carried out between *N*-*tert*-butyl-aryldiazene **145b** and aliphatic alkynes **10** under reaction conditions, neutral cinnoline derivatives **148** were obtained in 64–70% yields through isobutene removal. The reactions occurred by *N*-atom-directed *ortho*-C–H activation of azobenzene, followed by coordination and insertion of an alkyne molecule, with subsequent reductive elimination (Scheme 2.49) (Zhao et al. 2013). Moreover, Rh(III)- and Co(III)-catalyzed synthesis of cinnolinium salts were reported by dehydrogenative

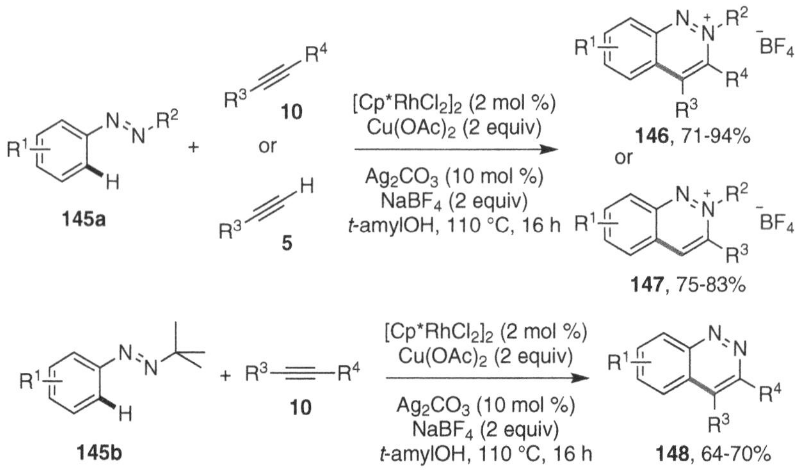

145a + 10 or 5 → [Cp*RhCl2]2 (2 mol %), Cu(OAc)2 (2 equiv), Ag2CO3 (10 mol %), NaBF4 (2 equiv), t-amylOH, 110 °C, 16 h → 146, 71-94% or 147, 75-83%

145b + 10 → [Cp*RhCl2]2 (2 mol %), Cu(OAc)2 (2 equiv), Ag2CO3 (10 mol %), NaBF4 (2 equiv), t-amylOH, 110 °C, 16 h → 148, 64-70%

SCHEME 2.49

coupling of azobenzenes with alkynes (Muralirajan and Cheng 2013, Prakash, Muralirajan, and Cheng 2016).

Synthesis of cinnoline **150** and cinnolinium salt **151** derivatives was achieved via Rh(III)-catalyzed cascade dehydrogenative coupling/cyclization reaction of Boc-arylhydrazines **149** with alkynes. Reacting an equimolar amount of Boc-arylhydrazines **149** with alkynes **10** in the presence of $[Cp^*RhCl_2]_2$ (5 mol%) and $Cu(OAc)_2{\cdot}H_2O$ (1.5 equiv) in $PhCl/CH_2Cl_2$ at 80°C for 24 h under Ar atmosphere resulted in the formation of cinnolines **150** in 56–95% yields. The reaction of Boc-arylhydrazines **149** with 2.5 equiv alkynes **10** catalyzed with $[Cp^*RhCl_2]_2$ (5 mol%) and $AgSbF_6$ (20 mol%) in the presence of $Cu(OAc)_2{\cdot}H_2O$ (4 equiv) as the external oxidant and $AgBF_4$ (2 equiv) in CH_3CN at 80°C for 24 h under Ar gave rise to cinnolinium salts **151** in 38–95% yields. The reaction was initiated by Cu(II)-induced oxidation of the Boc-phenylhydrazine **149** to the corresponding diazocompound **145c**, which underwent C–H bond activation with active species Cp^*RhX_2 (X = OAc, SbF_6) to generate intermediate **I**. Then, the coordination of alkyne **10** to intermediate **I**, followed by insertion into the Rh–C bond, led to the seven-membered rhodacycle intermediate **II**. Subsequently, reductive elimination of **II** afforded intermediate **III** and the Rh(I) species, which was oxidized to an Rh(III) species by a Cu(II) salt. Then, the sequential elimination of CO_2 and isobutene formed the intermediate **IV**, which was converted to five-membered cyclometalated intermediate **V** by C–H bond activation. By coordination and insertion of the second molecule of alkyne **10** into the Rh–C bond, the intermediate **VI** was generated, which transformed into cinnoline **151** upon protolysis. When $AgSbF_6$ and $AgBF_4$ were used, reductive elimination of **VI** and an exchange of the anion with $AgBF_4$ occurred rather than protolysis, giving cinnolinium salt **151** (Scheme 2.50) (Wang, Li, and Wang 2018).

Rh-catalyzed dehydrogenative annulation of *N*-Boc diazocompounds **145c** with α-diazo ketoesters **34** was accomplished using $[Cp^*RhCl_2]_2$ (2.5 mol%) and $AgSbF_6$ (20 mol%) in PivOH/DCE at 50°C for 24 h under air atmosphere, leading to

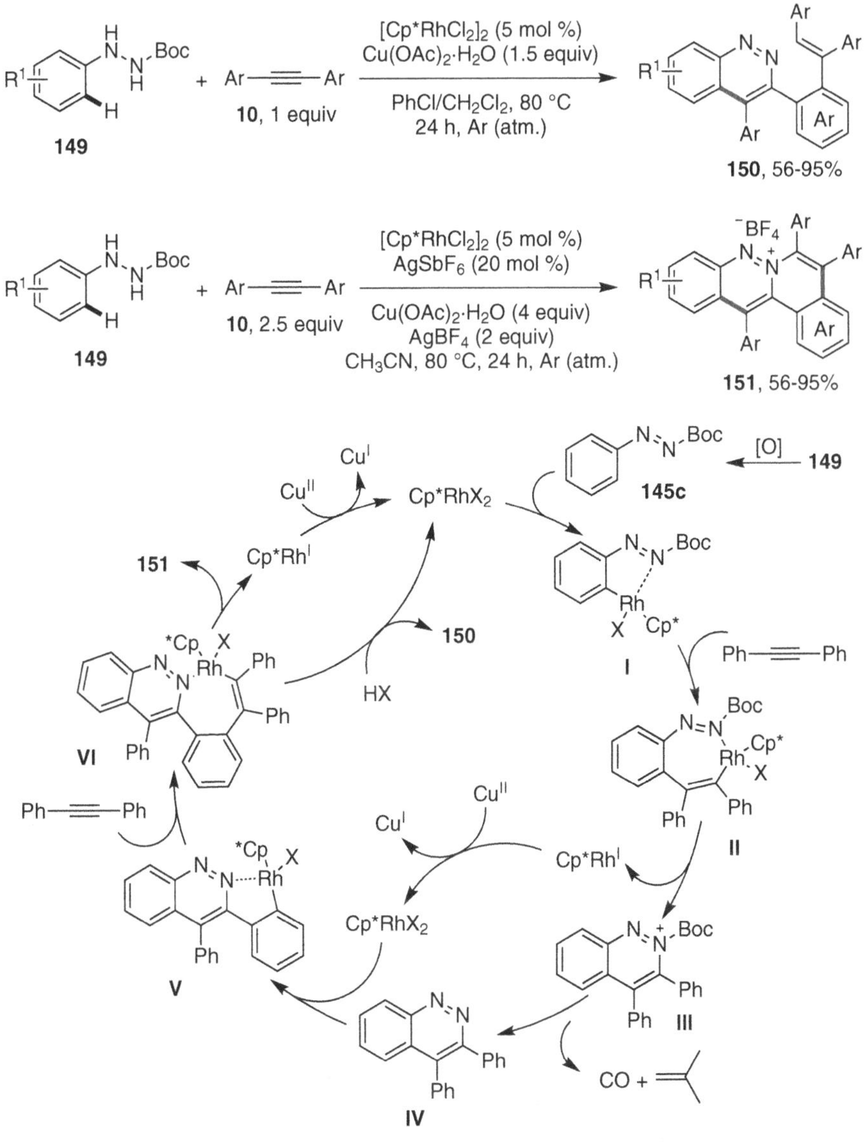

SCHEME 2.50

cinnoline-4-carboxylates **152** in 37–96% yields. Reaction with α-diazo diketone and α-diazo ketophosphonate resulted in the formation of corresponding cinnoline derivatives in 90% and 76% yields, respectively (Sun et al. 2016). The cinnoline scaffolds **154** were accessed via Rh(III)-catalyzed hydrazine-directed C–H functionalization of 1-alkyl-1-arylhydrazines **153** with α-diazo-β-ketoesters **34**. Reactions were performed by $[Cp^*RhCl_2]_2$ (2 mol%) and LiOAc (10 mol%) in MeOH at room temperature for 8 h, to give corresponding 3-substituted 1,4-dihydrocinnoline-4-carboxylates **154**

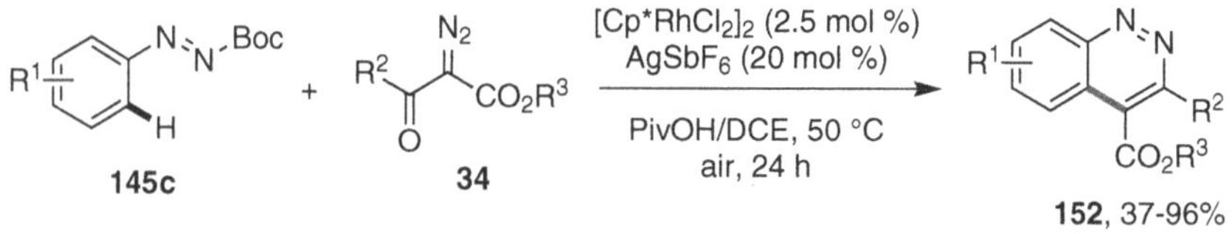

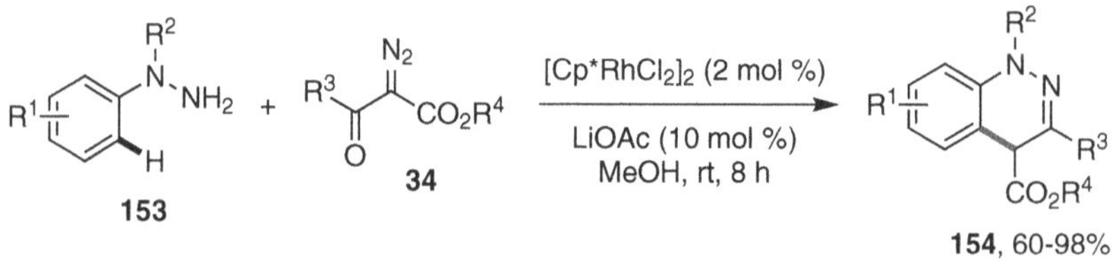

SCHEME 2.51

in 60–98% yields. The reaction took place by generation of Rh-carbenoid species by C–H activation of 1-phenylhydrazine **153** with Rh(III), followed by migratory insertion of the diazo moiety, with subsequent protonolysis to generate an *ortho*-alkylated product, which underwent intramolecular dehydrative condensation to achieve the final 1,4-dihydrocinnoline-4-carboxylate **154** (Scheme 2.51) (Song et al. 2016).

Pd(II)-catalyzed intramolecular oxidative C–H/C–H cross-coupling of *N'*-methylenebenzohydrazides **155** was reported by Matsuda, Tomaru, and Matsuda (2013) to obtain phthalazin-1(2*H*)-ones **156**. Reactions were conducted using $Pd(OAc)_2$ (10 mol%) and benzoquinone (1.2–2 equiv) in AcOH at 120°C, producing phthalazin-1(2*H*)-ones **156** in 24–85% yields. No product was obtained in the case of hydrazine, which lacked a carbonyl group. Reactions occurred via electrophilic *ortho*-palladation and subsequent *C*-arylation of the carbon–nitrogen double bond (Scheme 2.52).

2.9 QUINAZOLINE

Conversion of *N*-arylamidines **157** to 4-aminoquinazolines **158** occurred via Pd(II)-catalyzed C(sp^2)–H amidination by isonitrile insertion. Reactions were conducted by refluxing a mixture of *N*-arylamidines **157** and isonitrile (3 equiv) in the presence of $Pd(OAc)_2$ (5 mol%) and Cs_2CO_3 (1.5 equiv) in toluene under an O_2 balloon to afford corresponding 4-aminoquinazolines **158** in 42–97% yields. In the proposed reaction mechanism, Cs_2CO_3-assisted C–H bond activation of the amidine by isonitrile-coordinated Pd(II), followed by migratory insertion of the isonitrile into

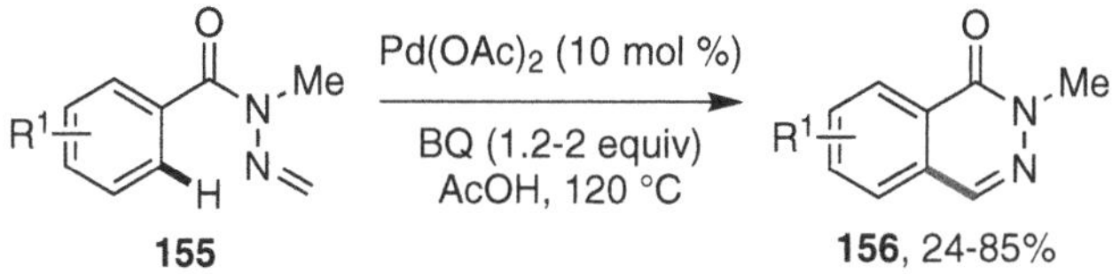

SCHEME 2.52

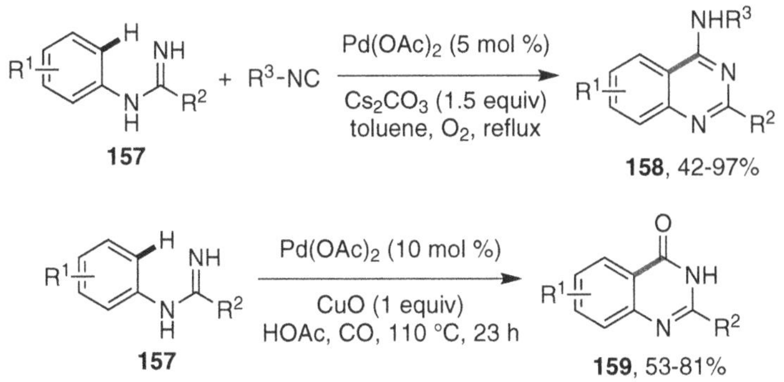

157 + R^3-NC → [$Pd(OAc)_2$ (5 mol %), Cs_2CO_3 (1.5 equiv), toluene, O_2, reflux] → **158**, 42-97%

157 → [$Pd(OAc)_2$ (10 mol %), CuO (1 equiv), HOAc, CO, 110 °C, 23 h] → **159**, 53-81%

SCHEME 2.53

the C–Pd bond, with subsequent reductive elimination, and then tautomerization, delivered the product **158** with concurrent formation of a Pd(0) species, which was reoxidized to Pd(II) by oxygen (Wang et al. 2011). Pd(II)-catalyzed intramolecular C(sp^2)–H carboxamidation of *N*-arylamidines **127** was developed in order to synthesize quinazolin-4(3*H*)-ones **159**. Reactions were carried out by heating a solution of *N*-arylamidines **157**, $Pd(OAc)_2$ (10 mol%), and CuO (1 equiv) under CO atmosphere in HOAc at 110°C for 23 h, leading to quinazolin-4(3*H*)-ones **159** in 53–81% yields. The reaction occurred by activation of the *ortho*-C–H bond with Pd(II), followed by coordination and insertion of CO into the C–Pd bond, with subsequent reductive elimination (Scheme 2.53) (Ma et al. 2011).

Synthesis of quinazolines **161** from amidines **157** and 1,6-diynes **160** was reported via Rh(III)-catalyzed cascade [5+1] annulation/5-*exo*-cyclization initiated by C–H activation. Reactions were performed using $[Cp^*RhCl_2]_2$ (5 mol%), Ag_2CO_3 (10 mol%), 2,6-dimethylbenzoic acid (1 equiv), and Li_2CO_3 (80 mol%) in *N,N*-dimethylacetamide (DMA)/EtOH/*N*-methyl-2-pyrrolidone (NMP) mixture in the presence of 4 equiv of water at 80°C, producing quinazoline derivatives **161** in 25–91% yields. The reaction was initiated by *ortho*-C–H activation to generate six-membered rhodacycle **I**, followed by ligand exchange with the terminal alkyne to give rhodacycle **II**, which underwent 5-*exo*-cyclization to **III**. Subsequent migratory insertion of **III** generated seven-membered rhodacycle intermediate **IV**, which was transformed into the final desired product **161** by reductive elimination to intermediate **V**, with subsequent isomerization, along with release of the Cp*Rh(I), which was reoxidized to the Cp*Rh(III) catalyst by Ag(I) to complete the catalytic cycle (Scheme 2.54) (Xu et al. 2018a).

Rh and Cu co-catalyzed aerobic oxidative [4+2] C–H annulation of aryl imidates **41** with benzyl azides **162** was reported by Wang and Jiao (2016) to synthesize quinazolines **163**. Reactions were conducted using $[Cp^*RhCl_2]_2$ (2.5 mol%), $AgSbF_6$ (10 mol%), CuI (20 mol%), and 4 Å MS under an atmosphere of O_2 in PhCl at 90°C for 16 h, leading to 2-aryl-4-alkoxyquinazolines **163** in 68–90% yields. Alkyl azide afforded the corresponding 2-alkyl-4-ethoxyquinazoline in low yield (38%) along with a related indazole derivative. The reaction proceeded by Rh(III)-catalyzed *ortho*-C–H amination of imidates via generation of a five-membered rhodacyclic intermediate,

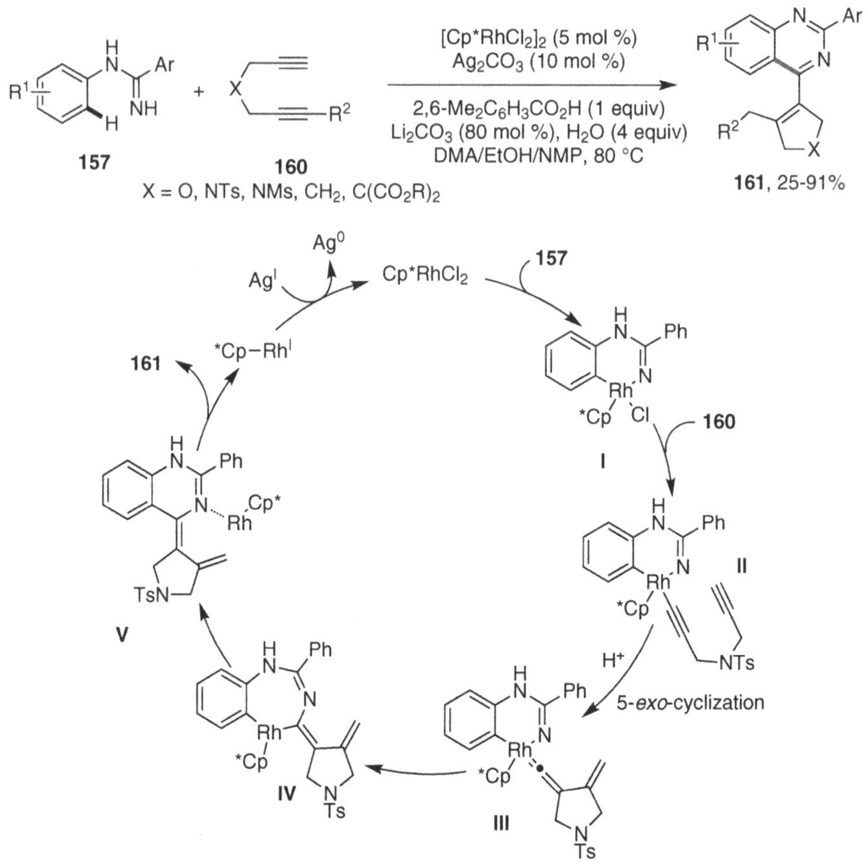

SCHEME 2.54

followed by coordination with benzyl azide, with subsequent migratory insertion by release of N_2, and then protonation. The amine **I** executes the Cu-catalyzed aerobic oxidation to produce the final quinazoline product **163** (Scheme 2.55).

Synthesis of quinazolines **163** was also accomplished by *ortho*-C–H activation and functionalization with dioxazolones **164**. In this context, reaction of benzimidates **41** and dioxazolones **164** was described under the catalytic redox-neutral $[Cp^*RhCl_2]_2/AgBF_4$ system, where dioxazolones **164** acted as an internal oxidant. Treatment of benzimidates **41** with dioxazolones **164** (1.5 equiv) in the presence of $[Cp^*RhCl_2]_2$ (2 mol%) and $AgBF_4$ (8 mol%) in DCE at 50°C afforded the corresponding quinazolines **163** in 66–96% yields. 3-aryl- as well as 3-alkyldioxazolones **164** were tolerated under reaction conditions, giving the expected 4-alkoxyquinazolines **163**. The plausible reaction mechanism involves the C–H bond activation of benzimidate **41** with in situ generated cationic [Cp*Rh(III)] to afford rhodacyclic complex **I**, followed by coordination and migratory insertion of dioxazol-5-one **164**, to give the Rh-amido species **III** with release of CO_2. Upon protonation the amido imidate **IV** was formed along with regeneration of the active [Cp*Rh(III)] catalyst.

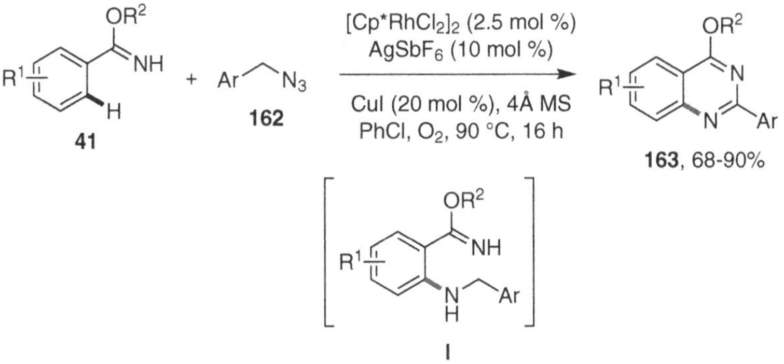

SCHEME 2.55

Dehydrative cyclization of **IV** afforded the quinazoline product **163** (Wang et al. 2016b). A similar approach was developed by Wang, Lerchen, and Glorius (2016) using [Cp*Co(CO)I_2] (10 mol%), $AgSbF_6$ (20 mol%), and NaOAc (50 mol%) in DCE at 60°C, resulting in the formation of quinazoline derivatives in 48–99% yields. Synthesis of quinazoline *N*-oxides **165** was achieved via an Rh(III)-catalyzed C–H activation–amidation of the ketoximes **67b** with 1,4,2-dioxazol-5-ones **164**, with subsequent Zn(II)-catalyzed cyclization (Scheme 2.56) (Wang et al. 2016c).

2.10 PHENAZINE

[3+3] Annulations of aromatic azides **114** with azobenzenes **145d** to give phenazines **166** were developed by an Rh(III)-catalyzed amination/cyclization/aromatization cascade. Reactions were performed using $[Cp^*RhCl_2]_2$ (5 mol%) and $AgB(C_6F_5)_4$ (20 mol%) in AcOH at 100°C for 24 h, affording phenazine derivatives **166** in 30–88% yields (Scheme 2.57). For the mechanism see Scheme 2.40 (Lian et al. 2013).

2.11 SEVEN-MEMBERED N-HETEROCYCLES

2.11.1 Benzazepine

An Rh(III)-catalyzed dehydrogenative cross-coupling reaction between benzylamines **53b** and Morita–Baylis–Hillman (MBH) adducts **167** is described. Subjecting benzylamines **53b** to 3 equiv of MBH adducts **167** in the presence of $[Cp^*RhCl_2]_2$ (2.5 mol%), $AgSbF_6$ (10 mol%), and $Cu(OAc)_2$ (70 mol%) in DCE under air at 110°C for 10 min resulted in the formation of 2-benzazepine derivatives **168** in 30–80% yields. In addition to α-unsubstituted benzyl amine, α-substituted and α,α-disubstituted benzyl amines were compatible with the reaction and gave rise to the formation of corresponding 2-benzazepine derivatives **168** in 30–70% and 45% yields, respectively. Tetrahydro-1-naphthylamine gave naphtho[1,8-*bc*]azepine-3-carboxylate **168a** in 31% yield. By C–H activation of secondary benzylamine **I**, derived from the reaction of benzylamine **53b** with MBH adduct **167** through the S_N2', with a cationic Rh(III) species, a rhodacycle intermediate **II** was generated,

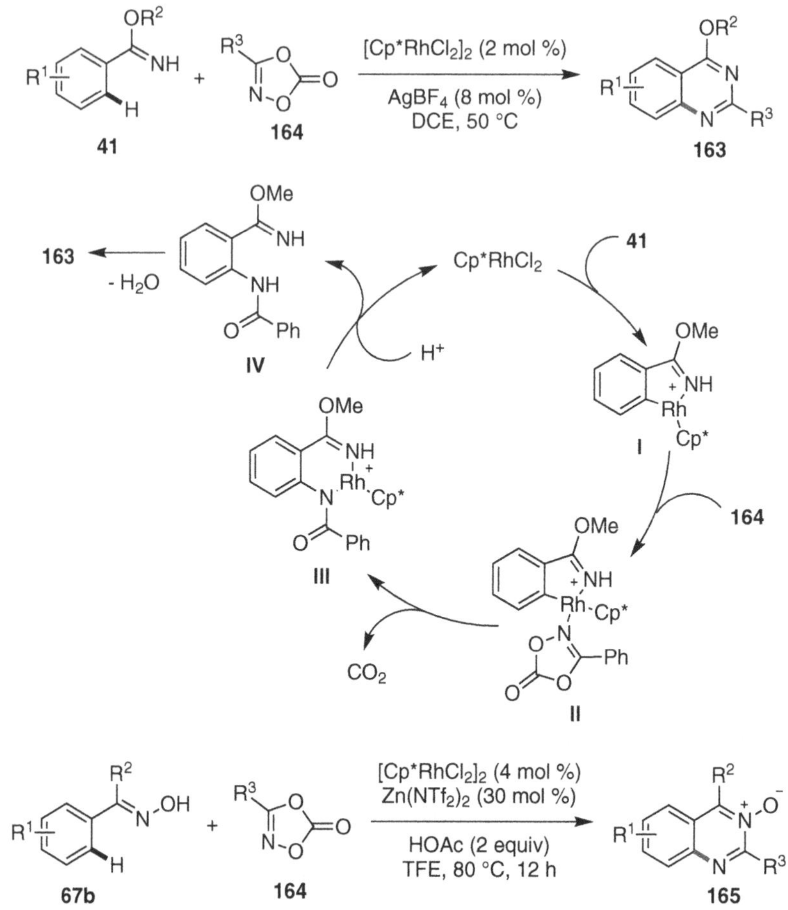

SCHEME 2.56

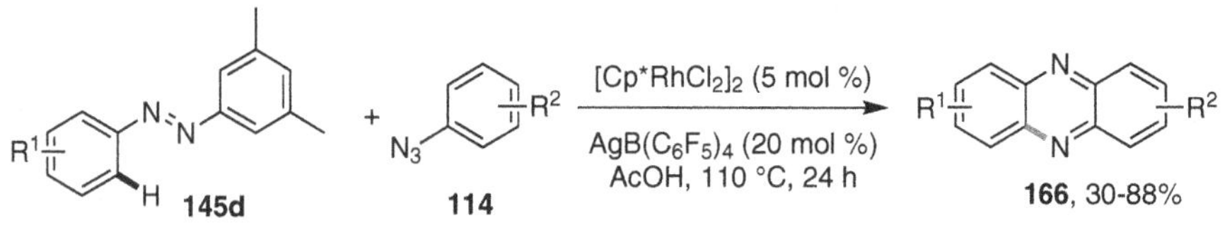

SCHEME 2.57

which underwent olefin coordination and migratory insertion to give seven-membered Rh(III) complex **III**. *β*-H elimination of **III** delivered the intermediate **IV** and the Rh(III) catalyst. Finally, the *N*-allylation of **IV** furnished the desired product **168** (Scheme 2.58) (Pandey et al. 2017).

Rh(III)-catalyzed dehydrogenative coupling of secondary benzamides **7a** with *α,β*-unsaturated aldehydes and ketones **169** was developed to synthesize benzoazepinone derivatives **170**. Reactions were performed using $[Cp^*RhCl_2]_2$ (2.5 mol%),

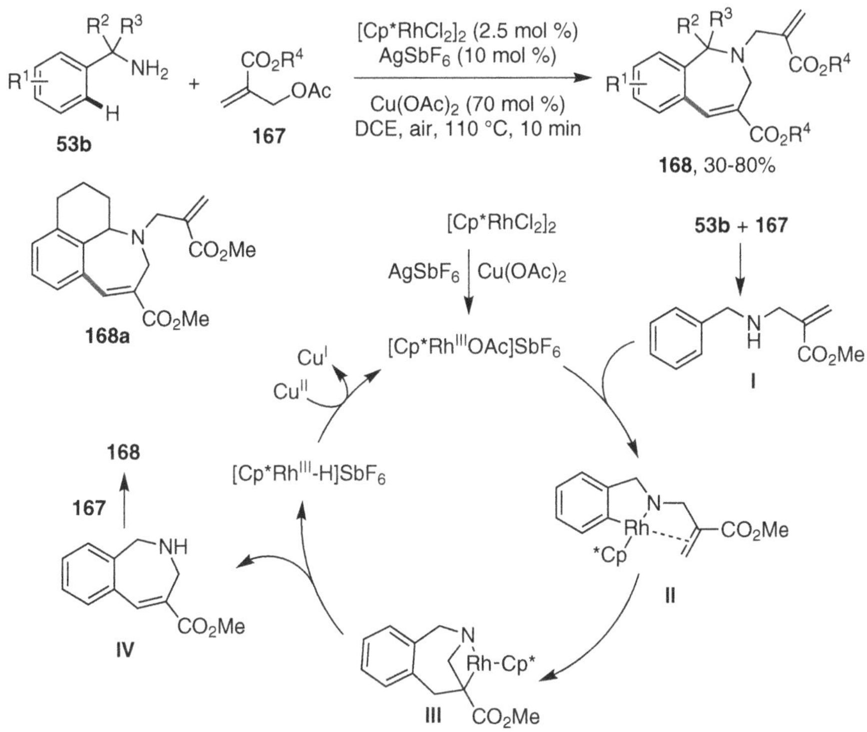

SCHEME 2.58

$AgSbF_6$ (10 mol%), and PivOH (2 equiv) in dioxane at 60°C under argon atmosphere for 12 h, which resulted in the formation of benzo[*c*]azepin-1-ones **170** in 32–88% yields. The primary benzamide and *N*-Ph- and *N*-allyl benzamides did not afford the expected products. Thieno[2,3-*c*]azepin-8(7*H*)-one **170a** was obtained in 60% yield, from *N*-methoxy thiophene-2-benzamide substrate under reaction conditions. The reaction proceeded by *N*-atom-directed *ortho*-C–H activation with rhodium, followed by C=C double bond coordination and insertion, with subsequent protonolysis to an *ortho*-alkylated intermediate, which underwent intramolecular cyclodehydration to achieve the final desired product (Shi, Grohmann, and Glorius 2013). Synthesis of 1-oxobenzo[*c*]azepine-5-carboxylate derivatives **172** was achieved by Rh(III)-catalyzed C–H activation/[4+3] cycloaddition of benzamides **73b** with vinylcarbenoid species **171**. Reactions were catalyzed with $[Cp^*RhCl_2]_2$ (2 mol%), in the presence of CsOAc (1 equiv) in CH_3CN at room temperature, affording 1-oxo-2,3-dihydro-1*H*-benzo[*c*]azepine-5-carboxylates **172** in 50–97% yields. α-diazoketone provided the corresponding 5-acetylbenzo[*c*]azepine in 43% yield. The reaction was initiated by amine-assisted C–H activation to generate five-membered rhodacycle **I**, which afforded Rh-carbene **II** by coordination of vinylcarbenoids **171**, followed by extrusion of N_2. Subsequent migratory insertion of the carbene moiety into the C–Rh bond generated η^3-Rh(III) species **III**, which underwent 1,3-allylic suprafacial migration to **IV**. Then, a C–N bond formation occurred to afford **V** along with N–O

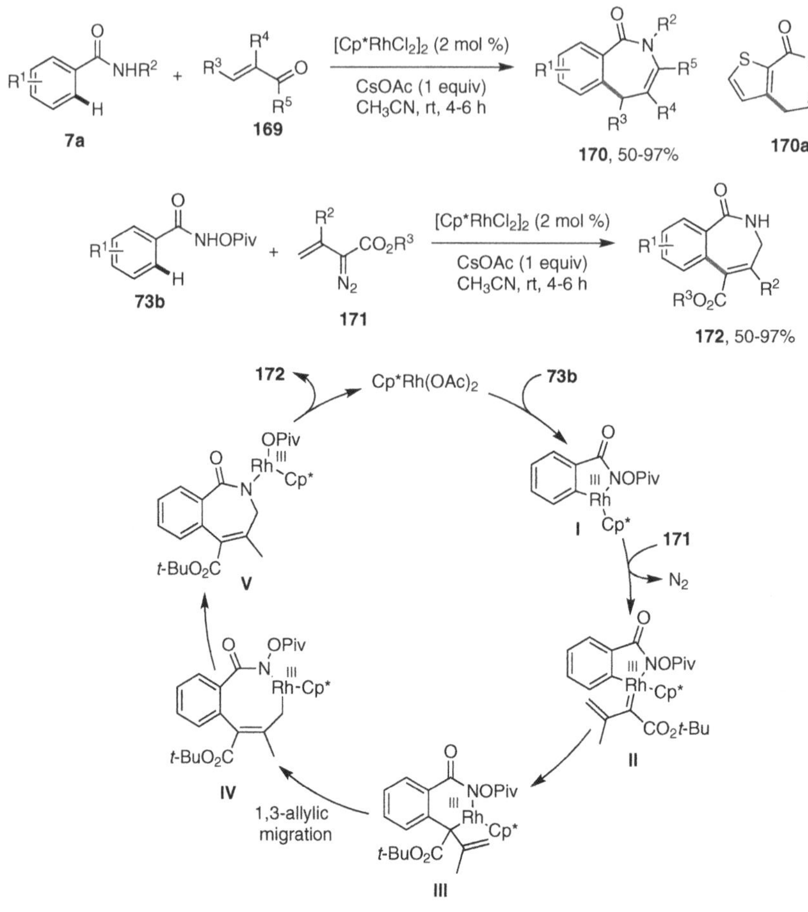

SCHEME 2.59

bond cleavage. The final desired product **172** was obtained upon protonolysis of **V** with concomitant Rh(III) catalysis regeneration (Scheme 2.59) (Cui et al. 2013).

Rh(III)-catalyzed C–H activation/cycloaddition of furan-2-carboxamides **173** and arylidene cyclopropanes **78** delivered furo[2,3-*c*]azepinone derivatives **174**. Reactions were performed using 2 equiv of arylidene cyclopropanes **78** in the presence of $[Cp^*RhCl_2]_2$ (2 mol%) and CsOAc (1 equiv) in MeOH at 60°C, and 4-arylidene-6,7-dihydro-4*H*-furo[2,3-*c*]azepin-8(5*H*)-ones **174** were obtained in 30–85% yields. Reaction with 2-cyclopropylideneacetate led to the expected 2-(8-oxo-5,6,7,8-tetra-hydro-4*H*-furo[2,3-*c*]azepin-4-ylidene)acetates **174a** in 45–54% yields. Benzofuran-2-carboxamide was also tolerated under the reaction conditions, and gave rise to the corresponding benzofuran-fused azepinone in 30% yield. Thiophene-2-carboxamide afforded the corresponding thieno[2,3-*c*]azepinone **174b** in 37% yield, along with the formation of spiro[cyclopropane-1,4'-thieno[2,3-*c*]pyridin]-7'(6'*H*)-one **174c** in 55% yield as the major product. In the proposed reaction mechanism, carboxylate-assisted C–H activation via a concerted metalation/deprotonation (CMD) pathway generated intermediate **I**, which underwent coordination and regioselective insertion of arylidene cyclopropane **78** into the C–Rh bond to afford rhodacycle **III**. A cyclopropylcarbinyl–butenyl rearrangement gave rhodacycle **IV**, which underwent subsequent C–N bond formation to afford **V** along with N–O bond cleavage. Protonation of **V** furnished product **174** with regeneration of the Rh(III) catalyst (Scheme 2.60) (Cui, Zhang, and Wu 2013).

Rh(III)-catalyzed synthesis of indeno[1,7-*cd*]azepine derivatives **176** was developed by [3+2]/[5+2] annulation of 4-aryl-1-tosyl-1,2,3-triazoles **175** with internal alkynes **10** through dual C(sp^2)–H activation. Reactions were conducted using 2.5 equiv of alkynes **10** in the presence of $[Cp^*RhCl_2]_2$ (5 mol%), $AgSbF_6$ (20 mol%), $Cu(OAc)_2{\cdot}H_2O$ (2 equiv), and H_2O (3 equiv) in DCE at 85°C under Ar atmosphere for 15 h, affording 1*H*-indeno[1,7-*cd*]azepin-1-ol derivatives **176** in 48–71% yields. No reaction occurred when *N*-benzyl 1,2,3-triazole was used. The reaction of unsymmetrical alkyl aryl alkynes led to corresponding 1-alkylidene-2,3-dihydro-1*H*-indeno[1,7-*cd*]azepines **177** in 37–67% yields. Reaction was initiated by reaction of triazole **175** with the active Cp^*RhX_2 species, generated in situ from $[Cp^*RhCl_2]_2$ and $AgSbF_6$, generating Rh(III)-carbenoid intermediate **I**, which attacked an alkyne **10** to afford intermediate **II**. Electrophilic cyclization of one of the phenyl groups

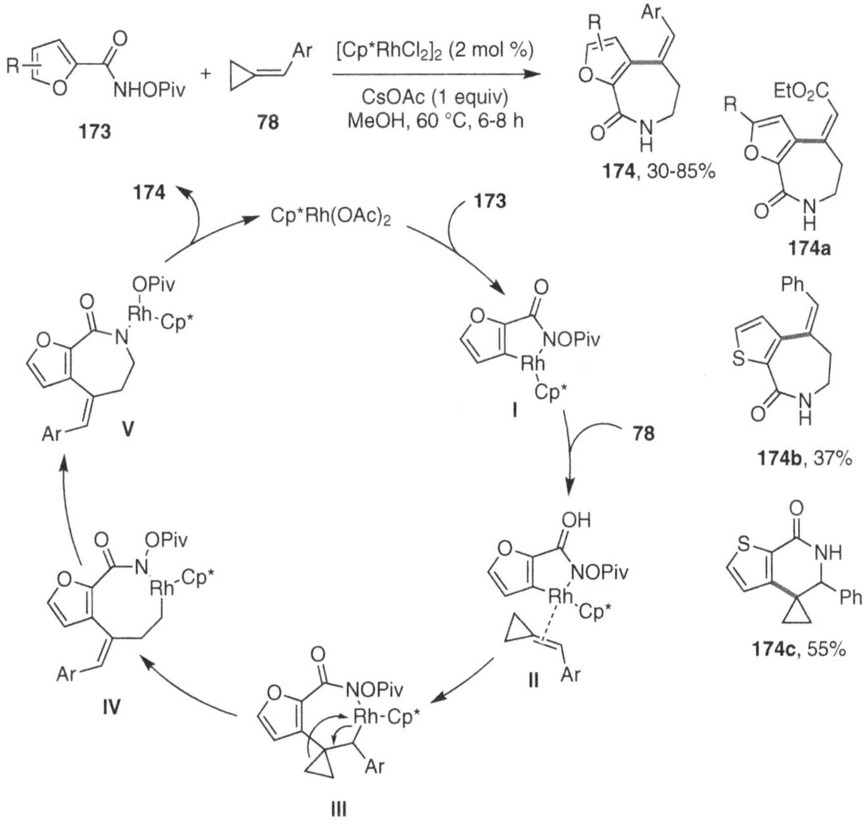

SCHEME 2.60

gave intermediate **III**, which underwent annulation with a second alkyne **10** to form the Cp*(H)Rh coordinated intermediate **V**. Coordination of Rh with the nitrogen atom led to transaddition to afford intermediate **VI**, which was transformed into **176** or **177** by cleavage of the C–Rh bond with the aid of $Cu(OAc)_2$, through hydration with H_2O by backside attack of the C–Rh bond (in the case of diphenylacetylene) or selective β-H elimination (in the case of phenylpropyne), respectively, along with regeneration of the active Cp^*RhX_2 species (Scheme 2.61) (Yang et al. 2015).

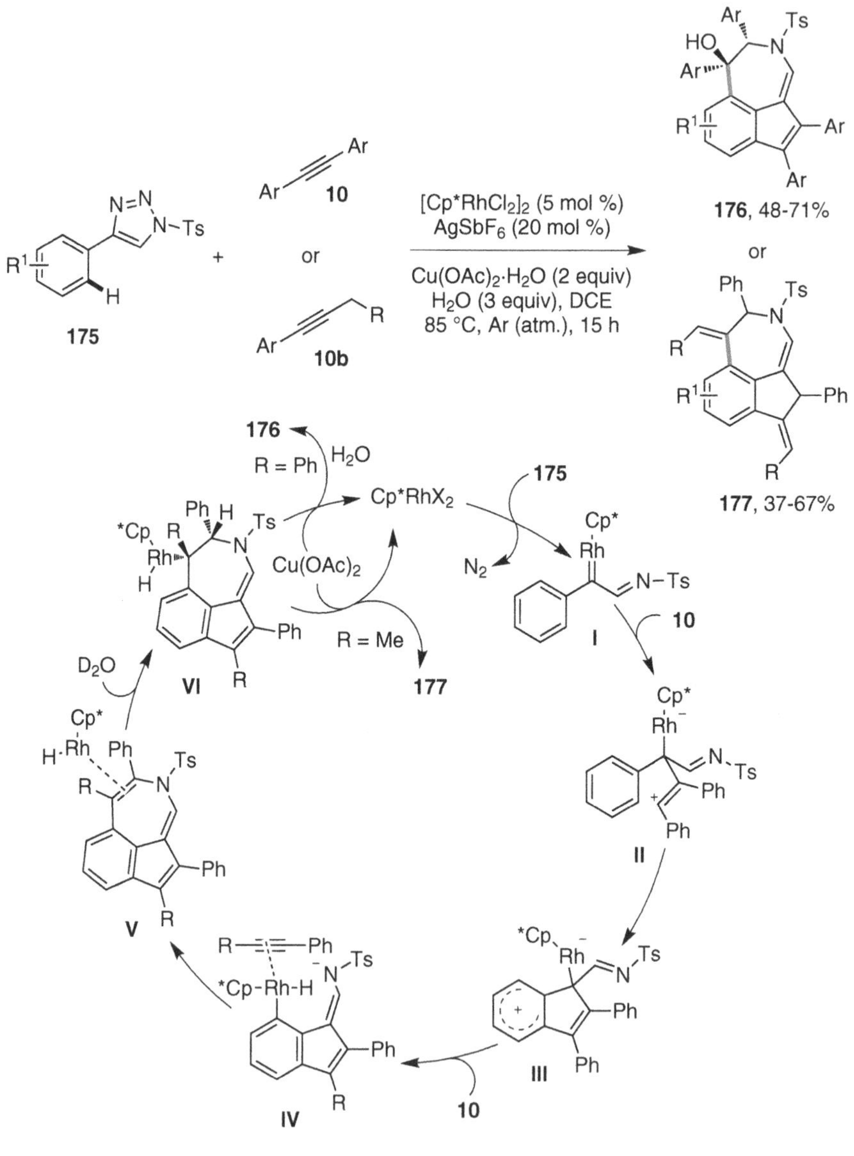

SCHEME 2.61

2.11.2 Dibenzazepine

The construction of dibenzo[*b,d*]azepines **179** in a highly diastereoselective fashion was described by a Pd(II)-catalyzed dehydrogenative cross-coupling reaction between *o*-arylanilines **127** with dienes **178**. Reactions were catalyzed with $Pd(TFA)_2$ (5 mol%) in the presence of $Cu(OAc)_2$ (2.1 equiv) as the oxidant in CH_3CN at 120°C for 36 h, giving dibenzo[*b,d*]azepines **179** in 29–92% yields. Aryl-substituted dienes **178** worked as well as carboxylate-group-functionalized dienes under reaction conditions. The reaction took place with complexation of Pd(II) species with substrate **127**, followed by the electrophilic palladation of the C–H bond to generate a six-membered palladium species **I**, which underwent coordination and migratory insertion of the 1,3-diene **178** to form an eight-membered palladacycle **III**. Finally, C–N reductive elimination delivered the desired product **179** and concomitantly regenerated the Pd(II) species with a Cu(II) oxidant to complete the catalytic cycle (Scheme 2.62) (Bai et al. 2017).

Also, cascade C–H functionalization/amidation reaction of *ortho*-aminobiaryls **180** with diazomalonates **54b** was developed under Rh(III) catalysis, affording dibenzo[*b,d*]azepin-6(7*H*)-one derivatives **181**. Reactions were carried out by heating a solution of *ortho*-aminobiaryls **180** and diazomalonates **54b** (2 equiv) in the presence of $[Cp^*RhCl_2]_2$ (2.5 mol%), $AgSbF_6$ (10 mol%), and AcOH (2.5 equiv) in EtOH at 60°C for 24 h, to furnish 6-oxo-6,7-dihydro-5*H*-dibenzo[*b,d*]azepine-7-carboxylates **181** in 53–94% yields. 2-Furylaniline and 2-thienylaniline gave the expected fused benzo[*b*]furo- and benzo[*b*]

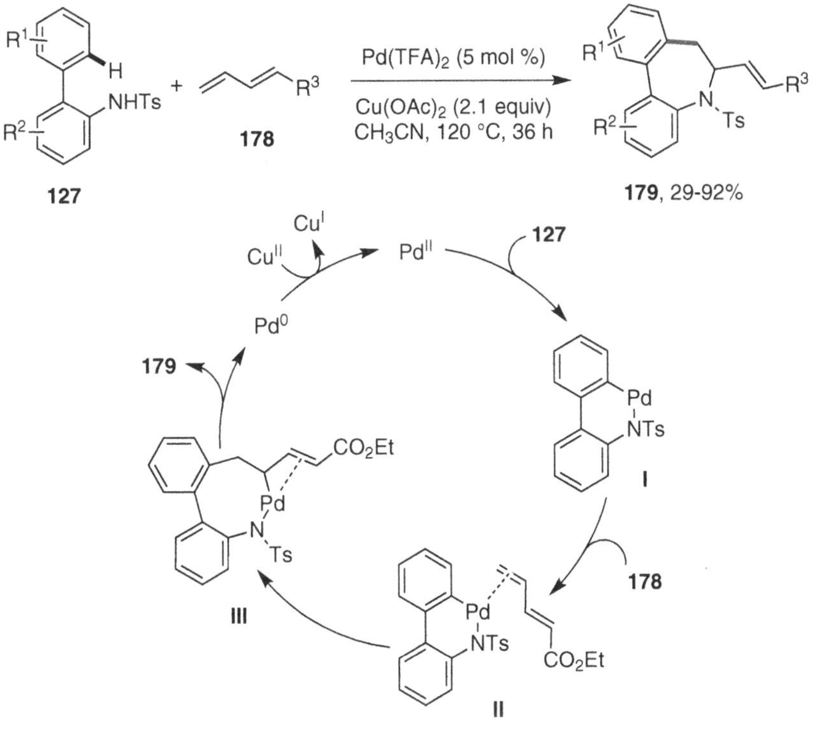

SCHEME 2.62

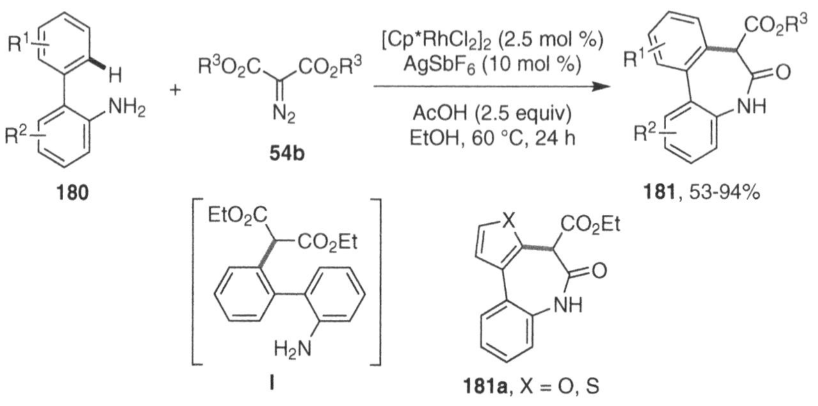

SCHEME 2.63

thieno[3,2-*d*]azepine-4-carboxylate derivatives **181a** in 80% and 48% yields, respectively. Amino-group-directed activation of the α-C(sp^2)–H bond of 2-aminobiaryl **180** with [Cp*Rh(III)]$^{2+}$ followed by coordination and release of N_2 gave the corresponding rhodium(III) carbene intermediate, which underwent migratory insertion of the carbene group into the Rh–C bond with subsequent protonolysis, and then intramolecular amidation via intermediate **I**, to achieve the final desired product (Scheme 2.63) (Bai et al. 2016).

Synthesis of 6,11-dihydro-5*H*-dibenzo[*b,e*]azepines **183** was developed by radical difluoromethylation of alkyne, followed by intramolecular cyclization under electrochemical conditions. Reactions between *N*-benzyl-*N*-(2-ethynylphenyl) amides **182a** and $CF_2HSO_2NHNHBoc$ as a CF_2H radical precursor were conducted in an undivided cell in the presence of Cp_2Fe (10 mol%), Et_4NBF_4, and Na_2HPO_4 in MeOH at 70°C under 13 mA, affording dibenzo[*b,e*]azepines **183** in 43–79% yields. Also, *N*-(2-ethynylbenzyl)-*N*-phenylacetamide **182b** was compatible with the reaction, giving the corresponding dibenzo[*b,e*]azepine **183b** in 61% yield. The reactions occurred by generation of F_2HC radicals upon Cp_2Fe-assisted anodic oxidation of $CF_2HSO_2NHNHBoc$, followed by addition to the alkyne moiety of **182** to give intermediate **I**, which underwent cyclization at the 7-*ortho* position to form cyclic radical intermediate **II**. Anodic dehydrogenative aromatization of **II** led to the dibenzo[*b,e*] azepine product **183** (Scheme 2.64) (Xiong et al. 2018).

N-sulfonyl-6,7-dihydro-5*H*-dibenzo[*c,e*]azepines **186** were accessed by Pd(II)-catalyzed domino *N*-benzylation/intramolecular arylation of *N*-Ts-benzyl amines **185** with 2-bromobenzyl bromides **120**. Reactions were performed using $Pd(OAc)_2$ (10 mol%), PPh_3 (20 mol%), and Cs_2CO_3 (2.5 equiv) in dioxane or toluene under N_2 atmosphere to give 6,7-dihydro-5*H*-dibenzo[*c,e*]azepines **186** in 14–76% yields (Laha et al. 2014). Moreover, Pd(II)-catalyzed homocoupling of benzamides **7** occurred via dual *ortho*-C–H bond activation, using amide as a directing group, which by the subsequent intramolecular condensation reaction resulted in the formation of 5*H*-dibenzo[*c,e*]azepine-5,7(6*H*)-diones **187** in 37–93% yields, in a one-pot fashion. Reactions were catalyzed with $Pd(OAc)_2$ (5 mol%) in the presence of $Na_2S_2O_8$ (2 equiv) as the oxidant in TFA at 130°C. [1,1'-Biphenyl]-2,2'-dicarboxamide **I** was proposed as the reaction intermediate (Kondapalli et al. 2017) (Scheme 2.65).

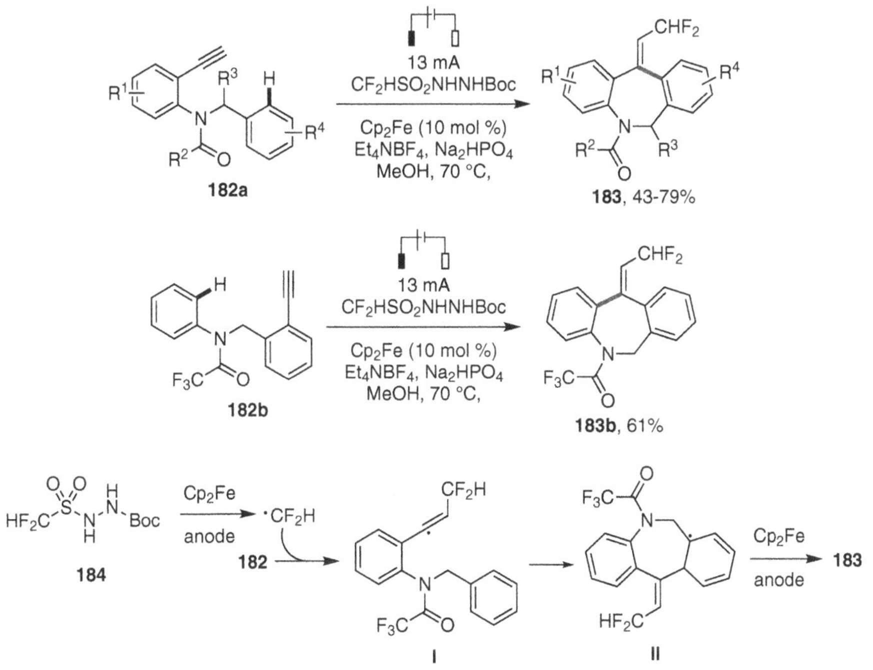

SCHEME 2.64

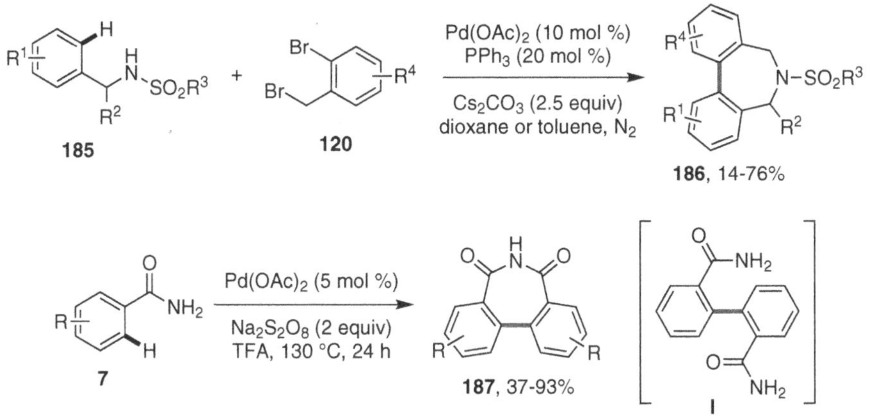

SCHEME 2.65

2.11.3 BENZODIAZEPINE

Rh(III)-catalyzed C–H activation of azomethine imines/dipolar addition with alkylidenecyclopropanes was accomplished to construct benzo[*d*]pyrazolo[1,2-*a*][1,2]diazepin-1(5*H*)-one scaffolds **189**. By subjecting azomethine imine **188** with alkylidenecyclopropanes **78** (2.5 equiv) in the presence of [Cp*Rh(OAc)$_2$] (8 mol%), and AgOAc (2.5 equiv) in TFE at 40°C, 10-benzylidene-tetrahydrobenzo[*d*] pyrazolo[1,2-*a*][1,2]diazepin-1(5*H*)-ones **189** were obtained in 45–84% yields. The proposed reaction mechanism is shown in Scheme 2.66. Coordination and C–H activation of **189** with Rh(III) produced a rhodacyclic intermediate **I**, which underwent coordination of alkylidenecyclopropanes **78** with subsequent migratory insertion to the Rh–aryl bond to provide Rh(III)-alkyl intermediate **II**. β-C elimination and ring scission delivered an Rh(III)-alkyl species **III**, which was converted to diene **IV** by subsequent β-H elimination, with generation of a Rh(III)-hydride species. The active catalyst was regenerated upon oxidation by AgOAc. Intramolecular [3+2] cycloaddition of diene **IV** afforded the desired product **189** (Scheme 2.66) (Bai et al. 2018).

The construction of ethyl 1,4-dialkyl-5*H*-benzo[*d*][1,2]diazepine-5-carboxylates **191** was developed by Rh(III)-catalyzed dehydrogenative cross-coupling of *N*-Boc hydrazones **190** and diazoketoesters **54**. Reactions were conducted using $[Cp^*RhCl_2]_2$ (2 mol%) in the presence of AcOH (50 mol%) in TFE at 80°C for 12 h, leading to the benzo[*d*][1,2]diazepine-5-carboxylate derivatives **191** in 37–90% yields. Reactions proceeded by C–H bond activation, followed by generation of carbene-Rh species by coordination of the diazo moiety and then migratory insertion, with subsequent protoderhodation to give *C*-arylated β-ketoester **I**, which underwent cyclocondensation and then release of the Boc moiety (Scheme 2.67) (Wang et al. 2017).

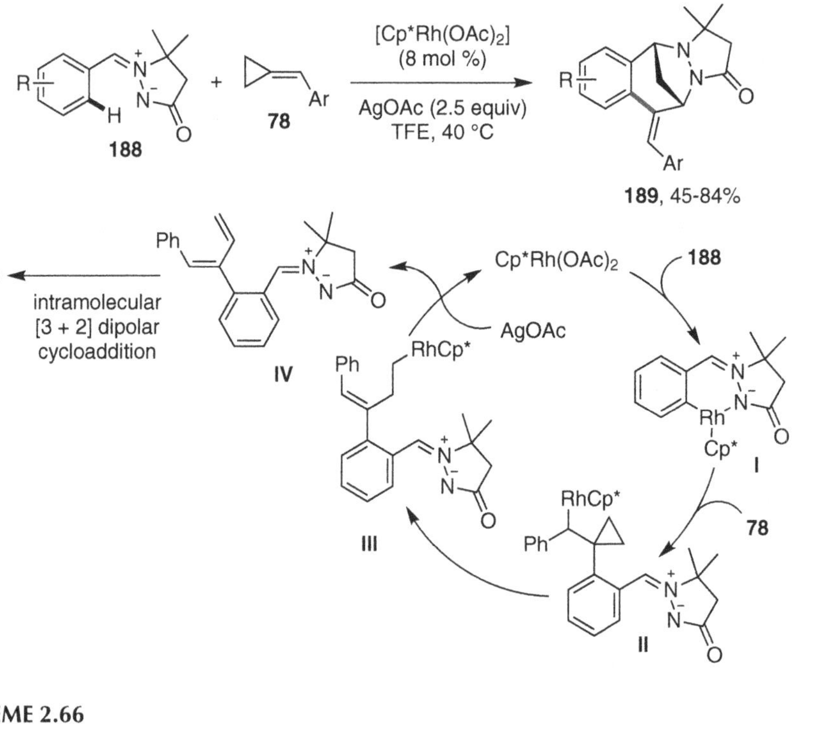

SCHEME 2.66

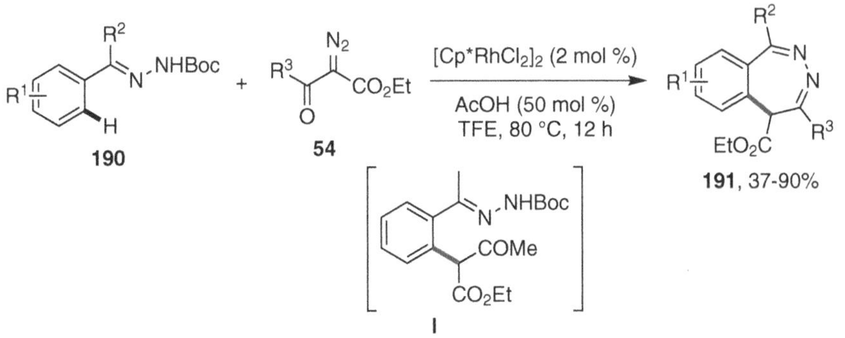

SCHEME 2.67

2.11.4 Benzazocine and Benzazonine

Rh(III)-catalyzed formal [4+2+2] cyclization of *N*-pivaloyloxybenzamides **73b** with 1,6-allene-enes **192** was developed to construction of benzo[*c*]cyclopenta[*f*]azocin-6(11*H*)-one derivatives **193**. Subjecting a solution of equimolar amounts of *N*-pivaloyloxybenzamides **73b** and 1,6-allene-enes **192** in MeOH to $[Cp^*RhCl_2]_2$ (2 mol%) and K_2CO_3 (30 mol%) at room temperature gave rise to the formation of benzo[*c*]cyclopenta[*f*]azocin-6(11*H*)-ones **193** in 23–68% yields. The plausible reaction mechanism involves the base-promoted C–H bond rhodation of **73b** with in situ formed $Cp^*Rh(CO_3)$ catalytic species to form rhodacyclic intermediate **I**, which underwent insertion of the allene moiety of **192**, affording seven-membered rhodacycle **II**. Subsequent cyclic intramolecular carborhodation of the C=C double bond gave nine-membered rhodacyclic intermediate **III**, which underwent reductive elimination to produce the desired [4+2+2] product **193** (Scheme 2.68) (Wu et al. 2015).

Two examples of dibenzoazocinone **194** were prepared by dehydrogenative cross-coupling between aryl aldehyde **38** and 3-methylanthranil **112b** using $[Cp^*RhCl_2]_2$ (2.5 mol%) and $AgSbF_6$ (10 mol%) in the presence of AcOH (1 equiv) in DCE at 110°C, in 32–36% yields. The reaction occurred by C–H bond functionalization of aromatic aldehyde **38** with anthranil **112b** with subsequent intramolecular aldol condensation (Scheme 2.69) (Kim et al. 2018a).

Rh(III)-catalyzed annulation of benzamides **73a** with quinone monoacetals **195** was developed for the one-pot synthesis of benzo[*c*]azonine-1,5(2*H*)-diones **196**. Reactions were performed using $[Cp^*RhCl_2]_2$ (5 mol%) and CsOAc (100 mol%) in trifluorotoluene at 70°C, leading to benzo[*c*]azonine-1,5(2*H*)-diones **196** in 46–93% yields. The reaction occurred by C–H bond activation and functionalization with quinone monoacetal **195** to generate intermediate **I**. Finally, intramolecular aza-Michael addition delivered the nine-membered aza-heterocycle **196** under basic conditions (Scheme 2.70) (Yang et al. 2017b).

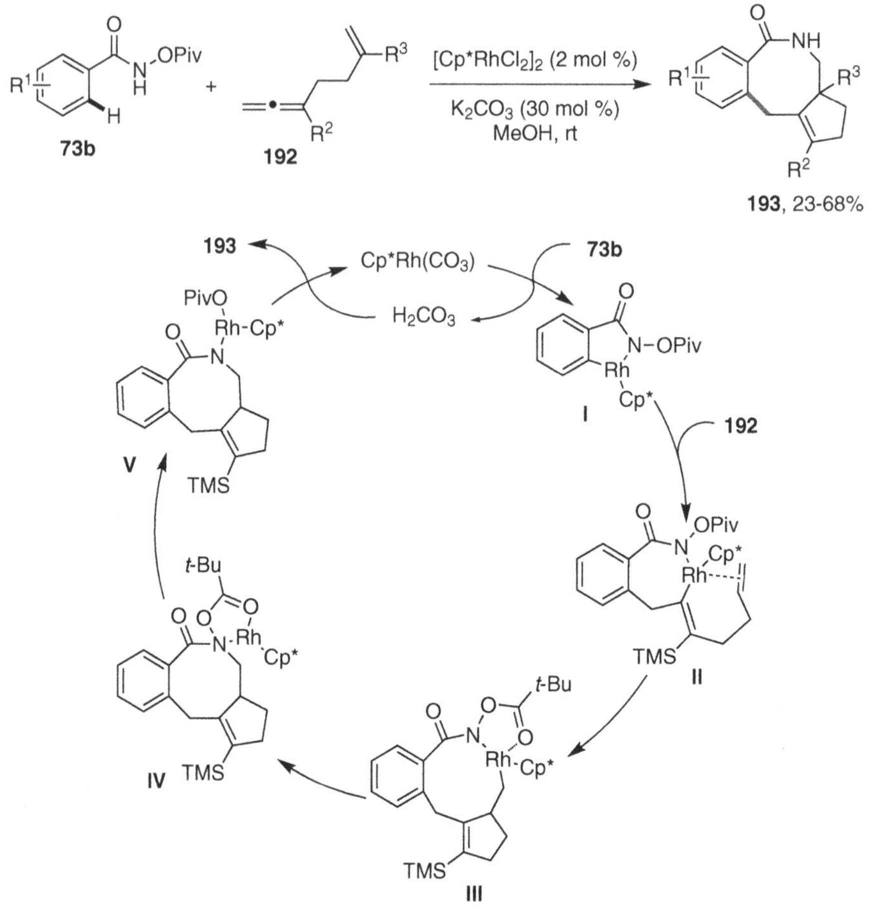

SCHEME 2.68

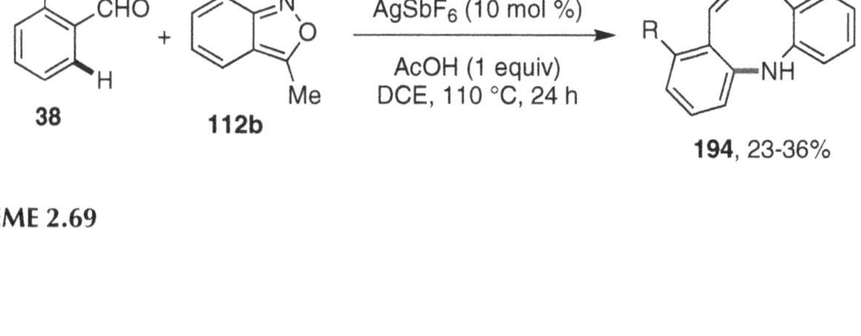

SCHEME 2.69

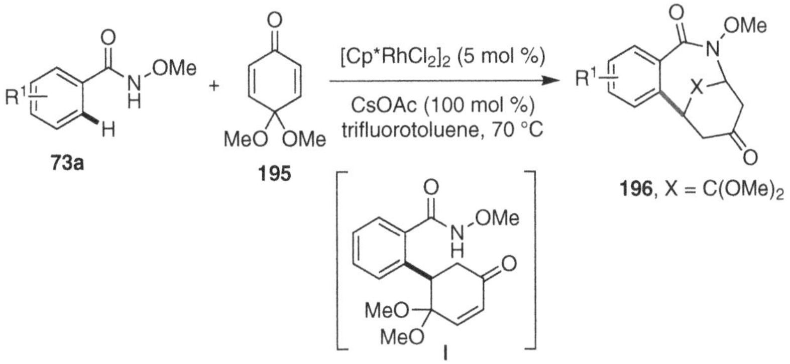

SCHEME 2.70

2.12 CONCLUSION

In summary, aromatic C(sp^2)–H dehydrogenative coupling processes are studied in depth for the preparation of six-membered *N*-heterocycles, such as quinolines, isoquinolines, acridines, phenanthiridines, cinnolines, and quinazolines. Dehydrogenative coupling was also developed for the synthesis of seven- and higher-membered benzoid *N*-heterocyclic compounds such as benzacepine and benzazocine. However, many different methods were developed for the construction of these types of heterocycles, due to the quantitative and one-step synthesis and a broad spectrum of substitution on the synthesized heterocycles, the aromatic C–H bond direct functionalization approach could be of interest in the synthesis of pharmaceutical, medicinal, and natural products. On the other hand, there is still room for further studies in the field of synthesis of seven-membered nitrogen-containing heterocyclic compounds via aromatic C–H dehydrogenative coupling reactions.

REFERENCES

Anukumar, Adapa, Masilamani Tamizmani, and Masilamani Jeganmohan. 2018. "Ruthenium (II)-catalyzed regioselective-controlled allenylation/cyclization of benzimides with propargyl alcohols." *The Journal of Organic Chemistry* no. 83(15):8567–8580.

Bai, Dachang, Teng Xu, Chaorui Ma, Xin Zheng, Bingxian Liu, Fang Xie, and Xingwei Li. 2018. "Rh (III)-catalyzed mild coupling of nitrones and azomethine imines with alkylidenecyclopropanes via C–H activation: Facile access to bridged cycles." *ACS Catalysis* no. 8(5):4194–4200.

Bai, Lu, Yan Wang, Yicong Ge, Jingjing Liu, and Xinjun Luan. 2017. "Diastereoselective synthesis of dibenzo [b, d] Azepines by Pd (II)-catalyzed [5+2] annulation of o-arylanilines with dienes." *Organic Letters* no. 19(7):1734–1737.

Bai, Peng, Xing-Fen Huang, Guo-Dong Xu, and Zhi-Zhen Huang. 2016. "Cascade C–H functionalization/amidation reaction for synthesis of azepinone derivatives." *Organic Letters* no. 18(13):3058–3061.

Bhowmik, Subhendu, Garima Pandey, and Sanjay Batra. 2013. "Substituent-guided switch between C-H activation and decarboxylative cross-coupling during Palladium/Copper-catalyzed cascade reactions of 2-aminobenzoates with 2-haloarylaldehydes." *Chemistry – A European Journal* no. 19(32):10487–10491.

Carral-Menoyo, Asier, Verónica Ortiz-de-Elguea, Mikel Martinez-Nunes, Nuria Sotomayor, and Esther Lete. 2017. "Palladium-catalyzed dehydrogenative coupling: An efficient synthetic strategy for the construction of the quinoline core." *Marine Drugs* no. 15(9):276.

Chen, Jinlei, Guoyong Song, Cheng-Ling Pan, and Xingwei Li. 2010. "Rh (III)-catalyzed oxidative coupling of N-Aryl-2-aminopyridine with alkynes and alkenes." *Organic Letters* no. 12(23):5426–5429.

Chu, Haoke, Peiran Xue, Jin-Tao Yu, and Jiang Cheng. 2016. "Rhodium-catalyzed annulation of primary benzylamine with α-diazo ketone toward isoquinoline." *The Journal of Organic Chemistry* no. 81(17):8009–8013.

Cui, Sunliang, Yan Zhang, Dahai Wang, and Qifan Wu. 2013. "Rh (III)-catalyzed C–H activation/[4+3] cycloaddition of benzamides and vinylcarbenoids: Facile synthesis of azepinones." *Chemical Science* no. 4(10):3912–3916.

Cui, Sunliang, Yan Zhang, and Qifan Wu. 2013. "Rh (III)-catalyzed C–H activation/cycloaddition of benzamides and methylenecyclopropanes: Divergence in ring formation." *Chemical Science* no. 4(9):3421–3426.

Eftekhari-Sis, Bagher, and Somayeh Mirdoraghi. 2016. "Graphene oxide-terpyridine conjugate: A highly selective colorimetric and sensitive fluorescence nano-chemosensor for Fe^{2+} in aqueous media." *Nanochemistry Research* no. 1(2):214–221.

Eftekhari-Sis, Bagher, Zahra Rezazadeh, Ali Akbari, and Mojtaba Amini. 2018. "8-Hydroxyquinoline functionalized graphene oxide: An efficient fluorescent nanosensor for Zn^{2+} in aqueous media." *Journal of Fluorescence* no. 28(5): 1173–1180.

Eftekhari-Sis, Bagher, Khadijeh Samadneshan, and Saleh Vahdati-Khajeh. 2018. "Design and synthesis of Nanosensor based on CdSe quantum dots functionalized with 8-hydroxyquinoline: A fluorescent sensor for detection of Al^{3+} in aqueous solution." *Journal of Fluorescence* no. 28(3):767–774.

Eftekhari-Sis, Bagher, Masoumeh Sarvari Karajabad, and Shiva Haqverdi. 2017. "Pyridylmethylaminoacetic acid functionalized Fe3O4 magnetic nanorods as an efficient catalyst for the synthesis of 2-aminochromene and 2-aminopyran derivatives." *Scientia Iranica* no. 24(6):3022–3031.

Eftekhari-Sis, Bagher, and Maryam Zirak. 2014. "Chemistry of α-oxoesters: A powerful tool for the synthesis of heterocycles." *Chemical Reviews* no. 115(1):151–264.

Eftekhari-Sis, Bagher, Maryam Zirak, and Ali Akbari. 2013. "Arylglyoxals in synthesis of heterocyclic compounds." *Chemical Reviews* no. 113(5):2958–3043.

Gandeepan, Parthasarathy, Thomas Müller, Daniel Zell, Gianpiero Cera, Svenja Warratz, and Lutz Ackermann. 2018. "3d transition metals for C–H activation." *Chemical Reviews* no. 119(4):2192–2452.

Gong, Shasha, Wanlin Xi, Zhenhua Ding, and Haiying Sun. 2017. "Synthesis of isoquinolines from benzimidates and alkynes via cobalt (III)-catalyzed C–H functionalization/cyclization." *The Journal of Organic Chemistry* no. 82(14):7643–7647.

Guimond, Nicolas, and Keith Fagnou. 2009. "Isoquinoline synthesis via rhodium-catalyzed oxidative cross-coupling/cyclization of aryl aldimines and alkynes." *Journal of the American Chemical Society* no. 131(34):12050–12051.

Han, Wenyong, Xiaojian Zhou, Siyi Yang, Guangyan Xiang, Baodong Cui, and Yongzheng Chen. 2015. "Palladium-catalyzed nucleophilic substitution/C–H activation/aromatization cascade reaction: One approach to construct 6-unsubstituted phenanthridines." *The Journal of Organic Chemistry* no. 80(22):11580–11587.

Hurst, Timothy E, and Richard JK Taylor. 2017. "A Cu-catalysed radical cross-dehydrogenative coupling approach to acridanes and related heterocycles." *European Journal of Organic Chemistry* no. 2017(1):203–207.

Inamoto, Kiyofumi, Tadataka Saito, Kou Hiroya, and Takayuki Doi. 2010. "Palladium-catalyzed intramolecular amidation of C (sp^2)–H bonds: Synthesis of 4-aryl-2-quinolinones." *The Journal of Organic Chemistry* no. 75(11):3900–3903.

Jayakumar, Jayachandran, Kanniyappan Parthasarathy, and Chien-Hong Cheng. 2012. "One-pot synthesis of isoquinolinium salts by rhodium-catalyzed C-H bond activation: Application to the total synthesis of oxychelerythrine." *Angewandte Chemie International Edition* no. 51(1):197–200.

Jiang, Yaojia, Gongtao Deng, Shuaishuai Zhang, and Teck-Peng Loh. 2018. "Directing group participated benzylic C (sp^3)–H/C (sp^2)–H cross-dehydrogenative coupling (CDC): Synthesis of azapolycycles." *Organic Letters* no. 20(3):652–655.

Kalsi, Deepti, Subhradeep Dutta, Nagaraju Barsu, Magnus Rueping, and Basker Sundararaju. 2018. "Room-temperature C–H bond functionalization by merging cobalt and photoredox catalysis." *ACS Catalysis* no. 8(9):8115–8120.

Kim, Byung Seok, Sun Young Lee, and So Won Youn. 2011. "Pd-catalyzed sequential C-C and C-N bond formations for the synthesis of N-heterocycles: Exploiting protecting group-directed C-H activation under modified reaction conditions." *Chemistry–An Asian Journal* no. 6(8):1952–1957.

Kim, Ju Hyun, Steffen Gressies, and Frank Glorius. 2016. "Cooperative Lewis acid/Cp* CoIII catalyzed C–H bond activation for the synthesis of isoquinolin-3-ones." *Angewandte Chemie International Edition* no. 55(18):5577–5581.

Kim, Saegun, Sang Hoon Han, Neeraj Kumar Mishra, Rina Chun, Young Hoon Jung, Hyung Sik Kim, Jung Su Park, and In Su Kim. 2018. "Dual role of anthranils as amination and transient directing group sources: Synthesis of 2-acyl acridines." *Organic Letters* no. 20(13):4010–4014.

Kim, Saegun, Amit Kundu, Rina Chun, Sang Hoon Han, Ashok Kumar Pandey, Sungin Yoo, Junghyun Park, Hyung Sik Kim, Jin-Mo Ku, and In Su Kim. 2018. "Direct synthesis of 2-Acyl acridines using aldimines and anthranils: Evaluation of cytotoxicity and anti-inflammatory activity." *Asian Journal of Organic Chemistry* no. 7(10):2069–2075.

Kondapalli, Vijayakumar, Xiaoqiang Yu, Yoshinori Yamamoto, and Ming Bao. 2017. "Synthesis of 5 H-Dibenzo [c, e] azepine-5, 7 (6 H)-diones from benzamides via palladium-catalyzed double C–H bond activation." *The Journal of Organic Chemistry* no. 82(4):2288–2293.

Kong, De-Shen, Yi-Fan Wang, Yi-Shuang Zhao, Qing-Hua Li, Yue-Xin Chen, Ping Tian, and Guo-Qiang Lin. 2018. "Bisannulation of benzamides and cyclohexadienone-tethered allenes triggered by Cp* Rh (III)-catalyzed C–H activation and relay ene reaction." *Organic Letters* no. 20(4):1154–1157.

Kong, Lingheng, Songjie Yu, Xukai Zhou, and Xingwei Li. 2016. "Redox-neutral couplings between amides and alkynes via cobalt (III)-catalyzed C–H activation." *Organic Letters* no. 18(3):588–591.

Kuai, Changsheng, Lianhui Wang, Bobin Li, Zhenhui Yang, and Xiuling Cui. 2017. "Cobalt-catalyzed selective synthesis of isoquinolines using picolinamide as a traceless directing group." *Organic Letters* no. 19(8):2102–2105.

Kumaran, Elumalai, and Weng Kee Leong. 2015. "[Cp* RhCl2] 2-catalyzed alkyne hydroamination to 1, 2-dihydroquinolines." *Organometallics* no. 34(9):1779–1782.

Kuppusamy, Ramajayam, Rajagopal Santhoshkumar, Ramadoss Boobalan, Hsin-Ru Wu, and Chien-Hong Cheng. 2018. "Synthesis of 1, 2-dihydroquinolines by Co (III)-catalyzed [3+3] annulation of anilides with benzylallenes." *ACS Catalysis* no. 8(3):1880–1883.

Laha, Joydev K, Neetu Dayal, Roli Jain, and Ketul Patel. 2014. "Palladium-catalyzed regiocontrolled domino synthesis of N-sulfonyl dihydrophenanthridines and dihydrodibenzo [c, e] azepines: Control over the formation of biaryl sultams in the intramolecular direct arylation." *The Journal of Organic Chemistry* no. 79(22):10899–10907.

Li, Bin, Jianfeng Ma, Nuancheng Wang, Huiliang Feng, Shansheng Xu, and Baiquan Wang. 2012. "Ruthenium-catalyzed oxidative C–H bond olefination of N-methoxybenzamides using an oxidizing directing group." *Organic Letters* no. 14(3):736–739.

Li, Jie, Mengyao Tang, Lei Zang, Xiaolei Zhang, Zhao Zhang, and Lutz Ackermann. 2016a. "Amidines for versatile cobalt (III)-catalyzed synthesis of isoquinolines through C–H functionalization with diazo compounds." *Organic Letters* no. 18(11): 2742–2745.

Li, Jie, Zhao Zhang, Mengyao Tang, Xiaolei Zhang, and Jian Jin. 2016b. "Selective synthesis of isoquinolines by rhodium (III)-catalyzed C–H/N–H functionalization with α-substituted ketones." *Organic Letters* no. 18(15):3898–3901.

Li, Xing Guang, Min Sun, Qiao Jin, Kai Liu, and Pei Nian Liu. 2016c. "Access to isoquinolines and isoquinolin-3-ols via Rh (III)-catalyzed coupling/cyclization cascade reaction of arylimidates and diazo compounds." *The Journal of Organic Chemistry* no. 81(9):3901–3910.

Li, Xinyao, Xinwei Li, and Ning Jiao. 2015. "Rh-Catalyzed construction of quinolin-2 (1 H)-ones via C–H bond activation of simple anilines with CO and alkynes." *Journal of the American Chemical Society* no. 137(29):9246–9249.

Lian, Yajing, Joshua R Hummel, Robert G Bergman, and Jonathan A Ellman. 2013. "Facile synthesis of unsymmetrical acridines and phenazines by a Rh (III)-catalyzed amination/cyclization/aromatization cascade." *Journal of the American Chemical Society* no. 135(34):12548–12551.

Liang, Dongdong, Ziwei Hu, Jiangling Peng, Jinbo Huang, and Qiang Zhu. 2013. "Synthesis of phenanthridinones via palladium-catalyzed C (sp 2)–H aminocarbonylation of unprotected o-arylanilines." *Chemical Communications* no. 49(2):173–175.

Liang, Zunjun, Jitan Zhang, Zhanxiang Liu, Kai Wang, and Yuhong Zhang. 2013. "Pd (II)-catalyzed C (sp^2)–H carbonylation of biaryl-2-amine: Synthesis of phenanthridinones." *Tetrahedron* no. 69(31):6519–6526.

Ling, Fei, Chaowei Zhang, Chongren Ai, Yaping Lv, and Weihui Zhong. 2018. "Metal oxidant free cobalt catalyzed C (sp^2)–H carbonylation of ortho-arylanilines: An approach towards free (NH)-phenanthridinones." *The Journal of Organic Chemistry* 83(10):5698–5706.

Liu, Kai, Shuang Chen, Xing Guang Li, and Pei Nian Liu. 2015. "Multicomponent Cascade synthesis of trifluoroethyl isoquinolines from alkynes and vinyl azides." *The Journal of Organic Chemistry* no. 81(1):265–270.

Lu, Qingquan, Suhelen Vásquez-Céspedes, Tobias Gensch, and Frank Glorius. 2016. "Control over organometallic intermediate enables Cp* Co (III) catalyzed switchable cyclization to quinolines and indoles." *ACS Catalysis* no. 6(4):2352–2356.

Luo, Shuang, Zhuang Xiong, Yongzhi Lu, and Qiang Zhu. 2018. "Enantioselective synthesis of planar chiral pyridoferrocenes via palladium-catalyzed imidoylative cyclization reactions." *Organic Letters* no. 20(7):1837–1840.

Ma, Bin, Yong Wang, Jiangling Peng, and Qiang Zhu. 2011. "Synthesis of quinazolin-4 (3 H)-ones via Pd (II)-catalyzed intramolecular C (sp^2)–H carboxamidation of N-arylamidines." *The Journal of Organic Chemistry* no. 76(15):6362–6366.

Malkov, Andrei V, Lenka Dufkova, Louis Farrugia, and Pavel Kočovský. 2003. "Quinox, a quinoline-Type N-oxide, as organocatalyst in the asymmetric allylation of aromatic aldehydes with allyltrichlorosilanes: The role of arene–arene interactions." *Angewandte Chemie International Edition* no. 42(31):3674–3677.

Mandal, Anup, Jayaraman Selvakumar, Suman Dana, Upasana Mukherjee, and Mahiuddin Baidya. 2018. "A cross-dehydrogenative annulation strategy towards synthesis of polyfluorinated phenanthridinones with copper." *Chemistry – A European Journal* no. 24(14):3448–3454.

Manikandan, Rajendran, and Masilamani Jeganmohan. 2014. "Ruthenium-catalyzed cyclization of anilides with substituted propiolates or acrylates: An efficient route to 2-quinolinones." *Organic Letters* no. 16(13):3568–3571.

Matsuda, Takanori, Yuki Tomaru, and Yoshiya Matsuda. 2013. "Synthesis of phthalazinones via palladium (ii)-catalysed intramolecular oxidative C–H/C–H cross-coupling of N′-methylenebenzohydrazides." Organic & Biomolecular Chemistry no. 11(13):2084–2087.

Mei, Ruhuai, Nicolas Sauermann, João CA Oliveira, and Lutz Ackermann. 2018. "Electroremovable traceless hydrazides for cobalt-catalyzed electro-oxidative C–H/N–H activation with internal alkynes." *Journal of the American Chemical Society.*

Meyer, Tjark H, João CA Oliveira, Samaresh Chandra Sau, Nate WJ Ang, and Lutz Ackermann. 2018. "Electrooxidative allene annulations by mild cobalt-catalyzed C–H activation." *ACS Catalysis* no. 8(10):9140–9147.

Miura, Masahiro, Takatoshi Tsuda, Tetsuya Satoh, Sommai Pivsa-Art, and Masakatsu Nomura. 1998. "Oxidative cross-coupling of N-(2'-phenylphenyl) benzene-sulfonamides or benzoic and naphthoic acids with alkenes using a Palladium–Copper catalyst system under air." *The Journal of Organic Chemistry* no. 63(15):5211–5215.

Muralirajan, Krishnamoorthy, and Chien-Hong Cheng. 2013. "Rhodium (III)-catalyzed synthesis of cinnolinium salts from azobenzenes and alkynes: Application to the synthesis of indoles and cinnolines." *Chemistry – A European Journal* no. 19(20):6198–6202.

Nageswar Rao, D, Sk Rasheed, and Parthasarathi Das. 2016. "Palladium/silver synergistic catalysis in direct aerobic carbonylation of C (sp^2)–H bonds using DMF as a carbon source: Synthesis of pyrido-fused quinazolinones and phenanthridinones." *Organic Letters* no. 18(13):3142–3145.

Pandey, Ashok Kumar, Sang Hoon Han, Neeraj Kumar Mishra, Dahye Kang, Suk Hun Lee, Rina Chun, Sungwoo Hong, Jung Su Park, and In Su Kim. 2017. "Synthesis of 2-benzazepines from benzylamines and MBH adducts under Rhodium (III) catalysis via C (sp^2)–H functionalization." *ACS Catalysis* no. 8(1):742–746.

Pearson, Ryan, Shuyu Zhang, Gang He, Nicola Edwards, and Gong Chen. 2013. "Synthesis of phenanthridines via palladium-catalyzed picolinamide-directed sequential C–H functionalization." *Beilstein Journal of Organic Chemistry* no. 9:891.

Peng, Xianglong, Weiguo Wang, Chao Jiang, Di Sun, Zhenghu Xu, and Chen-Ho Tung. 2014. "Strain-promoted oxidative annulation of arynes and cyclooctynes with benzamides: Palladium-catalyzed C–H/N–H activation for the synthesis of N-heterocycles." *Organic Letters* no. 16(20):5354–5357.

Phatake, Ravindra S, Pitambar Patel, and Chepuri V Ramana. 2015. "Ir (III)-catalyzed synthesis of isoquinoline N-oxides from aryloxime and α-diazocarbonyl compounds." *Organic Letters* no. 18 (2):292–295.

Phatake, Ravindra S, Pitambar Patel, and Chepuri V Ramana. 2016. "Ir (III)-catalyzed carbenoid functionalization of benzamides: Synthesis of N-methoxyisoquinolinediones and N-methoxyisoquinolinones." *Organic Letters* no. 18 (12):2828–2831.

Prakash, Sekar, Krishnamoorthy Muralirajan, and Chien-Hong Cheng. 2016. "Cobalt-catalyzed oxidative annulation of nitrogen-containing arenes with alkynes: An atom-economical route to heterocyclic quaternary ammonium salts." *Angewandte Chemie International Edition* no. 55(5):1844–1848.

Rajeshkumar, Venkatachalam, Tai-Hua Lee, and Shih-Ching Chuang. 2013. "Palladium-catalyzed oxidative insertion of carbon monoxide to N-sulfonyl-2-aminobiaryls through C–H bond activation: Access to bioactive phenanthridinone derivatives in one pot." *Organic Letters* no. 15(7):1468–1471.

Raju, Selvam, Pratheepkumar Annamalai, Pei-Ling Chen, Yi-Hung Liu, and Shih-Ching Chuang. 2017. "Palladium-catalyzed C–H bond activation by using iminoquinone as a directing group and an internal oxidant or a co-oxidant: Production of dihydrophenanthridines, phenanthridines, and carbazoles." *Organic Letters* no. 19(15):4134–4137.

Shaikh, Tanveer Mahamadali, and Fung-E Hong. 2016. "Recent developments in the preparation of N-heterocycles using Pd-catalysed C–H activation." *Journal of Organometallic Chemistry* no. 801:139–156.

Shi, Zhuangzhi, Christoph Grohmann, and Frank Glorius. 2013. "Mild rhodium (III)-catalyzed cyclization of amides with α, β-unsaturated aldehydes and ketones to azepinones: Application to the synthesis of the homoprotoberberine framework." *Angewandte Chemie International Edition* no. 52(20):5393–5397.

Song, Chao, Chen Yang, Feifei Zhang, Jinhu Wang, and Jin Zhu. 2016. "Access to the cinnoline scaffold via rhodium-catalyzed intermolecular cyclization under mild conditions." *Organic Letters* no. 18(18):4510–4513.

Song, Guoyong, Dan Chen, Cheng-Ling Pan, Robert H Crabtree, and Xingwei Li. 2010. "Rh-catalyzed oxidative coupling between primary and secondary benzamides and alkynes: Synthesis of polycyclic amides." *The Journal of Organic Chemistry* no. 75(21):7487–7490.

Sun, Peng, Youzhi Wu, Yue Huang, Xiaoming Wu, Jinyi Xu, Hequan Yao, and Aijun Lin. 2016. "Rh (III)-catalyzed redox-neutral annulation of azo and diazo compounds: One-step access to cinnolines." *Organic Chemistry Frontiers* no. 3(1):91–95.

Taneda, Hiroshi, Kiyofumi Inamoto, and Yoshinori Kondo. 2016. "Palladium-catalyzed highly chemoselective intramolecular C–H aminocarbonylation of phenethylamines to six-membered benzolactams." *Organic Letters* no. 18(11):2712–2715.

Tang, Shi, Sheng-Wen Yang, Hongwei Sun, Yali Zhou, Juan Li, and Qiang Zhu. 2018. "Pd-catalyzed divergent C (sp^2)–H activation/cycloimidoylation of 2-isocyano-2, 3-diarylpropanoates." *Organic Letters* no. 20(7):1832–1836.

Tian, Miaomiao, Guangfan Zheng, Xuesen Fan, and Xingwei Li. 2018. "Rhodium (III)-catalyzed redox-neutral synthesis of isoquinolinium salts via CH activation of imines." *The Journal of Organic Chemistry* no. 83(12):6477–6488.

Wang, Hui, Mélanie M Lorion, and Lutz Ackermann. 2017. "Domino C–H/N–H allylations of imidates by cobalt catalysis." *ACS Catalysis* no. 7(5):3430–3433.

Wang, Jie, Lili Wang, Shan Guo, Shanke Zha, and Jin Zhu. 2017. "Synthesis of 2, 3-benzodiazepines via Rh (III)-catalyzed C–H functionalization of N-Boc hydrazones with diazoketoesters." *Organic Letters* no. 19(13):3640–3643.

Wang, Jie, Shanke Zha, Kehao Chen, Feifei Zhang, Chao Song, and Jin Zhu. 2016a. "Quinazoline synthesis via Rh (III)-catalyzed intermolecular C–H functionalization of benzimidates with dioxazolones." *Organic Letters* no. 18(9):2062–2065.

Wang, Jie, Shanke Zha, Kehao Chen, Feifei Zhang, and Jin Zhu. 2016b. "Synthesis of isoquinolines via Rh-catalyzed C–H activation/C–N cyclization with diazodiesters or diazoketoesters as a C 2 source." *Organic & Biomolecular Chemistry* no. 14(21): 4848–4852.

Wang, Kongchao, Xia Chen, Ming Yuan, Meng Yao, Hucheng Zhu, Yongbo Xue, Zengwei Luo, and Yonghui Zhang. 2018. "Silver-mediated cyanomethylation of cinnamamides by direct C (sp^3)–H functionalization of acetonitrile." *The Journal of Organic Chemistry* no. 83(3):1525–1531.

Wang, Qiang, Fen Wang, Xifa Yang, Xukai Zhou, and Xingwei Li. 2016c. "Rh (III)-and Zn (II)-catalyzed synthesis of quinazoline N-oxides via C–H amidation–cyclization of oximes." *Organic Letters* no. 18(23):6144–6147.

Wang, Shao-Bo, Jun Zheng, and Shu-Li You. 2016. "Synthesis of ferrocene-based pyridinones through Rh (III)-catalyzed direct C–H functionalization reaction." *Organometallics* no. 35(10):1420–1425.

Wang, Xiaoming, Andreas Lerchen, and Frank Glorius. 2016. "A comparative investigation: Group 9 Cp* M (III)-catalyzed formal [4+2] cycloaddition as an atom-economic approach to quinazolines." *Organic Letters* no. 18(9):2090–2093.

Wang, Xiaoyang, and Ning Jiao. 2016. "Rh-and Cu-cocatalyzed aerobic oxidative approach to quinazolines via [4+2] C–H annulation with alkyl azides." *Organic Letters* no. 18(9):2150–2153.

Wang, Yanwei, Bin Li, and Baiquan Wang. 2018. "Synthesis of cinnolines and cinnolinium salt derivatives by Rh (III)-catalyzed cascade oxidative coupling/cyclization reactions." *The Journal of Organic Chemistry* no. 83(18):10845–10854.

Wang, Yong, Honggen Wang, Jiangling Peng, and Qiang Zhu. 2011. "Palladium-catalyzed intramolecular C (sp^2)–H amidination by isonitrile insertion provides direct access to 4-aminoquinazolines from N-arylamidines." *Organic Letters* no. 13(17):4604–4607.

Wei, Xiaohong, Miao Zhao, Zhengyin Du, and Xingwei Li. 2011. "Synthesis of 1-aminoisoquinolines via Rh (III)-catalyzed oxidative coupling." *Organic Letters* no. 13(17):4636–4639.

Weinstein, Adam B, and Jonathan A Ellman. 2016. "Convergent synthesis of diverse nitrogen heterocycles via Rh (III)-catalyzed C–H conjugate addition/cyclization reactions." *Organic Letters* no. 18(13):3294–3297.

Wen, Jiangwei, Shan Tang, Fan Zhang, Renyi Shi, and Aiwen Lei. 2016. "Palladium/copper Co-catalyzed oxidative C–H/C–H carbonylation of diphenylamines: A way to access acridones." *Organic Letters* no. 19(1):94–97.

Wu, Jiwei, Yuchen Zhou, Ting Wu, Yi Zhou, Chien-Wei Chiang, and Aiwen Lei. 2017. "From ketones, amines, and carbon monoxide to 4-quinolones: Palladium-catalyzed oxidative carbonylation." *Organic Letters* no. 19(23):6432–6435.

Wu, Shangze, Rong Zeng, Chunling Fu, Yihua Yu, Xue Zhang, and Shengming Ma. 2015. "Rhodium-catalyzed C–H functionalization-based approach to eight-membered lactams." *Chemical Science* no. 6(4):2275–2285.

Wu, Xiao-Lin, and Lin Dong. 2018. "Synthesis of α-ketone-isoquinoline derivatives via tandem ruthenium (II)-catalyzed C–H activation and annulation." *Organic Letters* no. 20(22):6990–6993.

Wu, Youzhi, Peng Sun, Kaifan Zhang, Tie Yang, Hequan Yao, and Aijun Lin. 2016. "Rh (III)-catalyzed redox-neutral annulation of primary benzamides with diazo compounds: Approach to isoquinolinones." *The Journal of Organic Chemistry* no. 81(5): 2166–2173.

Xie, Caixia, Zhen Dai, Yadi Niu, and Chen Ma. 2018. "Cascade one-pot method to synthesize isoquinolin-1 (2 H)-ones with α-bromo ketones and benzamides via Pd-catalyzed C–H activation." *The Journal of Organic Chemistry* no. 83(4):2317–2323.

Xie, Wucheng, Bin Li, Shansheng Xu, Haibin Song, and Baiquan Wang. 2014. "Palladium-catalyzed direct dehydrogenative annulation of ferrocenecarboxamides with alkynes in air." *Organometallics* no. 33(9):2138–2141.

Xiong, Peng, He-Huan Xu, Jinshuai Song, and Hai-Chao Xu. 2018. "Electrochemical difluoromethylarylation of alkynes." *Journal of the American Chemical Society* no. 140(7):2460–2464.

Xu, Fen, Wei-Fen Kang, Yang Wang, Chun-Sen Liu, Jia-Yue Tian, Rui-Rui Zhao, and Miao Du. 2018a. "Rhodium (III)-catalyzed cascade [5+1] annulation/5-exo-cyclization initiated by C–H activation: 1, 6-Diynes as one-carbon reaction partners.." *Organic Letters* 20(11):3245–3249.

Xu, Guo-Dong, and Zhi-Zhen Huang. 2017. "A cascade dehydrogenative cross-coupling/annulation reaction of benzamides with β-keto esters for the synthesis of isoquinolinone derivatives." *Organic Letters* no. 19(23):6265–6267.

Xu, Zhong-Qi, Chao Wang, Lin Li, Lili Duan, and Yue-Ming Li. 2018b. "Construction of 3, 4-dihydroisoquinolinones and indanones via DTBP-promoted oxidative coupling of N-allylbenzamides with aromatic aldehydes." *The Journal of Organic Chemistry* no. 83(17):9718–9728.

Yang, Fan, Jiaojiao Yu, Yun Liu, and Jin Zhu. 2017. "Cobalt (III)-catalyzed oxadiazole-directed C–H activation for the synthesis of 1-aminoisoquinolines." *Organic Letters* no. 19(11):2885–2888.

Yang, Wei, Jinhuan Dong, Jingyi Wang, and Xianxiu Xu. 2017. "Rh (III)-catalyzed diastereoselective annulation of amides with quinone monoacetals: Access to bridged nine-membered heterocycles via C–H activation." *Organic Letters* no. 19(3):616–619.

Yang, Wei, Jingyi Wang, Zhonglin Wei, Qian Zhang, and Xianxiu Xu. 2016. "Kinetic control of Rh (III)-catalyzed annulation of C–H bonds with quinones: Chemoselective synthesis of hydrophenanthridinones and phenanthridinones." *The Journal of Organic Chemistry* no. 81(4):1675–1680.

Yang, Xifa, Song Liu, Songjie Yu, Lingheng Kong, Yu Lan, and Xingwei Li. 2018. "Redox-neutral access to isoquinolinones via rhodium (III)-catalyzed annulations of O-pivaloyl oximes with ketenes." *Organic Letters* no. 20(9):2698–2701.

Yang, Yuan, Ming-Bo Zhou, Xuan-Hui Ouyang, Rui Pi, Ren-Jie Song, and Jin-Heng Li. 2015. "Rhodium (III)-catalyzed [3+2]/[5+2] annulation of 4-aryl 1, 2, 3-triazoles with internal alkynes through dual C (sp2)-H functionalization." *Angewandte Chemie International Edition* no. 54(22):6595–6599.

Yu, Xiaolong, Kehao Chen, Fan Yang, Shanke Zha, and Jin Zhu. 2016. "Oxadiazolone-enabled synthesis of primary azaaromatic amines." *Organic Letters* no. 18(20):5412–5415.

Zhai, Sheng-Xian, Shuxian Qiu, Xiaoming Chen, Cheng Tao, Yun Li, Bin Cheng, Huifei Wang, and Hongbin Zhai. 2018. "Trifunctionalization of allenes via cobalt-catalyzed MHP-assisted CH bond functionalization and molecular oxygen activation." *ACS Catalysis* 8(7):6645–6649.

Zhang, Honglin, Zhangxi Gu, Zhenyi Li, Changduo Pan, Weipeng Li, Hongwen Hu, and Chengjian Zhu. 2016. "Silver-catalyzed cascade radical cyclization: A direct approach to 3, 4-disubstituted dihydroquinolin-2 (1 H)-ones through activation of the P–H bond and functionalization of the C (sp2)–H bond." *The Journal of Organic Chemistry* no. 81(5):2122–2127.

Zhang, Shang-Shi, Jia-Qiang Wu, Xuge Liu, and Honggen Wang. 2014. "Tandem catalysis: Rh (III)-catalyzed C–H allylation/Pd (II)-catalyzed N-allylation toward the synthesis of vinyl-substituted N-heterocycles." *ACS Catalysis* no. 5(1):210–214.

Zhang, Xuan, Weili Si, Ming Bao, Naoki Asao, Yoshinori Yamamoto, and Tienan Jin. 2014. "Rh (III)-catalyzed regioselective functionalization of C–H bonds of naphthylcarbamates for oxidative annulation with alkynes." *Organic Letters* no. 16(18):4830–4833.

Zhao, Dongbing, Qian Wu, Xiaolei Huang, Feijie Song, Taiyong Lv, and Jingsong You. 2013. "A general method to diverse cinnolines and cinnolinium salts." *Chemistry–A European Journal* no. 19(20):6239–6244.

Zhou, Shi-Liu, Li-Na Guo, Shun Wang, and Xin-Hua Duan. 2014. "Copper-catalyzed tandem oxidative cyclization of cinnamamides with benzyl hydrocarbons through cross-dehydrogenative coupling." *Chemical Communications* no. 50(27):3589–3591.

Zirak, Maryam, and Bagher Eftekhari-Sis. 2015. "Kojic acid in organic synthesis." *Turkish Journal of Chemistry* no. 39(3):439–496.

Zuo, Youpeng, Xinwei He, Yi Ning, Yuhao Wu, and Yongjia Shang. 2018. "Selective synthesis of aminoisoquinolines via Rh (III)-catalyzed C–H/N–H bond functionalization of N-aryl amidines with cyclic 2-Diazo-1, 3-diketones." *The Journal of Organic Chemistry* no. 83(21):13463–13472.

3 *N*-Bridged Heterocycles

3.1 INTRODUCTION

Heterocycles with nitrogen-bridged frameworks are broadly present in a number of natural products and pharmaceuticals, possessing biological activities such as antitumor, antimalarial, antitubercular, antioxidant, antimicrobial, anticancer, anti-inflammatory, antiviral, and cytotoxicity through bacterial cell division and DNA alkylation activity, and have attracted the interest of many in the research fields of medicines and pharmaceuticals (Eftekhari-Sis, Zirak, and Akbari 2013, Eftekhari-Sis and Zirak 2014, Sutariya et al. 2015, Sharma and Kumar 2014). However, there are many reports on the construction of nitrogen-bridged heterocyclic compounds by multi-step synthesis in the literature; the development of new approaches to synthesis of these types of *N*-heterocycles with tolerance of different functional groups are of interest to synthetic and medicinal scientists.

Cross-dehydrogenative-coupling (CDC) (Gandeepan et al. 2018, Shaikh and Hong 2016) processes are widely used in the synthesis of various types of heterocyclic compounds. In this chapter, intramolecular and intermolecular CDC of aromatic $C(sp^2)$–H bonds are discussed for the construction of five-, six-, and seven-membered *N*-bridged heterocycles, including pyrrolo-indoles, indolo-indoles, isoindolo-indoles and indolo-carbazoles, imidazo-indoles, imidazo-isoindoles, isoindolo-isoquinolines, indolizines, imidazo-pyridines, and indazolo-phthalazines, etc.

3.2 FIVE-MEMBERED *N*-BRIDGED HETEROCYCLES

3.2.1 Pyrrolo-Indoles

In an interesting work, Piou, Neuville, and Zhu (2013) developed Pd(II)-catalyzed intramolecular aminopalladation/direct C–H arylation of substituted 6-(phenylamino)hex-2-ynoates **1** under aerobic conditions, to construct pyrrolo[1,2-*a*]indoles **2**. Reactions were carried out under oxygen atmosphere using $Pd(OAc)_2$ (10 mol%) in a *N,N*-dimethylacetamide (DMA)/PivOH mixture at 120°C, to give 2,3-dihydro-1*H*-pyrrolo[1,2-*a*]indole-9-carboxylates **2** in 18–83% yields. No reaction occurred when $PdCl_2$ was used as the catalyst. Electron-withdrawing and donating substituents at the *para* position were tolerated well under reaction conditions, resulting in corresponding pyrrolo[1,2-*a*]indoles. In the case of *ortho*-substituted aniline substrate, a low yield (18%) of the product was obtained. Unsymmetrical substituted *meta*-methylaniline and *β*-naphthylaniline derivatives led to two regioisomers. The presence of the electron-withdrawing carboxylate group at the terminal position of alkyne is mandatory, as the *N*-(5-phenylpent-4-yn-1-yl)aniline failed to produce the expected pyrrolo[1,2-*a*]indole derivative. The reaction occurred by coordination of substrate **1** with Pd(II) with subsequent intramolecular *syn*-aminopalladation of the alkyne, followed by activation of the neighboring aromatic C–H bond, and then reductive elimination. In 2011, Yip and Yang (2011) described the Pd(II)-catalyzed synthesis of pyrrolo[1,2-*a*]indol-3(2*H*)-one derivatives **4**, using O_2 as the sole oxidant. Reactions occurred via formation of new C–N and C–C bonds across an alkene through an intramolecular dehydrogenative coupling of *N*-arylpent-4-enamides **3**. Reactions were performed using $Pd(OAc)_2$ (10 mol%), Na_2CO_3 (1.5 equiv), and ethyl nicotinate (40 mol%) in dioxane at 70°C, under atmosphere of O_2, affording 9,9a-dihydro-1*H*-pyrrolo[1,2-*a*]indol-3(2*H*)-ones **4** in 57–96% yields. Different ligands including PPh_3, pyridine, quinoline, isoquinoline, acridine, 3-acetylpyridine, 2-cyanopyridine, 3-cyanopyridine, 4-cyanopyridine, and ethyl nicotinate were investigated, in which ethyl nicotinate led to a high yield of the product. Reactions with unsymmetrical *meta*-substituted anilides occurred regioselectively at the *para*-position of the substituent. In the proposed reaction mechanism, anilide **3** gave amidopalladation intermediate **I**, which underwent *syn*-amidopalladation to generate σ-alkylpalladium species **II**. The resulting Pd(II) center on intermediate **II** activated the *ortho*-C–H bond, leading to palladacycle **III**, which underwent reductive elimination to afford pyrrolo[1,2-*a*]indol-3(2*H*)-one **4** along with generation of a Pd(0) species. The catalytic cycle was completed by the regeneration of Pd(II) through oxidation of Pd(0) by molecular oxygen (Scheme 3.1).

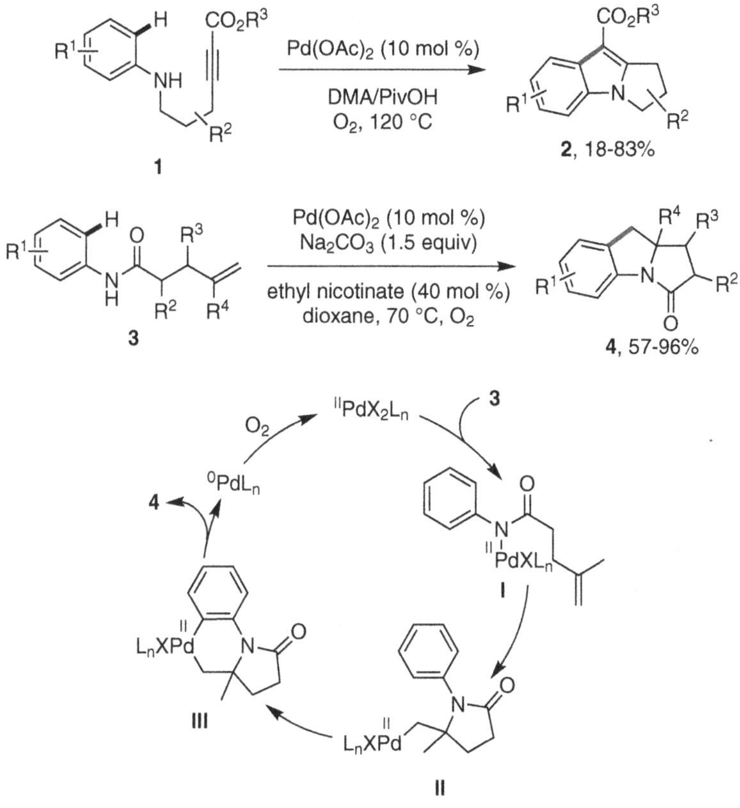

SCHEME 3.1

Also, the activation of C–H bonds of five-membered heteroaromatics was established in order to construct *N*-bridged heterocyclic compounds. Accordingly, 3*H*-pyrrolo[1,2-*a*]indol-3-ones **8** were accessed by Ru(II)-catalyzed redox-neutral [3+2] annulation of *N*-ethoxycarbamoyl indoles **5** with internal alkynes **6** via C–H bond activation. Reactions were performed using 1.3 equiv of an alkyne **6** in the presence of $[RuCl_2(p\text{-cymene})]_2$ (5 mol%) as the catalyst and CsOAc (1 equiv) as a base in dichloromethane (DCM) at 60°C under Ar atmosphere, and 3*H*-pyrrolo[1,2-*a*]indol-3-one derivatives **8** were obtained in 27–97% yields. The reaction failed to give the expected product with the terminal alkyne phenylacetylene. Reaction with unsymmetrical alkyne 1-arylbut-2-yn-1-ol **7** and alkyl aryl alkynes led regioselectively to 2-(aryl(hydroxy)methyl)-3-methyl-3*H*-pyrrolo[1,2-*a*]indol-3-ones **9** and 3-alkyl-2-aryl-3*H*-pyrrolo[1,2-*a*]indol-3-ones, respectively. The plausible reaction mechanism involves the generation of five-membered ruthenacycle **I** by coordination of *N*-carboxamides of indoles **5** with an active Ru(II) catalyst, generated in situ through ligand exchange with CsOAc, with subsequent C2-selective C–H bond activation with the aid of a base. By regioselective coordination and migratory insertion of alkyne **6** into the Ru–C bond of **I**, intermediate **II** was formed, which underwent intramolecular nucleophilic addition to give intermediate **III**, affording the desired product by release of ethoxyamine, with the regeneration of the Ru(II) catalyst (Scheme 3.2) (Xie et al. 2017). A similar approach was developed by Ikemoto et al. (2014) using a $[Cp^*Co(Ph)](PF_6)_2$ catalyst in 1,2-dichloroethane (DCE) at 130°C.

In 2017, Xu et al. (2017) reported the synthesis of 9*H*-pyrrolo[1,2-*a*]indoles **12** by Pd(II)-catalyzed [4+1] annulation reaction via formation of two different C–C bonds. Heating a solution of 1-(2-bromophenyl)-1*H*-pyrrole derivatives **10** and (trimethylsilyl)diazomethane **11** (1.2 equiv) in dioxane in the presence of $Pd(PPh_3)_4$ (5 mol%), K_2CO_3 (1 equiv), and KOAc (1 equiv) at 100°C under an atmosphere of N_2 for 24 h furnished the expected 9*H*-pyrrolo[1,2-*a*]indoles **12** in 65–90% yields. When 1-(2-bromophenyl)-1*H*-indole **10** was used, the corresponding 10*H*-indolo[1,2-*a*]indole **12** was obtained in 61% yield. First, an oxidative addition to the C–Br bond generated Pd(II) species **I**, which was transformed into palladacycle intermediate **IV** via two possible pathways. In path a, **I** reacted with diazo substrate **11** to form Pd(II) carbene species **V**, followed by migratory insertion to give intermediate **VI**, and then C–H bond activation. Alternatively, in path b, C–H bond activation occurred to generate palladacycle intermediate **II**, followed by carbene-metal formation by N_2 release and then migratory insertion. Finally, reductive elimination and desilylation afforded the desired product **12** (Scheme 3.3).

3.2.2 Indolo-Indole, Isoindolo-Indole and Indolo-Carbazole

Synthesis of indolo-indolone derivatives was accomplished by Ag-catalyzed intramolecular C-2 selective acylation via direct dehydrogenative coupling between aldehyde C–H and the $C(sp^2)$–H bonds of 2-(1*H*-indol-1-yl)benzaldehyde derivatives **13**. Various Ag salts in the presence of different oxidants were screened, from which AgOMs (4–7.5 mol%) and oxone (2–3 equiv) were selected as the best catalyst and oxidant, respectively. Therefore, reactions were carried out in dioxane or dioxane/DCE at 100°C under N_2 atmosphere, and 10*H*-indolo[1,2-*a*]indol-10-one derivatives

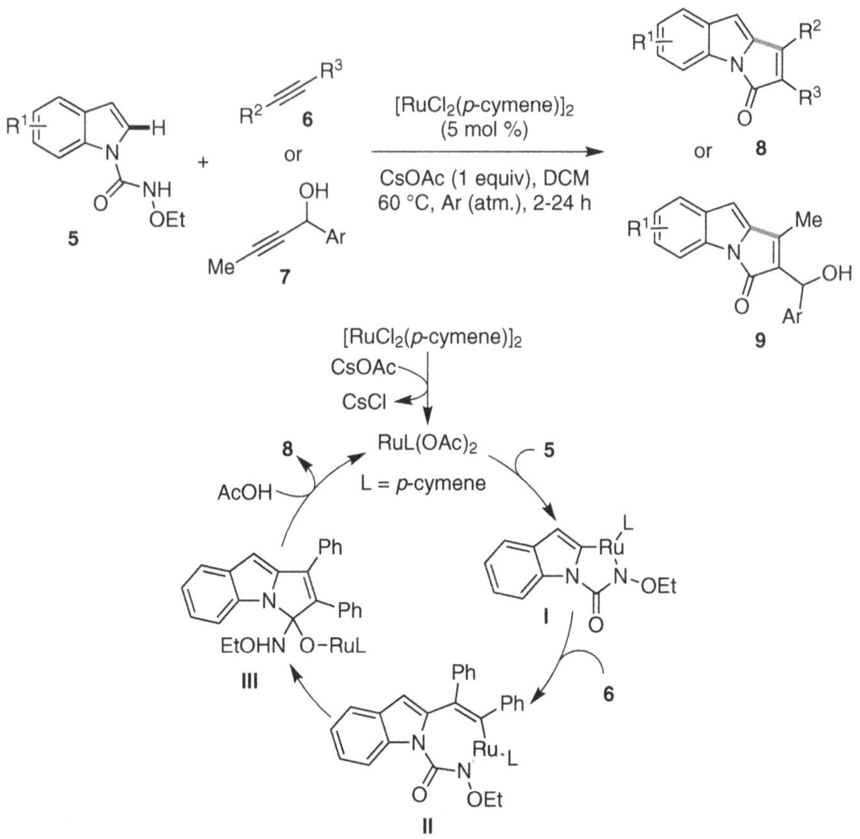

SCHEME 3.2

14 were obtained in 23–90% yields. The reaction of 3-methylindole derivatives led to poor yield of the product. The proposed reaction mechanism includes the generation of an Ag(II)OH species by disproportionates of a peroxymonosulfate anion into a hydroxide anion and a sulfate radical anion in the presence of an Ag(I) salt. Oxidation or abstraction of the aldehyde C–H bond in **13** with the Ag(II) species or the sulfate radical anion generated an acyl radical **I**, which underwent intramolecular radical cyclization to the indolyl C=C bond to give radical **II**. By direct oxidation with either the sulfate radical anion or the Ag(II) species and deprotonation of **II**, the final product **14** was obtained (Scheme 3.4) (Wang et al. 2016).

Intramolecular oxidative arylation of 7-azaindoles **15** bearing an *N*-benzyl substituent was described by Laha, Bhimpuria, and Hunjan (2017) Reactions were performed using $Pd(OAc)_2$ (10 mol%) and AgOAc (3 equiv) in PivOH at 130°C for 12 h, delivering aza-derivatives of 6*H*-isoindolo[2,1-*a*]indole 10*H*-pyrido[3',2':4,5]pyrrolo[2,1-*a*] isoindoles **16**, in 19–85% yields. A few examples of 5*H*-pyrrolo[2,1-*a*]isoindole derivatives were obtained in 67–79% yields by intramolecular dehydrogenative coupling reaction of *N*-benzylpyrroles under the same reaction conditions. The presence of electron-withdrawing groups on the pyrrole moiety is mandatory (Scheme 3.5).

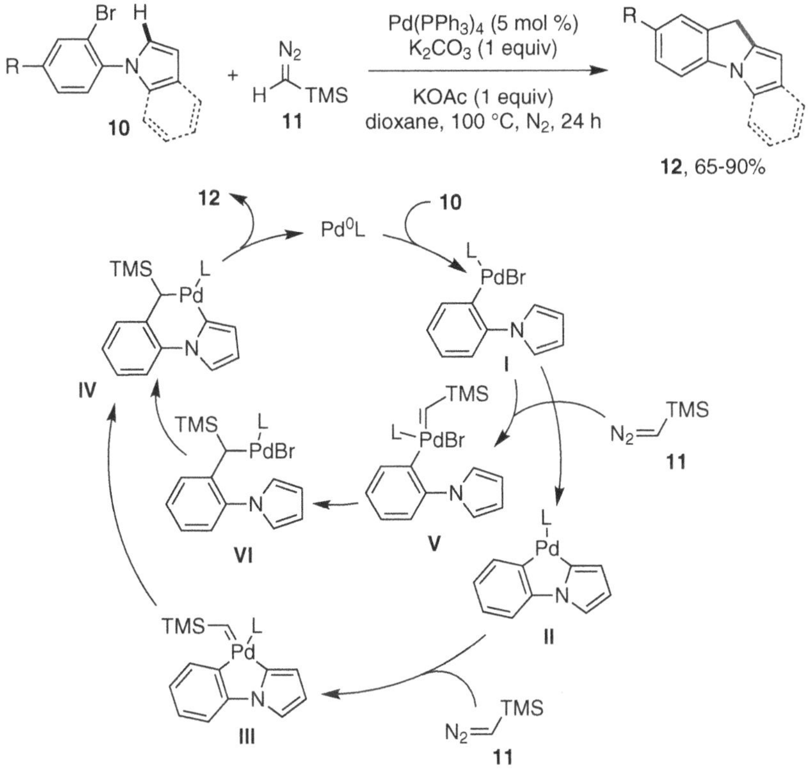

SCHEME 3.3

Synthesis of 6*H*-isoindolo[2,1-*a*]indol-6-ones **18** was achieved through Rh(III)-catalyzed NH-indole-directed C–H carbonylation of 2-arylindoles **17** with carbon monoxide. A variety of Rh(III) complexes, oxidants, and bases were investigated, from which $[Cp^*RhCl_2]_2$ (2.5 mol%), AgOAc (3 equiv), and K_2CO_3 (2 equiv) in xylene under CO atmosphere (1 atm) furnished 6*H*-isoindolo[2,1-*a*]indol-6-one derivatives **18** in 10–92% yields. No reactions occurred in the case of 2-pyridylindoles and 2-phenylbenzimidazole. 2-Phenylindoles bearing strong electron-withdrawing groups NO_2 and CN at the C-5 position afforded corresponding isoindolo-indolones in lower yields. The reaction was initiated by coordination and C–H bond activation to generate five-membered rhodacycle **I**, which underwent migratory insertion of CO into the Rh–N or Rh–C bond to give six-membered rhodacycle **IIa** or **IIb**, respectively. By reductive elimination the desired product **18** was obtained with generation of an Rh(I) species, which was oxidized with AgOAc to the catalytically active Rh(III) species (Huang et al. 2016). Also, two examples of 6*H*-isoindolo[2,1-*a*]indol-6-one **20** were prepared via intramolecular dehydrogenative coupling of *N*-benzoylindoles **19** by subjection to $Pd(OAc)_2$ (20 mol%) and CuOAc (1 equiv) in AcOH under atmosphere of O_2, in 29–82% yields (Scheme 3.6) (Dwight et al. 2007).

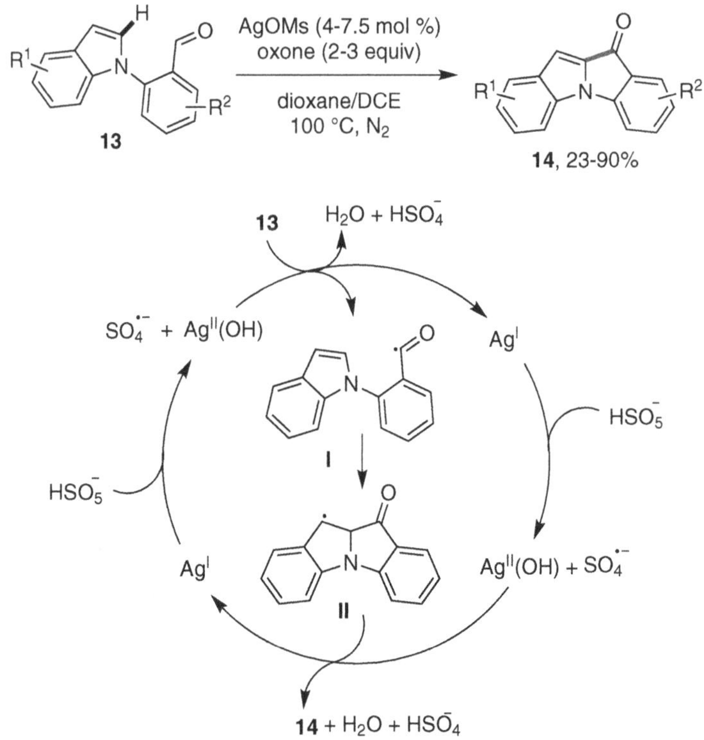

SCHEME 3.4

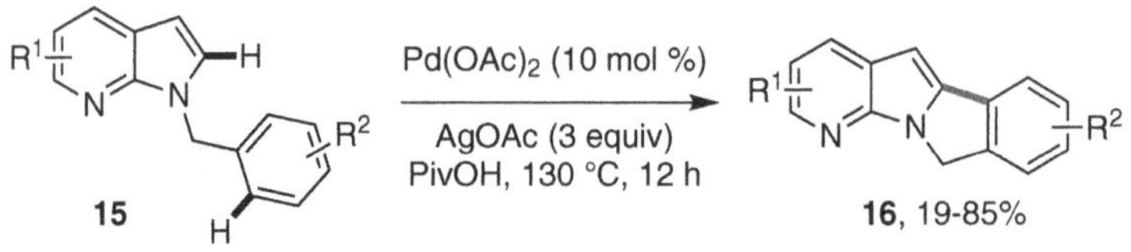

SCHEME 3.5

The dehydrogenative ring closure of arylcarbazoles **21** was developed by heating in the presence of $Pd(OPiv)_2$ (10 mol%), Ag_2O (1.2 equiv), and CuO (1.2 equiv) in PivOH under O_2 atmosphere at 130°C for 24 h. Corresponding indolo[3,2,1-*jk*]carbazoles **22** were obtained in 10–70% yields. Unsymmetrical substrates, whether at the *N*-phenyl moiety or at the carbazole moiety, afforded moderate regioselectivities, up to 1.6:1 (Scheme 3.7) (Jones, Louillat-Habermeyer, and Patureau 2015).

3.2.3 Isoindolo-Isoquinoline

Isoindolo[2,1-*b*]isoquinoline-7-carboxylate derivatives **25** were prepared by Rh(III)-catalyzed dehydrogenative [4+1] cycloaddition of isoquinolones **23** with diazoketoesters **24**. Treating 3,4-diarylisoquinolin-1(2*H*)-ones **23** with diazoketoesters

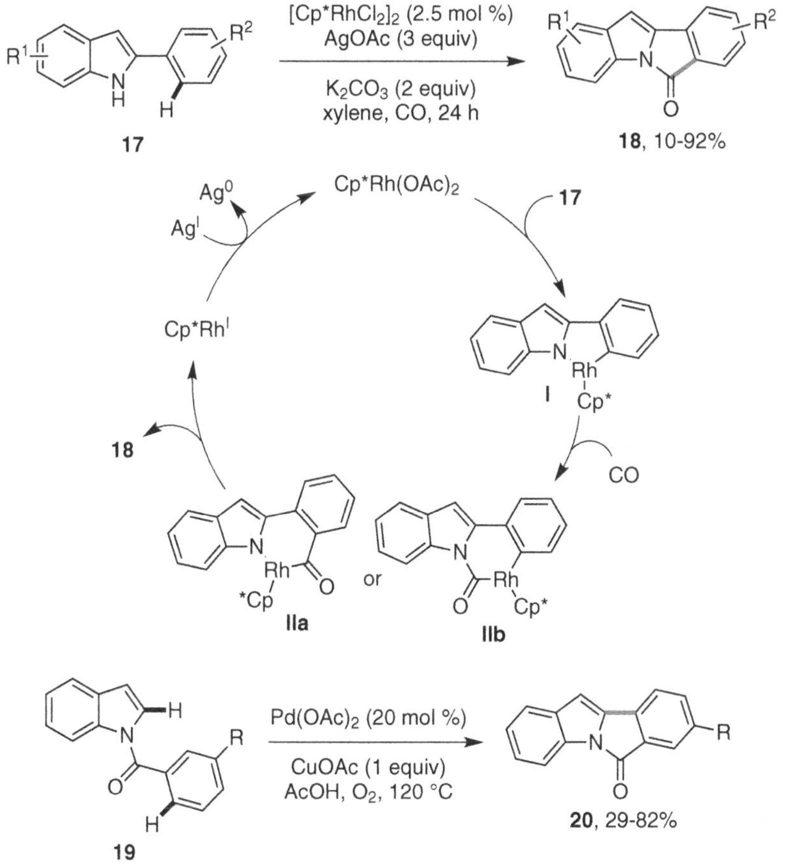

SCHEME 3.6

24 (1.5 equiv), in the presence of $[Cp^*RhCl_2]_2$ (2.5 mol%) and AgOAc (2 equiv) in dioxane at 140°C resulted in the formation of 5-oxo-5,7-dihydroisoindolo[2,1-*b*]isoquinoline-7-carboxylates **25** in 64–90% yields. Subjecting 4-methyl-3-phenylisoquinolin-1(2*H*)-one to a diazoketoester afforded the expected product in moderate yield (49%). The reaction did not occur with cyclic diazocompounds. In the proposed reaction mechanism, five-membered rhodacycle **I** was generated by Rh(III)-catalyzed N–H/C–H bond cleavage of **23**, which was transformed into rhodium carbene intermediate **II** upon reaction with **24** along with release of N_2. Migratory insertion of **II** gave six-membered rhodacycle **III**, which underwent reductive elimination to afford intermediate **IV** with regeneration of the Rh(I) species, which was reoxidized to the active Rh(III) catalyst by AgOAc. Finally, by hydrolysis of the benzoyl moiety of **IV**, the desired product **25** was obtained (Scheme 3.8) (Guo et al. 2018).

Synthesis of 2-(5-oxo-5,7-dihydroisoindolo[2,1-*b*]isoquinolin-7-yl)acetate **27** was achieved by Rh(III)-catalyzed dehydrogenative coupling reactions of isoquinolones **23** bearing a 3-aryl substituent with activated alkenes **26**. Reactions were performed

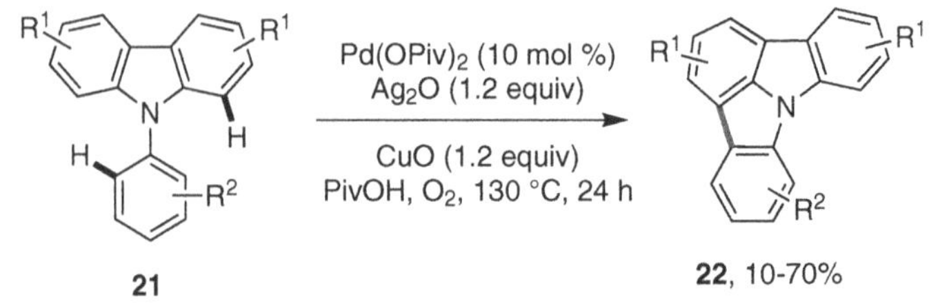

SCHEME 3.7

using 2 equiv of alkenes **26**, $[Cp^*RhCl_2]_2$ (2 mol%) as the catalyst, and anhydrous $Cu(OAc)_2$ (2.2 equiv) as an oxidant in MaCN at 115°C under N_2 atmosphere, giving isoindolo[2,1-*b*]isoquinoline derivatives **27** in 45–91% yields. The reaction was tolerated well using acrylates, acrylonitrile, and acrylamides. Reactions with diethyl maleate and fumarate occurred under reaction conditions to give the same corresponding product **27a** in 40–41% yields. Spiro-isoindolo[2,1-*b*]isoquinoline **27b** was obtained in 40% yield when *N*-methylmaleimide was reacted with 3-phenylisoquinolones **23** under the same reaction conditions. No coupling reaction was observed when the activated olefin was replaced by styrene. The reaction was initiated by addition of the NH group in **23** to the C=C double bond, activated by coordination to Rh(III) center **I**, to give a *trans*-amidorodation product **II**, which underwent *β*-hydride elimination to olefin intermediate **III**, followed by reinsertion into the C=C double bond, leading to the isomerization of the intermediate **III** to achieve an Rh(III) species **IV**. Subsequent cyclometalation of **IV** gave rise to a rhodacycle **V**,

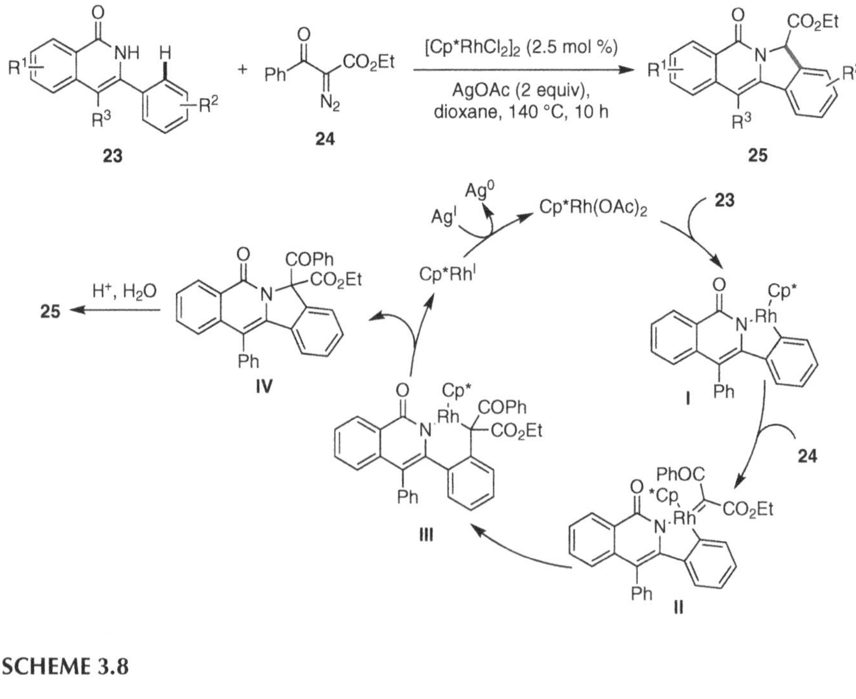

SCHEME 3.8

which afforded the final product **27** via reductive elimination, along with generation of an Rh(I) species, which was oxidized by Cu(II) to regenerate the Rh(III) species (Scheme 3.9) (Wang et al. 2011).

Ir(III)-catalyzed dehydrogenative coupling of isoquinolones **23** with benzoquinone **28** was developed by Zhou et al. (2015) to give spiro-isoindolo[2,1-*b*] isoquinoline derivatives **29**. Reactions were carried out by heating a solution of 3,4-diarylisoquinolin-1-one **23** and benzoquinone **28** (2.2 equiv) in the presence of $[Cp^*IrCl_2]_2$ (3 mol%) and NaOAc (50 mol%), in toluene at 100°C under Ar atmosphere for 12 h, producing 5'*H*-spiro[cyclohex[3]ene-1,7'-isoindolo[2,1-*b*] isoquinoline]-2,5,5'-triones **29** in 62–99% yields. Activation of $C(sp^3)$–H in 3,4-diethylisoquinolin-1-one to obtain a similar spiro compound did not occur. Reaction with 1,4-naphthoquinone afforded a low yield of the product. However, addition of $Cu(OAc)_2{\cdot}H_2O$ (2 equiv) as the oxidant improved the yield of corresponding spiro compounds up to 99% yield (Scheme 3.10). Also, similar spiro-isoindolo[2,1-*b*] isoquinoline scaffolds were prepared by Ru-catalyzed one-pot sequential dehydrogenative coupling of (hetero)arene-carboxamides with alkyne and then quinone (Mukherjee et al. 2018).

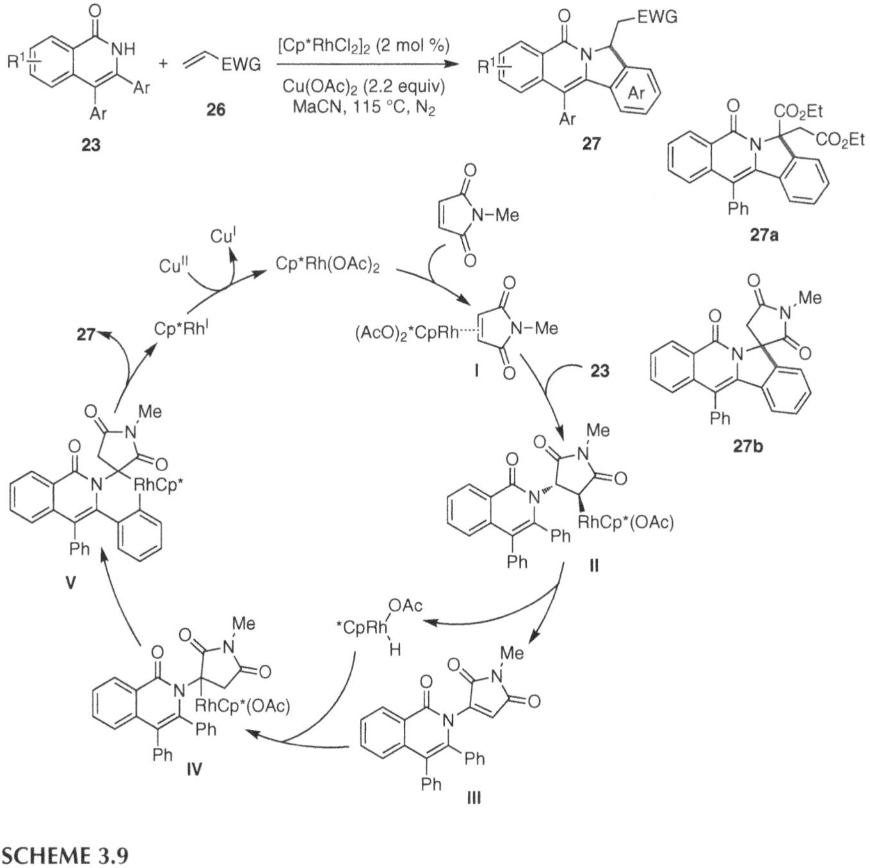

SCHEME 3.9

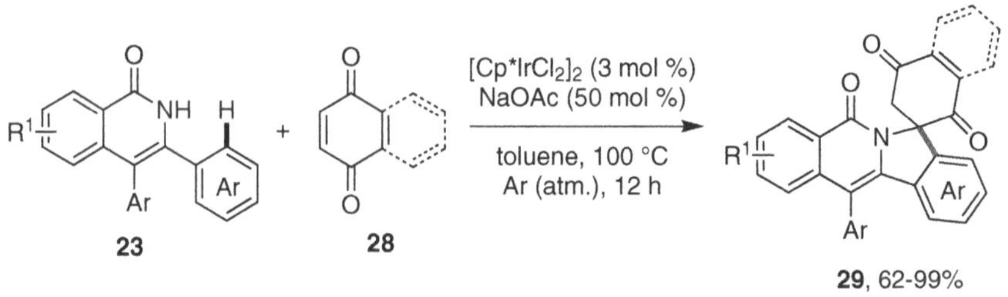

[Cp*IrCl2]2 (3 mol %)
NaOAc (50 mol %)
toluene, 100 °C
Ar (atm.), 12 h

23 28 29, 62-99%

SCHEME 3.10

3.2.4 Indolizine

An interesting approach to access ethyl indolizine-3-carboxylates **31** was reported via Rh(III)-catalyzed dehydrogenative annulation reactions of pyridinium trifluoromethanesulfonate salts **30** with alkynes **6** by cleavage of C(sp^2)–H/C(sp^3)–H bonds. Reactions were performed using $[Cp^*RhCl_2]_2$ (5 mol%) as the catalyst, $Cu(OAc)_2{\cdot}H_2O$ (1 equiv) as an oxidant, and KOAc (2 equiv) in DCE at 120°C under Ar atmosphere. Reaction of diaryl alkynes **6** with different substituted 1-(2-ethoxy-2-oxoethyl)pyridinium salts **30** gave corresponding ethyl 1,2-diarylindolizine-3-carboxylates **31** in 24–94% yields, in which strong electron-donating groups NMe_2 and OMe at the C-4 position of pyridinium salts led to lower yields of products, 24% and 55%, respectively. When the reaction was carried out using dialkyl alkynes **6**, corresponding indolizines **31** were obtained in moderate yields (52–68%). Unsymmetrical alkyl aryl alkynes afforded corresponding 1-alkyl-2-aryl-indolizine-3-carboxylates **31** in 59–75% yields. Dimethyl acetylenedicarboxylate was also investigated under reaction conditions, leading to 3-ethyl 1,2-dimethyl indolizine-1,2,3-tricarboxylate **31a** in 34% yield. When 1-(2-oxopropyl)pyridinium trifluoromethanesulfonate salt was used, the expected 1-(1,2-diphenylindolizin-3-yl)ethanone **31b** was obtained in 53% yield. The plausible reaction mechanism involves the formation of enolate-Rh intermediate **II** by generation of enolate **I** of the pyridinium salt in the presence of KOAc as the base, followed by reaction with $Cp^*Rh(OAc)_2$, generated in situ by ligand exchange. By C–H bond activation, intermediate **II** was converted to a six-membered rhodacyclic intermediate **III**, which underwent alkyne **6** coordination and migratory insertion into the Rh–C bond to afford eight-membered rhodacycle **IV**, which is in equilibrium with the six-membered rhodacycle **V**. Reductive elimination gave the intermediate **VI** with the release of the Rh(I) species, which was oxidized by $Cu(OAc)_2$ to the active Rh(III) catalyst for the next catalytic cycle. Finally, the intermediate **VI** was transferred to the desired product **31** by the base-induced loss of a proton (Scheme 3.11) (Shen, Li, and Wang 2016).

Co(III)-catalyzed [4+1] cycloaddition of 2-arylpyridines **32** with ethyl glyoxylate **35** via a C(sp^2)–H bond activation was accomplished to synthesize benzoindolizines pyrido[2,1-*a*]isoindoles **36**. Different salts and complexes of Co(III), including $Co(acac)_3$, $CoCl_3$, $[Cp^*Co(CO)I_2]$, and $[Cp^*Co(MeCN)_3(SbF_6)_2]$ were investigated, from which 5 mol% of $[Cp^*Co(CO)I_2]$ in the presence of $AgSbF_6$ (10 mol%) and $Cu(OAc)_2$ (20 mol%) in DCE at 110°C, under Ar atmosphere in a sealed tube,

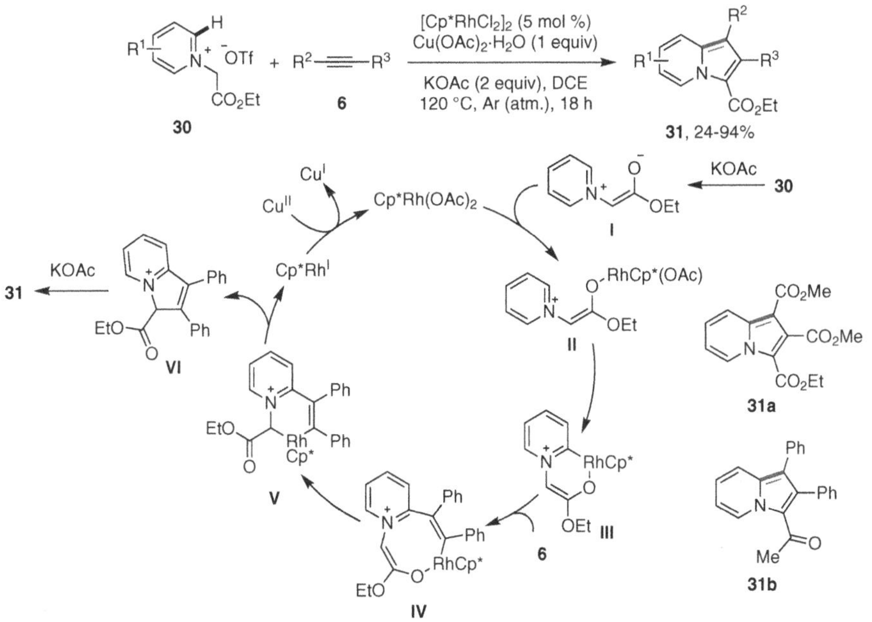

SCHEME 3.11

furnished ethyl pyrido[2,1-*a*]isoindole-6-carboxylate derivatives **36** in 53–90% yields. No reaction occurred with benzaldehyde and acetaldehyde. However, reaction with α-ketoaldehydes methyl and phenylglyoxal delivered the expected 6-acylpyrido[2,1-*a*] isoindoles **36a** in 21% and 19% yields, respectively. The proposed reaction mechanism is outlined in Scheme 3.12. The reaction was initiated by coordination and C–H bond activation of 2-arylpyridines **32** with $Cp^*Co(OAc)_2$ active species, generated in situ by $AgSbF_6$-assisted ligand exchange between $[Cp^*Co(CO)I_2]$ and $Cu(OAc)_2$, followed by insertion of the C=O bond of ethyl glyoxylate **35** to a Co–C bond, with subsequent protonation to deliver alcohol **I**, which converted to the final desired product through intramolecular nucleophilic addition of pyridine nitrogen to activated hydroxymethene carbon by a Co(III) Lewis acid, with release of the Co(III) catalyst (Chen et al. 2016). Rh-catalyzed dehydrogenative cross-coupling of 2-arylpyridines **32** with methyl trifluoroacrylate **33** was developed to construct benzoindolizine derivatives, methyl 2-oxo-2-(pyrido[2,1-*a*]isoindol-6-yl)acetates **34**. Reactions were catalyzed with $[Cp^*Rh(CH_3CN)_3](SbF_6)_2$ (5 mol%) in the presence of NaOAc (20 mol%), and 1 equiv of H_2O in DCE at 80°C for 24 h, affording corresponding pyrido[2,1-*a*] isoindoles **34** in 40–91% yields (Scheme 3.12) (Gong et al. 2018).

3.2.5 Imidazo-Indole and Imidazo-Isoindole

1,2-Dihydro-3*H*-imidazo[1,5-*a*]indol-3-one derivatives **37** having a quaternary carbon stereocenter were synthesized by Rh-catalyzed C–H annulation of *O*-pivaloyl 1-indolehydroxamic acid **5b** with donor/acceptor diazocompounds **24**. Treatment of

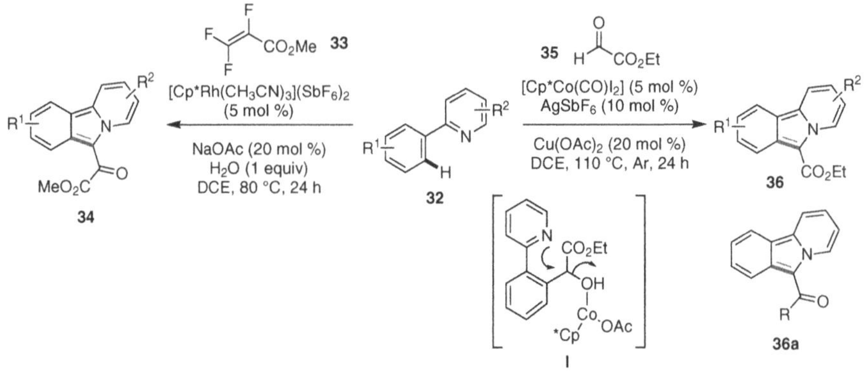

SCHEME 3.12

O-pivaloyl 1-indolehydroxamic **5b** with *α*-diazoesters **24** (1.5 equiv), in the presence of chiral rhodium catalyst **A** (2.5 mol%), $AgSbF_6$ (20 mol%), and CsOAc (1 equiv) in acetone, provided the expected 3*H*-imidazo[1,5-*a*]indol-3-ones **37** in 61–98% yields, with 84/16–98/2 enantioselectivities. The reaction occurred by activation of the indole C2–H bond by chiral Rh-catalyst **A**, followed by formation of carbene-metal species, generated by diastereoselective coordination of a diazoester and then release of N_2, which underwent migratory insertion of the carbene moiety to the Rh–C bond and then formation of the C–N bond. Subsequently the final product was obtained upon protonolysis. The obtained enantioselectivity could be attributed to the formation of metal-carbene species, preferably avoiding steric repulsion between the large ester substituent and the bulk pivalate group (Scheme 3.13) (Chen et al. 2017).

A dehydrogenative C(sp²)–N bond-forming strategy via Cu(I)-catalyzed intramolecular C–H/N–H coupling was developed in order to synthesize benzoimdazo[1,2-*a*]indoles **39**. Different salts of Cu, such as CuCl, CuBr, CuI, CuOAc, Cu_2O, $CuCl_2$, and $Cu(OAc)_2$ in the presence of $(t\text{-BuO})_2$ as the oxidant and various phosphine ligands were screened for intramolecular dehydrogenative coupling of *N*-(2-acetamidophenyl)indoles **38**, from which 10 mol% of Cu_2O and 1.5 equiv of $(t\text{-BuO})_2$

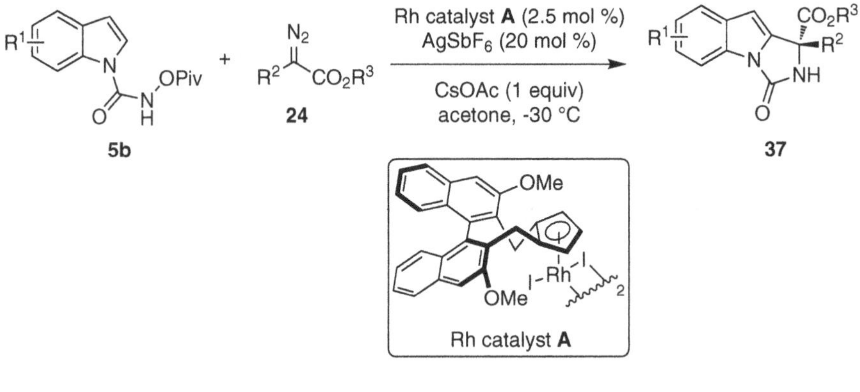

SCHEME 3.13

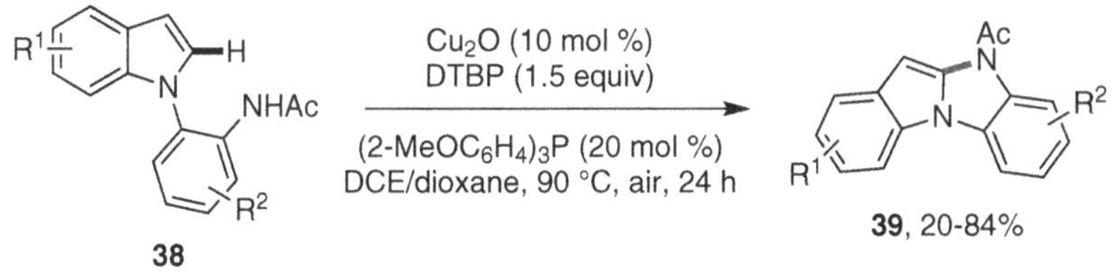

SCHEME 3.14

in the presence of a tris(2-methoxyphenyl)phosphine ligand (20 mol%) in DCE/dioxane at 90°C under air led to the desired benzoimdazo[1,2-*a*]indole products **39** in 20–84% yields. The reaction took place by abstraction of the H atom of the amide N–H bond, generating an anilidyl radical, followed by intramolecular radical cyclization, with subsequent direct oxidation by either Cu(II)(O*t*-Bu) complex or *tert*-BuO radical and then deprotonation (Scheme 3.14) (Wang et al. 2017).

Benzoimidazo[2,1-*a*]isoindole derivatives **41** were synthesized through Pd(II)/Cu(I)-catalyzed intramolecular dehydrogenative coupling (CDC) reaction of *N*-benzylbenzimidazoles **40**. Reactions were conducted using $Pd(OAc)_2$ (20 mol%), $Cu(OAc)_2{\cdot}H_2O$ (2 equiv), CuOAc (0.5 equiv), and CsOPiv (2.5 equiv) in dioxane at 150°C for 3 h under microwave heating, liberating 11*H*-benzo[4,5]imidazo[2,1-*a*]isoindoles **41** in 15–63% yields. Electron-withdrawing functionalities hindered coupling, affording the expected products in lower yields. The reaction occurred by transmetalation between the metalated benzimidazole C2–H bond with Cu(I) and the benzyl *ortho*-C–H bond activated by Pd(II), generating a six-membered palladacycle, followed by reductive elimination (Scheme 3.15) (Pereira, Porter, and DeBoef 2014).

3.2.6 Imidazo-Pyridine

A Cu(II)/Fe(III)-cocatalyzed intramolecular C–H amination reaction for the synthesis of benzo[4,5]imidazo[1,2-*a*]pyridine **43** from *N*-aryl-2-aminopyridines **42** was developed by heating a solution in *N,N*-dimethylformamide (DMF) in the presence of $Cu(OAc)_2$ (20 mol%) and $Fe(NO_3)_3{\cdot}9H_2O$ (10 mol%) and PivOH (5 equiv) under O_2 atmosphere at 130°C. Corresponding imidazo-pyridines **43** were obtained in 24–96% yields. The proposed reaction mechanism involves Fe(III)-induced *ortho*-C–H bond activation, with subsequent reductive elimination to achieve the desired product (Scheme 3.16) (Wang et al. 2010).

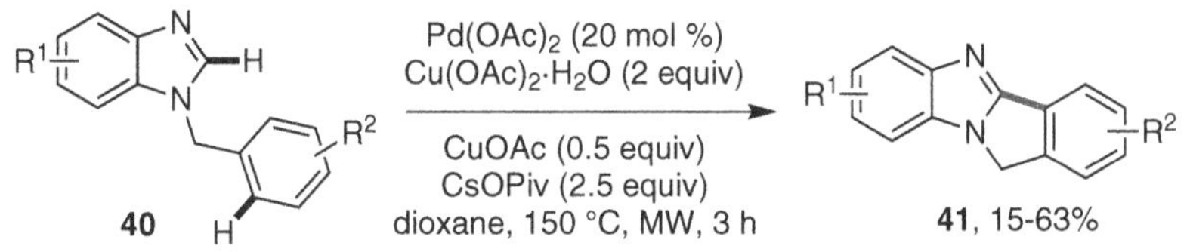

SCHEME 3.15

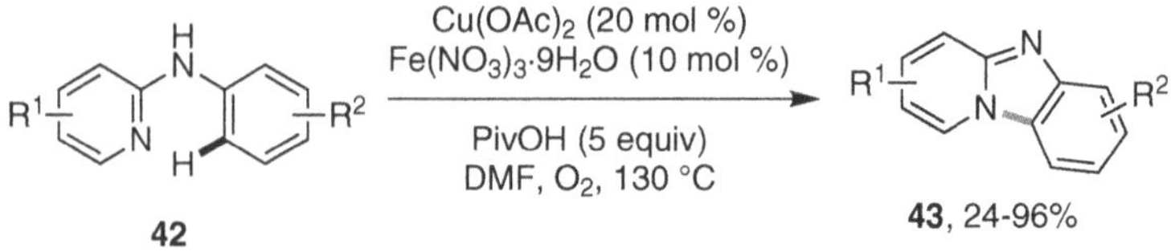

$Cu(OAc)_2$ (20 mol %)
$Fe(NO_3)_3 \cdot 9H_2O$ (10 mol %)
PivOH (5 equiv)
DMF, O_2, 130 °C

42 → **43**, 24-96%

SCHEME 3.16

3.2.7 Indazolo-Phthalazine

Gholamhosseyni and Kianmehr (2018) and Wu and Ji (2018) independently reported the synthesis of indazolo[1,2-*b*]phthalazine derivatives **45** by dehydrogenative coupling of *N*-aryl 2,3-dihydrophthalazine-1,4-diones **44** with acrylates **26a** and propargyl alcohols **7**, respectively. Reactions of *N*-aryl 2,3-dihydrophthalazine-1,4-diones **44** with acrylate **26a** (3 equiv) were conducted using $[RuCl_2(p\text{-cymene})]_2$ (5 mol%) as the catalyst and $Cu(OAc)_2 \cdot H_2O$ (1 equiv) as the oxidant, in the presence of KPF_6 (10 mol%) in water at 120°C for 24 h, leading to corresponding 2-(6,11-dioxo-11,13-dihydro-6*H*-indazolo[1,2-*b*]phthalazin-13-yl)acetate derivatives **45** in 66–99% yields. The reaction of *N*-aryl 1,2-dihydropyridazine-3,6-diones **44** with acrylates **26a** under the same reaction conditions led to the expected dihydro-6*H*-pyridazino[1,2-*a*] indazoles **45** in 71–98% yields. The reaction proceeded by N–H-assisted C–H bond activation of the substrate **44** generating ruthenacycle complex **I**, which underwent coordination with subsequent migratory insertion of the acrylate **26a** to give intermediate **II**. β-hydride elimination led to intermediate **III**, which afforded alkenylated product **IV** by reductive elimination. Finally, the indazolo[1,2-*b*]phthalazine product **45** was obtained by an intramolecular aza-Michael addition reaction of **IV** (Scheme 3.17). In Wu and Ji's work, reaction of *N*-aryl 2,3-dihydrophthalazine-1,4-diones **44** with propargyl alcohols **7** was performed in the presence of $[Cp^*RhCl_2]_2$ (5 mol%) and NaOAc (1 equiv) in PhCl at 90°C under air atmosphere, to furnish the expected 13-acylmethyl substituted 6*H*-indazolo[1,2-*b*]phthalazine-6,11(13*H*)-diones **46** in 25–83% yields. The C–H bond activation of substrate **44** by in situ generated active catalyst $Cp^*Rh(OAc)_2$ through the ligand exchange with sodium acetate gave a five-membered rhodacycle **I**, which underwent coordination of propargyl alcohol **7**, followed by migratory insertion to form intermediate **II**. Abstraction of the allylic proton produced allene intermediate **III**, which underwent reductive elimination and enol–keto tautomerization to produce the desired product **46**, along with generation of a Cp*Rh(I) complex, which was reoxidized to the active catalyst Cp*Rh(III) under air to complete the catalytic cycle (Scheme 3.18).

Also, synthesis of 13-hydroxy-6*H*-indazolo[1,2-*b*]phthalazine-13-carboxylate derivatives **47** was achieved in 68–86% yields by Rh(III)-catalyzed dehydrogenative coupling between *N*-arylphthalazine-1,4-diones **44** and α-diazoketoesters **24**, by subjection to $[Cp^*RhCl_2]_2$ (2.5 mol%) in the presence of CsOAc (50 mol%) in DCE at room temperature under air conditions. The reaction failed in the case of *N*-arylphthalazine-1,4-diones bearing electron-withdrawing groups on the aryl moiety. When reactions were carried out using $AgSbF_6$ (10 mol%) instead of CsOAc at 110°C under an atmosphere of N_2, phthalazino[2,3-*a*]cinnoline-5-carboxylate

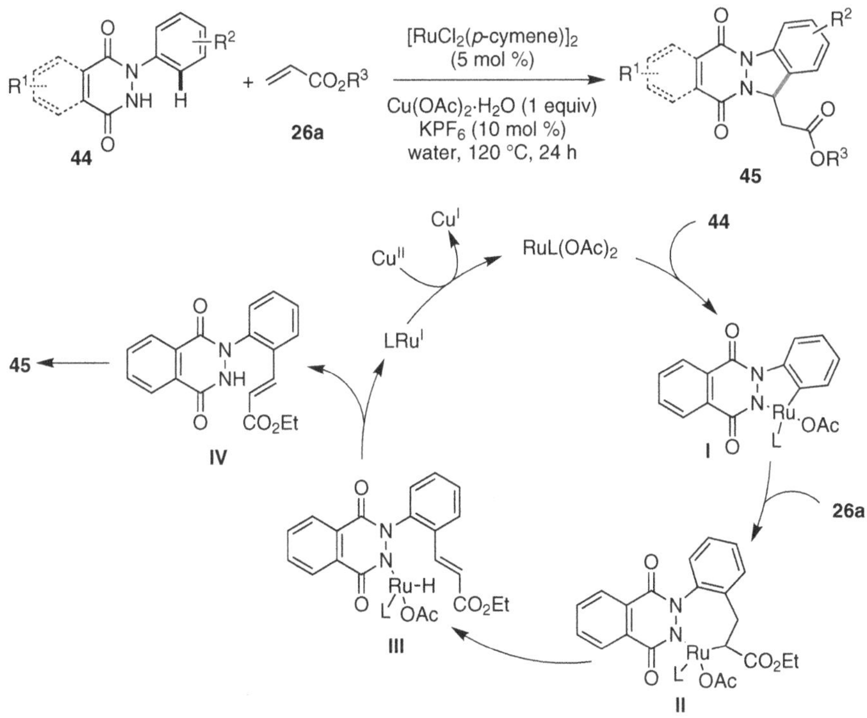

SCHEME 3.17

derivatives **48** were obtained in 65–82% yields. The reaction occurred by the formation of *ortho*-alkylated product **I**, via activation of the *ortho*-C–H bond followed by formation of carbene-Rh species with subsequent migratory insertion and then protonolysis. Intermediate **I** was converted to indazolo[1,2-*b*]phthalazine **47** or phthalazino[2,3-*a*]cinnoline **48** by CsOAc-induced proton abstraction, oxygen insertion, elimination of acyl, and intramolecular nucleophilic addition sequences, or $AgSbF_6$ (Lewis acid)-promoted cyclocondensation, respectively (Scheme 3.19) (Karishma et al. 2018).

3.3 SIX-MEMBERED *N*-BRIDGED HETEROCYCLES

3.3.1 Pyrrolo-Quinoline and Pyrrolo-Isoquinoline

Rh(III)-catalyzed carbocyclization reactions of 3-(indolin-1-yl)-3-oxopropanenitriles **49** with alkynes **6** and alkenes **26** were developed to form 1,7-fused indolines **50–51** through C–H activation. Reactions with alkynes **6** were performed using $[Cp^*RhCl_2]_2$ (3–5 mol%) and NaOAc (50 mol%) in DMF at 80 or 100°C under Ar atmosphere to produce 1*H*-pyrrolo[3,2,1-*ij*]quinoline derivatives **51** in 66–97% yields. Dialkyl alkynes as well as diaryl alkynes afforded corresponding products under the reaction conditions. Unsymmetrical alkyl aryl alkynes afforded 2-(6-alkyl-5-aryl-1*H*-pyrrolo[3,2,1-*ij*]quinolin-4(2*H*)-ylidene)acetonitriles **51**, regioselectively.

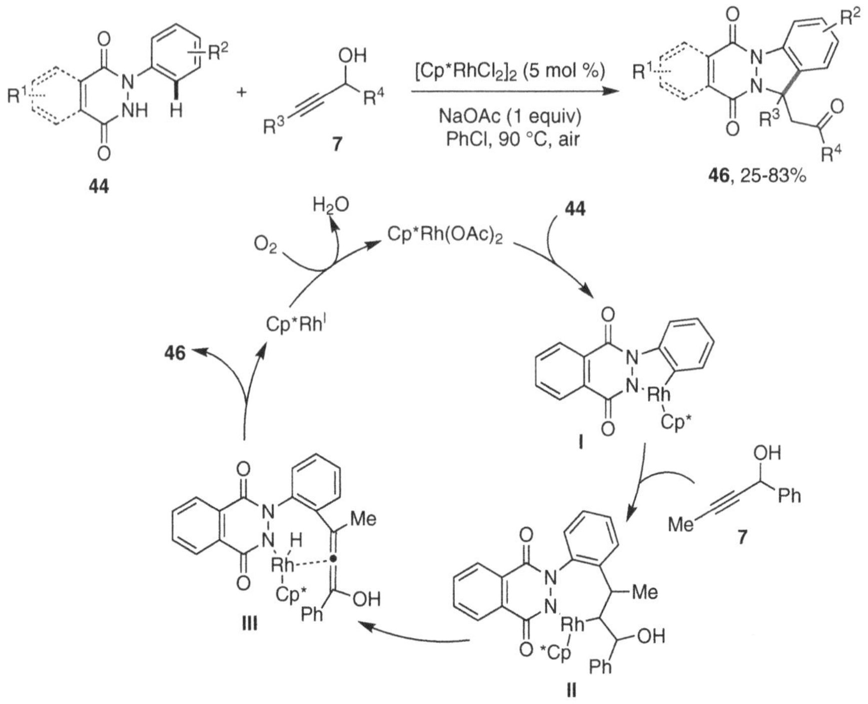

SCHEME 3.18

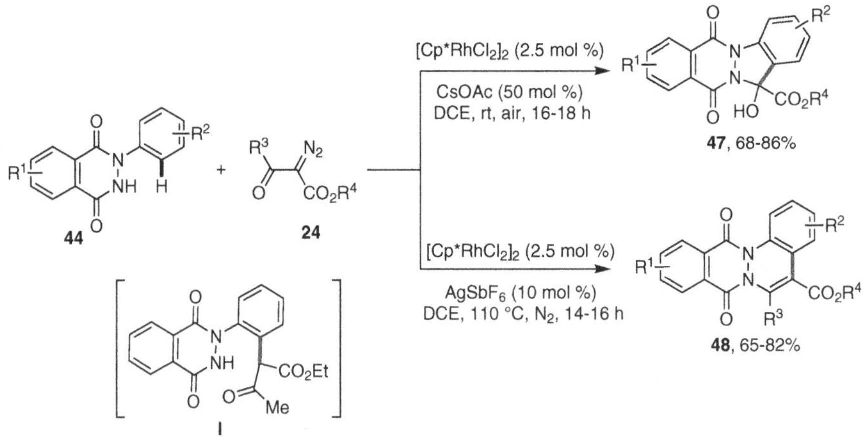

SCHEME 3.19

Reactions with alkenes **26** were carried out by heating a solution of 3-(indolin-1-yl)-3-oxopropanenitriles **49** with acrylates **26** (2 equiv) in the presence of $[Cp^*RhCl_2]_2$ (5 mol%), $Cu(OAc)_2{\cdot}H_2O$ (2 equiv), and CsOAc (50 mol%) in DMF at 100°C, leading to 2-(5-cyano-4-oxo-2,4-dihydro-1*H*-pyrrolo[3,2,1-*ij*]quinolin-6-yl)acetates **50** in 58–90% yields. *N,N*-dimethylacrylamide and (vinylsulfonyl)benzene furnished the desired product in 74% and 37% yields, respectively. Reaction with acrylic acid resulted in the formation of a decarboxylation product in 35% yield. The proposed reaction mechanism involves the generation of six-membered rhodacycle intermediate **I** via C-7 C(sp^2)–H activation, which underwent coordination and insertion of an alkyne **6** to give the alkenyl-Rh intermediate **VI**. By intramolecular nucleophilic addition of the C–Rh bond into the carbamoyl group, the intermediate **VII** was formed, delivering the final expected product **51** upon protonation with AcOH, along with regeneration of active catalytic species Cp*Rh$(OAc)_2$. Similarly, insertion of acrylate C=C double bond into the C–Rh bond of **I** afforded the eight-membered rhodacycle **II**, which was converted to alkenylated product **III** through *β*-hydrogen elimination. C(sp^2)–H and C(sp^3)–H activation delivered intermediates **V**, which underwent reductive elimination with a subsequent double bond shift and gave the final product **50** with generation of Cp*Rh(I), which was oxidized by $Cu(OAc)_2{\cdot}H_2O$ to Cp*Rh$(OAc)_2$ for the next catalytic cycle (Zhou et al. 2016). A similar methodology was developed by reacting *N*-carbamoyl indolines **52** with internal alkynes **6** in the presence of $[Cp^*RhCl_2]_2$ (5.5 mol%), $AgSbF_6$ (22 mol%), and $Zn(OTf)_2$ (30 mol%) in DCE at 100–120°C, leading to fused-ring pyrroloquinolinones **53** in 32–79% yields. Reaction with dialkyl alkyne, 3-hexyne, led to the corresponding pyrroloquinolinone in lower yield (44%). Unsymmetrical alkyl aryl alkynes afforded 6-alkyl-5-aryl-1*H*-pyrrolo[3,2,1-*ij*]quinolin-4(2*H*)-ones, regioselectively. The reaction occurred by the activation of the C–H bond of the indoline at the C-7 position and the C–N bond of the urea motif (Scheme 3.20) (Yang, Hu, and Loh 2015). An $[RuCl_2(p\text{-cymene})]_2$ catalyst was also applied to the similar reaction between *N*-carbamoyl indolines with alkynes (Manoharan and Jeganmohan 2015).

Pyrrolo[1,2-*b*]isoquinolines **55** were synthesized by Rh(III)-catalyzed intramolecular annulation of alkyne-tethered benzamides **54**. Reactions were carried out by $[Cp^*RhCl_2]_2$ (2.5 mol%) in the presence of $Cu(OAc)_2$ (2 equiv) in *t*-AmOH at 110°C, producing 2,3-dihydropyrrolo[1,2-*b*]isoquinolin-5(1*H*)-ones **55** in 58–98% yields. No reaction occurred in the case of terminal alkyne-tethered substrates. 1*H*-pyrido[1,2-*b*]isoquinolin-6(2*H*)-one (**55**, n=2) and azepino[1,2-*b*]isoquinolin-5(7*H*)-one (**55**, n=3) were produced in 67% and 62% yields, respectively, when a longer carbon-tethered alkyne was investigated. The reaction occurred through *ortho*-C–H activation by Rh(III), followed by intramolecular alkyne coordination and migratory insertion into the C–Rh bond, with subsequent reductive elimination (Quinones et al. 2013). Synthesis of 1*H*-pyrrolo[1,2-*b*]isoquinolin-4-ium salts **58** was achieved in 75–91% yields by heating a solution of 5-phenylpent-4-yn-1-amine **57** and aromatic aldehydes **56** (1.2 equiv) in the presence of $[Cp^*RhCl_2]_2$ (2 mol%) and $Cu(BF_4)_2{\cdot}6H_2O$ (1.125 equiv) under O_2 atmosphere in *t*-BuOH at 90°C. 4-Phenylbut-3-yn-1-amine did not afford the corresponding azeto[1,2-*b*]isoquinolin-3-ium salt under the reaction conditions, while 6-phenylhex-5-yn-1-amine and 7-phenylhept-6-yn-1-amine were tolerated in the reaction, giving the expected pyrido[1,2-*b*]isoquinolin-5-ium (**58**, n=2) and 7*H*-azepino[1,2-*b*]isoquinolin-6-ium salts (**58**, n=3) in 86–93% and 88% yields, respectively. Heterocyclic aldehydes, furan-3-carbaldehyde, and thiophen-3-carbaldehyde were also investigated, to afford the corresponding 5*H*-furo[2,3-*f*]indolizin-8-ium and 5*H*-thieno[2,3-*f*]indolizin-8-ium salts **58a** in 75% and 84% yields, respectively. The reaction was initiated by in situ generated

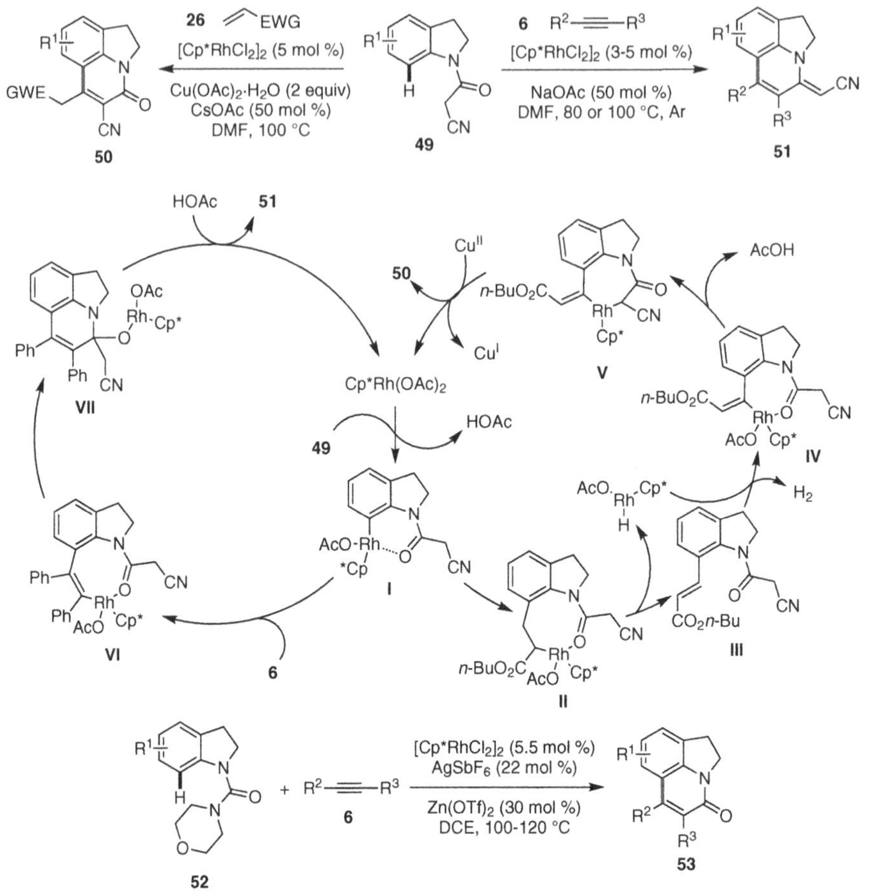

SCHEME 3.20

imine-directed C–H activation, followed by intramolecular migratory insertion of the alkyne and then reductive elimination (Scheme 3.21) (Upadhyay, Jayakumar, and Cheng 2017).

3.3.2 Indolo-Quinoline and Indolo-Isoquinoline

5,6-Diarylindolo[1,2-*a*]quinolones **60** were synthesized by Pd(II)-catalyzed dehydrogenative coupling of *N*-arylindole-3-carboxylic acids **59** with alkynes **6**. Reactions were performed using 1.5 equiv of alkyne **6** in the presence of $Pd(OAc)_2$ (10 mol%), $Cu(OAc)_2{\cdot}3H_2O$ (2 equiv), and LiOAc (6 equiv) in DMA at 120°C under N_2 atmosphere, to furnish indolo[1,2-*a*]quinoline derivatives **60** in 25–55% yields. The reaction was accomplished by coordination of the carboxyl group to $Pd(OAc)_2$, followed by palladation at the C-2 position to give a palladacycle intermediate **I**, which underwent alkyne **6** insertion (intermediate **II**) and decarboxylation to generate a five-membered palladacycle intermediate **III**. Seven-membered palladacycle **V** was generated by protonolysis

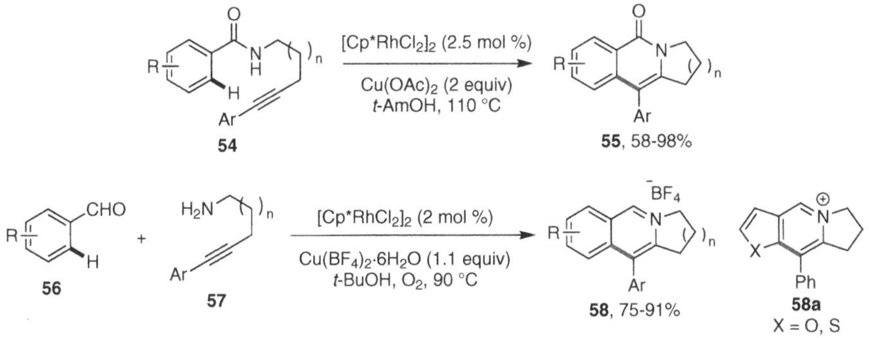

SCHEME 3.21

of the indolyl–Pd bond in **III** to vinylpalladium intermediate **IV**, followed by cyclopalladation on the phenyl group. Finally, reductive elimination afforded the product **60** (Yamashita et al. 2009). A similar approach to obtaining indolo[1,2-*a*]quinolines was developed using an Rh(III) catalyst (Okada et al. 2018). Also, cross-dehydrogenative coupling of *N*-pyridylindoles (X=CH) **61** with alkynes **6** provided indolo[1,2-*a*][1,8] naphthyridine derivatives **62**. Reactions were catalyzed with $[Cp^*RhCl_2]_2$ (2 mol%) in the presence of AgOAc (3 equiv) as the oxidant in PhCl at 140°C under N_2 atmosphere, and indolo-naphthyridines **62** were obtained in 51–94% yields. Unsymmetrical 1-phenyl-1-hexyne gave a mixture of regioisomers. Reaction with *N*-(2-pyridyl)benzimidazole (X=N) **61** led to 5,6-diphenylbenzo[4,5]imidazo[1,2-*a*][1,8]naphthyridine **62** in 95% yield. The reaction occurred by pyridine nitrogen-directed C–H activation of the C-2 position of the indole with Rh, followed by insertion of the alkyne with subsequent reductive elimination (Scheme 3.22) (Morioka et al. 2015).

Synthesis of indolo[2,1-*a*]isoquinoline derivatives **63** was developed by Rh(III)-catalyzed dehydrogenative coupling/cyclization of 2-arylindoles **17** with alkynes **6a** via C–H and N–H bond cleavages. Reactions were carried out using an equivalent amount of 2-arylindoles **17** and diaryl alkynes **6a** in the presence of $[Cp^*RhCl_2]_2$ (2 mol%), $Cu(OAc)_2{\cdot}H_2O$ (10 mol%), and Na_2CO_3 (2 equiv) in *o*-xylene at 100°C under N_2-air, to give 5,6-diarylindolo[2,1-*a*]isoquinolines **63** in 64–94% isolated yields. Reaction with dialkyl alkynes **6b**, 4-octyne, and 8-octadecyne gave unexpected products 6-alk yl-5-alkylidene-5,6-dihydroindolo[2,1-*a*]isoquinolines **64** in 62–75% yields, along with 8–9% production of the expected indolo[2,1-*a*]isoquinolines **63**. Unsymmetrical alkyne, 2-methyl-4-phenyl-3-butyn-2-ol, resulted in the formation of 2-(5-phenylindolo[2,1-*a*] isoquinolin-6-yl)propan-2-ol **63a** in 30% yield, regioselectively. When AgOAc was used instead of $Cu(OAc)_2$, the yield improved to 76%. The reaction proceeded by indole-directed activation of the *ortho*-C–H bond of the aryl, followed by migratory insertion of the alkyne, with subsequent reductive elimination (Scheme 3.23) (Morimoto et al. 2010).

3.3.3 Isoquinolino-Isoquinoline

Rh(III)-catalyzed double dehydrogenative coupling between primary benzamides **65** and alkynes **6** was described by Song et al. (2010). Subjecting benzamides **65** to 2

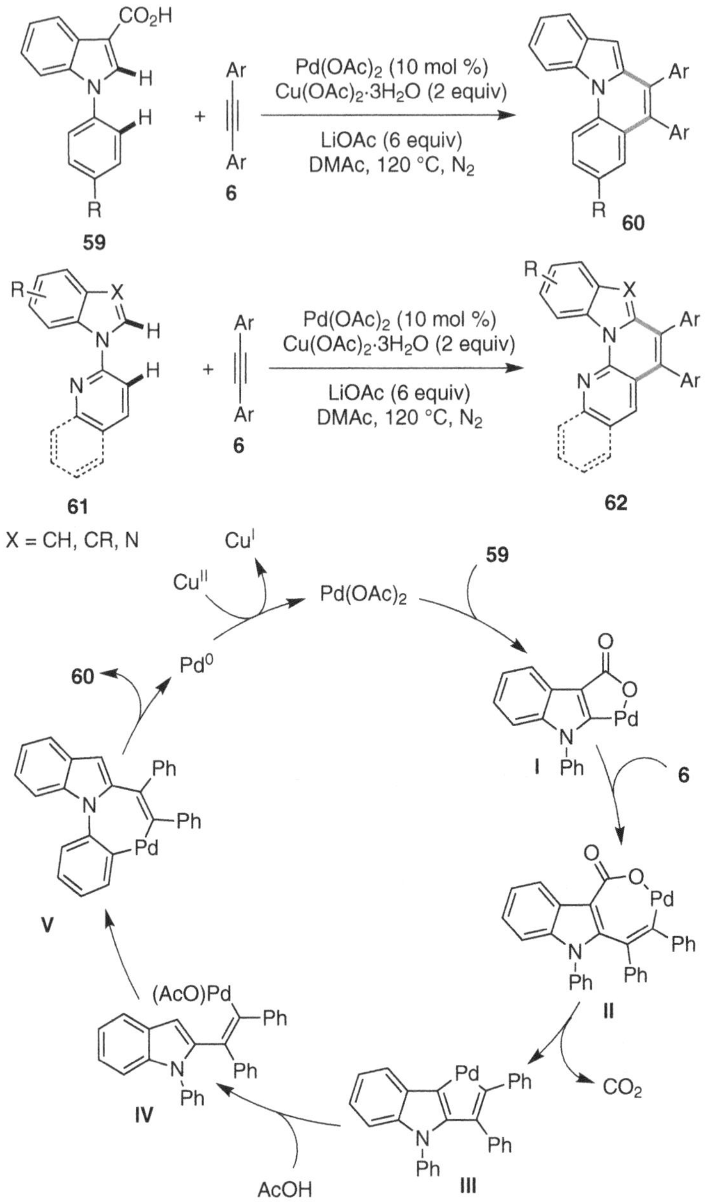

SCHEME 3.22

equiv of diphenylacetylene **6** in the presence of $[Cp^*RhCl_2]_2$ (4 mol%) and Ag_2CO_3 (3 equiv) in CH_3CN at 115°C for 12 h delivered 5,6,13-triphenyl-8*H*-isoquinolino[3,2-*a*]isoquinolin-8-ones **66** in 71–93% yields. From four possible regioisomeric products, only 5,13-dimethyl-6-phenyl-8*H*-isoquinolino[3,2-*a*]isoquinolin-8-one **66a** was obtained (83%) when 1-phenyl-1-propyne was reacted with benzamide **65** under the

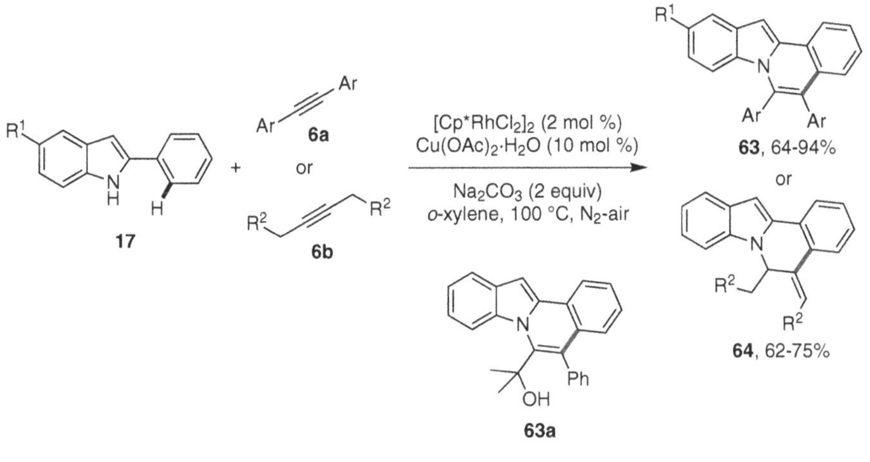

SCHEME 3.23

same reaction conditions (Scheme 3.24). Also, a methodology using 10 mol% of $[RuCl_2(p\text{-cymene})]_2$ was developed for the construction of similar heterocyclic-fused scaffolds (Shankar et al. 2016).

3.3.4 Pyrazolo-Isoquinoline and Pyrazolo-Cinnoline

Pyrazole-directed C–H bond activation of 5-aryl-1*H*-pyrazoles **67** was developed in order to construct 5,6-diarylpyrazolo[5,1-*a*]isoquinolines **68** via Rh-catalyzed dehydrogenative coupling with alkynes **6**. $[Cp^*RhCl_2]_2$ (2 mol%) in the presence of $Cu(OAc)_2$ (2.2 equiv) as the oxidant in acetone at 110°C afforded pyrazolo[5,1-*a*] isoquinoline derivatives **68** in 25–95% yields. In the case of unsymmetrical methyl phenyl acetylene, 6-methyl-5-phenylpyrazolo[5,1-*a*]isoquinoline was regioselectively obtained in 61% yield. The C–H activation of 5-(furan-2-yl)-1*H*-pyrazole and 5-(thiophen-2-yl)-1*H*-pyrazole under the same reaction conditions produced the corresponding furo[2,3-*c*]pyrazolo[1,5-*a*]pyridine and pyrazolo[1,5-*a*]thieno[2,3-*c*] pyridine **68a** in lower yields (13% and 25%, respectively). When reactions were carried out with 2.2 equiv of acrylates, 8*H*-pyrazolo[5,1-*a*]isoindole derivatives **69** were obtained in 46–80% yields (Scheme 3.25) (Li and Zhao 2011).

Synthesis of pyrazolo[1,2-*a*]cinnolines **71** was achieved through an Rh-catalyzed dehydrogenative coupling of *N*-aryl-1*H*-pyrazol-5(4*H*)-ones **70** with internal

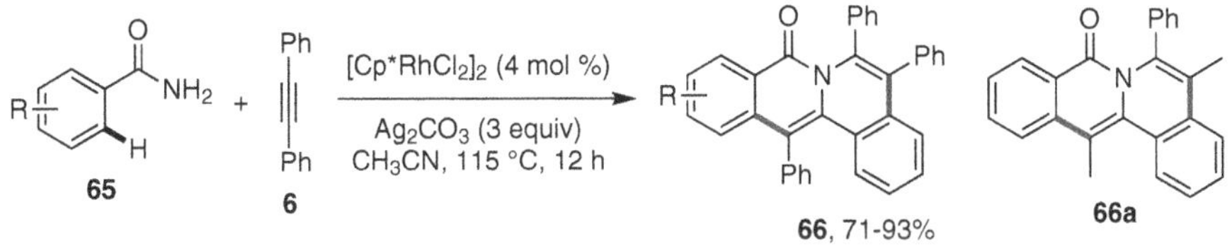

SCHEME 3.24

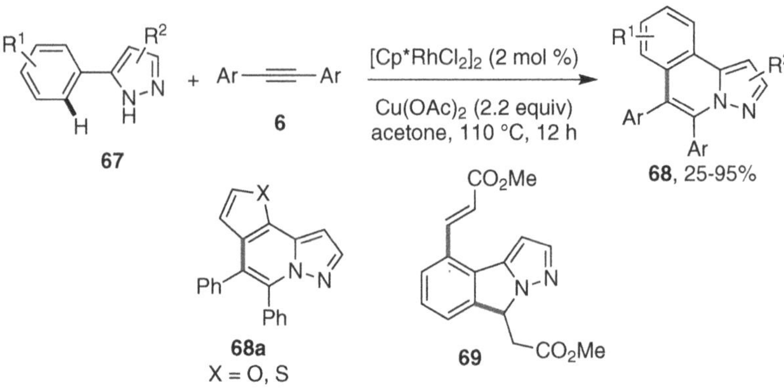

SCHEME 3.25

alkynes **6**, by heating with $[Cp^*RhCl_2]_2$ (2.5 mol%), $Cu(OAc)_2{\cdot}H_2O$ (1 equiv), and $AgSbF_6$ (10 mol%) in hexafluoroisopropanol (HFIP) at 70°C under air conditions. After 4 h, 1*H*-pyrazolo[1,2-*a*]cinnolin-1-one derivatives **71** were obtained in 41–96% yields. Reaction with dialkyl alkyne, di(*n*-propyl)acetylene, furnished the expected pyrazolo[1,2-*a*]cinnolin-1-one **71** in 84% yield. When unsymmetrical alkyl aryl alkynes were treated with *N*-phenyl-1*H*-pyrazol-5(4*H*)-one **70**, 6-alkyl-5-phenyl-1*H*-pyrazolo[1,2-*a*]cinnolin-1-ones **71** were obtained in 83–90% yields, regioselectively (Xing et al. 2014). Benzo[*c*]pyrazolo[1,2-*a*]cinnolin-1-ones **73** were synthesized by Pd-catalyzed dual C–H activation to construct C–C/C–N bonds. Reacting *N*-aryl-1*H*-pyrazol-5(4*H*)-ones **70** with aryl iodides **72** under a $Pd(OAc)_2$ (10 mol%) and AgOAc (1.5 equiv) catalytic system in trifluoroacetic acid (TFA) at 120°C led to *N*-biaryl-1*H*-pyrazol-5(4*H*)-ones **I**, which were converted to the final benzo[*c*]pyrazolo[1,2-*a*]cinnolin-1-one derivative **73** by addition of $Pd(OAc)_2$ (10 mol%), $K_2S_2O_8$ (1.5 equiv), and K_2CO_3 (2 equiv) in a one-pot manner (Scheme 3.26) (Fan et al. 2014).

Liu, Liu, and Dong (2018) reported an Ir-catalyzed cascade annulation of pyrazolones **70** and sulfoxonium ylides **74** to access pyrazolo[1,2-*α*]cinnoline derivatives **75**. $[Cp^*IrCl_2]_2$ (5 mol%) in the presence of $AgSbF_6$ (0.2 equiv), $Zn(OTf)_2$ (0.3 equiv), and TsOH (1 equiv) in TFE at 120°C under air produced 17–96% yield of pyrazolo[1,2-*α*]cinnolines **75**. Alkyl-substituted sulfoxonium ylides, as well as aryl sulfoxonium ylides, were compatible with the reaction conditions, leading to the expected products **75** in 44–82% yields. However, sulfoxonium ylides bearing a 2-furyl substituent led to the corresponding pyrazolo[1,2-*α*]cinnoline in 17% yield. Due to the steric effects, no reaction occurred with substrate bearing an *ortho*-substituent on the phenyl ring. The proposed reaction mechanism involves the generation of five-membered species **I** by the C–H activation of **70** with an active Ir(III) catalyst, generated in situ by ligand exchange of $[Cp^*IrCl_2]_2$ with $AgSbF_6$. Coordination of **74** with the Ir center of intermediate **I** gave the intermediate **II**, which was transformed to Ir-*α*-oxo carbene species **III** by release of dimethyl sulfoxide (DMSO). Migratory insertion of the carbene into the C–Ir bond afforded a six-membered intermediate

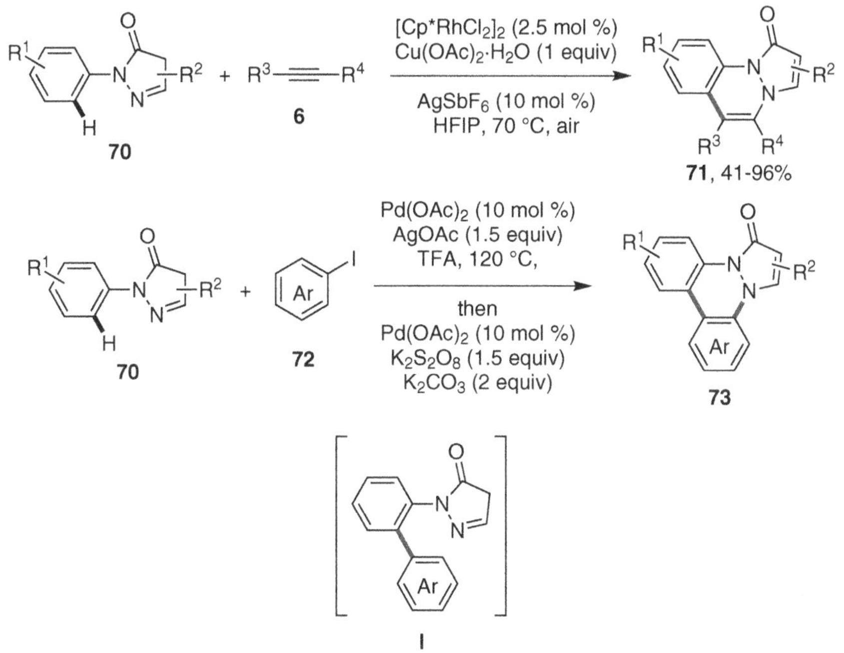

SCHEME 3.26

IV, which underwent protonolysis to release acylmethylated intermediate **V** together with the regeneration of the active catalyst. The acylmethylated compound **V** underwent cyclization in the presence of TsOH or Lewis acid to form the desired product **75** (Scheme 3.27).

Pyrazolo[1,2-*a*]cinnolines **77** was also obtained by Rh(III)-catalyzed redox-neutral annulation of pyrazolidinones **76** with diazoketoesters **24**. Various catalytic systems were studied, from which $[Cp^*RhCl_2]_2$ (5 mol%) in the presence of $AgSbF_6$ (10 mol%) in DCE at 40°C under air atmosphere gave corresponding 3-oxo-2,3-dihydro-1*H*-pyrazolo[1,2-*a*]cinnoline-6-carboxylate derivatives **77** in 44–84% yields. α-diazo-β-diketones afforded the expected 6-acyl-1*H*-pyrazolo[1,2-*a*]cinnolin-3(2*H*)-ones **77a** in 45–68% yields. The reaction occurred by pyrazole-nitrogen-assisted *ortho*-C–H activation, generating the corresponding five-membered rhodacycle, followed by Rh-carbene formation by addition of diazocompounds with N_2 release, with subsequent migratory insertion and then protonolysis to achieve the alkylated intermediate. This underwent cyclocondensation to obtain the final product (Li et al. 2018). Pyrazolo[1,2-*a*]cinnoline scaffolds **79** were also prepared by Rh(III)-catalyzed C–H functionalization of *N*-arylpyrazolidinones **76** with α-*O*-mesyl arylketones **78**. Treatment of pyrazolidinone **76** with α-OMs arylketones **78** (1.5 equiv) in MeOH solution including $[Cp^*RhCl_2]_2$ (5 mol%) and NaOPiv (1.4 equiv) at 60°C under Ar atmosphere resulted in construction of pyrazolo[1,2-*a*]cinnolines **79** in 50–81% yields. When reactions were carried out using NaOCN instead of NaOPiv, 5-hydroxy-5-aryl-5,6-dihydro-1*H*-pyrazolo[1,2-*a*]cinnolin-3(2*H*)-one derivatives **80**

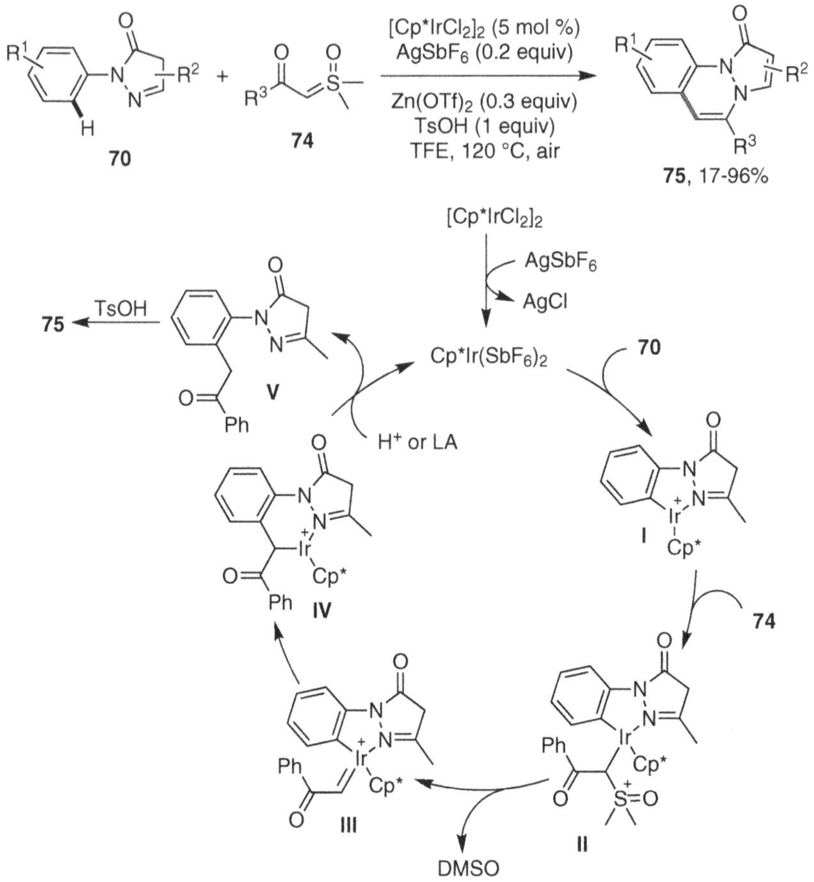

SCHEME 3.27

were obtained in 62–92% yields. Reaction of α-OMs acetone provided the expected 5-hydroxypyrazolo[1,2-*a*]cinnolin-3(2*H*)-one in 61% yield (Scheme 3.28) (Yang et al. 2017).

3.3.5 Imidazo-Fused Heterocycles

Pd(II)-catalyzed C–H arylation/dehydrogenative C–H amination was described for the construction of benzo[4,5]imidazo[1,2-*f*]phenanthridine scaffolds **82**. Reacting 2-arylbenzimidazoles **81** with 3 equiv of arylhalides **72** in the presence of $PdCl_2$ (10 mol%), Xphos ligand (20 mol%), and K_2CO_3 (3 equiv) in DMF at 160°C under air gave benzoimidazo-phenanthridine derivatives **82** in 27–90% yields. Generally, the yield for the arylhalide substrates followed the order ArI > ArBr > ArCl. No reaction occurred when an imidazopyridine substrate was used. The proposed reaction mechanism involves imidazole-nitrogen-directed Pd(II)-catalyzed *ortho*-C–H activation on the aryl moiety and functionalization with the aryl halide to give *ortho*-arylated

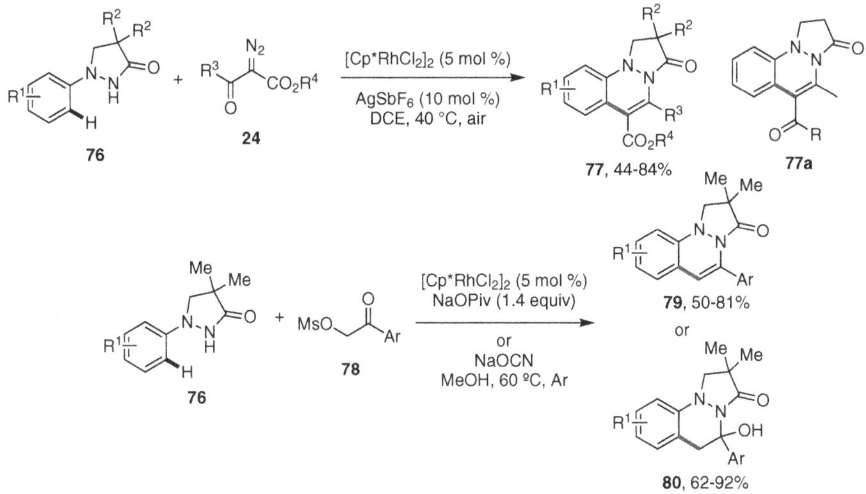

SCHEME 3.28

intermediate **I**, which was then transformed into a seven-membered palladacycle **III** through C–H activation (**II**), with subsequent K_2CO_3-promoted deprotonation of the N–H of the imidazole. Reductive elimination gave the desired product **82** and a Pd(0) species, which was oxidized by O_2 to regenerate the active Pd(II) species to complete the catalytic cycle (Scheme 3.29) (Zhao et al. 2015a).

Imidazo[2',1':3,4]pyrazino[1,2-*a*]indole derivatives **84** were prepared by a Cu(II)-promoted intramolecular dehydrogenative cross-coupling reaction between the indole-C2 and imidazole-C2 moieties of 1-(2-(1*H*-imidazol-1-yl)ethyl)-1*H*-indole **83**. Reactions were performed using $Cu(OAc)_2$ (1 equiv), a 1,10-phenanthroline ligand (2 equiv), Ag_2CO_3 (2 equiv) and K_2CO_3 (2 equiv) in xylene at 140°C for 12 h, resulting in the generation of imidazo[2',1':3,4]pyrazino[1,2-*a*]indoles **84** in 35–84% yields. Similarly, reaction with 1-(2-(1*H*-indol-1-yl)ethyl)-1*H*-benzo[*d*]imidazole afforded the corresponding benzo[4',5']imidazo[2',1':3,4]pyrazino[1,2-*a*]indoles **84a** in 40–63% yields. 1-(2-(1*H*-pyrazol-1-yl)ethyl)-1*H*-indole and 1-(2-(1*H*-1,2,4-triazol-1-yl)ethyl)-1*H*-indoles were also tolerated, affording corresponding fused heterocycles **84b,c** in 20–68% yields (Ray et al. 2015). $Pd(OAc)_2$ was also used as a catalyst for a similar intramolecular dehydrogenative coupling reaction under air conditions (Mantenuto et al. 2017). Moreover, benzo[4,5]imidazo[1,2-*a*]pyrrolo[2,1-*c*]pyrazine derivatives were synthesized via Cu-catalyzed intramolecular dehydrogenative coupling of 3-substituted pyrrole-benzimidazole systems (Tripathi, Ray, and Singh 2017). Also, a one-pot synthesis of imidazo[1,2-*a*][1,2,3]triazolo[5,1-*c*]quinoxaline derivatives **88** was reported by Bharathimohan et al. (2014) through Cu-catalyzed decarboxylative azide-alkyne cycloaddition/Pd-catalyzed intramolecular dehydrogenative cross-coupling reaction sequence. By treatment of 1-(2-azidophenyl)-1*H*-benzo[*d*]imidazole or 1-(2-azidophenyl)-1*H*-imidazole **85** with 2-alkynoic acid **86** (1.2 equiv) in the presence of $Cu(OAc)_2{\cdot}H_2O$ (10 mol%) and sodium ascorbate (20 mol%) in toluene at 80°C for 2 h (producing compounds **87** in situ), followed by addition of $Cu(OAc)_2{\cdot}H_2O$ (2 equiv), $Pd(OAc)_2$ (5 mol%), and PivOH (2.6 equiv),

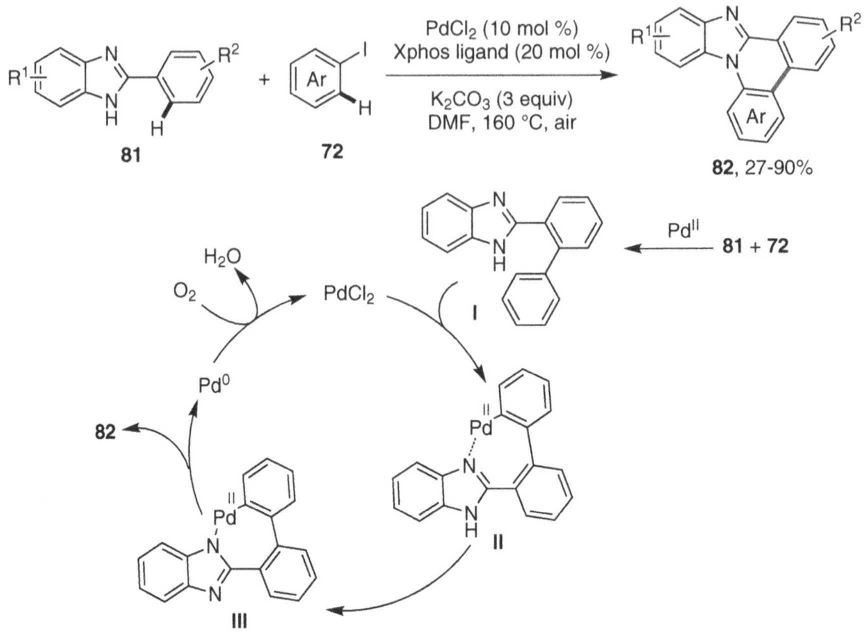

SCHEME 3.29

and then refluxing at 120°C for 3 h, the desired imidazo[1,2-*a*][1,2,3]triazolo[5,1-*c*] quinoxalines **88** were obtained in 60–97% yields (Scheme 3.30).

3.3.6 Pyrimido-Indole

Rh(III)-catalyzed C2–H bond carbenoid insertion/cyclization of *N*-amidoindoles **5a** with α-diazoketoesters **24** was accomplished to synthesize 2*H*-pyrimido[1,6-*a*]indol-1-ones **89**. The reaction was studied under different catalytic systems, such as Ir(III), Co(III), Ru(III), Pd(II), and Rh(III) catalysts, from which $[Cp^*RhCl_2]_2$ (2.5 mol%) in the presence of $AgSbF_6$ (10 mol%) and PivOH (1 equiv) in DCE under an atmosphere of argon at 100°C afforded the corresponding 2*H*-pyrimido[1,6-*a*]indol-1-one derivatives **89** in 48–94% yields. α-acyl diazoketone and α-acyl diazosulfone were also tolerated as the substrate in the reaction, affording the desired 4-acetyl and 4-sulfonyl-2*H*-pyrimido[1,6-*a*]indol-1-ones in 57% and 81% yields, respectively. The reaction was initiated by *N*-coordination and C2–H bond activation of the indole, followed by coordination of the diazocompound to the Rh(III) center with subsequent denitrogenation to generate an Rh-carbene species, which was transformed into the final expected product by migratory insertion and protonolysis/intramolecular cyclization (Scheme 3.31) (Chen et al. 2015).

3.3.7 Pyrido-Quinazoline

The construction of 11*H*-pyrido[2,1-*b*]quinazolin-11-one scaffolds **91** was achieved through Pd(II)-catalyzed C(sp²)–H pyridocarbonylation of *N*-aryl-2-aminopyridines

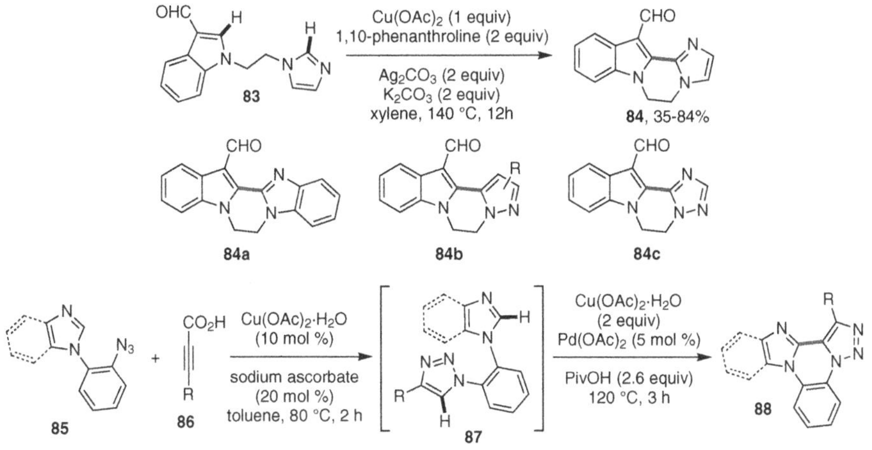

SCHEME 3.30

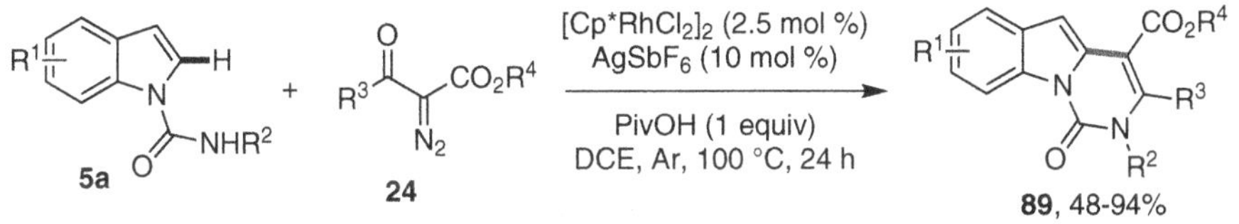

SCHEME 3.31

90. Reactions were performed by $Pd(OAc)_2$ (5 mol%) as the catalyst, in the presence of $K_2S_2O_8$ (3 equiv) as the oxidant, under a CO balloon (1 atm) in TFA at 70°C, delivering 11*H*-pyrido[2,1-*b*]quinazolin-11-ones **91** in 46–86% yields. *N*-phenyl-2-aminopyrimidine was also compatible with the reaction conditions, affording the corresponding 6*H*-pyrimido[2,1-*b*]quinazolin-6-one **91a** in 40% yield (Liang, He, and Zhu 2014). Also, the Pd/Ag-catalyzed pyrido carbonylation of *N*-aryl-2-aminopyridines **90** was described using DMF as the CO source. Reactions were carried out using $Pd(OAc)_2$ (10 mol%) and AgOTf (0.4 equiv) in DMF at 140°C under an atmosphere of O_2, to furnish the expected pyrido[2,1-*b*]quinazolinone structural motifs **91** in 70–91% yields. *N*-(3-thienyl)-2-aminopyrdine and *N*-phenyl-2-aminopyrimidine provided the corresponding pyrido[1,2-*a*]thieno[3,2-*d*]pyrimidin-10-one **91b** and pyrimido[2,1-*b*]quinazolin-6-one **91c** in 68% and 82% yields, respectively, while no desired product was obtained when *N*-(pyridin-3-yl)pyridin-2-amine was investigated. Reaction with *N*-phenyl isoquinolin-1-amine under reaction conditions gave 8*H*-isoquinolino[1,2-*b*]quinazolin-8-one **91d** in 81% yield. The reaction proceeded by insertion of *N*,*N*-dimethylmethyleneammonium salt **I**, generated in situ from DMF via a sequential decarbonylation, nucleophilic addition, and elimination process under silver catalysis with oxygen as the oxidant, into the C–Pd bond in the *ortho*-C–H activated intermediate, followed by reductive elimination (Scheme 3.32) (Nageswar Rao, Rasheed, and Das 2016).

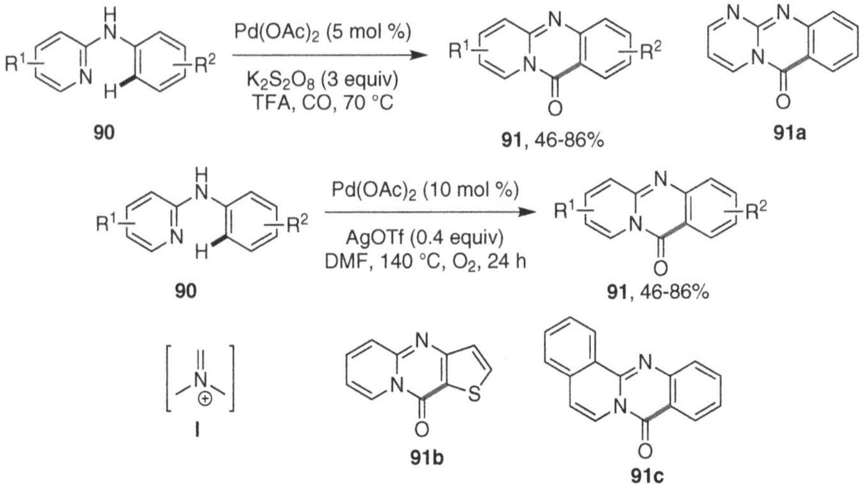

$Pd(OAc)_2$ (5 mol %)
$K_2S_2O_8$ (3 equiv)
TFA, CO, 70 °C

90 → 91, 46-86% 91a

$Pd(OAc)_2$ (10 mol %)
AgOTf (0.4 equiv)
DMF, 140 °C, O_2, 24 h

90 → 91, 46-86%

I 91b 91c

SCHEME 3.32

3.3.8 Isoquinolino-Quinazoline and Quinazolino-Phenanthridine

5,6-Diaryl-8*H*-isoquinolino[1,2-*b*]quinazolin-8-one derivatives **93** were synthesized by Ru(II)-catalyzed dehydrogenative cross-coupling/annulations of quinazolones **92** with internal alkynes **6**. $[RuCl_2(p\text{-cymene})]_2$ (5 mol%) in the presence of $Cu(OAc)_2$ (2.2 equiv) and Na_2CO_3 (2 equiv) in toluene at 90°C for 16 h furnished isoquinolino[1,2-*b*]quinazolin-8-ones **93** in 37–98% yields. Reaction with an unsymmetrical alkyne, 1-phenylpropyne, afforded 5-methyl-6-phenyl-8*H*-isoquinolino[1,2-*b*]quinazolin-8-one in 79% yield, as a major regioisomer (Lu et al. 2014). Also, Pd(II)-catalyzed synthesis of 3*aH*-furo[3',2':3,4]isoquinolino[1,2-*b*]quinazolin-13(14*aH*)-one derivatives **94** was achieved by aerobic dehydrogenative coupling reaction of quinazolinones **92** and alkynes **6** through sequential [4+2]/[3+2] cycloadditions. Subjecting quinazolinones **92** and 3 equiv of alkyne **6** in DMF/DMA solution to $Pd(CH_3CN)_2Cl_2$ (10 mol%), CuBr (0.5 equiv), and TFA (4 equiv) under O_2 (1 atm) led to the construction of furo[3',2':3,4]isoquinolino[1,2-*b*]quinazolines **94** in 41–78% yields. Reactions initiated by Pd(II)-promoted C–H activation generated isoquinolino[1,2-*b*]quinazolin-8-ones **93**, which then underwent O_2-incorporated [3+2] cycloaddition with another alkyne, to afford the final desired product **94** (Scheme 3.33) (Feng et al. 2017).

Pd-catalyzed intramolecular C–H bond activation and arylation was developed in order to prepare 14*H*-quinazolino[3,2-*f*]phenanthridin-14-one derivatives **96**. The reactions were catalyzed with $Pd(OAc)_2$ (10 mol%) in the presence of Na_2CO_3 (2.5 equiv) in DMF under N_2 atmosphere at 100°C, in which both 3-(2-bromophenyl)-2-arylquinazolin-4(3*H*)-ones **95a** and 2-(2-bromoaryl)-3-arylquinazolin-4(3*H*)-ones **95b** were converted to corresponding quinazolino-phenanthridines **96** in 62–75% and 73–82% yields, respectively (Scheme 3.34) (Gupta et al. 2015).

Also, 1*H*-quinazolino[3,2-*f*]phenanthridine-4,14-dione scaffolds **98** were synthesized by the amide-directed Ru(II)-catalyzed dehydrogenative coupling reaction

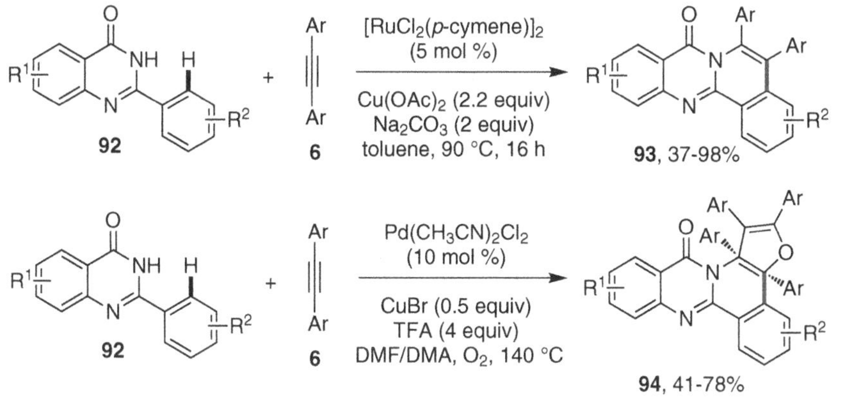

SCHEME 3.33

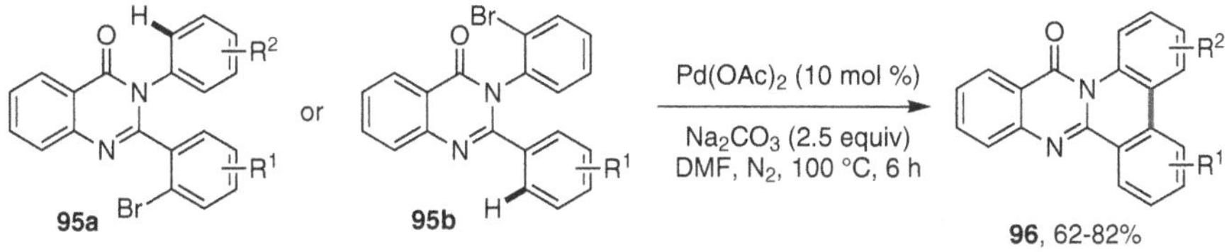

SCHEME 3.34

of 2-arylquinazolin-4(3*H*)-one **97** with cyclic 2-diazo-1,3-diketones **24b**. Reactions were carried out using $[RuCl_2(p\text{-cymene})]_2$ (3 mol%) and $AgNTf_2$ (30 mol%) in *t*-BuOH at 90°C, providing 1*H*-quinazolino[3,2-*f*]phenanthridine-4,14-diones **98** in 35–74% yields. Ir(III)-catalyzed construction of benzo[*c*]phthalazino[2,3-*a*]cinnoline-4,10,15(1*H*)-trione derivatives **100** was also developed by the dehydrogenative coupling reaction of 2-aryl-2,3-dihydrophthalazine-1,4-diones **99** with cyclic 2-diazo-1,3-diketones **24b**, using $[Cp^*IrCl_2]_2$ (2 mol%) and $AgSbF_6$ (20 mol%) in DCE at 100°C. The expected phthalazino[2,3-*a*]cinnoline products **100** were obtained in 75–95% yields. The reactions took place by Ru(II)/Ir(III)-catalyzed quinazolin- or phthalazine-nitrogen-directed C–H bond activation, followed by carbene coordination and migratory insertion, with subsequent protonolysis to the alkylated product, which was converted into the final desired fused heterocyclic systems by isomerization/intramolecular condensation (Scheme 3.35) (Cai et al. 2018).

Mayakrishnan et al. (2016) reported an Rh-catalyzed dehydrogenative C–H/N–H functionalization to obtain phthalazino[2,3-*a*]cinnolines **101** by reacting equivalent amounts of *N*-aryl phthalazine **99** with internal alkynes **6** in the presence of $[Cp^*RhCl_2]_2$ (2.5 mol%), $AgSbF_6$ (10 mol%), and $Cu(OAc)_2{\cdot}H_2O$ (1 equiv) in *t*-AmOH at 100°C under air, in 65–98% yields. Dialkyl acetylenes led to higher yields of products (96–98%) than diarylacetylene substrates (70–94%). Unsymmetrical alkyl phenyl alkynes were also compatible with the reaction, affording 5-alkyl-6-phenylphthalazino[2,3-*a*]cinnoline-8,13-diones **101** in 65–85% yields, regioselectively. The methodology was also extended to the construction of indazolo[1,2-*a*]

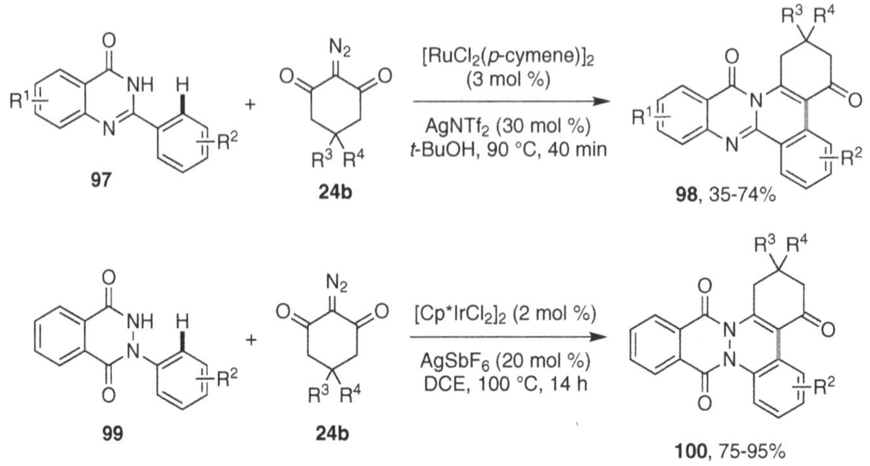

SCHEME 3.35

cinnolines **103** in 75–94% yields, by reacting *N*-phenylindazole **102** with alkynes **6** under similar reaction conditions (Scheme 3.36).

5-Alkyl-6-arylphthalazino[2,3-*a*]cinnoline-8,13-diones **104** were also synthesized by Ru-catalyzed C–H bond activation of phthalazinones, accomplished by the unusual deoxygenation of propargyl alcohols. Reactions were conducted at 110°C in acetic acid using $[RuCl_2(p\text{-cymene})]_2$ (5 mol%), $Cu(OAc)_2$ (2 equiv), and KPF_6 (30 mol%) catalytic systems, and phthalazino[2,3-*a*]cinnolines **104** were obtained unexpectedly in 70–92% yields. No reaction occurred when *ortho*-substituted aryl phthalazinones **99** were used. Secondary propargyl alcohols **7** were also tolerated under the reaction conditions. Moreover, *N*-arylpyridazinones were also compatible with the reaction, delivering pyridazino[1,2-*a*]cinnoline-1,4-diones in 75–94% yields. The proposed reaction mechanism includes the regioselective insertion of the

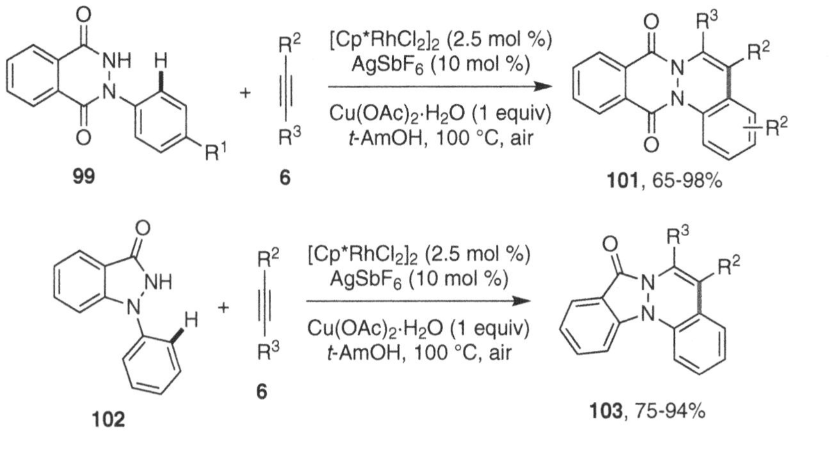

SCHEME 3.36

in situ formed propargylic ester **7a** into ruthenacycle **II**, generated by C–H activation of *N*-arylphthalazinones **99**, to give seven-membered Ru(II) species **IV**, which underwent isomerization to intermediate **VI** under acidic condition. By rearrangement of **VI** to Ru(II) species **VII**, a seven-membered Ru(II) species **IX** was generated with the removal of acetate, converting to the final product **104** by oxidative elimination in the presence of Cu(II) (Scheme 3.37) (Rajkumar et al. 2016).

3.3.9 Cationic Six-Membered *N*-Bridged Heterocycles

Zhao et al. (2015b) reported a synthesis of mesoionic triazolo[5,1-*a*]isoquinolium **106** via Rh-catalyzed annulation of 4-aryltriazoles **105** with 1.2 equiv dialkyl acetylenes **6** using $[Cp^*RhCl_2]_2$ (5 mol%), in the presence of $Cu(OAc)_2{\cdot}H_2O$ (2 equiv) and

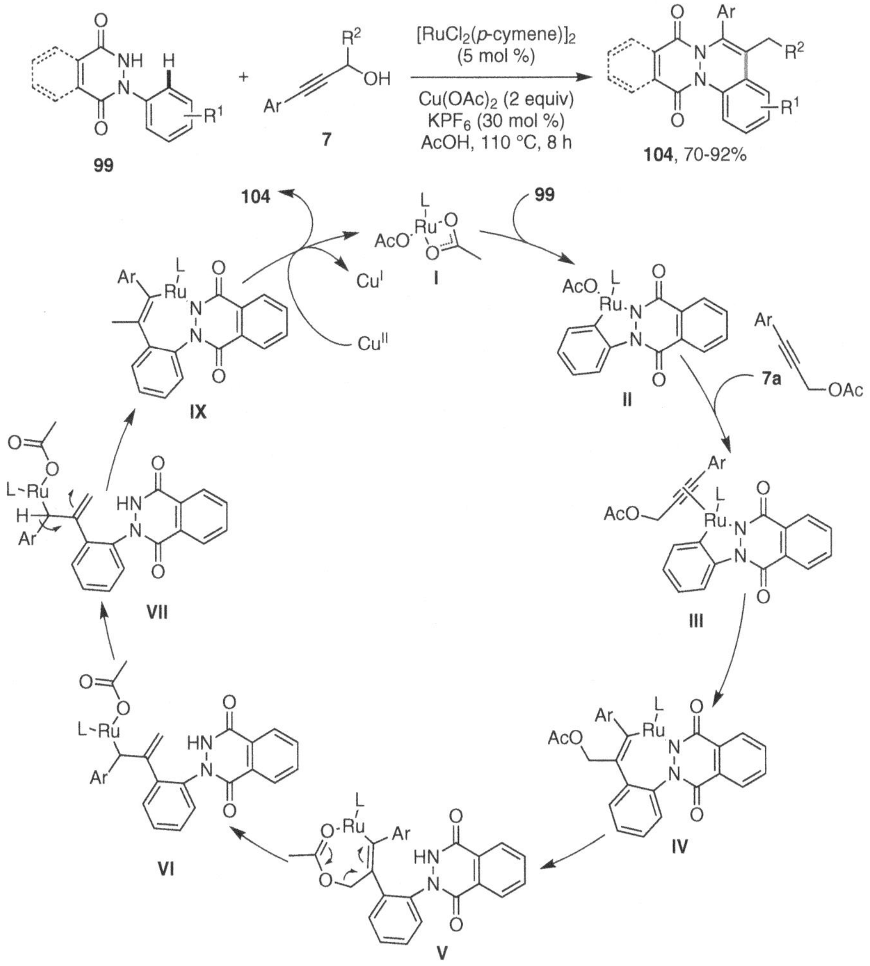

SCHEME 3.37

Na_2O_2 (2 equiv) in DCE at 100°C under air. Corresponding triazolo[5,1-*a*]isoquinolin-4-ium-1-olates **106** were obtained in 30–62% yields. Unsymmetrical alkynes led to two regioisomers. Reactions of aryl alkynes were carried out using di-*tert*-butyl peroxide (DTBP, 2 equiv) instead of Na_2O_2, leading to the expected mesoionic triazolo[5,1-*a*]isoquinoliums **106** in 29–56% yields. Reaction with 1-phenylpropyne afforded corresponding products in 62–66% yields, with low regioselectivity. The reaction involves sequential triazole-directed C–H activation and C–C, C–N, and C–O bond formation processes. Also, mesoionic oxadiazolo[3,4-*a*]quinolinium-3-olate derivatives **108** were obtained by Rh(III)-catalyzed C–H activation of sydnones **107** and dehydrogenative coupling with internal arylalkynes **6**. Reactions were catalyzed with $[Cp^*RhCl_2]_2$ (1 mol%), and $Cu(OAc)_2$ (2 equiv) in CH_3CN at 25°C, giving [1,2,3]oxadiazolo[3,4-*a*]quinolin-10-ium-3-olates **108** in 37–97% yields. Reaction with 3-hexyne led to corresponding mesoionic oxadiazolo[3,4-*a*]quinolinium-3-olate in 79% yields. The reaction of unsymmetrical alkyl aryl alkynes led to corresponding mesoionic oxadiazolo[3,4-*a*]quinolinium-3-olates **108** in 85–94% yields, with low regioselectivity. The reaction was initiated by cyclometalation of the sydnone and rollover C–H activation, with subsequent insertion of an alkyne into the Rh–C bond, followed by reductive elimination to obtain product **108** (Scheme 3.38) (Li et al. 2016).

The reaction between 2,3-dimethyl-1-(pyridin-2-yl)-1*H*-imidazol-3-iums **109** (1.1 equiv) and internal alkynes **6** in the presence of $[Cp^*RhCl_2]_2$ (3 mol%) as the catalyst, NaOAc (5 equiv) as the base, and AgOTf (2.5 equiv) as the oxidant in DCE at 110°C under Ar atmosphere led to annulated products imidazo[1,5-*a*][1,8]naphthyridiniums **110** in 60–92% yields. However, diaryl acetylenes afforded higher yields of products rather than dialkyl acetylenes. Unsymmetrical 1-phenylpropyne afforded the expected 5-methyl-6-phenylimidazo[1,5-*a*][1,8]naphthyridinium **110** in 80% yield, regioselectively (Dutta, Ghorai, and Choudhury 2018). Dual C(sp²)–H bond activation/annulation of *N*-aryl imidazolium **111** with internal alkynes **6** was reported to construct benzo[*ij*]imidazo[2,1,5-*de*]quinolizinium scaffolds **112**. Reactions were catalyzed with $[Cp^*RhCl_2]_2$ (5 mol%) in the presence of AgOTf (6 equiv) as

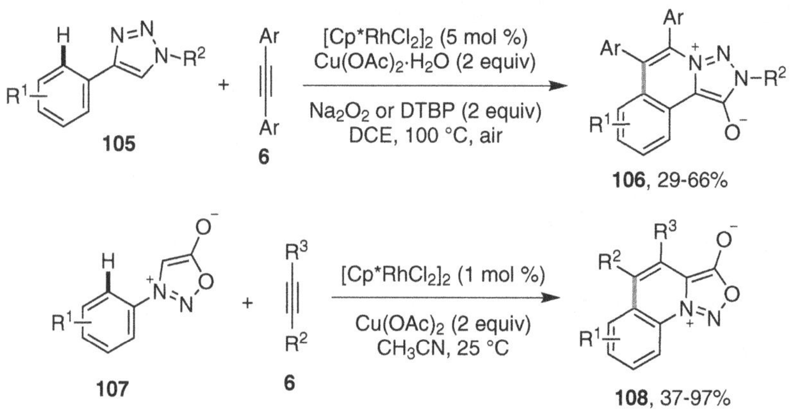

105 + 6 (Ar≡Ar) → [$[Cp^*RhCl_2]_2$ (5 mol %), $Cu(OAc)_2 \cdot H_2O$ (2 equiv); Na_2O_2 or DTBP (2 equiv), DCE, 100 °C, air] → **106**, 29-66%

107 + 6 (R^3≡R^2) → [$[Cp^*RhCl_2]_2$ (1 mol %); $Cu(OAc)_2$ (2 equiv), CH_3CN, 25 °C] → **108**, 37-97%

SCHEME 3.38

an oxidant and NaOAc (8 equiv) as a base in DCE under N_2 atmosphere at 100°C. Benzo[*ij*]imidazo[2,1,5-*de*]quinoliziniums **112** were obtained in 57–95% yields. The reaction with unsymmetrical alkynes afforded 4,8-dialkyl-3,9-diphenylbenzo[*ij*]imidazo[2,1,5-*de*]quinolizinium derivatives **112** in 57–83% yields, regioselectively (Ghorai and Choudhury 2015, Ge et al. 2015). The Rh(III)-catalyzed reaction of 2-arylbenzimidazoles **113** with alkynes **6** (2 equiv) afforded two C–N bonds formation through a double-dehydrogenative C–H/N–H annulation, which resulted in the formation of imidazo[1,2,3-*ij*][1,8]naphthyridinium derivatives **114** (Scheme 3.39) (Villar et al. 2017).

The dehydrogenative C–H activation/annulation of 2-phenylpyridine **115** and internal alkynes **6** was catalyzed with $[Cp^*Rh(H_2O)_3](OTf)_2$ (1 mol%), in the presence of TfOH (1 equiv) in MeOH at 120°C under O_2 atmosphere, leading to pyrido[2,1-*a*]isoquinolinium derivatives **116** in 49–99% yields. Reaction with dialkyl acetylene, 4-octyne, resulted in the formation of the expected pyrido[2,1-*a*]isoquinolinium **116** in 78% yield. Unsymmetrical alkyl aryl alkynes were also tolerated under reaction conditions, resulting in the formation of 7-alkyl-6-phenylpyrido[2,1-*a*]isoquinoliniums **116** in 77–79% yields, with 6:1 regioselectivity. Benzo[*de*]pyrido[3,2,1-*ij*]quinoliniums **116** and isoquinolino[2,1-*a*]quinoliniums **116a** were obtained in 71–87% and

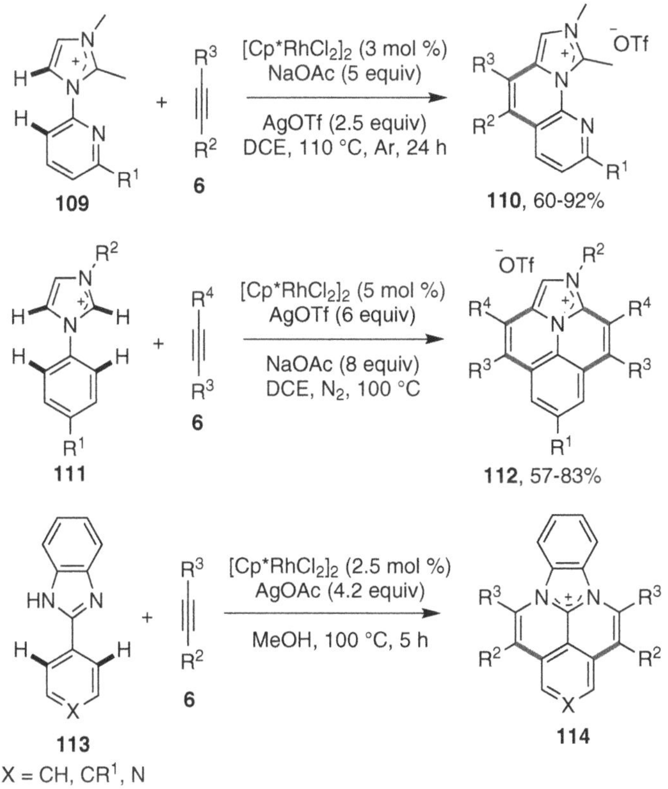

109 + 6 → 110, 60-92%
$[Cp^*RhCl_2]_2$ (3 mol %), NaOAc (5 equiv), AgOTf (2.5 equiv), DCE, 110 °C, Ar, 24 h

111 + 6 → 112, 57-83%
$[Cp^*RhCl_2]_2$ (5 mol %), AgOTf (6 equiv), NaOAc (8 equiv), DCE, N_2, 100 °C

113 + 6 → 114
$[Cp^*RhCl_2]_2$ (2.5 mol %), AgOAc (4.2 equiv), MeOH, 100 °C, 5 h
X = CH, CR¹, N

SCHEME 3.39

92–99% yields when benzo[*h*]quinolines and 2-phenylquinolines were treated with diphenyl acetylene under reaction conditions, respectively (Zhang et al. 2013). A similar approach was also developed using a Co(III) catalyst (Prakash, Muralirajan, and Cheng 2016). Also, pyrido[1,2-*a*]quinolinium derivatives **118** were prepared by Rh-catalyzed dehydrogenative coupling of 1-phenylpyridinium trifluoromethanesulfonates **117** with internal alkynes **6** using $[Cp^*RhCl_2]_2$ (5 mol%) as the catalyst in the presence of $Cu(OAc)_2$ (2 equiv) as an oxidant in CH_3CN at 100°C. The corresponding pyrido[1,2-*a*]quinolinium **118** was obtained in 27–94% yields. An amount of 2 equiv of alkynes afforded double annulation products quinolizino[3,4,5,6-*ija*]quinoliniums **119** in 52–94% yields (Scheme 3.40) (Ge et al. 2016).

The construction of pyrido[2,1-*a*]isoquinolinium salts **121** was reported by Rh(III)-catalyzed cascade double *N*-annulation reactions between allylamines **120** and diarylacetylenes **6** in the presence of HBF_4. Reactions were performed using 3 equiv of alkyne catalyzed with $[Cp^*RhCl_2]_2$ (5 mol%) in the presence of $Cu(OAc)_2{\cdot}H_2O$ (5 equiv) and HBF_4 (1.5 equiv) in MeOH at 130°C, and pyrido[2,1-*a*]isoquinoliniums **121** were obtained in 37–97% yields. 2,3-Diphenylpyridine **I** was proposed as the reaction intermediate, as it underwent pyridine-nitrogen-directed *ortho*-C–H activation of the 2-phenyl ring, followed by alkyne coordination and insertion, with subsequent reductive elimination to the final desired product **121**. Also, 2,3-diphenylpyridines were subjected to the reaction with alkynes under reaction conditions, to furnish the expected pyrido[2,1-*a*]isoquinoliniums **121** in 93–99% yields (Scheme 3.41) (Han et al. 2017).

A cascade of double C–H annulation of aldoximes **122** with alkynes **6** was developed by using $[Cp^*Rh(OAc)_2]_2$ (2.5 mol%) as the catalyst with oxygen as the oxidant in DCE at 100°C, leading to the construction of isoquinolino[3,2-*a*]isoquinolinum salts **123**. Reactions were carried out in the presence of $Zn(OTf)_2$ (50 mol%) and AcOH (1 equiv), and isoquinolino[3,2-*a*]isoquinolinums **123** were delivered in 42–98% yields. Thiophene-2-carbaldoxime afforded the corresponding thieno[3',2':4,5]pyrido[2,1-*a*]isoquinolinium **123a** in 81% yield. However, thieno[2',3':3,4]pyrido[1,2-*b*]isoquinolinium **123b** was obtained in 56% yield when benzaldoxime was treated with di(2-thienyl)acetylene under reaction conditions. Reaction with unsymmetrical alkyl aryl

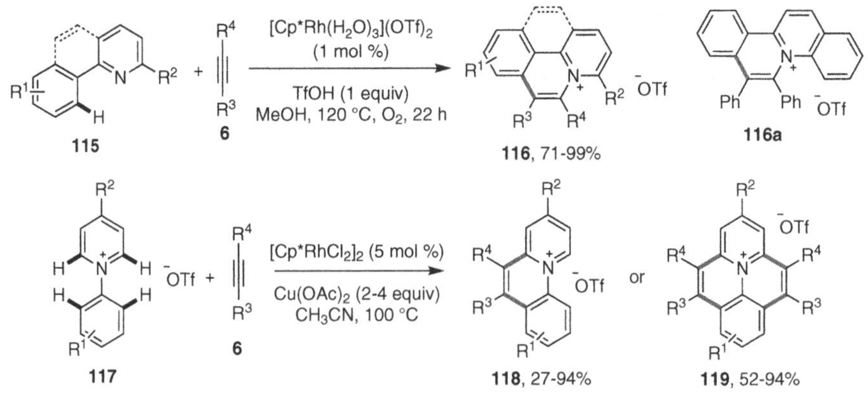

SCHEME 3.40

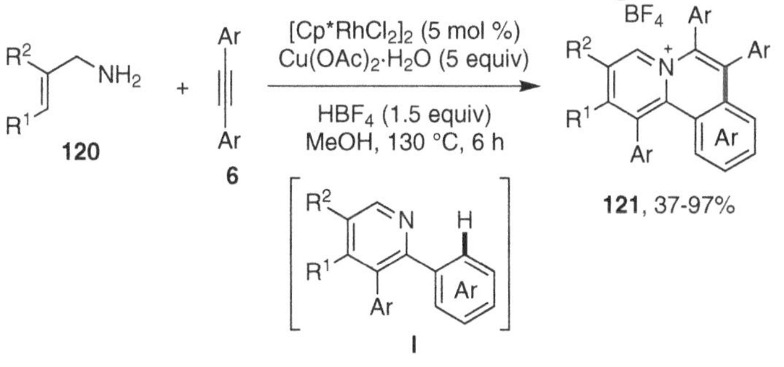

SCHEME 3.41

alkyne 1-phenylpropyne led to 5,13-dimethyl-6-phenylisoquinolino[3,2-*a*]isoquinolin-7-ium **123c** in 73% yield, with 20:1 regioselectivity. The reaction proceeded by *ortho*-C–H activation of aldoxime and functionalization with alkyne to generate intermediate 3,4-diarylisoquinoline **I**, which underwent a second alkyne insertion and annulation to obtain the final desired product. Subjecting 3,4-diphenylisoquinoline **I** to an alkyne under reaction conditions led to the expected isoquinolino[3,2-*a*]isoquinolinum **123**, revealing the generation of 3,4-diphenylisoquinoline **I** as the reaction intermediate. Dialkyl acetylene and phenylacetylene carboxylate, as well as diaryl alkynes, were compatible in the reaction with 3,4-diphenylisoquinoline **I**, giving corresponding isoquinolino[3,2-*a*]isoquinolinums **123** (Tang et al. 2017). Moreover, isoquinolino[2,1-*a*]quinolinium salts **125** were prepared by Rh-catalyzed cascade C–H activation and N–C annulation reaction of imines, generated in situ from benzaldehydes **56** and anilines **124**, with alkynes **6**. Reactions were carried out using $[Cp^*RhCl_2]_2$ (5 mol%), $AgBF_4$ (20 mol%), $Cu(OAc)_2$ (4 equiv), and NaOAc (2 equiv) in DCE at 140°C under a N_2 atmosphere, to give isoquinolino[2,1-*a*]quinoliniums **125** in 45–85% yields. Thiophene-2-carbaldehyde afforded thieno[3',2':3,4]pyrido[1,2-*a*]quinolinium when reacted with aniline and diphenyl acetylene under the reaction conditions. 2-Phenylisoquinolin-2-ium **II** was proposed as the reaction intermediate (Scheme 3.42) (Kadam et al. 2018).

Synthesis of isoquinolino[2,1-*b*]cinnolinium salts **127** was achieved by an Rh(III)-catalyzed cascade of aromatic $C(sp^2)$–H dehydrogenative coupling/cyclization reaction of Boc-arylhydrazines **126** and alkynes **6**. Heating a solution of Boc-arylhydrazines **126** and alkynes **6** (2.5 equiv) in CH_3CN in the presence of $[Cp^*RhCl_2]_2$ (5 mol%), $AgSbF_6$ (20 mol%), $Cu(OAc)_2{\cdot}H_2O$ (4 equiv), and $AgBF_4$ (2 equiv) at 80°C for 24 h under an atmosphere of argon produced isoquinolino[2,1-*b*]cinnolinium derivatives **127** in 54–95% yields. Thieno[2',3':3,4]pyrido[1,2-*b*]cinnolin-6-ium was obtained in 38% yield when di(2-thienyl)acetylene was reacted with Boc-arylhydrazine **126**. The plausible reaction mechanism involves the *ortho*-C–H bond cleavage of diazocompound **I**, derived in situ from Boc-phenylhydrazine **126** through oxidation with a Cu(II) salt, to generate intermediate **II**, which underwent an alkyne **6** coordination and insertion into the Rh–C bond, leading to the seven-membered rhodacycle intermediate **III**. Reductive elimination afforded intermediate **IV** along with an Rh(I) species, which was oxidized to the catalytic active

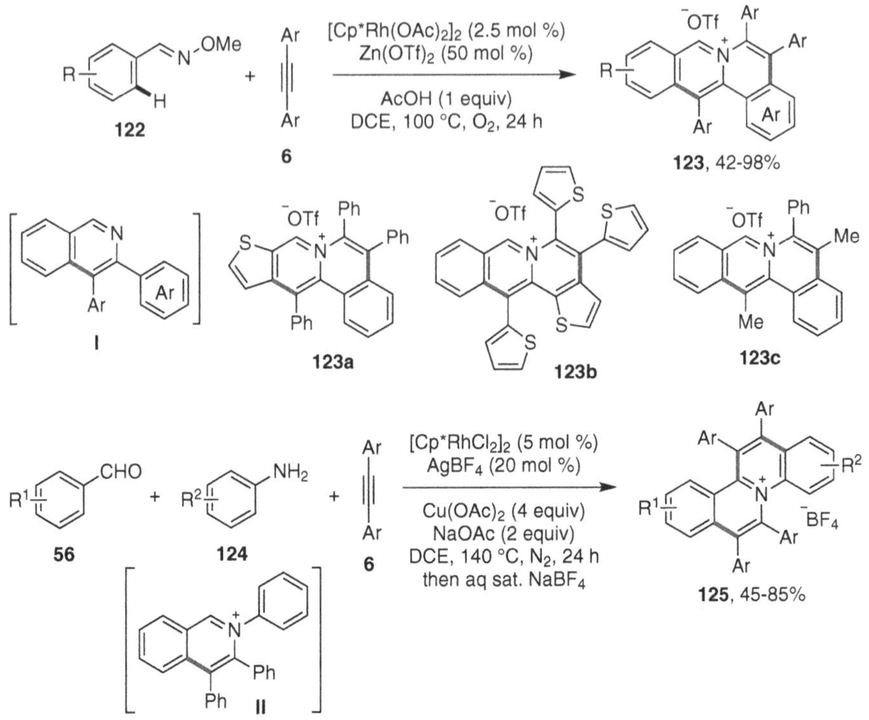

SCHEME 3.42

Rh(III) species by the Cu(II) salt. Intermediate **IV** was transformed into 3,4-diphenylcinnoline intermediate **V** by the elimination of CO_2 and an *iso*-butylene. The C–H bond activation of **V** gave five-membered rhodacycle intermediate **VI**, which underwent coordination and insertion of the second molecule of alkyne **6** to form the intermediate **VII**. Finally, reductive elimination of **VII** and an exchange of anion with $AgBF_4$ led to cinnolinium salt **127**, along with an Rh(I) species (Scheme 3.43) (Wang, Li, and Wang 2018). Rh(III)-catalyzed dehydrogenative coupling of *N-tert*-butyl-aryldiazene with diaryl acetylene was also described in order to construct isoquinolino[2,1-*b*]cinnolinium salt derivatives (Zhao et al. 2013).

3.4 SEVEN-MEMBERED *N*-BRIDGED HETEROCYCLES

Pd(II)-catalyzed intramolecular dehydrogenative coupling of indoles (**128**, **130** and **132**) with aryls and hetaryls tethered at the N-1 position was developed in order to synthesize seven-membered *N*-fused heterocyclic compounds. Reactions were carried out using 10 mol% of $Pd(OAc)_2$, K_2CO_3 (1 equiv), and $Cu(OAc)_2$ (3 equiv) in DMA at 90°C for 16 h, to give 5*H*-benzo[3,4]azepino[1,2-*a*]indole derivatives **129** in 60–95% yields. Benzo[5,6][1,3]oxazepino[3,4-*a*]indoles, benzo[6,7][1,4]oxazepino[4,5-*a*] indoles, and 5*H*-benzo[5,6][1,4]diazepino[1,7-a]indoles **129** were obtained in 63–70% and 78–87% yields when oxygen- and nitrogen-containing tethered substrates were

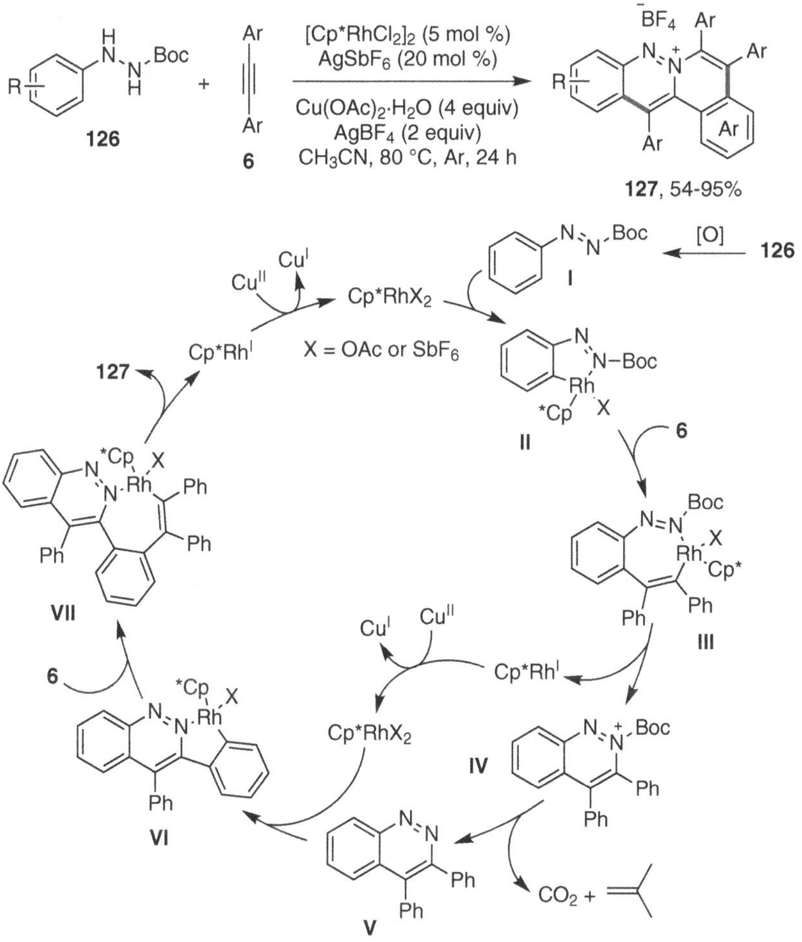

SCHEME 3.43

used, respectively. The reaction of 1-(*N*-hetarylpropyl)indoles **130** led to diazepino[1,2-*a*]indole derivatives **131** in 43–91% yields. Also, diazocino[1,8-*a*]indole derivatives **133** were prepared in 51–62% yields via intramolecular dehydrogenative coupling of *N*-benzyl-2-(1*H*-indol-1-yl)ethanamines **132** under reaction conditions. The reaction was initiated by activation of the C2–H of the indole, followed by base-mediated *ortho*-palladation of the aryl moiety, with subsequent reductive elimination to obtain the final products (Scheme 3.44) (Pintori and Greaney 2010).

Construction of azepino[3,2,1-*hi*]indole-1,2-dione derivatives **135** was accomplished by Pd-catalyzed dehydrogenative cycloaddition between isatin **134** and internal alkynes **6** through C-H/N-H activation. By treatment of isatin **134** with internal alkynes **6** (5 equiv) in the presence of $Pd(OAc)_2$ (10 mol%) as the catalyst and AgOAc (2 equiv) as an oxidant, in a mixture of CH_3CN/dioxane at 100°C under an atmosphere of N_2, azepino[3,2,1-*hi*]indole-1,2-diones **135** were obtained in 55–96%

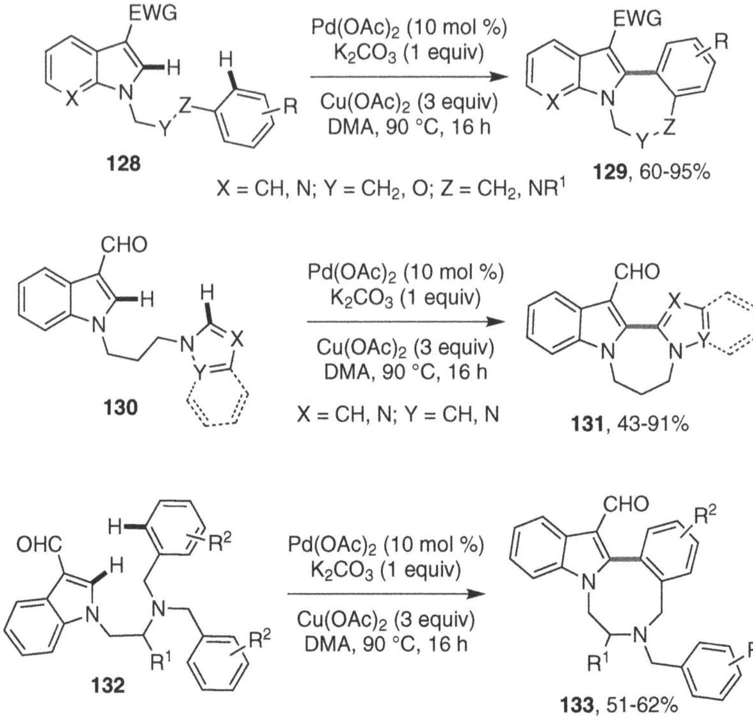

SCHEME 3.44

yields. The reaction of unsymmetrical alkyne, 1-phenylpropyne, afforded the corresponding 4,6-dimethyl-5,7-diphenylazepino[3,2,1-*hi*]indole-1,2-dione **135** in 60% yield, with moderate regioselectivity (13:7). Reaction with methyl phenylacetylene carboxylate led to corresponding azepino[3,2,1-*hi*]indole-5,7-dicarboxylate **135a** in 63% yield, regioselectively. The reaction occurred by N–H activation and functionalization with an alkyne to alkenyl palladium species **I**, followed by insertion of a second alkyne molecule (intermediate **II**) and then *ortho*-C–H bond activation (**III**) with subsequent reductive elimination (Scheme 3.45) (Wang et al. 2013).

Rh(III)-catalyzed redox-neutral C7-selective C–H activation/annulation of *N*-methoxycarbamoyl-protected indoline **136** with diaryl alkynes **6** was developed to access diazepino[6,7,1-*hi*]indol-1(2*H*)-ones **137**. Reactions were performed using $[Cp^*RhCl_2]_2$ (2.5 mol%) in the presence of NaOAc (2 equiv) in MeOH at 80°C, and the desired diazepino-indoles **137** were furnished in 52–80% yields. Reaction with alkyl aryl alkynes led to corresponding 4-alkyl-3-phenyl-8,9-dihydro-[1,3]diazepino[6,7,1-*hi*]indol-1(2*H*)-ones **137** in 52–54% yields, regioselectively. Interestingly, reaction with aliphatic alkynes resulted in the formation of pyrrolo[3,2,1-*ij*]quinolin-4(2*H*)-one scaffolds **138** in 40–60% yields. The proposed reaction mechanism involves the *N*-methoxycarbamoyl-directed activation of the C7–H bond to generate a six-membered rhodacycle intermediate **I**, which underwent coordination and insertion of an alkyne molecule to give intermediate **II**. Utilizing *N*-methoxycarbamoyl as

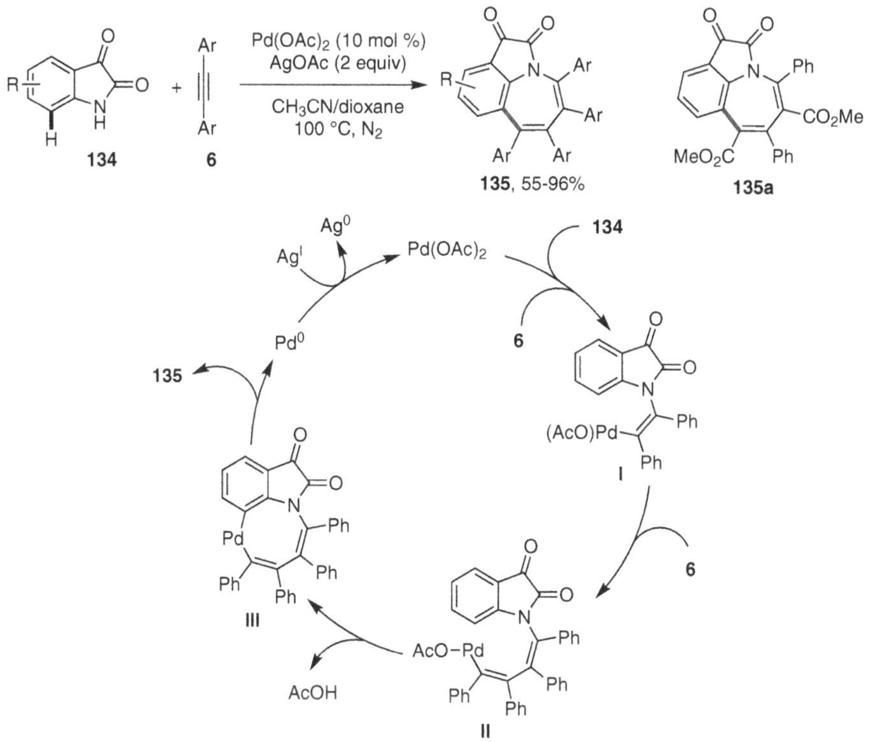

SCHEME 3.45

an oxidizing–directing group led to reductive elimination to intermediate **III** in the case of aryl alkynes, along with formation of an Rh(I) species, which underwent oxidation with intermediate **III** to regenerate the Rh(III) active catalyst species while releasing the final desired product **137**. In the case of aliphatic alkynes, intermediate **II** underwent intramolecular nucleophilic addition to intermediate **IV**, followed by release of methoxyamine, to give product **138** with the regeneration of the Rh(III) catalyst (Scheme 3.46) (Wang et al. 2015).

3.5 CONCLUSION

In summary, aromatic C(sp^2)–H dehydrogenative coupling processes are studied for the preparation of five-, six-, and seven-membered *N*-bridged heterocycles, such as pyrrolo-indoles, indolo-indoles, isoindolo-indoles and indolo-carbazoles, imidazo-indoles, imidazo-isoindoles, isoindolo-isoquinolines, indolizines, imidazo-pyridines, and indazolo-phthalazines. Cationic *N*-bridged heterocycles were also prepared by dehydrogenative coupling reactions. However, many different methods were developed for the construction of these types of heterocycles, due to the quantitative and one-step synthesis and a broad spectrum of substitution on the synthesized heterocycles, the aromatic C–H bond direct functionalization approach could be of interest in the synthesis of pharmaceutical, medicinal, and natural products.

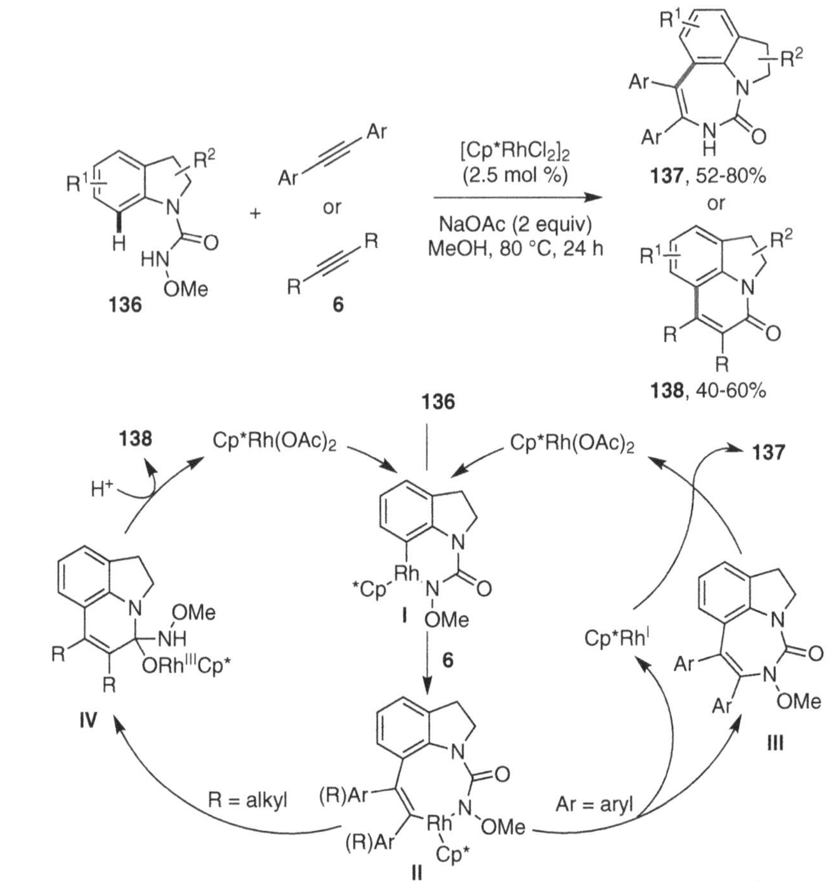

SCHEME 3.46

REFERENCES

Bharathimohan, Kuppusamy, Thanasekaran Ponpandian, A Jafar Ahamed, and Nattamai Bhuvanesh. 2014. "Sequential decarboxylative azide–alkyne cycloaddition and dehydrogenative coupling reactions: One-pot synthesis of polycyclic fused triazoles." *Beilstein Journal of Organic Chemistry* no. 10:3031.

Cai, Panyuan, Enshen Zhang, Yinsong Wu, Taibei Fang, Qianqian Li, Chen Yang, Jian Wang, and Yongjia Shang. 2018. "Ru (II)/Ir (III)-catalyzed C–H bond activation/annulation of cyclic amides with 1, 3-diketone-2-diazo compounds: Facile access to 8 H-isoquinolino [1, 2-b] quinazolin-8-ones and phthalazino [2, 3-a] cinnoline-8, 13-diones." *ACS Omega* no. 3(11):14575–14584.

Chen, Xiaohong, Shujing Yang, Helong Li, Bo Wang, and Guoyong Song. 2017. "Enantioselective C–H annulation of indoles with diazo compounds through a chiral Rh (III) catalyst." *ACS Catalysis* no. 7(4):2392–2396.

Chen, Xun, Xinwei Hu, Siyi Bai, Yuanfu Deng, Huanfeng Jiang, and Wei Zeng. 2015. "Rh (III)-catalyzed [4+ 2] annulation of indoles with diazo compounds: Access to pyrimido [1, 6-a] indole-1 (2 H)-ones." *Organic Letters* no. 18(2):192–195.

Chen, Xun, Xinwei Hu, Yuanfu Deng, Huanfeng Jiang, and Wei Zeng. 2016. "A [4+ 1] cyclative capture access to indolizines via cobalt (III)-catalyzed Csp^2–H bond functionalization." *Organic Letters* no. 18(18):4742–4745.

Dutta, Champak, Debasish Ghorai, and Joyanta Choudhury. 2018. "To "rollover" or not? Stereoelectronically guided C–H functionalization pathways from rhodium–abnormal NHC intermediates." *ACS Omega* no. 3(2):1614–1620.

Dwight, Timothy A, Nicholas R Rue, Dagmara Charyk, Ryan Josselyn, and Brenton DeBoef. 2007. "C–C bond formation via double C–H functionalization: Aerobic oxidative coupling as a method for synthesizing heterocoupled biaryls." *Organic Letters* no. 9(16):3137–3139.

Eftekhari-Sis, Bagher, and Maryam Zirak. 2014. "Chemistry of α-oxoesters: A powerful tool for the synthesis of heterocycles." *Chemical Reviews* no. 115(1):151–264.

Eftekhari-Sis, Bagher, Maryam Zirak, and Ali Akbari. 2013. "Arylglyoxals in synthesis of heterocyclic compounds." *Chemical Reviews* no. 113(5):2958–3043.

Fan, Zhoulong, Kui Wu, Li Xing, Qizheng Yao, and Ao Zhang. 2014. "Palladium-catalyzed double C–H activation: One-pot synthesis of benzo [c] pyrazolo [1, 2-a] cinnolin-1-ones from 5-pyrazolones and aryl iodides." *Chemical Communications* no. 50(14):1682–1684.

Feng, Yadong, Nian Tian, Yudong Li, Chunqi Jia, Xuening Li, Lianhui Wang, and Xiuling Cui. 2017. "Construction of fused polyheterocycles through sequential [4+ 2] and [3+ 2] cycloadditions." *Organic Letters* no. 19(7):1658–1661.

Gandeepan, Parthasarathy, Thomas Müller, Daniel Zell, Gianpiero Cera, Svenja Warratz, and Lutz Ackermann. 2018. "3d transition metals for C–H activation." *Chemical Reviews* no. 119(4):2192–2452.

Ge, Qingmei, Yang Hu, Bin Li, and Baiquan Wang. 2016. "Synthesis of conjugated polycyclic quinoliniums by rhodium (III)-catalyzed multiple C–H activation and annulation of arylpyridiniums with alkynes." *Organic Letters* no. 18(10):2483–2486.

Ge, Qingmei, Bin Li, Haibin Song, and Baiquan Wang. 2015. "Rhodium (iii)-catalyzed cascade oxidative annulation reactions of aryl imidazolium salts with alkynes involving multiple C–H bond activation." *Organic & Biomolecular Chemistry* no. 13(28):7695–7710.

Gholamhosseyni, Maral, and Ebrahim Kianmehr. 2018. "A ruthenium-catalyzed alkenylation–annulation approach for the synthesis of indazole derivatives via C–H bond activation." *Organic & Biomolecular Chemistry* no. 16(33):5973–5978.

Ghorai, Debasish, and Joyanta Choudhury. 2015. "Rhodium (III)–N-heterocyclic carbene-driven cascade C–H activation catalysis." *ACS Catalysis* no. 5(4):2692–2696.

Gong, Tian-Jun, Meng-Yu Xu, Shang-Hai Yu, Chu-Guo Yu, Wei Su, Xi Lu, Bin Xiao, and Yao Fu. 2018. "Rhodium (III)-catalyzed directed C–H coupling with methyl trifluoroacrylate: Diverse synthesis of fluoroalkenes and heterocycles." *Organic Letters* no. 20(3):570–573.

Guo, Shenghai, Lincong Sun, Fang Wang, Xinying Zhang, and Xuesen Fan. 2018. "Rh (III)-catalyzed oxidative annulation of isoquinolones with diazoketoesters featuring an in situ deacylation: Synthesis of isoindoloisoquinolones and their transformation to rosettacin analogues." *The Journal of Organic Chemistry* no. 83(19):12034–12043.

Gupta, Puneet K, Nisha Yadav, Subodh Jaiswal, Mohd Asad, Ruchir Kant, and Kanchan Hajela. 2015. "Palladium-catalyzed synthesis of phenanthridine/benzoxazine-fused quinazolinones by intramolecular C-H bond activation." *Chemistry–A European Journal* no. 21(38):13210–13215.

Han, Ye Ri, Su-Hyang Shim, Dong-Su Kim, and Chul-Ho Jun. 2017. "Synthesis of benzoquinolizinium salts by Rh (III)-catalyzed cascade double N-annulation reactions of allylamines, diarylacetylenes, and HBF4." *Organic Letters* no. 19(11): 2941–2944.

Huang, Qiufeng, Qingshuai Han, Shurong Fu, Zizhu Yao, Lv Su, Xiaofeng Zhang, Shen Lin, and Shengchang Xiang. 2016. "Rhodium-catalyzed NH-indole-directed C–H carbonylation with carbon monoxide: Synthesis of 6 H-isoindolo [2, 1-a] indol-6-ones." *The Journal of Organic Chemistry* no. 81(24):12135–12142.

Ikemoto, Hideya, Tatsuhiko Yoshino, Ken Sakata, Shigeki Matsunaga, and Motomu Kanai. 2014. "Pyrroloindolone synthesis via a Cp* CoIII-catalyzed redox-neutral directed C–H alkenylation/annulation sequence." *Journal of the American Chemical Society* no. 136(14):5424–5431.

Jones, Alexander W, Marie-Laure Louillat-Habermeyer, and Frederic W Patureau. 2015. "Strained dehydrogenative ring closure of phenylcarbazoles." *Advanced Synthesis & Catalysis* no. 357(5):945–949.

Kadam, Vilas D, Boya Feng, Xingyu Chen, Wenbo Liang, Fulin Zhou, Yanhong Liu, Ge Gao, and Jingsong You. 2018. "Cascade C–H annulation reaction of benzaldehydes, anilines, and alkynes toward dibenzo [a, f] quinolizinium salts: Discovery of photostable mitochondrial trackers at the nanomolar level." *Organic Letters* no. 20(22):7071–7075.

Karishma, Pidiyara, Chikkagundagal K Mahesha, Devesh S Agarwal, Sanjay K Mandal, and Rajeev Sakhuja. 2018. "Additive-driven rhodium-catalyzed [4+ 1]/[4+ 2] annulations of N-arylphthalazine-1, 4-dione with α-diazo carbonyl compounds." *The Journal of Organic Chemistry* no. 83(19):11661–11673.

Laha, Joydev K, Rohan A Bhimpuria, and Mandeep Kaur Hunjan. 2017. "Intramolecular oxidative arylations in 7-azaindoles and pyrroles: Revamping the synthesis of fused N-heterocycle tethered fluorenes." *Chemistry–A European Journal* no. 23(9):2044–2050.

Li, Lei, He Wang, Xifa Yang, Lingheng Kong, Fen Wang, and Xingwei Li. 2016. "Rhodium-catalyzed oxidative synthesis of quinoline-fused sydnones via 2-fold C–H bond activation." *The Journal of Organic Chemistry* no. 81(23):12038–12045.

Li, Panpan, Xiaoying Xu, Jiayi Chen, Hequan Yao, and Aijun Lin. 2018. "Rh (III)-catalyzed synthesis of pyrazolo [1, 2-a] cinnolines from pyrazolidinones and diazo compounds." *Organic Chemistry Frontiers* no. 5(11):1777–1781.

Li, Xingwei, and Miao Zhao. 2011. "Rhodium (III)-catalyzed oxidative coupling of 5-aryl-1 H-pyrazoles with alkynes and acrylates." *The Journal of Organic Chemistry* no. 76(20):8530–8536.

Liang, Dongdong, Yimiao He, and Qiang Zhu. 2014. "Palladium-catalyzed C (sp²)–H pyridocarbonylation of N-aryl-2-aminopyridines: Dual function of the pyridyl moiety." *Organic Letters* no. 16(10):2748–2751.

Liu, Chen-Fei, Man Liu, and Lin Dong. 2018. "Iridium (III)-catalyzed tandem annulation synthesis of pyrazolo [1, 2-α] cinnolines from pyrazolones and sulfoxonium ylides." *The Journal of Organic Chemistry* no. 84(1):409–416.

Lu, Hui, Qin Yang, Yirong Zhou, Yanqin Guo, Zhihong Deng, Qiuping Ding, and Yiyuan Peng. 2014. "Cross-coupling/annulations of quinazolones with alkynes for access to fused polycyclic heteroarenes under mild conditions." *Organic & Biomolecular Chemistry* no. 12(5):758–764.

Manoharan, Ramasamy, and Masilamani Jeganmohan. 2015. "Ruthenium-catalyzed cyclization of N-carbamoyl indolines with alkynes: An efficient route to pyrroloquinolinones." *Organic & Biomolecular Chemistry* no. 13(35):9276–9284.

Mantenuto, Serena, Cecilia Ciccolini, Simone Lucarini, Giovanni Piersanti, Gianfranco Favi, and Fabio Mantellini. 2017. "Palladium (II)-catalyzed intramolecular oxidative C–H/C–H cross-coupling reaction of C3, N-linked biheterocycles: Rapid access to polycyclic nitrogen heterocycles." *Organic Letters* no. 19(3):608–611.

Mayakrishnan, Sivakalai, Yuvaraj Arun, Chandrasekar Balachandran, Nobuhiko Emi, Doraiswamy Muralidharan, and Paramasivan Thirumalai Perumal. 2016. "Synthesis of cinnolines via Rh (III)-catalysed dehydrogenative C–H/N–H functionalization:

Aggregation induced emission and cell imaging." *Organic & Biomolecular Chemistry* no. 14(6):1958–1968.

Morimoto, Keisuke, Koji Hirano, Tetsuya Satoh, and Masahiro Miura. 2010. "Rhodium-catalyzed oxidative coupling/cyclization of 2-phenylindoles with alkynes via C–H and N–H bond cleavages with air as the oxidant." *Organic Letters* no. 12(9): 2068–2071.

Morioka, Ryosuke, Kazunori Nobushige, Tetsuya Satoh, Koji Hirano, and Masahiro Miura. 2015. "Synthesis of indolo [1, 2-a][1, 8] naphthyridines by rhodium (III)-catalyzed dehydrogenative coupling via rollover cyclometalation." *Organic Letters* no. 17(12):3130–3133.

Mukherjee, Kallol, Majji Shankar, Koushik Ghosh, and Akhila K Sahoo. 2018. "An orchestrated unsymmetrical annulation episode of C (sp^2)–H bonds with alkynes and quinones: Access to spiro-isoquinolones." *Organic Letters* no. 20(7):1914–1918.

Nageswar Rao, D, Sk Rasheed, and Parthasarathi Das. 2016. "Palladium/silver synergistic catalysis in direct aerobic carbonylation of C (sp^2)–H bonds using DMF as a carbon source: Synthesis of pyrido-fused quinazolinones and phenanthridinones." *Organic Letters* no. 18(13):3142–3145.

Okada, Takeshi, Asumi Sakai, Tomoaki Hinoue, Tetsuya Satoh, Yoshihiro Hayashi, Susumu Kawauchi, Kona Chandrababunaidu, and Masahiro Miura. 2018. "Rhodium (III)-catalyzed oxidative coupling of N-phenylindole-3-carboxylic acids with alkenes and alkynes via C4–H and C2–H/C2′–H bond cleavage." *The Journal of Organic Chemistry* no. 83(10):5639–5649.

Pereira, Kyle C, Ashley L Porter, and Brenton DeBoef. 2014. "Intramolecular arylation of benzimidazoles via Pd (II)/Cu (I) catalyzed cross-dehydrogenative coupling." *Tetrahedron Letters* no. 55(10):1729–1732.

Pintori, Didier G, and Michael F Greaney. 2010. "Intramolecular oxidative C–H coupling for medium-ring synthesis." *Journal of the American Chemical Society* no. 133(5):1209–1211.

Piou, Tiffany, Luc Neuville, and Jieping Zhu. 2013. "Pd (II)-catalyzed intramolecular aminopalladation/direct C–H arylation under aerobic conditions: Synthesis of pyrrolo [1, 2-a] indoles." *Tetrahedron* no. 69(22):4415–4420.

Prakash, Sekar, Krishnamoorthy Muralirajan, and Chien-Hong Cheng. 2016. "Cobalt-catalyzed oxidative annulation of nitrogen-containing arenes with alkynes: An atom-economical route to heterocyclic quaternary ammonium salts." *Angewandte Chemie International Edition* no. 55(5):1844–1848.

Quinones, Noelia, Andres Seoane, Rebeca García-Fandiño, Jose Luis Mascarenas, and Moises Gulias. 2013. "Rhodium (III)-catalyzed intramolecular annulations involving amide-directed C–H activations: Synthetic scope and mechanistic studies." *Chemical Science* no. 4(7):2874–2879.

Rajkumar, Subramani, S Antony Savarimuthu, Rajendran Senthil Kumaran, CM Nagaraja, and Thirumanavelan Gandhi. 2016. "Expedient synthesis of new cinnoline diones by Ru-catalyzed regioselective unexpected deoxygenation-oxidative annulation of propargyl alcohols with phthalazinones and pyridazinones." *Chemical Communications* no. 52(12):2509–2512.

Ray, Devalina, T Manikandan, Arup Roy, Krishna N Tripathi, and Ravi P Singh. 2015. "Ligand-promoted intramolecular dehydrogenative cross-coupling using a Cu catalyst: Direct access to polycyclic heteroarenes." *Chemical Communications* no. 51(32):7065–7068.

Shaikh, Tanveer Mahamadali, and Fung-E Hong. 2016. "Recent developments in the preparation of N-heterocycles using Pd-catalysed C–H activation." *Journal of Organometallic Chemistry* no. 801:139–156.

Shankar, Majji, Koushik Ghosh, Kallol Mukherjee, Raja K Rit, and Akhila K Sahoo. 2016. "Ru-catalyzed one-pot diannulation of heteroaryls: Direct access to π-conjugated polycyclic amides." *Organic Letters* no. 18(24):6416–6419.
Sharma, Vikas, and Vipin Kumar. 2014. "Indolizine: A biologically active moiety." *Medicinal Chemistry Research* no. 23(8):3593–3606.
Shen, Bingxue, Bin Li, and Baiquan Wang. 2016. "Rh (III)-catalyzed oxidative annulation leading to substituted indolizines by cleavage of C (sp^2)–H/C (sp^3)–H bonds." *Organic Letters* no. 18(12):2816–2819.
Song, Guoyong, Dan Chen, Cheng-Ling Pan, Robert H Crabtree, and Xingwei Li. 2010. "Rh-catalyzed oxidative coupling between primary and secondary benzamides and alkynes: Synthesis of polycyclic amides." *The Journal of Organic Chemistry* no. 75(21):7487–7490.
Sutariya, Tushar R, Balvantsingh M Labana, Narsidas J Parmar, Rajni Kant, Vivek K Gupta, Gabriela B Plata, and Jose M Padron. 2015. "Efficient synthesis of some new antiproliferative N-fused indoles and isoquinolines via 1, 3-dipolar cycloaddition reaction in an ionic liquid." *New Journal of Chemistry* no. 39(4):2657–2668.
Tang, Junbin, Shiqing Li, Zheng Liu, Yinsong Zhao, Zhijie She, Vilas D Kadam, Ge Gao, Jingbo Lan, and Jingsong You. 2017. "Cascade C–H annulation of aldoximes with alkynes using O2 as the sole oxidant: One-pot access to multisubstituted protoberberine skeletons." *Organic Letters* no. 19(3):604–607.
Tripathi, Krishna N, Devalina Ray, and Ravi P Singh. 2017. "Synthesis of pyrrole-annulated heterocycles through copper-catalyzed site-selective dehydrogenative cross-coupling." *European Journal of Organic Chemistry* no. 2017(38):5809–5813.
Upadhyay, Nitinkumar Satyadev, Jayachandran Jayakumar, and Chien-Hong Cheng. 2017. "Facile one-pot synthesis of 2, 3-dihydro-1 H-indolizinium derivatives by rhodium (iii)-catalyzed intramolecular oxidative annulation via C–H activation: Application to ficuseptine synthesis." *Chemical Communications* no. 53(16):2491–2494.
Villar, José M, Jaime Suárez, Jesús A Varela, and Carlos Saá. 2017. "N-doped cationic PAHs by Rh (III)-catalyzed double C–H activation and annulation of 2-arylbenzimidazoles with alkynes." *Organic Letters* no. 19(7):1702–1705.
Wang, Fen, Guoyong Song, Zhengyin Du, and Xingwei Li. 2011. "Oxidative coupling of NH isoquinolones with olefins catalyzed by Rh (III)." *The Journal of Organic Chemistry* no. 76(8):2926–2932.
Wang, Honggen, Yong Wang, Changlan Peng, Jiancun Zhang, and Qiang Zhu. 2010. "A direct intramolecular C–H amination reaction cocatalyzed by copper (II) and iron (III) as part of an efficient route for the synthesis of pyrido [1, 2-a] benzimidazoles from N-aryl-2-aminopyridines." *Journal of the American Chemical Society* no. 132(38):13217–13219.
Wang, Lei, Jiayao Huang, Shiyong Peng, Hui Liu, Xuefeng Jiang, and Jian Wang. 2013. "Palladium-catalyzed oxidative cycloaddition through C-H/N-H activation: Access to benzazepines." *Angewandte Chemie International Edition* no. 52(6):1768–1772.
Wang, Xiaoxia, Na Li, Zhongfeng Li, and Honghua Rao. 2017. "Copper-catalyzed dehydrogenative C (sp^2)–N bond formation via direct oxidative activation of an anilidic N–H bond: Synthesis of benzoimidazo [1, 2-a] indoles." *The Journal of Organic Chemistry* no. 82(19):10158–10166.
Wang, Xiaoxia, Zhongfeng Li, Shengli Cao, and Honghua Rao. 2016. "Silver-catalyzed intramolecular C-2 selective acylation of indoles with aldehydes: An atom-economical entry to indole–indolone scaffolds." *Advanced Synthesis & Catalysis* no. 358(13):2059–2065.
Wang, Xuan, Huanyu Tang, Huijin Feng, Yuanchao Li, Yaxi Yang, and Bing Zhou. 2015. "Access to six-and seven-membered 1, 7-fused indolines via Rh (III)-catalyzed redox-neutral C7-selective C–H functionalization of indolines with alkynes and alkenes." *The Journal of Organic Chemistry* no. 80(12):6238–6249.

Wang, Yanwei, Bin Li, and Baiquan Wang. 2018. "Synthesis of cinnolines and cinnolinium salt derivatives by Rh (III)-catalyzed cascade oxidative coupling/cyclization reactions." *The Journal of Organic Chemistry* no. 83(18):10845—10854.

Wu, Xiaowei, and Haitao Ji. 2018. "Rhodium-catalyzed [4+1] cyclization via C–H activation for the synthesis of divergent heterocycles bearing a quaternary carbon." *The Journal of Organic Chemistry* no. 83(8):4650–4656.

Xie, Yanan, Xiaowei Wu, Chunpu Li, Jiang Wang, Jian Li, and Hong Liu. 2017. "Ruthenium (II)-catalyzed redox-neutral [3+2] annulation of indoles with internal alkynes via C–H bond activation: Accessing a pyrroloindolone scaffold." *The Journal of Organic Chemistry* no. 82(10):5263–5273.

Xing, Li, Zhoulong Fan, Chengyu Hou, Guoping Yong, and Ao Zhang. 2014. "Synthesis of pyrazolo [1, 2-a] cinnolines via a rhodium-catalyzed oxidative coupling approach." *Advanced Synthesis & Catalysis* no. 356(5):972–976.

Xu, Shuai, Ri Chen, Zihao Fu, Qi Zhou, Yan Zhang, and Jianbo Wang. 2017. "Palladium-catalyzed formal [4+1] annulation via metal carbene migratory insertion and C (sp^2)–H bond functionalization." *ACS Catalysis* no. 7(3):1993–1997.

Yamashita, Mana, Hakaru Horiguchi, Koji Hirano, Tetsuya Satoh, and Masahiro Miura. 2009. "Fused ring construction around pyrrole, indole, and related compounds via palladium-catalyzed oxidative coupling with alkynes." *The Journal of Organic Chemistry* no. 74(19):7481–7488.

Yang, Chao, Feifei Song, Jiean Chen, and Yong Huang. 2017. "Construction of pyridazine analogues via rhodium-mediated C-H activation." *Advanced Synthesis & Catalysis* no. 359(20):3496–3502.

Yang, Xiao-Fei, Xu-Hong Hu, and Teck-Peng Loh. 2015. "Expedient synthesis of Pyrroloquinolinones by Rh-catalyzed annulation of N-carbamoyl indolines with alkynes through a directed C–H functionalization/C–N cleavage sequence." *Organic Letters* no. 17(6):1481–1484.

Yip, Kai-Tai, and Dan Yang. 2011. "Pd (II)-catalyzed intramolecular amidoarylation of alkenes with molecular oxygen as sole oxidant." *Organic Letters* no. 13(8):2134–2137.

Zhang, Guoying, Lei Yang, Yanyu Wang, Yinjun Xie, and Hanmin Huang. 2013. "An efficient Rh/O2 catalytic system for oxidative C–H activation/annulation: Evidence for Rh (I) to Rh (III) oxidation by molecular oxygen." *Journal of the American Chemical Society* no. 135(24):8850–8853.

Zhao, Dongbing, Qian Wu, Xiaolei Huang, Feijie Song, Taiyong Lv, and Jingsong You. 2013. "A general method to diverse cinnolines and cinnolinium salts." *Chemistry – A European Journal* no. 19(20):6239–6244.

Zhao, Gongyuan, Chunxia Chen, Yixia Yue, Yanhan Yu, and Jinsong Peng. 2015a. "Palladium (II)-catalyzed sequential C–H arylation/aerobic oxidative C–H amination: One-pot synthesis of benzimidazole-fused phenanthridines from 2-arylbenzimidazoles and aryl halides." *The Journal of Organic Chemistry* no. 80(5):2827–2834.

Zhao, Shixian, Ruicheng Yu, Wanzhi Chen, Miaochang Liu, and Huayue Wu. 2015b. "Efficient approach to mesoionic triazolo [5, 1-a] isoquinolium through rhodium-catalyzed annulation of triazoles and internal alkynes." *Organic Letters* no. 17(11):2828–2831.

Zhou, Tao, Liubo Li, Bin Li, Haibin Song, and Baiquan Wang. 2015. "Ir (III)-catalyzed oxidative coupling of NH isoquinolones with benzoquinone." *Organic Letters* no. 17(17):4204–4207.

Zhou, Tao, Yanwei Wang, Bin Li, and Baiquan Wang. 2016. "Rh (III)-catalyzed carbocyclization of 3-(Indolin-1-yl)-3-oxopropanenitriles with alkynes and alkenes through C–H activation." *Organic Letters* no. 18(19):5066–5069.

4 Five-Membered O-Heterocycles

4.1 INTRODUCTION

Five-membered oxygen heterocycles are broadly found in a number of biologically active compounds, natural products, and pharmaceuticals, exhibiting antifungal, antiarrhythmic, uricosuric, vasodilator, antimigraine, anti-inflammatory, antioxidant, antitumor, anti-HIV, antiplatelet, and analgesic activities (Kamal, Shakya, and Jawaid 2011, Bhargava and Rathore 2017). They have attracted the interest of many in the research fields of medicines and pharmaceuticals (Eftekhari-Sis, Zirak, and Akbari 2013, Eftekhari-Sis and Zirak 2014, Bhargava and Rathore 2017). However, there are many reports on the creation of five-membered *O*-heterocyclic compounds by multi-step synthesis in the literature; the development of new approaches to synthesis of these types of *O*-heterocycles with tolerance of different functional groups are of interest to synthetic and medicinal scientists.

Cross-dehydrogenative-coupling (CDC) (Gandeepan et al. 2018, Shaikh and Hong 2016) processes are widely used in the synthesis of various types of heterocyclic compounds. In this chapter, intramolecular and intermolecular CDC of aromatic $C(sp^2)$–H bonds are discussed to construct five-membered *O*-heterocycles, including dihydrobenzofurans, benzo- and dibenzofurans, isobenzofurans, and also fused furans, etc.

4.2 DIHYDROBENZOFURANS

Pd(II)-catalyzed hydroxyl-directed C–H activation/C–O cyclization was developed to synthesize dihydrobenzofuran derivatives **2**. Intramolecular dehydrogenative coupling of tertiary α-benzyl substituted alcohols **1** with 5 mol% of $Pd(OAc)_2$ in the presence of $PhI(OAc)_2$ (1.5 equiv), and Li_2CO_3 (1.5 equiv) in C_6F_6 at 100°C led to the formation of dihydrobenzofurans **2** in 42–91% yields. The reaction of 1-benzylcyclohexanol and 1-benzyldecahydronaphthalen-1-ol afforded corresponding spiro-dihydrobenzofurans **2a** in 85% and 83% yields, respectively. The reaction occurred by Pd(II)-catalyzed C–H cleavage (**I**) followed by oxidation of Pd(II) to a higher oxidation state, Pd(IV) (**II**), which underwent reductive elimination of the C–O bond (Wang et al. 2010). Also, Pd-catalyzed lactonization of 2-arylacetic acids **3** to benzofuranone derivatives **4** through a C–H activation/C–O formation was reported by Yang et al. (2013). Reactions were performed using $Pd(OAc)_2$ (10 mol%), $PhI(OAc)_2$ (2 equiv), AgOAc (0.5 equiv), CsOAc (0.5 equiv), and NaOAc (0.5 equiv) in PhCl/*t*-BuOH mixture at 100°C, to give benzofuranones **4** in 47–89% yields. The α,α-disubstitution is essential for dehydrogenative cyclization, as the reaction did not occur in the presence of the α-hydrogen of 2-arylacetic acids. 1-Arylcycloalkancarboxylic acids were

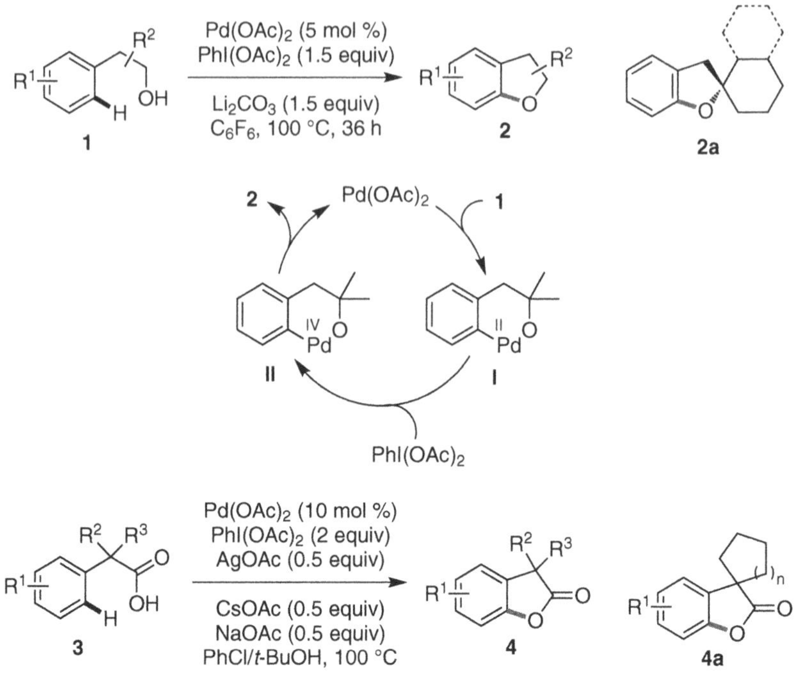

SCHEME 4.1

also compatible with the reaction, giving corresponding spiro-benzofuranones **4a** in 35–80% yields (Scheme 4.1). In addition, Pd(II)-catalyzed intramolecular dehydrogenative coupling of C(sp^2)–H and C(sp^3)–H bonds of phenyl alkyl ethers was described for the creation of dihydrobenzofuran derivatives (Shi et al. 2017).

An Fe(III)-catalyzed dehydrogenative cross-coupling reaction between phenols **5** and conjugated alkenes **6** was developed for the synthesis of 2,3-dihydrobenzofuran motif **7**. Reactions were carried out using $FeCl_3$ (20 mol%) in the presence of di-*tert*-butyl peroxide (DTBP) (2 equiv) in 1,2-dichloroethane (DCE) at 80°C. Styrene, α-alkyl- and α-arylstyrenes, β-alkyl styrenes, and stilbenes were tolerated under reaction conditions, giving corresponding dihydrobenzofurans **7** in 28–94% yields. The reaction of β-alkyl styrenes led to 3-alkyl-2-aryldihydrobenzofurans, regioselectively. The reaction was initiated by generation of Fe-chelated species **I**, which was oxidized to give the electrophilic phenol species **II**. An additional reaction with an alkene and then tautomerization afforded the intermediate **IV**, which underwent reductive elimination to achieve the final desired product via an Fe-chelated benzylic carbocation species **V** (Kshirsagar et al. 2013). A similar approach was developed for the preparation of 2,6-dihydro-1*H*-furo[3,2-*e*]indole derivatives **9** by oxidative coupling of 1-tosylindol-5-ol **8** and styrenes **6a** in the presence of $FeCl_3$ and 2,3-dichloro-5,6-dicyano-1,4-benzoquinone (DDQ) in toluene/dichloromethane (DCM) mixture (Scheme 4.2) (Liang et al. 2016).

Rh(III)-catalyzed redox-neutral C–H functionalization/cyclization of *N*-phenoxyacetamides **10** with propargyl carbonates **11** was accomplished to synthesize

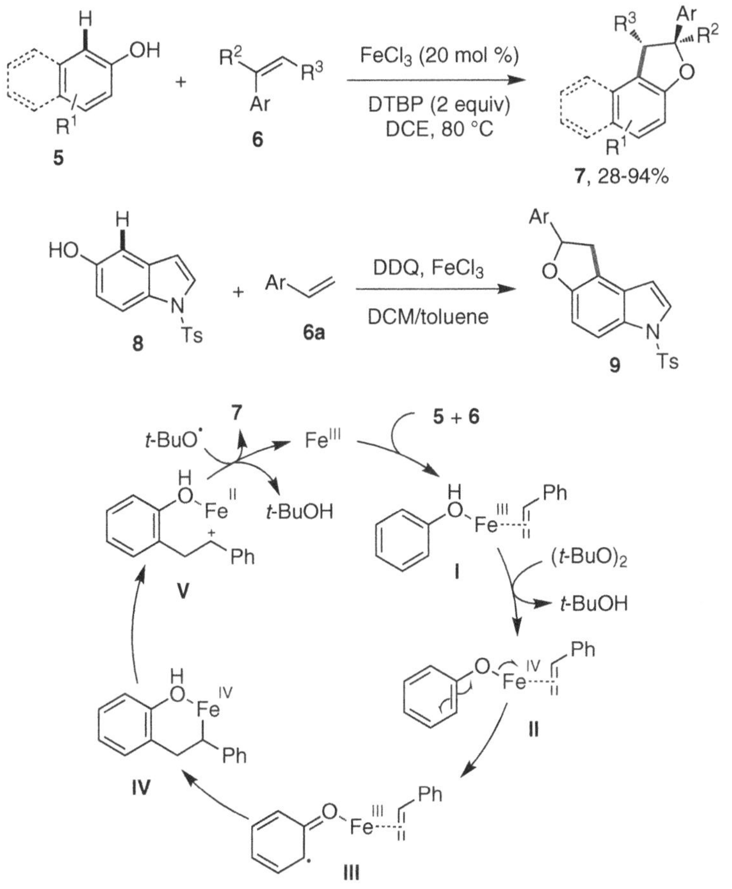

SCHEME 4.2

3-alkylidene dihydrobenzofuran derivatives **12**. Subjecting *N*-phenoxyacetamides **10** to propargyl carbonates **11** (1.2 equiv), in the presence of $[Cp^{*}RhCl_2]_2$ (2.5 mol%), CsOAc (0.5 equiv), and K_2CO_3 (0.5 equiv) in MeOH at room temperature for 24 h under air resulted in the formation of corresponding 3-alkylidene dihydrobenzofurans **12** in 35–90% yields. The reaction failed in the case of tertiary 2-alkynylic carbonate. In the proposed reaction mechanism, the N–H/C–H cleavage gave a five-membered rhodacycle intermediate **I**, which underwent regioselective insertion of alkyne **11** to generate intermediate **II**. Reductive elimination gave intermediate **III** and Rh(I) species. Subsequently, Rh(I)-mediated O–N bond cleavage (**IV**) and then protonolysis formed the enamide **V** along with regeneration of active Rh(III) species. Intramolecular substitution proceeded to afford the final desired product **12** with simultaneous formation of an exocyclic double bond (Scheme 4.3) (Zhou et al. 2015).

Also, Rh(III)-catalyzed redox-neutral cascade [3+2] annulation of *N*-phenoxyacetamides **10** with propiolates **13** was reported, accessing benzofuran-2(3*H*)-ones **14** via C–H functionalization/isomerization/lactonization. Subjecting

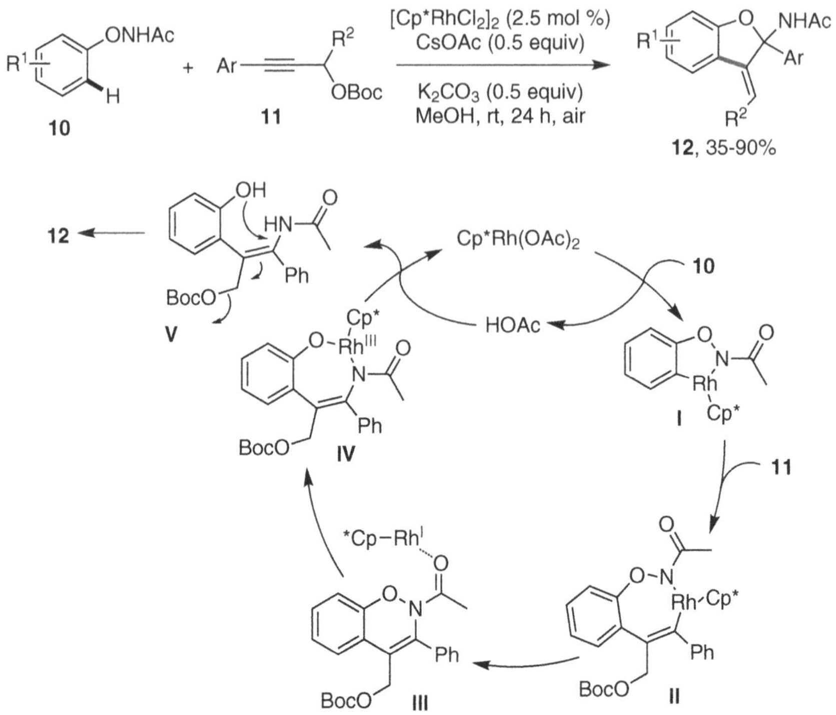

SCHEME 4.3

equimolar amounts of *N*-phenoxyacetamides **10** and propiolates **13** to $[Cp^*RhCl_2]_2$ (3.5 mol%) and KOAc (0.25 equiv) in CH_3CN at room temperature for 24 h furnished *N*-((2-oxobenzofuran-3(2*H*)-ylidene)(phenyl)methyl)acetamide derivatives **14** in 22–78% yields. The plausible reaction mechanism involves the generation of rhodacycle intermediate **I** by sequential N–H deprotonation and C–H cleavage, which underwent coordination and regioselective insertion of an alkyne moiety to give seven-membered intermediate **II**. The Rh-nitrenoid **III** was formed by N–O cleavage, which led to cyclic enamine intermediate **IV** by C–N bond formation through 1,2-alkenyl migration. By protonation of **IV**, intermediate **V** was obtained, which had a free phenol moiety with Rh(III) bonding to the enamine moiety. The enamine–imine isomerization with subsequent protonation of the amino moiety in **VI** with AcOH delivered intermediate **VII**, which underwent intramolecular lactonization to the final desired product **14** (Scheme 4.4) (Pan et al. 2018).

4.3 SPIRO-DIHYDROBENZOFURANS

Pd-catalyzed intramolecular spirocyclization of 2-aryloxymethyl-1,4-naphthoquinones **15** was described by Claes et al. (2013) for the construction of 2*H*,3'*H*-spiro[benzofuran-3,2'-naphthoquinones] **16**. Reactions were catalyzed with $Pd(OAc)_2$ (10 mol%) in the presence of dichloropyridine (15 mol%) and trifluoroacetic acid (TFA) (5 mol%) in AcOH at 110°C, providing corresponding

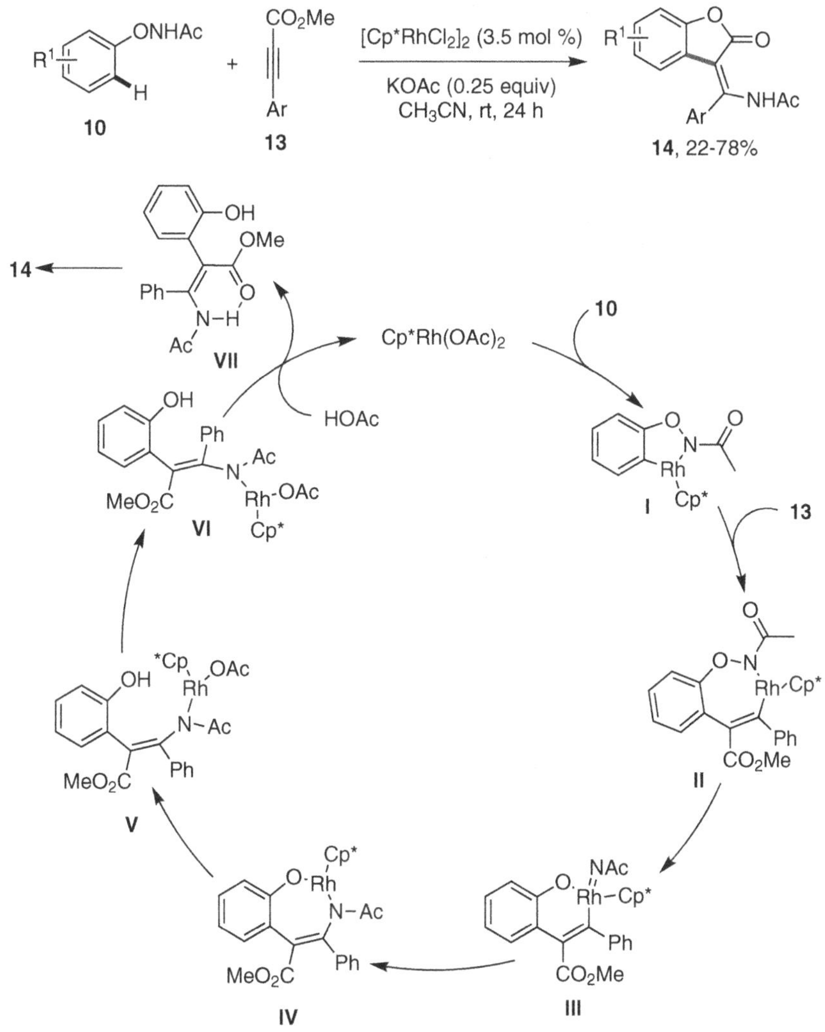

SCHEME 4.4

spiro-dihydrobenzofurans **16** in 21–71% yields. Little or no conversion was observed in the case of *ortho*-substituted aryloxy substrates. 2*H*,3'*H*-Spiro[benzofuran-3,2'-benzoquinones] were also synthesized by dehydrogenative spirocyclization of aryloxymethylbenzoquinones under reaction conditions, in low yield (7–20%). The reaction proceeded by coordination of the palladium with the naphthoquinone moiety, followed by C–H cleavage to the organopalladium intermediate **I**, which underwent an intramolecular Michael addition, leading to palladium enolate **III**. Upon protonolysis with subsequent tautomerization, **III** was converted to spiroquinone **16**, along with regeneration of the Pd(II) catalyst for the next catalytic cycle (Scheme 4.5).

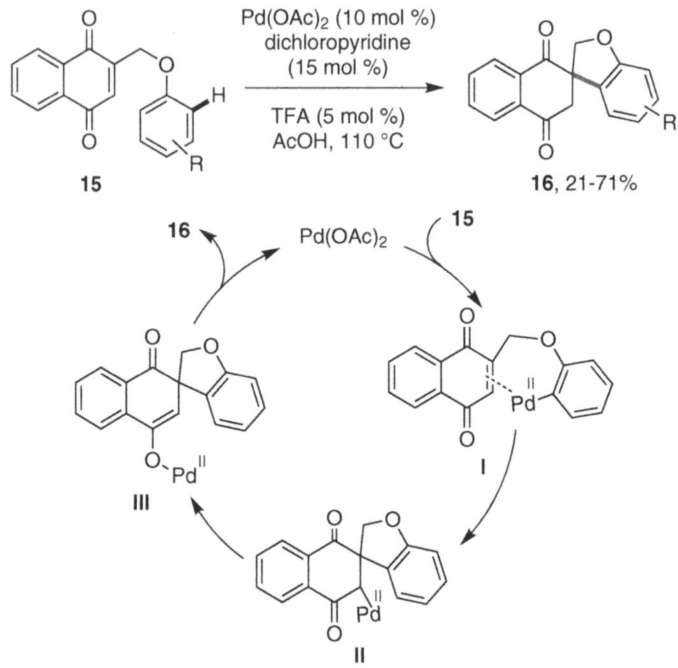

SCHEME 4.5

4.4 BENZOFURANS

The Pd-catalyzed intramolecular C–H activation/cyclization of 3-phenoxyacrylates **17** was developed to synthesize substituted benzofurans **18**. (*E*)-3-phenoxyacrylates **17** were transformed into benzofurans **18** in 51–85% yields when treated with $Pd(OAc)_2$ (5 mol%), PPh_3 (5 mol%), and CF_3CO_2Ag (2 equiv) in benzene at 110°C under air conditions. A low yield of the expected benzofuran product was obtained when the (*Z*) isomer was used. The reaction occurred through generation of a vinylpalladium intermediate **I** by the electrophilic palladation of (*E*)-3-phenoxyacrylates **17** with $Pd(OAc)_2$, followed by intramolecular electrophilic aromatic palladation through C–H activation to give intermediate **II**, which underwent reductive elimination to afford benzofuran product **18** with generation of Pd(0), oxidizing with CF_3CO_2Ag to regenerate the Pd(II) active species to complete the catalytic cycle (Li et al. 2011). Intramolecular Pd-catalyzed dehydrogenative C–H bond functionalization of (*Z*)-2-bromovinyl phenyl ethers **I** was described using $PdCl_2$ (5 mol%) and K_2CO_3 (2 equiv) in *N,N*-dimethylformamide (DMF) at 130°C, leading to 2-substituted benzofurans **20** in 81–92% yields. Also, one-pot synthesis of benzofuran derivatives **20** based on the addition/palladium-catalyzed C–H functionalization of phenols **5** with bromoalkynes **19** was investigated, in which benzofurans **20** were obtained in 41–91% yields (Scheme 4.6) (Wang et al. 2011).

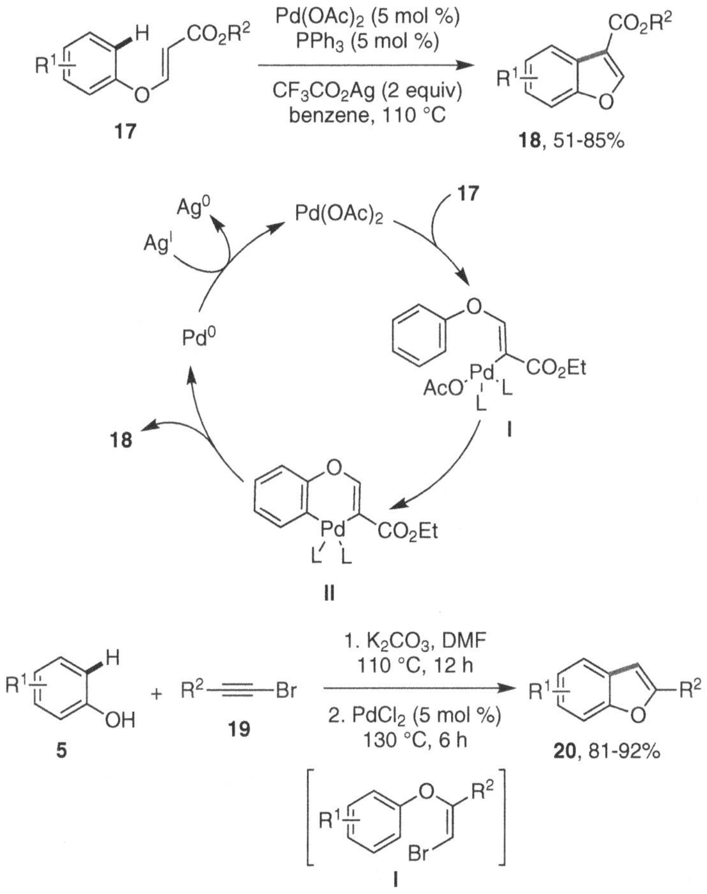

SCHEME 4.6

Synthesis of benzofurans **22** was achieved through Pd-catalyzed dehydrogenative coupling/annulation of phenols **5** and unactivated internal alkynes **21**. Reactions were performed using 2 equiv of phenol **5** and 1 equiv of alkyne **21** in the presence of $Pd_2(dba)_3$ (5 mol%), 1,10-phenanthroline (10 mol%), NaOAc (5 equiv), and $Cu(OAc)_2{\cdot}H_2O$ (2 equiv) in 1,4-dioxane at 130°C, giving 2,3-diarylbenzofurans **22** in 22–92% yields. Aliphatic alkyne, 4-octyne, was compatible with the reaction to give 2,3-dipropylbenzofuran **22** in 63% yield. Unsymmetrical alkyl aryl alkynes led to corresponding 3-alkyl-2-arylbenzofurans **22** in 64–81% yields. The proposed reaction mechanism involves the coordination of the 1,10-phenantroline bidentate ligand with $[Pd_2(dba)_3]$, followed by oxidation with $Cu(OAc)_2$ to active Pd(II) species **I**, which underwent an attack of phenol to generate Pd(II)-phenoxide species **II**. By subsequent coordination and insertion of an alkyne molecule **21** into the Pd–O bond, intermediate **III** was formed, which underwent base-assisted intramolecular *ortho*-C–H activation, leading to palladacycle **IV**. Finally, reductive elimination delivered

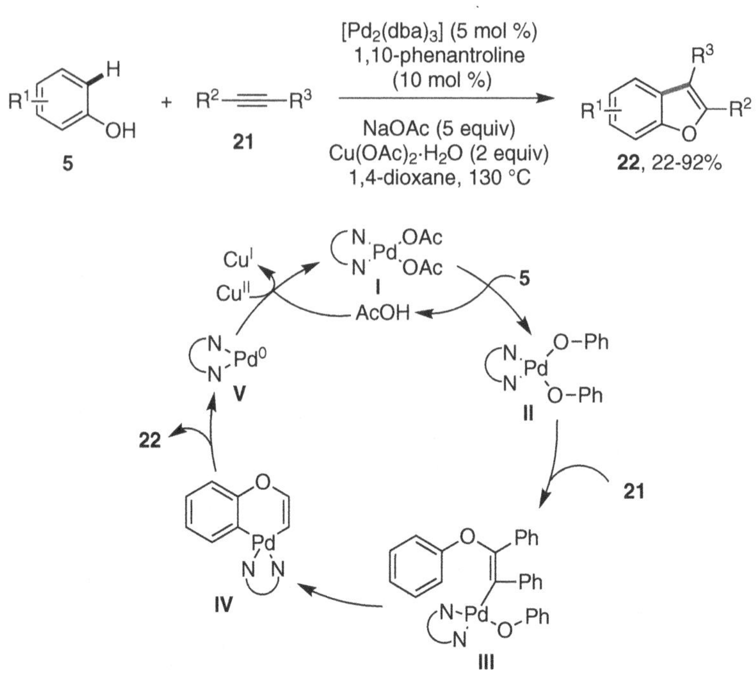

SCHEME 4.7

the benzofuran **22** and regenerated the Pd(0) species for the next catalytic cycle (Scheme 4.7) (Kuram, Bhanuchandra, and Sahoo 2013). A similar approach was developed using [Cp*Rh(MeCN)$_3$](SbF$_6$)$_2$ (5 mol%), Cu(OTf)$_2$ (1 equiv), AgPF$_6$ (2.5 equiv), an acetanilide ligand (10 mol%), and dodecyl trimethyl ammonium chloride (DTAC) (0.5 equiv) in decaline at 120°C, resulting in the formation of corresponding benzofurans in 29–86% yields (Zhu, Wei, and Shi 2013).

Rh(III)-catalyzed redox-neutral coupling of *N*-phenoxyacetamides **10** and alkynes **21** was described by Liu et al. (2013), providing 2,3-disubstituted benzofurans **22**. Treating *N*-phenoxyacetamides **10** (1.2 equiv) with alkynes **21** in the presence of [Cp*RhCl$_2$]$_2$ (2.5 mol%), CsOAc (25 mol %), and HOAc (1.2 equiv) in DCM at room temperature for 12–48 h under air pressure led to benzofuran derivatives **22** in 64–90% yields. The reaction of alkyl aryl alkynes led to the corresponding 3-alkyl-2-arylbenzofuran **22**, regioselectively. Alkylacetylene carboxylates afforded the corresponding 3-alkylbenzofuran-2-carboxylates. The proposed reaction mechanism involves nitrogen-directed *ortho*-C–H activation (**I**) followed by alkyne insertion into the Rh–C bond to give seven-membered rhodacycle **II**, with subsequent protonolysis of the Rh–N bond to intermediate **III**, which underwent an intramolecular substitution to form the C–O bond, along with N–O bond cleavage with the aid of acetic acid (**IV**) (Scheme 4.8). A similar methodology was used to access 3-arylbenzofuran-2-ylphosphines, by Rh(III)-catalyzed redox-neutral cyclization of *N*-phenoxyacetamides with 1-alkynylphosphine sulfides and oxides followed by reduction (Wang, Wang, and Li 2017).

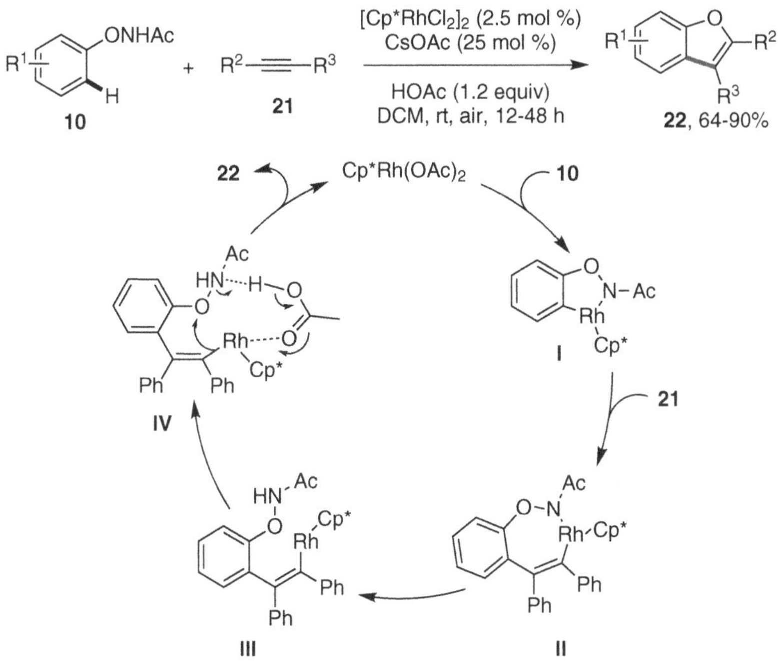

SCHEME 4.8

4.5 DIBENZOFURANS

Cu-catalyzed dehydrogenative C(sp^2)–H cycloetherification of *ortho*-arylphenols **23** was developed for the synthesis of dibenzofurans **24**. Reactions were conducted in dimethyl sulfoxide (DMSO) at 140°C under air conditions using CuBr (30 mol%), PivOH (1 equiv), and Cs_2CO_3 (0.5 equiv), generating dibenzofuran derivatives **24** in 44–72% yields. By coordination of Cu(II) species with the phenolic hydroxyl group, followed by C–H activation via the concerted metalation/deprotonation process and then subsequent reductive elimination, the expected product was obtained with concurrent formation of Cu(0) species, and was oxidized by O_2 to regenerate active Cu(II) species (Scheme 4.9) (Zhao et al. 2012). Also, construction of dibenzofuran scaffolds was reported by Pd(II)-catalyzed phenol-directed C–H activation/C–O cyclization (Xiao et al. 2011).

ZrO_2-supported $Pd(OH)_2$ was reported as the catalyst for the dehydrogenative intramolecular couplings of diarylethers **25** to dibenzofurans **24** via dual aryl C–H bond functionalizations. Treatment of diarylethers **25** with 10 wt% $Pd(OH)_2/ZrO_2$ in the presence of AcOH in dioxane at 130–140°C under pressure of O_2 resulted in the creation of dibenzofurans **24** in 9–84% gas chromatography (GC) yields. Reaction proceeded by double *ortho*-C–H activation via electrophilic substitution by Pd(II) species to give six-membered palladacycle **I**, followed by reductive elimination to dibenzofuran **24** with generation of Pd(0) species, which regenerated active Pd(II) by oxidation with O_2 (Ishida et al. 2014). Similarly, bisbenzofuro[2,3-*b*:3',2'-*e*]

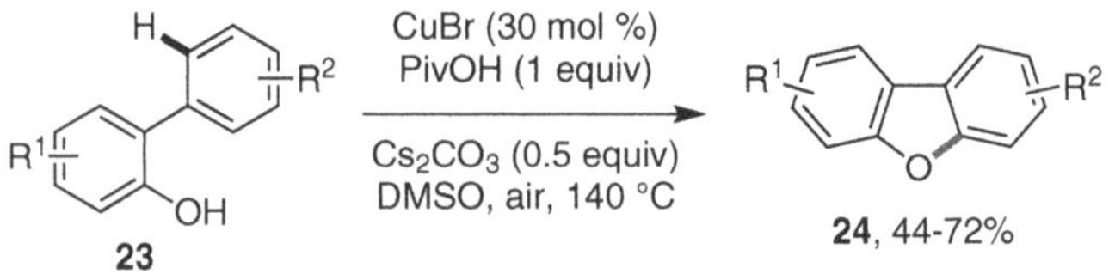

SCHEME 4.9

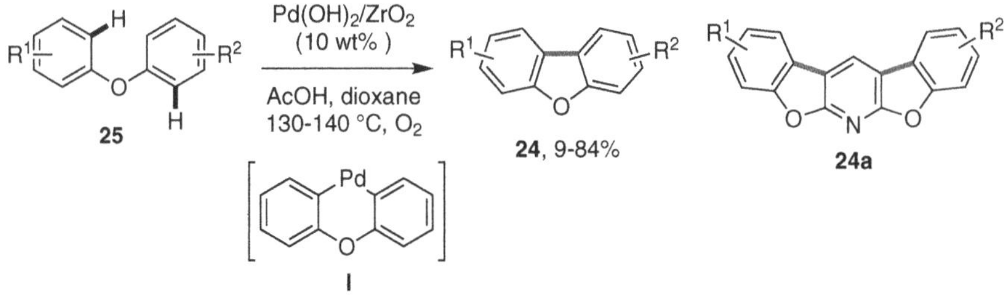

SCHEME 4.10

pyridines **24a** were produced by Pd-catalyzed intramolecular dehydrogenative C–H/C–H coupling of 2,6-diaryloxypyridines, in 31–74% yields (Scheme 4.10) (Kaida et al. 2017).

4.6 BENZOFUROCHROMENS

In 2016, Cheng et al. (2016) and Mackey et al. (2016) independently reported the Pd(II)-catalyzed intramolecular cross-dehydrogenative coupling of 4-aryloxy coumarins **26** as a route to access 6*H*-benzofuro[3,2-*c*]chromen-6-one scaffolds **27**. In Cheng et al.'s work, $Pd(OAc)_2$ (10 mol%) in the presence of AgOAc (2 equiv) as an oxidant and CsOAc (2 equiv) as a base in PivOH at 100°C led to 6*H*-benzofuro[3,2-*c*]chromen-6-ones **27** in 83–99% yields. Mackey et al. used Ag_2O (1.5 equiv) as an oxidant in the presence of NaO*t*-Bu (0.2 equiv), leading to the expected 6*H*-benzofuro[3,2-*c*]chromen-6-ones **27** in 55–99% yields. The reaction of *meta*-substituted 4-aryloxy coumarins occurred with 67:33–98:2 regioselectivity, favoring coupling at the *para* position of the substitution. Also, 1*H*-pyrano[4,3-*b*]benzofuran-1-one derivatives **27** were obtained in 20–65% yields, by intramolecular cross-dehydrogenative coupling of 4-aryloxy-2-pyrones under the same reaction conditions. The reaction was initiated by palladation at the C-3 position, followed by a concerted metalation/deprotonation at the *ortho* position of the aryl moiety to form a six-membered palladacycle, which delivered the final product by reductive elimination, along with the generation of Pd(0) species, which were reoxidized with Ag(I) to achieve active Pd(II) species for the next catalytic cycle (Scheme 4.11).

Also, 6*H*-benzofuro[3,2-*c*]chromen-6-one structural motifs **30** were obtained by Pd(II)-catalyzed C–H bond activation/C–C and C–O bond formation reaction between 4-hydroxycoumarins **28** and arynes, generated in situ from *ortho*-silyl aryl triflates **29**. Reacting 4-hydroxycoumarin **28** with *ortho*-silyl aryl triflate **29** (1.5 equiv) under $Pd(OAc)_2$ (5 mol%), $Cu(OAc)_2{\cdot}H_2O$ (1.2 equiv), CsF (2 equiv), and

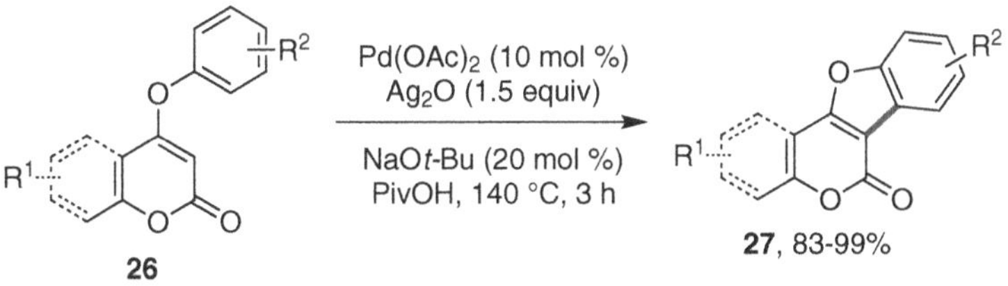

SCHEME 4.11

NaOAc (1.2 equiv) catalytic system in CH_3CN at 120°C afforded 6*H*-benzofuro[3,2-*c*]chromen-6-ones **30** in 57–87% yields. The plausible reaction mechanism involves the activation of C–H at the C-3 position to generate intermediate **I**, followed by carbopalladation of the aryne **II** to give intermediate **III**. Base-mediated reductive elimination afforded the desired product **30** with Pd(0) species, which were reoxidized to Pd(II) by $Cu(OAc)_2$ to complete the catalytic cycle (Scheme 4.12) (Neog, Borah, and Gogoi 2016).

4.7 THIENOBENZOFURANS

The synthesis of thieno[3,2-*b*]benzofurans **32** was reported by intramolecular C–H/C–H coupling of 3-aryloxythiophenes **31** in the presence of $Pd(TFA)_2$ (10 mol%) as catalyst and AgOAc (2 equiv) as an oxidant. Reactions were performed in

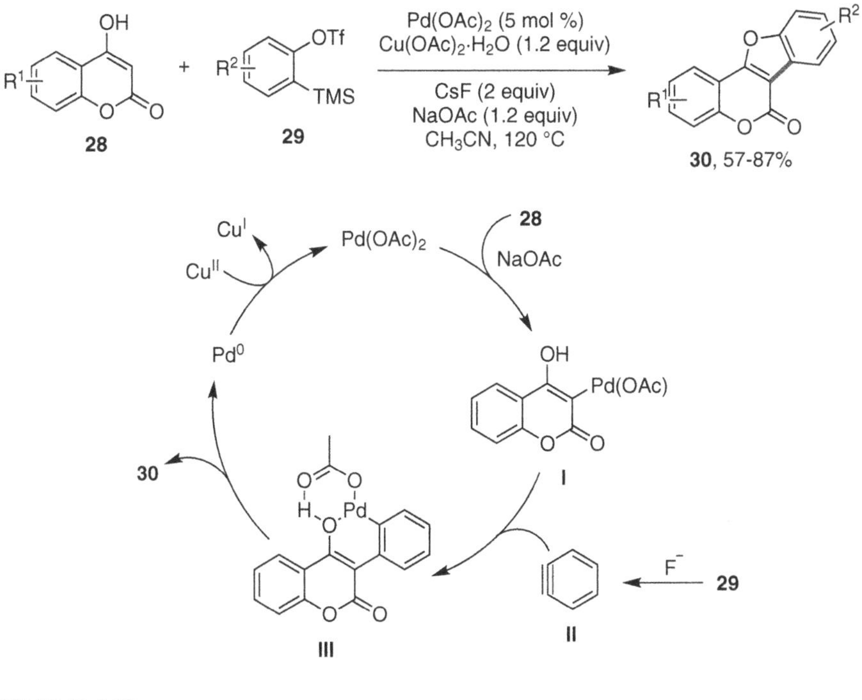

SCHEME 4.12

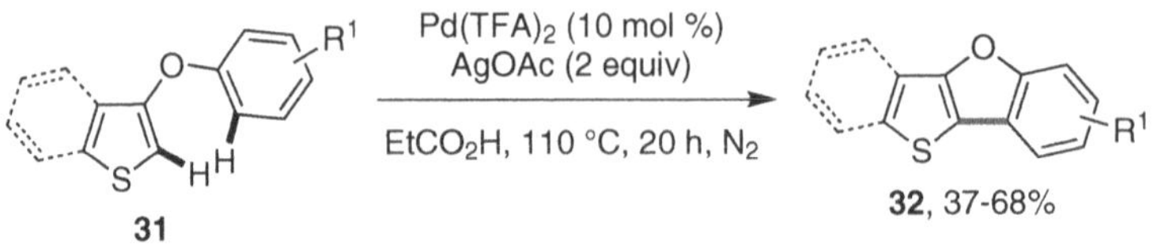

Pd(TFA)$_2$ (10 mol %)
AgOAc (2 equiv)
EtCO$_2$H, 110 °C, 20 h, N$_2$
31
32, 37-68%

SCHEME 4.13

propionic acid at 110°C under an atmosphere of N_2, and thieno[3,2-*b*]benzofurans **32** were produced in 37–68% yields. Also, [1]benzothieno[3,2-*b*]benzofurans **32** were obtained from 3-aryloxybenzo[*b*]thiophenes in 76–89% yields (Scheme 4.13) (Kaida et al. 2015).

4.8 ISOBENZOFURANS

Pd(II)-catalyzed dehydrogenative coupling of tertiary benzyl alcohols **33** with CO_2 was developed for the synthesis of isobenzofuran-1(3*H*)-one derivatives **34**. Reactions were performed using 10 mol% of Pd(TFA)$_2$ in the presence of LiO*t*-Bu (7 equiv), 4-bromo-*N,N*-dimethylbenzamide (BDB, 2 equiv), and Cs_2CO_3 (1 equiv) in DMF at 140°C under pressure of CO_2, leading to isobenzofuran-1-one derivatives **34** in 44–91% yields. The reaction occurred by base-mediated coordination of benzyl alcohol **33** to the Pd center, followed by C–H activation to generate five-membered palladacycle **I**, which underwent CO_2 insertion into the Pd–O bond to generate intermediate **II**. Insertion of a second CO_2 molecule into the Pd–O bond afforded nine-membered palladacycle **III**, which was converted to spiroheterocyclic intermediate **IV** via intramolecular nucleophilic addition. Finally, the decarbonization of **IV** delivered the isobenzofuran-1-one product **34** (Scheme 4.14) (Song et al. 2018).

In 2011, Ackermann and Pospech (2011) described Ru-catalyzed cross-dehydrogenative C–H bond alkenylations providing isobenzofuran-1-one derivatives **37a**. Treatment of benzoic acids **38** with 2 equiv of either acrylates or acrylonitrile **36** in the presence of [RuCl$_2$(*p*-cymene)]$_2$ (2 mol%) and Cu(OAc)$_2$·H_2O (2 equiv) in water at 80°C furnished corresponding 2-(3-oxo-1,3-dihydroisobenzofuran-1-yl) acetates or 2-(3-oxo-1,3-dihydroisobenzofuran-1-yl)acetonitriles **38** in 66–90% or 51–97% yields, respectively. Also, a similar methodology was developed by Zhao et al. (2015) using polyethylene glycol (PEG)/water as a reaction medium. The reaction was initiated by carboxylic group-induced C–H activation, followed by olefin insertion and then β-hydride elimination to give olefinated intermediate **I**, which underwent an intramolecular oxa-Michael addition reaction to afford the desired product **37a**. [(COD)RhCl]$_2$ catalyzed the dehydrogenative coupling of benzoic acids **35** with ethyl vinyl ketone yielded phthalides. Subjecting benzoic acids to 2 equiv of ethyl vinyl ketone in the presence of [(COD)RhCl]$_2$ (8 mol%), AgOTf (24 mol%), dicyclopentadiene (DCPD) (32 mol%), and Cu(OAc)$_2$·H_2O (4 equiv) in PhCl at 120°C for 48 h under air resulted in creation of 3-(2-oxobutyl)isobenzofuran-1(3*H*)-one derivatives **37b** in 15–75% yields. The reaction of acrylate and acrylamide afforded the expected products **37a** in 66% and 34% yields, respectively (Renzetti, Nakazawa, and Li 2016). Synthesis of 3-(2-oxoalkyl)-substituted isobenzofuran-1(3*H*)-ones **37b**

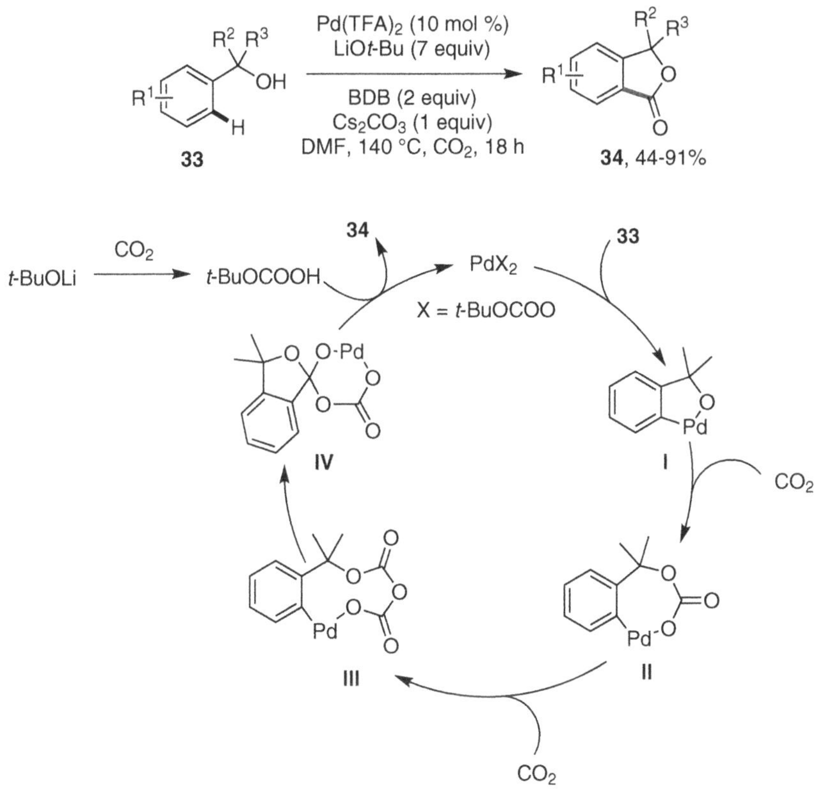

SCHEME 4.14

was also achieved by Ru-catalyzed *ortho*-C–H alkenylation of benzoic acids **35** with allyl alcohols **38**. Reactions took place through generation of a five-membered ruthenocycle through C–H activation followed by coordination and insertion of a vinyl ketone, generated in situ by Ru-catalyzed oxidation of allyl alcohol (Scheme 4.15) (Kumar et al. 2018).

An approach was developed for the creation of 3-((phenylsulfonyl)methyl)-substituted isobenzofuran-1(3*H*)-ones **37c** via $[RuCl_2(p\text{-cymene})]_2$ (5 mol%)-catalyzed dehydrogenative coupling of benzoic acids **35** with phenyl vinyl sulfone **36b** (1.2 equiv), in the presence of $Cu(OAc)_2{\cdot}H_2O$ (1 equiv) in CH_3CN under air conditions. Corresponding isobenzofuran-1(3*H*)-ones **37c** were obtained in 56–92% yields. Treating the obtained products with 1,8-diazabicyclo[5.4.0]undec-7-ene (DBU) (5 equiv) in the same reaction tube at 100°C led to corresponding 3-methyleneisobenzofuran-1(3*H*)-ones **39** in 67–72% yields (Mandal et al. 2018). Also, an unexpected 3-((fluorosulfonyl)methyl)-substituted isobenzofuran-1(3*H*)-one **37d** was obtained in the Rh-catalyzed dehydrogenative coupling of *N*-methoxy-*N*-methylbenzamides **40** with vinylsulfonyl fluoride. The reaction may occur through hydrolysis of *N*-methoxy-*N*-methylbenzamide **37d** to benzoic acid **35,** followed by dehydrogenative alkenylation with subsequent cyclization (Scheme 4.16) (Wang et al. 2018).

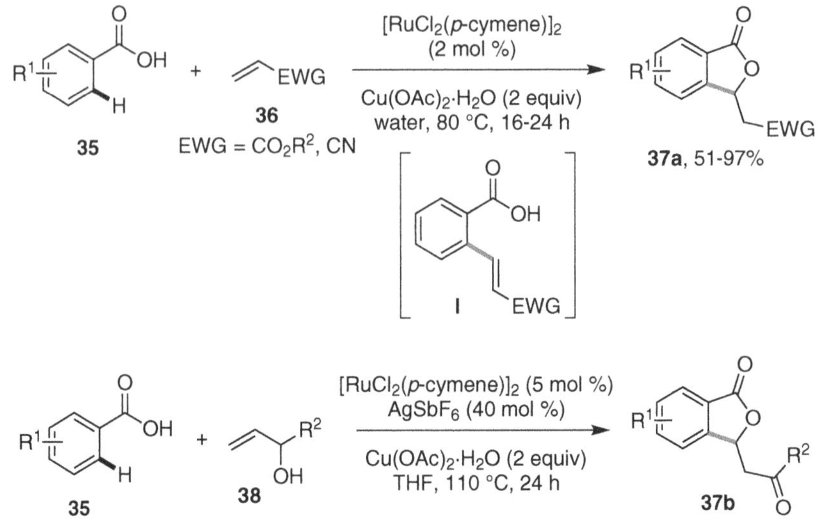

SCHEME 4.15

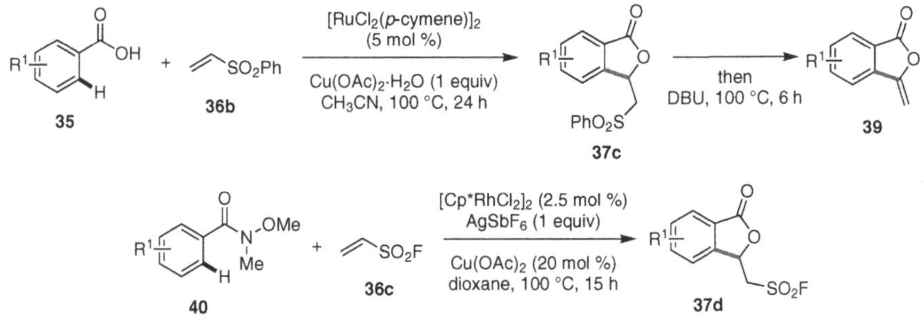

SCHEME 4.16

$Pd(OAc)_2$ catalyzed the dehydrogenative coupling of benzoic acids **35** and styrenes **6a** to yield 3-benzylidenephthalide **41**. Reactions were performed by equimolar amounts of benzoic acids **35** and styrenes **6a** in the presence of $Pd(OAc)_2$ (5 mol%) and Ag_2O (1 equiv) in DMF at 110°C, to give (*Z*)-3-benzylidene isobenzofuran-1(3*H*)-ones **41** in 47–80% yields. The proposed reaction mechanism involves the coordination of benzoic acid **35** to Pd(II), followed by C–H activation to generate five-membered palladacycle intermediate **I**, which underwent coordination and insertion of an alkene moiety to give seven-membered palladacycle intermediate **II**. An intramolecular attack of the coordinated O atom on C-1 of the alkene group afforded intermediate **IV**, which furnished the desired product **41** via subsequent *β*–H elimination. The existence of an electron-donating group at the *ortho* position of benzoic acid is essential to obtain benzylidene isobenzofuran-1-ones **41** (Scheme 4.17) (Nandi et al. 2013).

(*Z*)-3-ethylidenephthalides **43** were synthesized by a Ru(II)-catalyzed, redox-free, twofold aromatic/allylic C–H bond activation of benzoic acids **35** with allyl

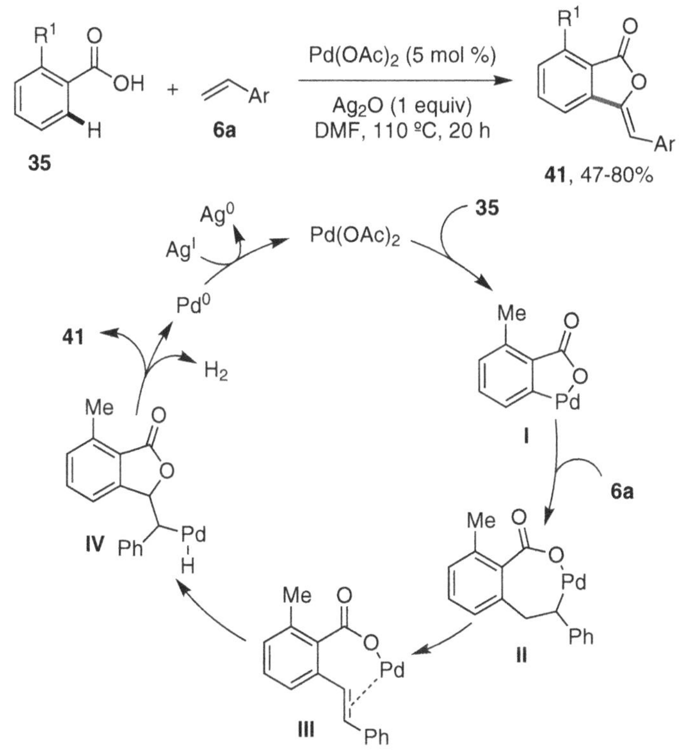

SCHEME 4.17

acetate **42**. Reactions were catalyzed with $[RuCl_2(p\text{-cymene})]_2$ (5 mol%) in the presence of K_2CO_3 (2 equiv) in DMF at 100°C for 12 h, and (Z)-3-ethylidenephthalides **43** were obtained in 33–90% yields. In the proposed reaction mechanism, the active catalyst species, generated in situ by ligand exchange, underwent coordination and C–H activation by *ortho*-metalation via a concerted deprotonation pathway to give a five-membered ruthenacycle intermediate **I**. By coordinative insertion of the C=C double bond of allyl acetate **42** into the Ru–C bond, intermediate **II** was formed, which underwent β-acetate elimination, giving *ortho*-allylated benzoate–ruthenium adduct **III**. AcO-mediated deprotonation at the allylic $C(sp^3)$–H bond provided the π-allylruthenium intermediate **IV**, which converted to isomerized *ortho*-vinylated ruthenium species **V** upon protonation of the η^1 C–Ru bond by RCO_2H. An intramolecular Wacker-type addition of the oxygen atom of CO_2–Ru at the double bond led to intermediate **VI**, which by β-hydride elimination gave cyclized product **43** along with a ruthenium hydride species, which reacted with RCO_2H to regenerate the active catalyst, while liberating H_2 gas (Scheme 4.18) (Jambu, Tamizmani, and Jeganmohan 2018).

$[Cp^*RhCl_2]_2$ catalyzed the dehydrogenative coupling/annulation of benzoic acids **35** with terminal alkynes **44** to deliver 3-arylidenephthalides **43b**. Reactions were conducted by heating a mixture of benzoic acid **35** with a terminal alkyne **44** (2 equiv)

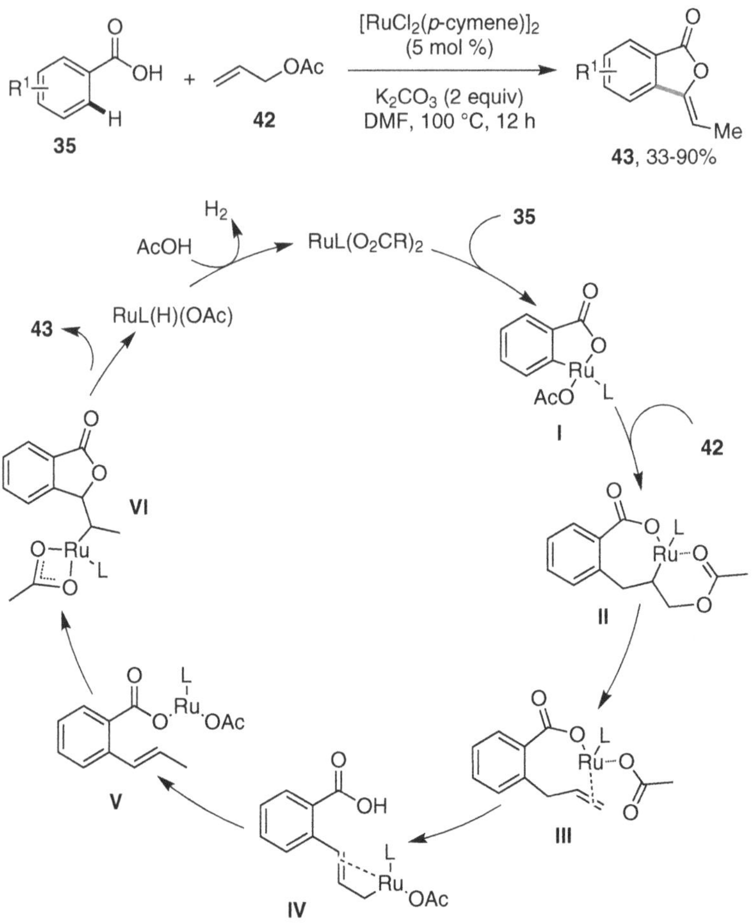

SCHEME 4.18

in the presence of $[Cp^*RhCl_2]_2$ (2.5 mol%), $AgSbF_6$ (10 mol%), PivOH (1 equiv), and Ag_2O (2 equiv) in *o*-xylene at 100°C, and 3-arylidenephthalides **43b** were obtained in 50–64% yields. An alkyl acetylene, 3,3-dimethyl-1-butyne, was also compatible with the reaction, leading to the corresponding 3-*t*-alkylidenephthalide **34b** in 66% yield. Tri(*iso*-propyl)silylacetylene was also tolerated under reaction with various substituted benzoic acids, affording (Z)-3-((tri(*iso*-propyl)silyl)methylene)phthalides **43b** in 31–91% yields. In the proposed reaction mechanism, five-membered rhodacycle **I** was generated by coordination of benzoic acid carboxylic oxygen with the Rh(III) center, with subsequent *ortho*-C–H activation. The reaction of **I** with the alkyne radical **II** formed in situ gave the Rh(IV) complex **III**, which underwent reductive elimination to produce the alkynylated product **IV**. A subsequent metal-catalyzed cyclization afforded the desired product **43b** (Scheme 4.19) (Liu et al. 2016).

Reacting benzoic acid **35** with 2,2-difluorovinyl tosylate **45** in the presence of $[Cp^*Rh(CH_3CN)_3](SbF_6)_2$ (5 mol%), Na_2CO_3 (1 equiv), and $AgSbF_6$ (10 mol%) in

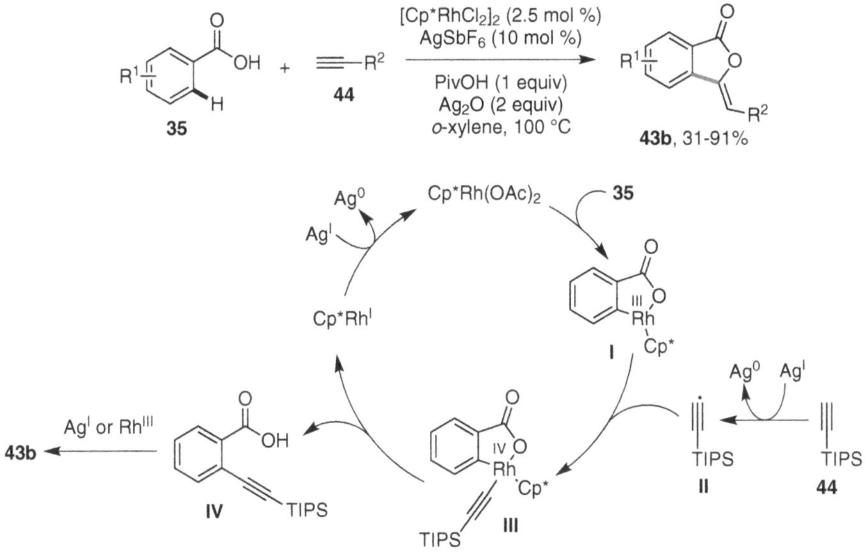

SCHEME 4.19

hexafluoroisopropanol (HFIP) at 60°C for 16 h afforded 3-alkylidene isobenzofuranones **46** in 40–82% yields. The reaction occurred through a redox-neutral bimetallic Rh(III)/Ag(I) relay catalysis (Scheme 4.20) (Ji et al. 2017).

Rh(III)-amine dual catalysis for dehydrogenative coupling of aldehydes **47** was developed for the synthesis of 3-arylsubstituted phthalides **48** by Tan, Juwaini, and Seayad (2013). Subjecting an aromatic aldehyde **47** to $[Cp^{*}RhCl_2]_2$ (1.25 mol%), $AgBF_4$ (5 mol%), Ag_2CO_3 (0.5 equiv), and 4-CF_3-aniline in diglyme at 90°C resulted in formation of phthalides **48** in 21–81% yields. Heterocoupling of benzaldehydes with an aldehyde without an *ortho*-C–H bond or benzaldehydes possessing electron-withdrawing F or CF_3 functional groups was also investigated, leading to corresponding 3-arylsubstituted phthalides **48** in 32–82% yields. Reaction of 4-methyl benzaldehyde with aliphatic aldehydes furnished the expected 3-alkyl phthalides **48** in low yields (21–22%). The reaction proceeded by generation of five-membered rhodacycle **II**, by *ortho*-C–H activation of imine **I**, generated in situ from aldehyde **47** and aniline, followed by insertion of the second molecule of the aldehyde C=O bond into Rh–C, giving intermediate **III**, which underwent intramolecular nucleophilic attack of the alkoxy on the electrophilic imine carbon leading to **IV**. β-hydride

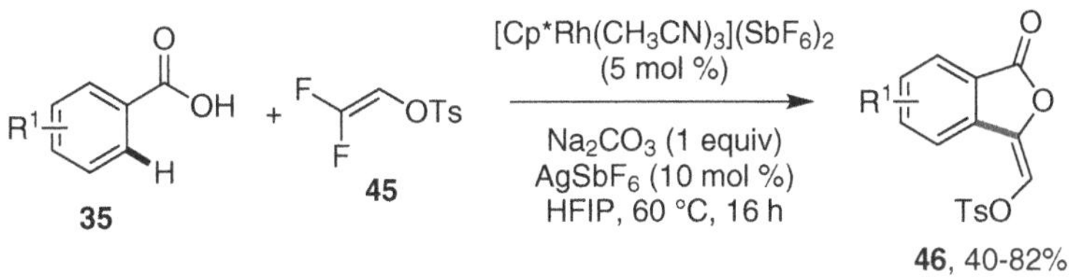

SCHEME 4.20

elimination afforded intermediate **V** along with formation of the Rh(III)hydride species, which regenerated the Rh(III) active species by reductive elimination of HBF_4 followed by oxidation by Ag_2CO_3. Finally, imine intermediate **V** was hydrolyzed to form the phthalide product **48**. Also, 3-alkylphthalide **51** derivatives were accessed by $Mn(CO)_5$-catalyzed dehydrogenative coupling of benzoates **49** with mono-substituted epoxides **50**. Electron-withdrawing as well as electron-donating substituted aromatic esters **49** were compatible with the reaction, affording isobenzofuranones **51** in 24–84% yields. The reactions occurred using $Mn(CO)_5$ (5 mol%) in the presence of Ph_3B (1 equiv) in DCE/hexane mixture at 150°C, in which Ph_3B acted as a Lewis acid for isomerization of the epoxide to the corresponding aldehyde, and then Mn acted as the catalyst for dehydrogenative coupling to the final 3-alkylphthalides **51**. 1,1-Disubstituted oxirane led to the corresponding 3-diphenylmethyl substituted phthalide **51** in 23% yield (Scheme 4.21) (Sueki, Wang, and Kuninobu 2016).

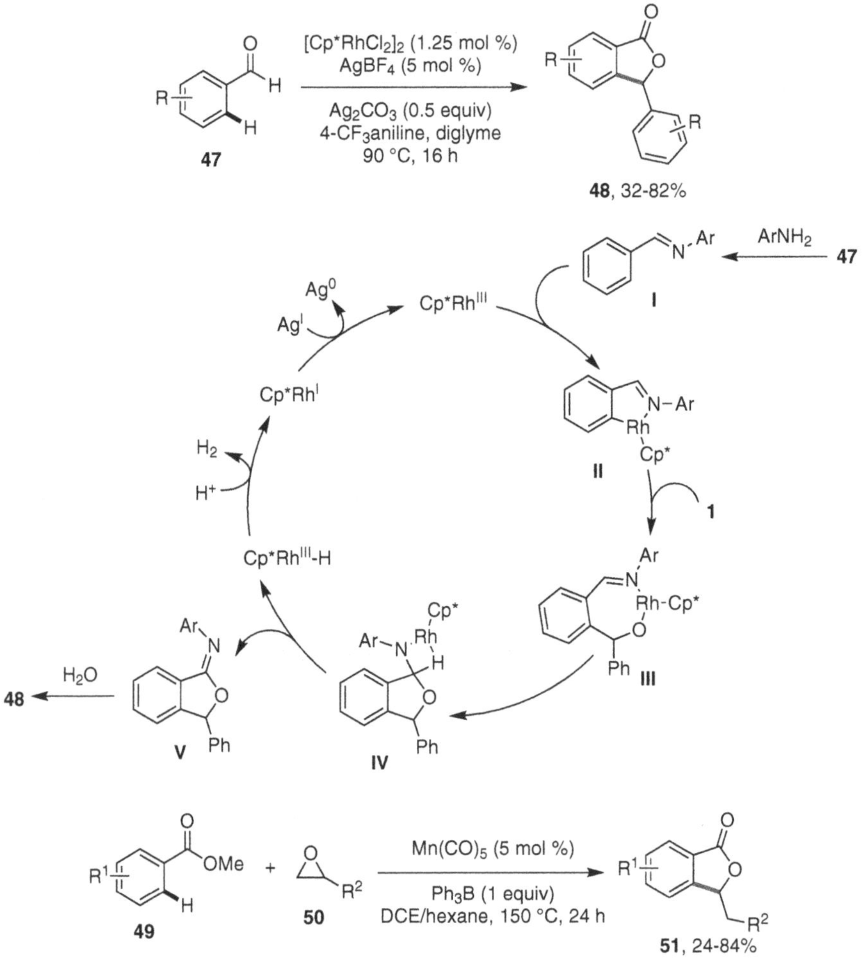

SCHEME 4.21

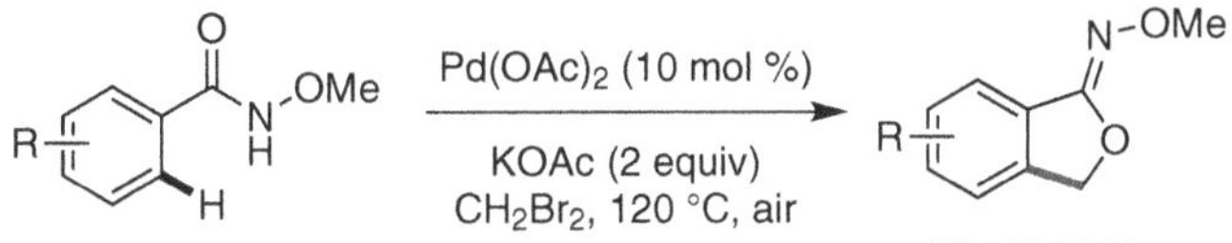

SCHEME 4.22

Pd-catalyzed synthesis of (*Z*)-isobenzofuran-1(3*H*)-one *O*-methyl oxime **53** was developed by dehydrogenative coupling/annulation of *N*-methoxy aryl amides **52** with CH_2Br_2. Reactions were performed in CH_2Br_2 as solvent in the presence of $Pd(OAc)_2$ (10 mol%) as catalyst and KOAc (2 equiv) as a base, at 120°C under air, and interestingly delivering isobenzofuran derivatives **53** in 56–78% yields. The proposed reaction mechanism involves *N*-atom coordinative C–H activation, followed by base-mediated insertion of CH_2Br_2 into the C–Pd bond to give an *ortho*-bromomethylated intermediate, which underwent base-assisted intramolecular S_N2 attack (Scheme 4.22) (Rao et al. 2018).

4.9 CONCLUSION

In summary, aromatic C(sp^2)–H dehydrogenative coupling processes are described for the synthesis of five-membered *O*-heterocycles, such as dihydrobenzofurans, benzo- and dibenzofurans, isobenzofurans, and also fused furans. Spiro-dihydrobenzofuran derivatives were also prepared by the dehydrogenative coupling reactions. However, many different methods were developed for the construction of these types of heterocycles, due to the quantitative and one-step synthesis and a broad spectrum of substitution on the synthesized heterocycles, the aromatic C–H bond direct functionalization approach could be of interest in the synthesis of pharmaceutical, medicinal, and natural products.

REFERENCES

Ackermann, Lutz, and Jola Pospech. 2011. "Ruthenium-catalyzed oxidative C–H bond alkenylations in water: Expedient synthesis of annulated lactones." *Organic Letters* no. 13(16):4153–4155.

Bhargava, Sangeeta, and Deepti Rathore. 2017. "Synthetic routes and biological activities of benzofuran and its derivatives: A review." *Letters in Organic Chemistry* no. 14(6):381–402.

Cheng, Chao, Wen-Wen Chen, Bin Xu, and Ming-Hua Xu. 2016. "Intramolecular cross dehydrogenative coupling of 4-substituted coumarins: Rapid and efficient access to coumestans and indole [3, 2-c] coumarins." *Organic Chemistry Frontiers* no. 3(9):1111–1115.

Claes, Pieter, Jan Jacobs, Bart Kesteleyn, Tuyen Nguyen Van, and Norbert De Kimpe. 2013. "Palladium (II)-Catalyzed Synthesis of 2 H, 3′ H-Spiro [benzofuran-3, 2′-naphthoquinones]." *The Journal of Organic Chemistry* no. 78(17):8330–8339.

Eftekhari-Sis, Bagher, and Maryam Zirak. 2014. "Chemistry of α-oxoesters: A powerful tool for the synthesis of heterocycles." *Chemical Reviews* no. 115(1):151–264.

Eftekhari-Sis, Bagher, Maryam Zirak, and Ali Akbari. 2013. "Arylglyoxals in synthesis of heterocyclic compounds." *Chemical Reviews* no. 113(5):2958–3043.

Gandeepan, Parthasarathy, Thomas Müller, Daniel Zell, Gianpiero Cera, Svenja Warratz, and Lutz Ackermann. 2018. "3d transition metals for C–H activation." *Chemical Reviews* no. 119(4):2192–2452.

Ishida, Tamao, Ryosuke Tsunoda, Zhenzhong Zhang, Akiyuki Hamasaki, Tetsuo Honma, Hironori Ohashi, Takushi Yokoyama, and Makoto Tokunaga. 2014. "Supported palladium hydroxide-catalyzed intramolecular double CH bond functionalization for synthesis of carbazoles and dibenzofurans." *Applied Catalysis B: Environmental* no. 150:523–531.

Jambu, Subramanian, Masilamani Tamizmani, and Masilamani Jeganmohan. 2018. "Ruthenium (II)-catalyzed cyclization of aromatic acids with allylic acetates via redox-free two-fold aromatic/allylic C–H activations: Combined experimental and DFT studies." *Organic Letters* no. 20(7):1982–1986.

Ji, Wei-Wei, E Lin, Qingjiang Li, and Honggen Wang. 2017. "Heteroannulation enabled by a bimetallic Rh (iii)/Ag (I) relay catalysis: Application in the total synthesis of aristolactam BII." *Chemical Communications* no. 53(41):5665–5668.

Kaida, Hiroyuki, Tsuyoshi Goya, Yuji Nishii, Koji Hirano, Tetsuya Satoh, and Masahiro Miura. 2017. "Construction of bisbenzofuro [2, 3-b: 3′, 2′-e] pyridines by palladium-catalyzed double intramolecular oxidative C–H/C–H coupling." *Organic Letters* no. 19(5):1236–1239.

Kaida, Hiroyuki, Tetsuya Satoh, Koji Hirano, and Masahiro Miura. 2015. "Synthesis of thieno [3, 2-b] benzofurans by palladium-catalyzed intramolecular C–H/C–H coupling." *Chemistry Letters* no. 44(8):1125–1127.

Kamal, Mehnaz, Ashok K Shakya, and Talha Jawaid. 2011. "Benzofurans: A new profile of biological activities." *International Journal of Medical & Pharmaceutical Sciences* no. 1:1–15.

Kshirsagar, Umesh A, Clil Regev, Regev Parnes, and Doron Pappo. 2013. "Iron-catalyzed oxidative cross-coupling of phenols and alkenes." *Organic Letters* no. 15(12):3174–3177.

Kumar, Gangam Srikanth, Tapasi Chand, Diksha Singh, and Manmohan Kapur. 2018. "Ruthenium-catalyzed C–H functionalization of benzoic acids with allyl alcohols: A controlled reactivity switch between C–H alkenylation and C–H alkylation pathways." *Organic Letters* no. 20(16):4934–4937.

Kuram, Malleswara Rao, M Bhanuchandra, and Akhila K Sahoo. 2013. "Direct access to benzo [b] furans through palladium-catalyzed oxidative annulation of phenols and unactivated internal alkynes." *Angewandte Chemie* no. 125(17):4705–4710.

Li, Chengliang, Yicheng Zhang, Pinhua Li, and Lei Wang. 2011. "Palladium-catalyzed oxidative cyclization of 3-phenoxyacrylates: An approach to construct substituted benzofurans from phenols." *The Journal of Organic Chemistry* no. 76(11):4692–4696.

Liang, Kangjiang, Jing Yang, Xiaogang Tong, Wenbin Shang, Zhiqiang Pan, and Chengfeng Xia. 2016. "Biomimetic synthesis of moschamine-related indole alkaloids via iron-catalyzed selectively oxidative radical coupling." *Organic Letters* no. 18(6):1474–1477.

Liu, Guixia, Yangyang Shen, Zhi Zhou, and Xiyan Lu. 2013. "Rhodium (III)-catalyzed redox-neutral coupling of N-phenoxyacetamides and alkynes with tunable selectivity." *Angewandte Chemie International Edition* no. 52(23):6033–6037.

Liu, Yang, Yudong Yang, Yang Shi, Xiaojie Wang, Luoqiang Zhang, Yangyang Cheng, and Jingsong You. 2016. "Rhodium-catalyzed oxidative coupling of benzoic acids with terminal alkynes: An efficient access to 3-ylidenephthalides." *Organometallics* no. 35(10):1350–1353.

Mackey, Katrina, Leticia M Pardo, Aisling M Prendergast, Marie-T Nolan, Lorraine M Bateman, and Gerard P McGlacken. 2016. "Cyclization of 4-phenoxy-2-coumarins and 2-pyrones via a double C–H activation." *Organic Letters* no. 18(11):2540–2543.

Mandal, Anup, Suman Dana, Deepan Chowdhury, and Mahiuddin Baidya. 2018. "RuII-catalyzed annulative coupling of benzoic acids with vinyl sulfone via weak

carboxylate-assisted C–H bond activation." *Asian Journal of Organic Chemistry* no. 7(7):1302–1306.
Nandi, Debkumar, Debalina Ghosh, Shih-Ji Chen, Bing-Chiuan Kuo, Nancy M Wang, and Hon Man Lee. 2013. "One-step synthesis of isocoumarins and 3-benzylidenephthalides via ligandless pd-catalyzed oxidative coupling of benzoic acids and vinylarenes." *The Journal of Organic Chemistry* no. 78(7):3445–3451.
Neog, Kashmiri, Ashwini Borah, and Pranjal Gogoi. 2016. "Palladium (ii)-catalyzed C–H bond activation/C–C and C–O bond formation reaction cascade: Direct synthesis of coumestans." *The Journal of Organic Chemistry* no. 81(23):11971–11977.
Pan, Jin-Long, Peipei Xie, Chao Chen, Yu Hao, Chang Liu, He-Yuan Bai, Jun Ding, Li-Ren Wang, Yuanzhi Xia, and Shu-Yu Zhang. 2018. "Rhodium (III)-catalyzed redox-neutral cascade [3+2] annulation of N-phenoxyacetamides with propiolates via C–H functional ization/isomerization/lactonization." *Organic Letters* no. 20(22):7131–7136.
Rao, Wei-Hao, Li-Li Jiang, Jin-Xiao Zhao, Xin Jiang, Guo-Dong Zou, Yu-Qiang Zhou, and Lin Tang. 2018. "Selective O-cyclization of N-methoxy aryl amides with CH2Br2 or 1,2-DCE via palladium-catalyzed C–H activation." *Organic Letters* no. 20(19):6198–6201.
Renzetti, Andrea, Hiroshi Nakazawa, and Chao-Jun Li. 2016. "Rhodium-catalysed tandem dehydrogenative coupling–Michael addition: Direct synthesis of phthalides from benzoic acids and alkenes." *RSC Advances* no. 6(47):40626–40630.
Shaikh, Tanveer Mahamadali, and Fung-E Hong. 2016. "Recent developments in the preparation of N-heterocycles using Pd-catalysed C–H activation." *Journal of Organometallic Chemistry* no. 801:139–156.
Shi, Jiang-Ling, Ding Wang, Xi-Sha Zhang, Xiao-Lei Li, Yu-Qin Chen, Yu-Xue Li, and Zhang-Jie Shi. 2017. "Oxidative coupling of sp 2 and sp 3 carbon–hydrogen bonds to construct dihydrobenzofurans." *Nature Communications* no. 8(1):238.
Song, Lei, Lei Zhu, Zhen Zhang, Jian-Heng Ye, Si-Shun Yan, Jie-Lian Han, Zhu-Bao Yin, Yu Lan, and Da-Gang Yu. 2018. "Catalytic lactonization of unactivated aryl C–H bonds with CO2: Experimental and computational investigation." *Organic Letters.*
Sueki, Shunsuke, Zijia Wang, and Yoichiro Kuninobu. 2016. "Manganese-and borane-mediated synthesis of isobenzofuranones from aromatic esters and oxiranes via C–H bond activation." *Organic Letters* no. 18(2):304–307.
Tan, Peng Wen, Nur Asyikin Binte Juwaini, and Jayasree Seayad. 2013. "Rhodium (III)-amine dual catalysis for the oxidative coupling of aldehydes by directed C–H activation: Synthesis of phthalides." *Organic Letters* no. 15(20):5166–5169.
Wang, Huanan, Baiquan Wang, and Bin Li. 2017. "Synthesis of 3-arylbenzofuran-2-ylphosphines via rhodium-catalyzed redox-neutral C–H activation and their applications in palladium-catalyzed cross-coupling of aryl chlorides." *The Journal of Organic Chemistry* no. 82(18):9560–9569.
Wang, Shihua, Pinhua Li, Lin Yu, and Lei Wang. 2011. "Sequential and one-pot reactions of phenols with bromoalkynes for the synthesis of (Z)-2-bromovinyl phenyl ethers and benzo [b] furans." *Organic Letters* no. 13(22):5968–5971.
Wang, Shi-Meng, Chen Li, Jing Leng, Syed Nasir Abbas Bukhari, and Hua-Li Qin. 2018. "Rhodium (iii)-catalyzed oxidative coupling of N-methoxybenzamides and ethenesulfonyl fluoride: A C–H bond activation strategy for the preparation of 2-aryl ethenesulfonyl fluorides and sulfonyl fluoride substituted γ-lactams." *Organic Chemistry Frontiers* no. 5(9):1411–1415.
Wang, Xisheng, Yi Lu, Hui-Xiong Dai, and Jin-Quan Yu. 2010. "Pd (II)-catalyzed hydroxyl-directed C–H activation/C–O cyclization: Expedient construction of dihydrobenzofurans." *Journal of the American Chemical Society* no. 132(35):12203–12205.
Xiao, Bin, Tian-Jun Gong, Zhao-Jing Liu, Jing-Hui Liu, Dong-Fen Luo, Jun Xu, and Lei Liu. 2011. "Synthesis of dibenzofurans via palladium-catalyzed phenol-directed

C–H activation/C–O cyclization." *Journal of the American Chemical Society* no. 133(24):9250–9253.

Yang, Mingyu, Xingyu Jiang, Wen-Juan Shi, Qi-Lei Zhu, and Zhang-Jie Shi. 2013. "Direct lactonization of 2-arylacetic acids through Pd (II)-catalyzed C–H activation/C–O formation." *Organic Letters* no. 15(3):690–693.

Zhao, Hong, Tinli Zhang, Tao Yan, and Mingzhong Cai. 2015. "Recyclable and reusable [RuCl2 (p-cymene)] 2/Cu (OAc) 2/PEG-400/H2O system for oxidative C–H bond alkenylations: Green synthesis of phthalides." *The Journal of Organic Chemistry* no. 80(17):8849–8855.

Zhao, Jiaji, Yong Wang, Yimiao He, Lanying Liu, and Qiang Zhu. 2012. "Cu-catalyzed oxidative C (sp^2)–H cycloetherification of o-arylphenols for the preparation of dibenzofurans." *Organic Letters* no. 14(4):1078–1081.

Zhou, Zhi, Guixia Liu, Yan Chen, and Xiyan Lu. 2015. "Cascade synthesis of 3-alkylidene dihydrobenzofuran derivatives via rhodium (III)-catalyzed redox-neutral C–H functionalization/cyclization." *Organic Letters* no. 17(23):5874–5877.

Zhu, Ruyi, Jiangbo Wei, and Zhangjie Shi. 2013. "Benzofuran synthesis via copper-mediated oxidative annulation of phenols and unactivated internal alkynes." *Chemical Science* no. 4(9):3706–3711.

5 Six- and Seven-Membered O-Heterocycles

5.1 INTRODUCTION

Due to the presence of six- and seven-membered oxygen-containing heterocycles in a number of biologically important molecules, numerous natural products and pharmaceuticals, they exhibit anti-HIV, antifungal, antimicrobial, antiviral, anti-inflammatory, antidiabetic, and antioxidant properties (Mohammadi et al. 2019, Eftekhari-Sis, Sarvari Karajabad, and Haqverdi 2017). In addition, they have a wide range of uses as building blocks in organic synthesis (Eftekhari-Sis, Zirak, and Akbari 2013, Eftekhari-Sis and Zirak 2014). However, various types of one-step or multi-step synthesis of these types of heterocycles have been reported in the literature. The development of new methodologies to construct the six-membered *O*-heterocycles with different functional groups are of interest to synthetic and medicinal scientists.

Cross-dehydrogenative-coupling (CDC) (Gandeepan et al. 2018, Shaikh and Hong 2016) processes are widely used in the synthesis of various types of heterocyclic compounds. In this chapter, intramolecular and intermolecular CDC of aromatic $C(sp^2)$–H bonds are discussed to construct six-membered *O*-heterocyclic compounds, including chromenes, coumarins, isochromenes, isocoumarins, fused coumarins and isocoumarins, benzochromenes, and xanthenes, etc.

5.2 CHROMENE

Ir(III)-catalyzed carbocarbation of alkynes **2** through sequential arene $C(sp^2)$–H and methoxy $C(sp^3)$–H bond activation of anisoles **1** was developed to create 3-(1-fluoroalken-1-yl)-2*H*-chromene derivatives **3**. Reactions were carried out using [mod-$Cp^*IrCl_2]_2$ (5 mol%) as the catalyst in the presence of cyclic sulfoxide **L2** (12 mol%) and dimethyl diacetoxysilane **Si2** (1 equiv), 1-$AdCO_2Cs$, and $Cu(OAc)_2$ (25 mol%) in *p*-xylene at 85°C under O_2 pressure; corresponding chromene derivatives **3** were obtained in 63–89% yields. The reaction was initiated by the generation of active catalyst species cationic $[Cp^*Ir(OAc)]^+$ **I**, followed by anisole **1** $C(sp^2)$–H activation by concerted metalation/deprotonation (CMD) to give Ir(III)-aryl species **II**, which underwent sulfoxide ligand exchange with acetate for the second CMD event via transition state **III**, leading to methoxy $C(sp^3)$–H bond cleavage to formation of the five-membered iridacycle **IV**. Coordination with subsequent migratory insertion of alkyne **2** into the $C(sp^2)$–Ir bond afforded the seven-membered iridacycle **V**, which underwent *β*–F elimination, giving metal-bound fluoroallene **VI**. A second migratory insertion of the allene into the C(sp^3)–Ir bond (**VII**) followed by *β*–H elimination liberated the chromene product **3**, with formation of Ir(I) species, which reoxidized to Ir(III) active species with $Cu(OAc)_2/O_2$ and cleavage of the Ir–F bond with **Si2** (Scheme 5.1) (Romanov-Michailidis et al. 2018).

In 2014, an Rh(III)-catalyzed synthesis of 2*H*-chromene scaffolds **6** from *N*-phenoxyacetamides **4** and cyclopropenes **5** (1.3 equiv) was developed using $[Cp^*RhCl_2]_2$ (2.5 mol%) as a catalyst and CsOPiv (0.25 equiv) as a base in MeOH at 25°C under air. Reactions occurred at 5 min, and corresponding 2,2-disubstituted 2*H*-chromenes **6** were obtained in 53–95% yields. *Meta*-substituted *N*-phenoxyacetamide substrate **4** gave the expected 2*H*-chromene product **6** via

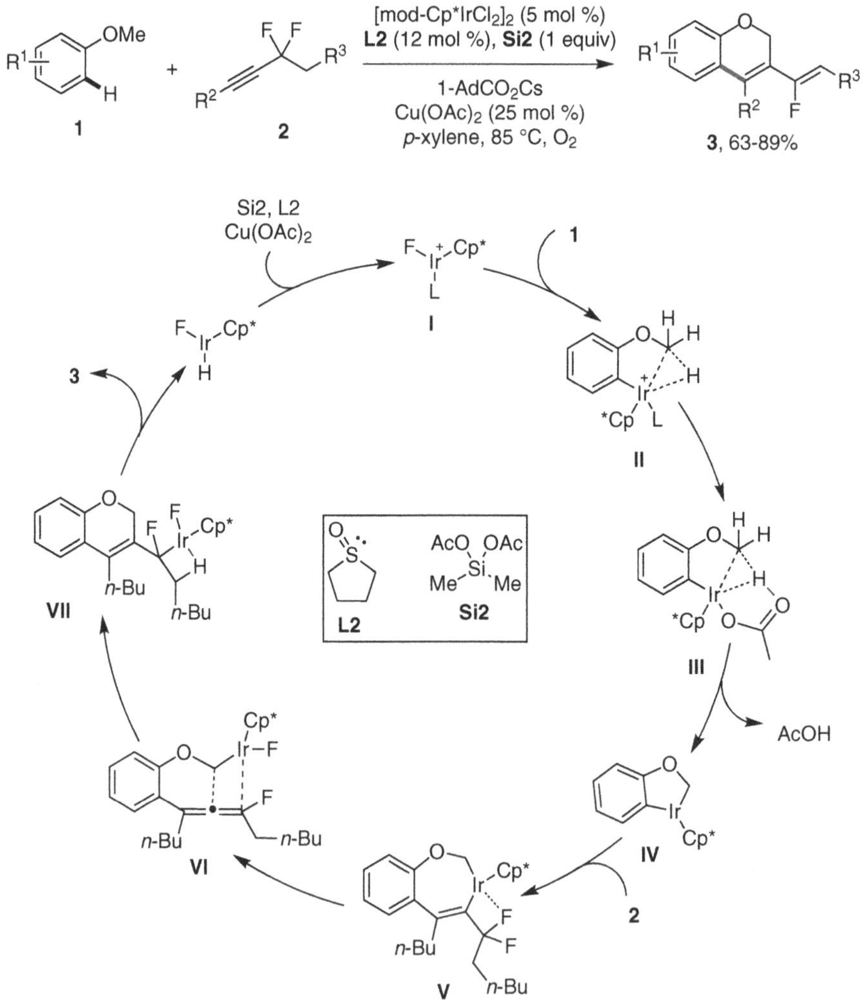

SCHEME 5.1

activation of the C–H bond in the *para* position to the acetamido group, resulting from reaction at the less hindered site. Spirocyclopropenes were also tolerated under reaction conditions, affording related spirochromenes **6a** in 59–69% yields. The proposed reaction mechanism involves the nitrogen coordination and *ortho*-C–H activation of *N*-phenoxyacetamide **4** with Cp*Rh(OPiv)$_2$ active species, generated in situ by the ligand exchange of [Cp*RhCl$_2$]$_2$ with CsOPiv, to five-membered rhodacyclic intermediate **I**, which activated cyclopropene to generate the intermediate **II**. The ring opening generated the Rh(III) carbene **III**, which underwent migratory insertion of the carbene into the C–Rh bond to afford six-membered rhodacyclic intermediate **IV**. 1,3-Allylic migration gave the eight-membered rhodacyclic intermediate **V**, which was converted to the final product **6** by intramolecular substitution to form the C–O bond and cleave the N–O bond (**VI**), along with the regeneration of the Rh(III) catalyst for the next catalytic cycle (Scheme 5.2) (Zhang et al. 2014).

2*H*-Chromene-3-carboxylic acids **8** were synthesized from *N*-phenoxyacetamides **4** and methyleneoxetanones **7** via Rh(III)-catalyzed C–H activation/an unusual [3+3] annulation sequence. Reactions were performed using 1.5 equiv of methyleneoxetanones

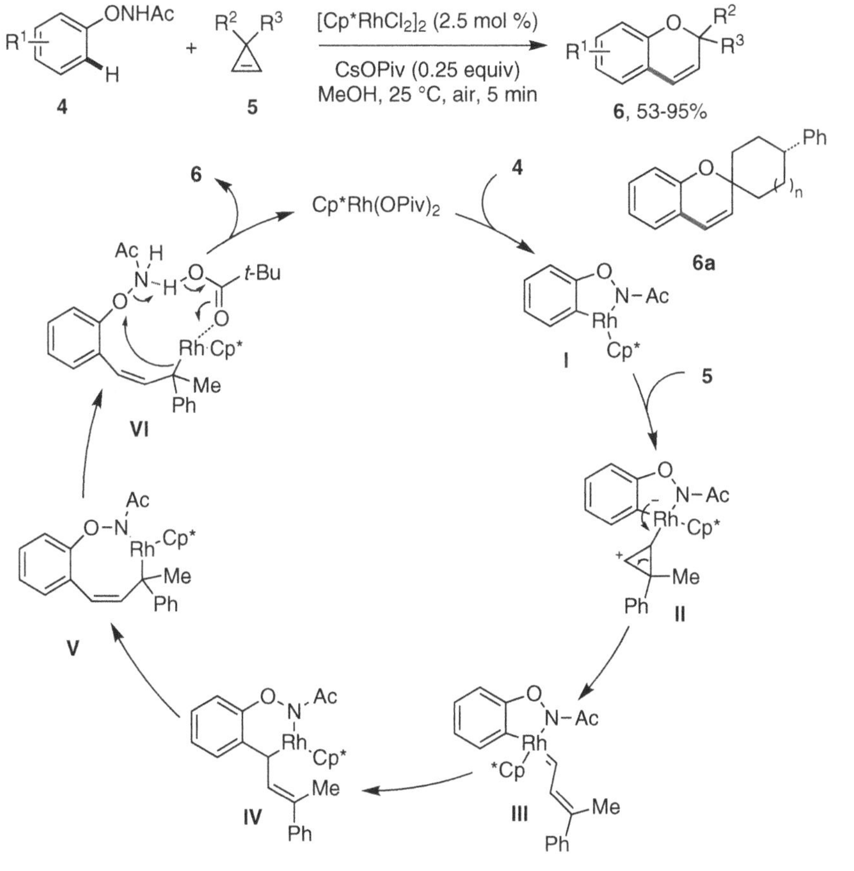

SCHEME 5.2

7, $[Cp^*RhCl_2]_2$ (2.5 mol%), and CsOAc (1 equiv) in CH_3CN at 60°C, providing 2*H*-chromene-3-carboxylic acids **8** in 23–85% yields. The reaction proceeded by coordination of the N atom and *ortho*-C–H activation of *N*-phenoxyacetamides **4** to give five-membered rhodacycle intermediate **I**, which underwent regioselective migratory insertion of the α-methylene moiety of **7** into the Rh–C bond, forming the seven-membered intermediate **II**. Oxidative addition of Rh(III) into the O–N bond generated the six-membered Rh(V) species **III**, converting to the intermediate **IV** by AcOH-assisted addition and subsequent *anti*-β–H elimination. Finally, protonation of **V** delivered the intermediate **VI**, regenerating the active catalyst. An intramolecular S_N2-type nucleophilic substitution reaction along with the alkyl C–O bond cleavage of **V** released the product **8** (Scheme 5.3) (Zhou et al. 2018).

5.3 COUMARIN

Synthesis of 3-functionalized coumarin derivatives **11** was achieved by visible-light-promoted dual C–C bond formations of alkynoates **9**. Reactions occurred via a

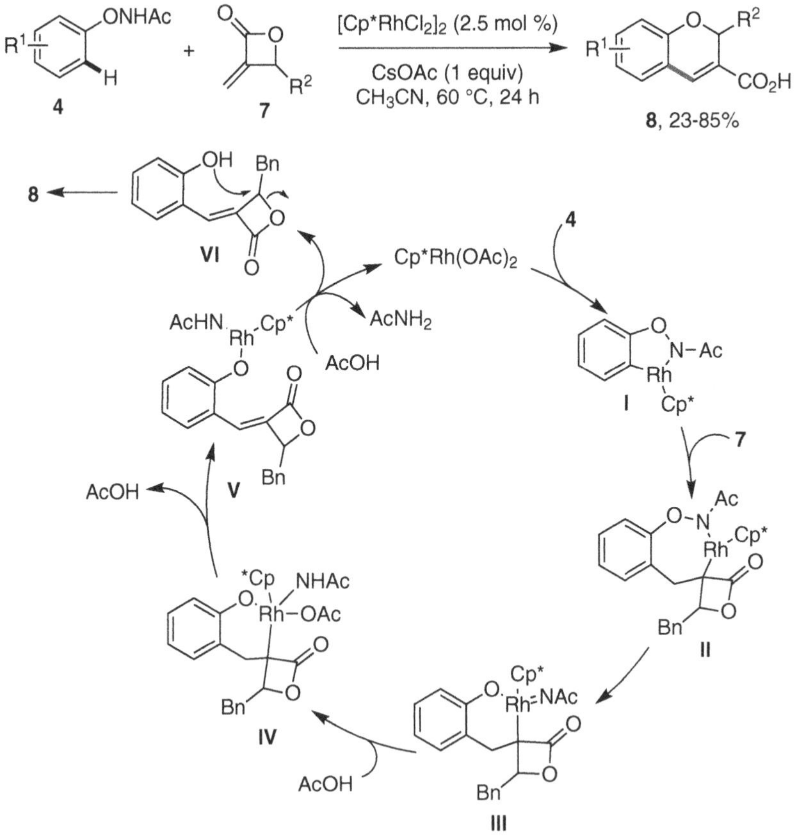

SCHEME 5.3

domino radical addition/cyclization sequence. Irradiation of a solution of aryl 3-arylpropiolates **9** in an ether/CH_3CN mixture solvent in the presence of [Ru(bpy)$_3$Cl$_2$] (5 mol%) and *tert*-butyl hydroperoxide (TBHP) (4 equiv) by 34 W blue LED strip at room temperature under argon for 40–48 h gave 3-etherified coumarins **11** in 51–82% yields. The reaction with 1,3-dioxolane occurred at the C-2 position to give the corresponding 3-(1,3-dioxolan-2-yl)-2*H*-chromen-2-one in 42% yield. 1,2-Dimethoxyethane, glyme, afforded a mixture of two regioisomers in 60% total yield with a ratio of 1:0.65 for CH_3:CH_2 functionalization, respectively. A free-radical-type process was proposed for the reaction mechanism, in which α-oxo radical **I** was generated through abstraction of a hydrogen atom from tetrahydrofuran by a *t*-BuO radical generated in situ via a single electron transfer between TBHP and Ru(II) species in an excited state under irradiation by the blue LED. Selective addition of the radical **I** to the α-position of the C=O bond in alkynoate **9** formed the vinyl radical **II**, which cyclized to form the radical intermediate **III**. Subsequently, oxidation of **III** with the Ru(III) species generated the carbocation intermediate **IV**, which was transformed into the desired 3-etherified coumarin product **11** upon deprotonation (Scheme 5.4) (Feng et al. 2016).

Coumarins **14** were accessed by Rh-catalyzed dehydrogenative annulation of aryl thiocarbamates **12** with internal alkynes **13**. Reacting aryl thiocarbamates **12** with 1.5 equiv of alkyne **13** in the presence of [Cp*RhCl$_2$]$_2$ (2.5 mol%), AgOTf (10 mol%), and Cu(OAc)$_2$ (1 equiv) in *t*-AmOH at 120°C under an atmosphere of argon

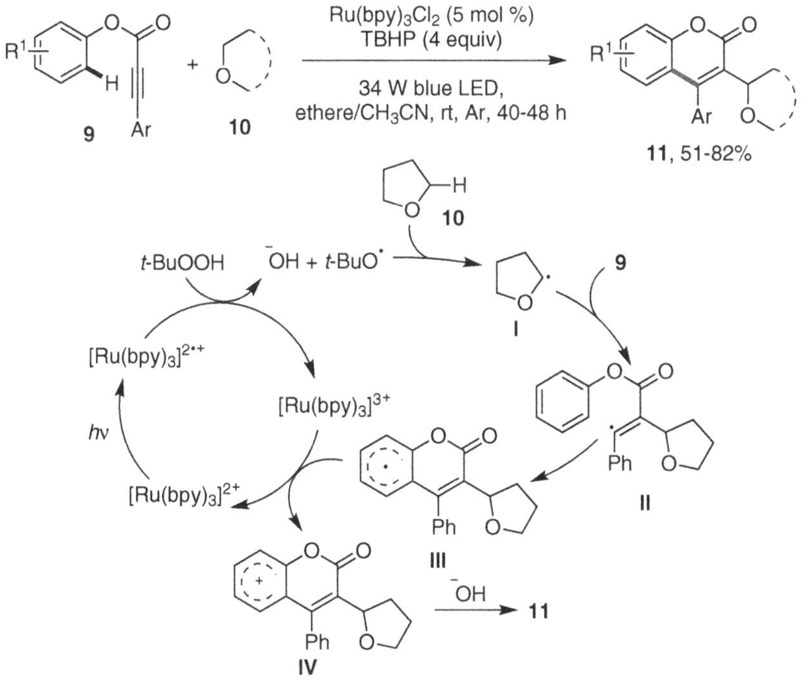

SCHEME 5.4

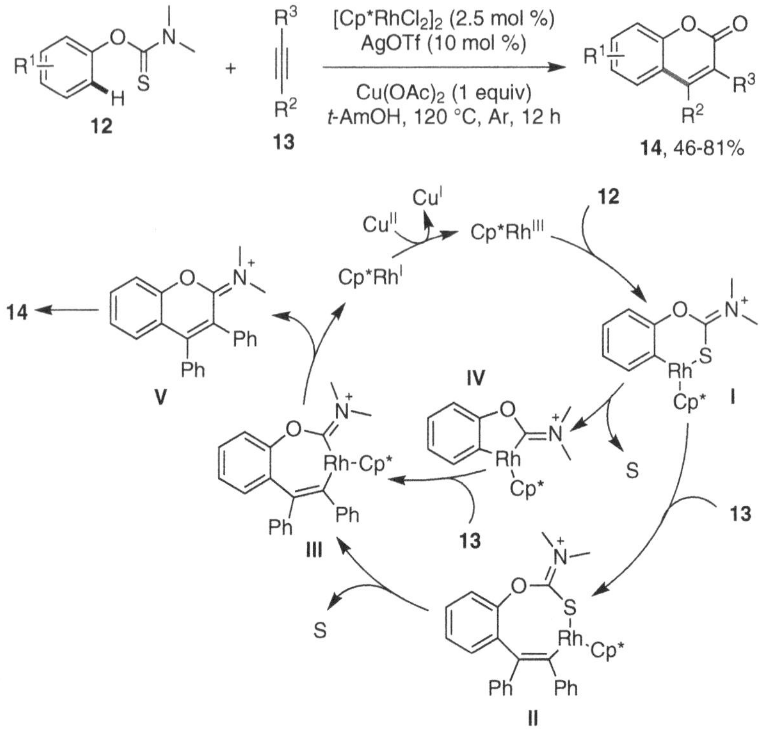

SCHEME 5.5

furnished coumarin derivatives **14** in 46–81% yields. Dialkyl acetylene was also compatible with the reaction, as 5-decyne afforded 3,4-dibutylcoumarin **14** in 61% yield. Unsymmetrical alkyl aryl alkyne, 1-phenylpropyne, led to 4-methyl-3-phenylcoumarin in 68% yield, regioselectively. No reaction occurred when phenyl dimethylcarbamate or *S*-phenyl dimethylcarbamothioate was used under reaction conditions. As postulated in the reaction mechanism, cationic [Cp*Rh(III)], generated in situ in the presence of AgOTf, was coordinated with the soft sulfur atom in the thiocarbamate **12**, followed by *ortho* C–H activation to afford rhodacyclic complex **I**, which was transformed into seven-membered rhodacycle intermediate **III** either by desulfurization followed by migratory insertion of an alkyne (**IV**) or by alkyne insertion to afford the eight-membered ring **II** followed by desulfurization sequences. Finally, reductive elimination released the iminium salt **V** and Rh(I) species, which was reoxidized to an Rh(III) active species by $Cu(OAc)_2$ to complete the catalytic cycle. The nucleophilic attack of the acetate anion on the iminium carbon of **V** followed by C–N bond cleavage led to the desired product **14** (Scheme 5.5) (Zhao et al. 2015).

5.4 ISOCHROMENE

(Isochroman-1-ylidene)acetate derivatives **17** were prepared by Pd(II)-catalyzed dehydrogenative coupling of tertiary 2-arylethanols **15** with acrylates **16**. Reactions

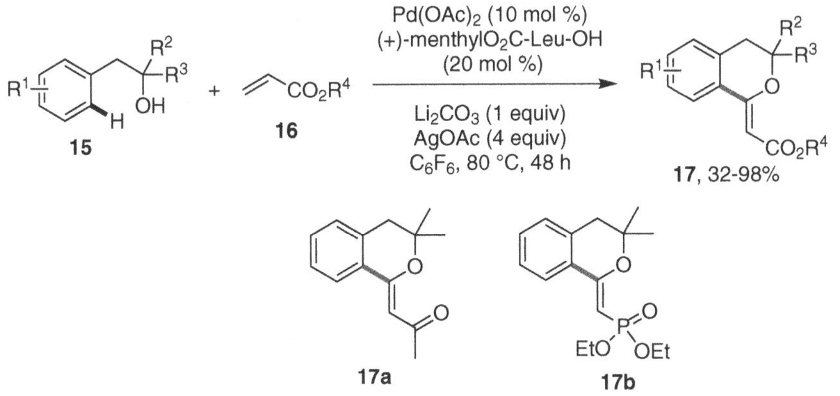

$Pd(OAc)_2$ (10 mol %)
(+)-menthylO_2C-Leu-OH
(20 mol %)
Li_2CO_3 (1 equiv)
AgOAc (4 equiv)
C_6F_6, 80 °C, 48 h

15 + **16** → **17**, 32-98%

17a **17b**

SCHEME 5.6

were catalyzed with 10 mol% of $Pd(OAc)_2$ in the presence of (+)-menthylO_2C-Leu-OH (20 mol%), Li_2CO_3 (1 equiv), and AgOAc (4 equiv) in C_6F_6 at 80°C, and 2-(3, 3-dialkylisochroman-1-ylidene)acetate derivatives **17** were obtained in 32–98% yields. Reaction of 1-phenyl-2-methylpropan-2-ol with methyl vinyl ketone and diethyl vinylphosphonate under reaction conditions afforded 1-(3,3-dimethylisoch roman-1-ylidene)propan-2-one **17a** and diethyl ((3,3-dimethylisochroman-1-ylidene) methyl)phosphonate **17b** in 53% and 89% yields, respectively. No expected cyclization product was obtained in the case of vinyl sulfone and vinyl sulfonate. Reaction with primary and secondary arylethanol substrates was also studied under reaction conditions, affording corresponding (isochroman-1-ylidene)acetates in 28–63% yields. Pd(II)-catalyzed *ortho*-C–H olefination followed by oxidative cyclization was proposed as the reaction mechanism for the construction of isochroman scaffolds (Scheme 5.6) (Lu et al. 2010).

Rh-catalyzed dehydrogenative coupling of α,α-disubstituted benzyl alcohols **18** with alkynes **13** was developed in order to synthesize isochromene structural motifs **19**. Treatment of 3 equiv of benzyl alcohols **18** with alkynes **13** in the presence of $[Cp^*Rh(MeCN)_3][SbF_6]_2$ (4 mol%) and $Cu(OAc)_2 \cdot H_2O$ (2 equiv) in refluxing dioxane under N_2 for 6 h resulted in the formation of 1,1-disubstituted 3,4-diaryl-1*H*-isochromenes **19** in 45–89% yields. Reaction of aliphatic alkyne produced the expected isochromene in 36% yield. However, unsymmetrical alkyl aryl alkyne, 1-phenylpropyne, led to the corresponding isochromene in 67% yield, with 82:18 regioselectivity, favoring the 4-methyl-3-phenyl isochromene regioisomer. The reaction occurred by O-atom-directed *ortho*-C-H activation to generate a five-membered rhodacycle, followed by alkyne insertion with subsequent reductive elimination (Scheme 5.7) (Morimoto et al. 2011).

5.5 ISOCOUMARIN

Dehydrogenative coupling of benzoic acids **20** with styrene derivatives **21** was described by Nandi et al. (2013). Reactions were catalyzed with $Pd(OAc)_2$ (5 mol%) and Ag_2O (1 equiv) in *N,N*-dimethylformamide (DMF) at 110°C to obtain 3-aryl

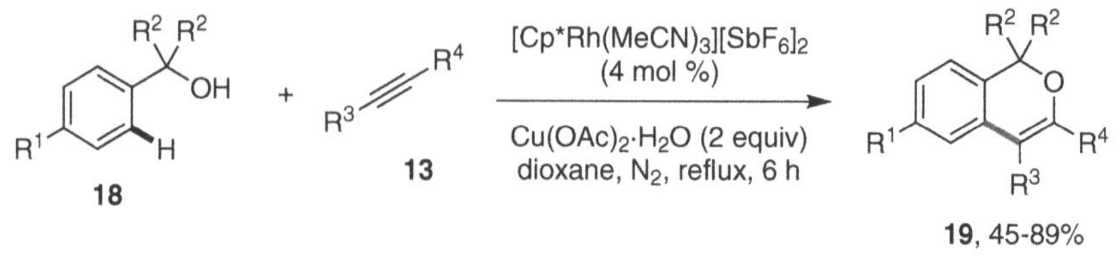

18 + 13 → 19, 45-89%

$[Cp^*Rh(MeCN)_3][SbF_6]_2$ (4 mol %)

$Cu(OAc)_2 \cdot H_2O$ (2 equiv) dioxane, N_2, reflux, 6 h

SCHEME 5.7

substituted isocoumarins **22** in 51–76% yields. Seven-membered ring lactones **24** were obtained when 9-anthracenecarboxylic acid **23** was reacted with styrenes **21** in 64–68% yields. The reaction was initiated by carboxylic acid *O*-atom-directed C–H activation of benzoic acid to give five-membered palladacycle **I**, followed by coordination and insertion of alkene into the Pd–C bond generating seven-membered intermediate **II**. *β*–H elimination at the C-1 position of the vinyl group produced the intermediate **III**, which underwent intramolecular attack of the coordinated O atom at the C-2 position of the vinyl moiety, leading to intermediate **IV**. Finally, *β*–H elimination of **IV** gave isocoumarin product **22**. Also, synthesis of 3-substituted isocoumarins **22** was achieved by Rh(III)-catalyzed dehydrogenative coupling of benzoic acids **20** with 1-arylvinyl acetates **25**. Reactions were performed using 2.5 equiv of vinyl acetates **25** in the presence of $[Cp^*RhCl_2]_2$ (5 mol%), CuO (2 equiv), KOAc (2 equiv), LiCl (1 equiv), and KI (20 mol%) in PhMe/*t*-AmOH at 120°C, giving 3-substituted isocoumarins **22** in 35–78% yields. In the case of 1-alkylvinyl acetates, in addition to the expected 3-substituted isocoumarins **22**, corresponding 3-alkylideneisobenzofuran-1(3*H*)-ones **26** were obtained as minor products. Reaction occurred by activation of the C–H bond with Rh, followed by insertion of the vinyl moiety and then reductive elimination with subsequent release of AcOH (Scheme 5.8) (Zhang et al. 2014).

Moreover, Pd-catalyzed sequential nucleophilic addition/dehydrogenative annulation of bromoalkynes **27** with benzoic acids **20** was developed by Jiang et al. (2017) to construct 3-substituted isocoumarins **22**. $Pd(TFA)_2$ (10 mol%), in the presence of a DPEPhos ligand (15 mol%) and K_2CO_3 (2 equiv) in dimethyl sulfoxide (DMSO)/EtOH at 120°C under N_2, catalyzed the reaction and led to 3-substituted isocoumarins **22** in 26–86% yields. Alkyl bromoalkynes, as well as aryl bromoalkynes, worked under reaction conditions. The reaction was proposed to proceed by *cis*-nucleophilic addition of benzoic acids **20** to bromoalkynes **27** to give the intermediate **I**, followed by oxidative addition to Pd(0) species to generate intermediate **II**, which underwent *ortho*-C(sp^2)–H activation to afford seven-membered palladacyclic intermediate **III**. Finally, reductive elimination delivered the coupling product **22** along with regeneration of the active Pd(0) species for further catalytic cycles (Scheme 5.9).

Dehydrogenative cross-coupling of benzoic acids with alkynes were reported using Rh(III), Ir(III), Ru(II), and Co(III) catalysts in order to obtain isocoumarins. Ueura, Satoh, and Miura (2007a) developed an Rh-catalyzed coupling reaction of benzoic acids **20** with diaryl acetylenes **13** using 1 mol% of $[Cp^*RhCl_2]_2$ in the presence of $Cu(OAc)_2 \cdot H_2O$ (5 mol%) under air as the oxidant in DMF at 120°C, leading to isocoumarin derivatives **28** in 81–97% yields. However, dialkyl acetylenes

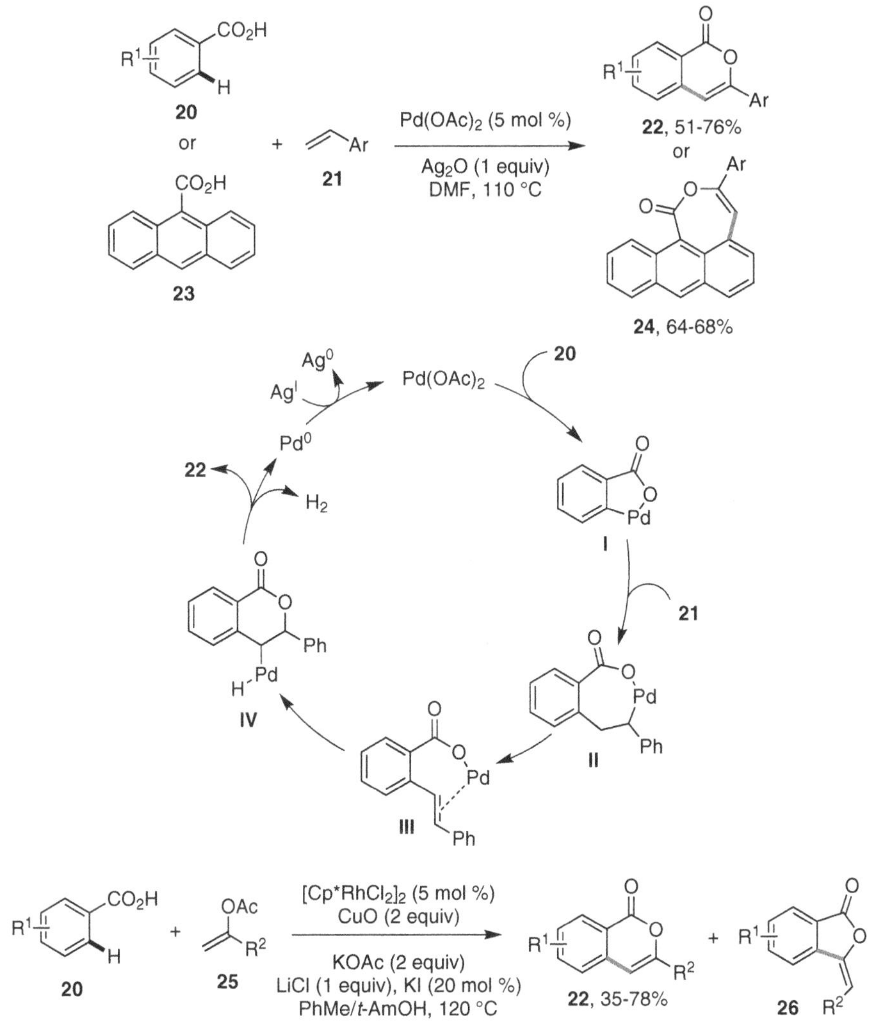

SCHEME 5.8

were tolerated well in the reaction to produce 3,4-dialkylisocoumarin **28** in 83–97% yields. Also, an Rh catalyst was used in the presence of 2 equiv of $Cu(OAc)_2$ under N_2 atmosphere (Ueura, Satoh, and Miura 2007b, Satoh, Ueura, and Miura 2008). Frasco et al. (2013) used $[Cp^*IrCl_2]_2$ (2.5 mol%) as the catalyst in the presence of AgOAc (2 equiv) as the oxidant in MeOH at 60°C for dehydrogenative coupling of benzoic acids with alkynes. Diaryl alkynes and dialkyl alkynes afforded corresponding isocoumarins **28** in 28–81% and 81–99% yields, respectively. In Ackermann's work (Ackermann et al. 2012), $[RuCl_2(p\text{-cymene})]_2$ (2.5 mol%) and KPF_6 (20 mol%) were utilized as the catalytic system for the dehydrogenative coupling between benzoic acids **20** with alkynes **13** in the presence of $Cu(OAc)_2{\cdot}H_2O$ (2 equiv) as an oxidant in *t*-AmOH at 120°C, delivering the expected 3,4-diaryl- and 3,4-dialkyl-isocoumarins

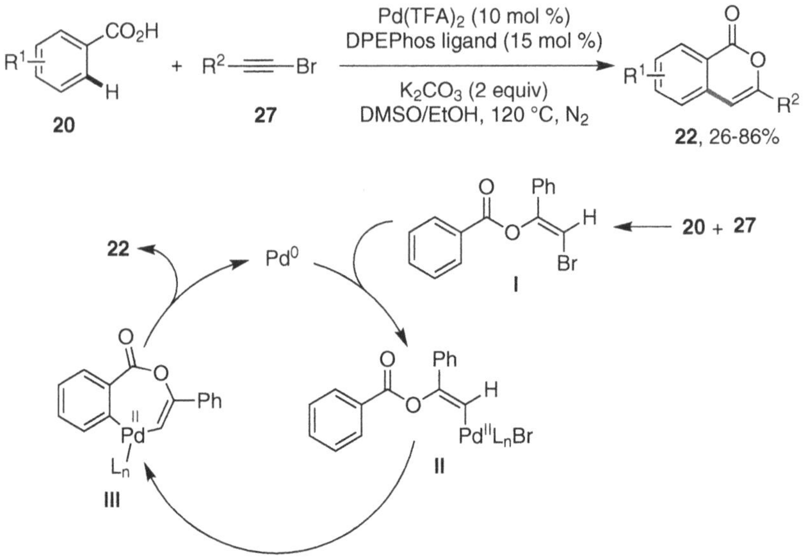
CO2H
R1
H
20
R2—≡—Br
27
Pd(TFA)2 (10 mol %)
DPEPhos ligand (15 mol %)
K2CO3 (2 equiv)
DMSO/EtOH, 120 °C, N2
R1
O
O
R2
22, 26-86%
O
Ph
H
O
Br
I
20 + 27
22
Pd0
O
O
Ph
Pd
II
Ln
III
O
Ph
H
O
PdIILnBr
II

SCHEME 5.9

28 in 60–87% and 58–85% yields, respectively. 4-Alkyl-3-arylisocoumarins **28** were obtained in 46–65% yields, regioselectively, when unsymmetrical alkyl aryl acetylenes were subjected to benzoic acids under reaction conditions. Indol-3-carboxylic acid was also compatible with the reaction to furnish pyrano[4,3-*b*]indol-1(5*H*)-ones **29** in 65–83% yields. Moreover, [Cp*Co(CO)I_2] (10 mol%) in the presence of NaOAc (20 mol%) and CuO (2 equiv) was used as the catalytic system in 2,2,2-trifluoroethanol (TFE) at 80°C for dehydrogenative coupling of benzoic acids **20** with internal alkynes **13**. Diaryl acetylenes delivered the expected isocoumarins **28** in 64–97% yields. Dialkyl acetylenes afforded related isocoumarins in 71–96% yields. Unsymmetrical alkyl aryl acetylenes led to the production of a mixture of two regioisomers, with a majority of 4-alkyl-3-arylisocoumarins (Sen et al. 2017). In all cases, the reaction was initiated by carboxylic acid *O*-atom-directed *ortho*-C–H activation by the metal center of the catalyst, followed by coordination and insertion of an alkyne into the C-metal bond, with subsequent reductive elimination (Scheme 5.10).

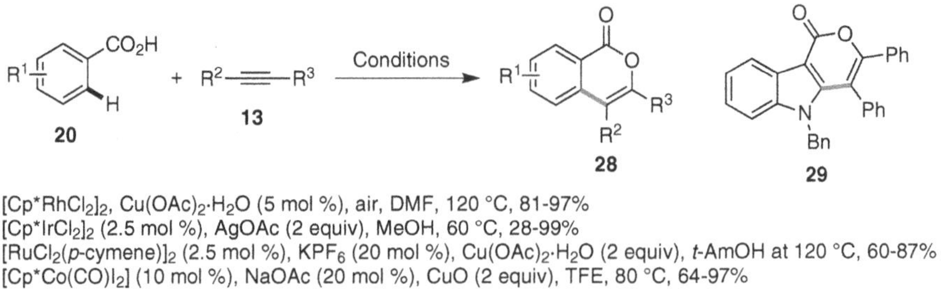
CO2H
R1
H
20
R2—≡—R3
13
Conditions
R1
O
O
R3
R2
28
O
O
Ph
N
Ph
Bn
29
[Cp*RhCl2]2, Cu(OAc)2·H2O (5 mol %), air, DMF, 120 °C, 81-97%
[Cp*IrCl2]2 (2.5 mol %), AgOAc (2 equiv), MeOH, 60 °C, 28-99%
[RuCl2(p-cymene)]2 (2.5 mol %), KPF6 (20 mol %), Cu(OAc)2·H2O (2 equiv), t-AmOH at 120 °C, 60-87%
[Cp*Co(CO)I2] (10 mol %), NaOAc (20 mol %), CuO (2 equiv), TFE, 80 °C, 64-97%

SCHEME 5.10

Ru-catalyzed synthesis of 4-benzylideneisochroman-1-one derivatives **32** was reported by reacting benzoic acids **20** with propargylic alcohols **31** (2.5 equiv) in the presence of $[RuI_2(p\text{-cymene})]_2$ (2 mol%), guanidine carbonate (0.5 equiv), and AcOH (1 equiv) in *t*-AmOH at 100°C. Primary and secondary propargylic alcohols were tolerated in the reaction, giving 4-benzylideneisochroman-1-ones **32** in 51–90% yields (Huang et al. 2016). Also, Rh(III)-catalyzed C–H activation and annulation of 1-benzoylpyrrolidine **30** with propargyl alcohols **31** was developed for the synthesis of 4-benzylideneisochroman-1-one scaffolds **32**. Reactions were performed using $[Cp^*RhCl_2]_2$ (4 mol%), $AgSbF_6$ (16 mol%), and AgOAc (20 mol%) in the presence of AcOH (2 equiv) in 1,4-dioxane at 80°C under an atmosphere of argon, leading to 4-benzylideneisochroman-1-ones **32** in 34–85% yields. The proposed reaction mechanism involves the generation of five-membered rhodacycle intermediate **I** via amide *O*-atom-directed *ortho*-C–H activation, followed by coordination and insertion of the alkyne moiety of propargyl alcohol **31** into the C–Rh bond in a regioselective manner to give seven-membered rhodacycle **II**. Protonolysis gave an alkenylation intermediate **III**, which underwent Rh- or Ag-catalyzed lactonization to achieve the final product **32** with the release of a pyrrolidine (Scheme 5.11) (Wang et al. 2013).

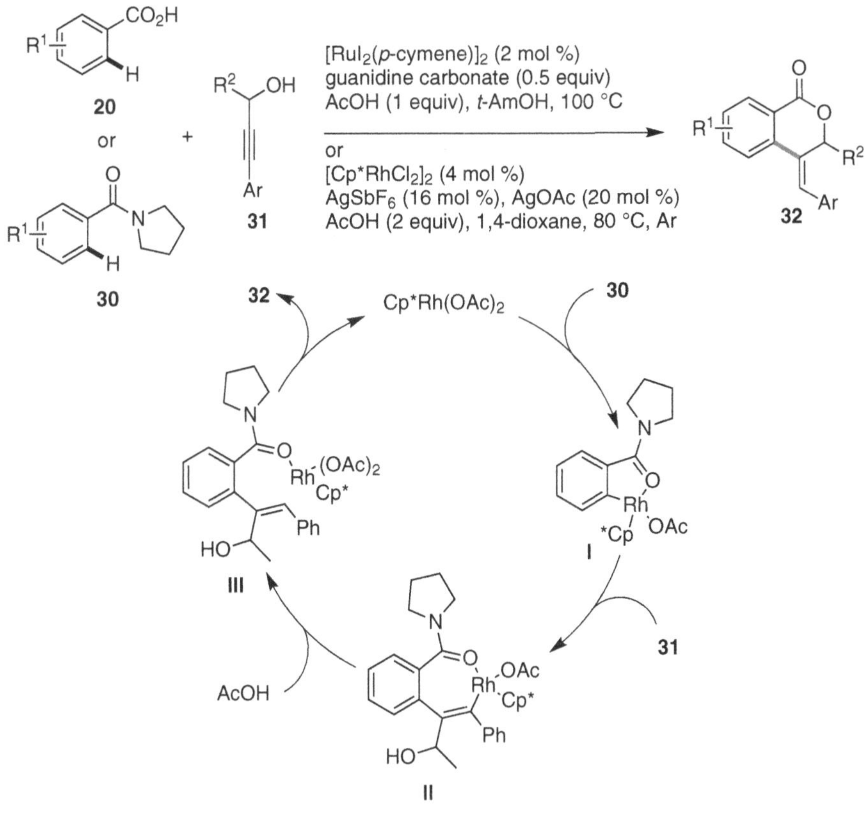

SCHEME 5.11

5.6 HETEROCYCLE-FUSED COUMARIN AND ISOCOUMARIN

Rh-catalyzed synthesis of 3-oxo-3,5-dihydropyrano[4,3-*b*]indole-4-carboxylates **35** was developed via dehydrogenative coupling of indoles **33** with diazo esters **34**. Reacting 3-acyl indoles **33** with α-diazo malonates **34** (2.5 equiv) in the presence of $[Cp^*RhCl_2]_2$ (2.5 mol%), $AgSbF_6$ (10 mol%), $Zn(OTf)_2$ (0.5 equiv), and AcOH (2 equiv) in 1,2-dichloroethane (DCE) at 100°C for 10 h produced 3-oxo-3,5-dihydropyrano[4,3-*b*]indole-4-carboxylate derivatives **35** in 32–81% yields. In the proposed reaction mechanism, a five-membered rhodacyclic intermediate **I**, generated in situ via C2–H activation of indole **33**, underwent coordination of the diazo moiety by denitrogenation to afford an Rh–carbene species **II**. Subsequent migratory insertion of the carbene moiety into the Rh–aryl bond gave the intermediate **III**, which converted to 2-alkylated intermediate **IV** upon protonolysis. By tautomerization of **IV** to **V**, followed by Rh- or Zn-assisted intramolecular nucleophilic attack, the annulated product **35** was obtained, together with regeneration of the active Rh(III) catalyst (Scheme 5.12) (Chen et al. 2017).

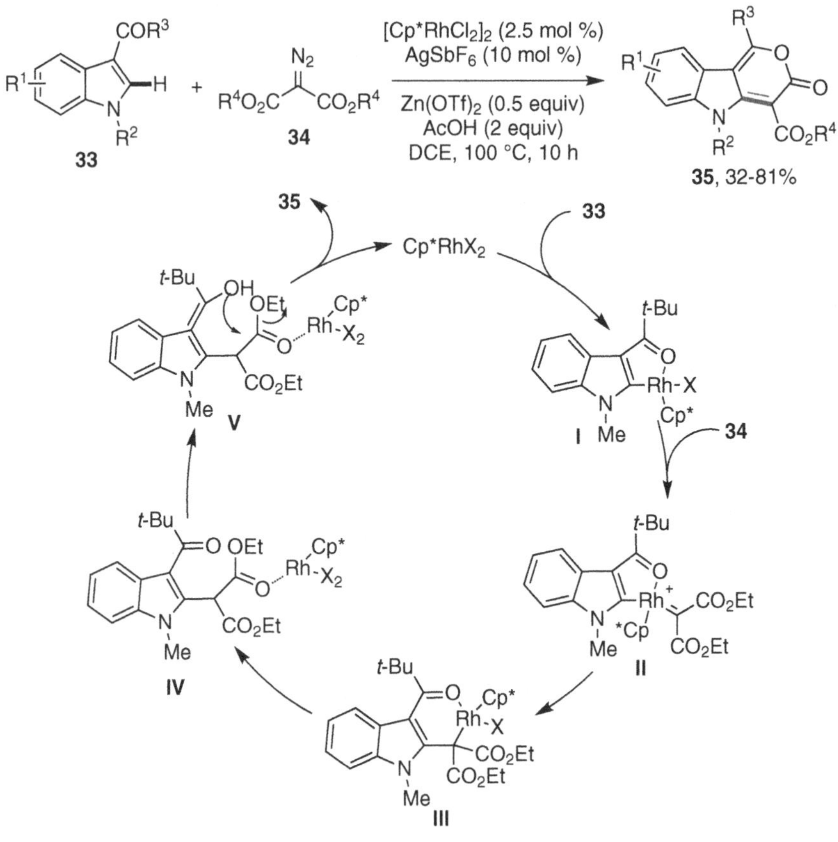

SCHEME 5.12

Indolo[3,2-*c*]coumarins **37** were prepared by Pd-catalyzed intramolecular dehydrogenative coupling of phenyl 1*H*-indole-3-carboxylates **36**. Reactions were carried out using $Pd(OAc)_2$ (10 mol%), AgOAc (3 equiv), PivOH (1 equiv), and K_2CO_3 (2 equiv) in *N,N*-dimethylacetamide (DMA) under N_2 at 120°C, and indolo[3,2-*c*] coumarins **37** were obtained in 52–86% yields. The starting substrates were prepared by Pd-catalyzed C–H carbonylation of indoles with aryl formats (Wu et al. 2014). 5*H*-Benzo[4,5]thieno[3,2-*c*]isochromen-5-one **40** was obtained in 75% yield when 2-(benzo[*b*]thiophen-2-yl)benzoic acid **38** was treated with $Pd(OAc)_2$ (10 mol%) and $PhI(OAc)_2$ (2 equiv) in the presence of *N*-acetylglycine (Ac-Gly-OH, 20 mol%) and KOAc (2 equiv) in *t*-BuOH at 120°C for 24 h. The reaction occurred by direct intramolecular dehydrogenative C–H/O–H coupling. A similar methodology was developed for the construction of 4*H*-benzo[4,5]thieno[3,2-*b*]thieno[2,3-*d*] pyran-4-one and 5*H*-dithieno[3,2-*b*:2',3'-*d*]pyran-5-one **39** in 69% and 60% yields, respectively (Qin et al. 2015). Also, benzothieno[3,2-*c*]isochromen-5-one derivatives **40** were obtained by Ir(III)-catalyzed oxidative annulation of phenylglyoxylic acids **41** with benzo[*b*]thiophenes **42**. Reactions were carried out by heating a mixture of phenylglyoxylic acids **41** and benzo[*b*]thiophenes **42** (2 equiv) in the presence of $[Cp^*IrCl_2]_2$ (5 mol%), $AgNTf_2$ (20 mol%), Ag_2O (4 equiv), and 1-AdCOOH (1 equiv) in hexafluoroisopropanol (HFIP) at 90°C under an atmosphere of N_2, producing benzothieno[3,2-*c*]isochromen-5-ones **40** in 29–90% yields. The reaction proceeded by Ir(III)-catalyzed coupling of phenylglyoxylic acids **41** with benzo[*b*]thiophenes **42**, followed by Ag-catalyzed decarboxylative oxidation to 2-(benzothien-2-yl)benzoic acid with subsequent Ir(III)-catalyzed intramolecular dehydrogenative coupling (Scheme 5.13) (Wang, Yang, and Yang 2018).

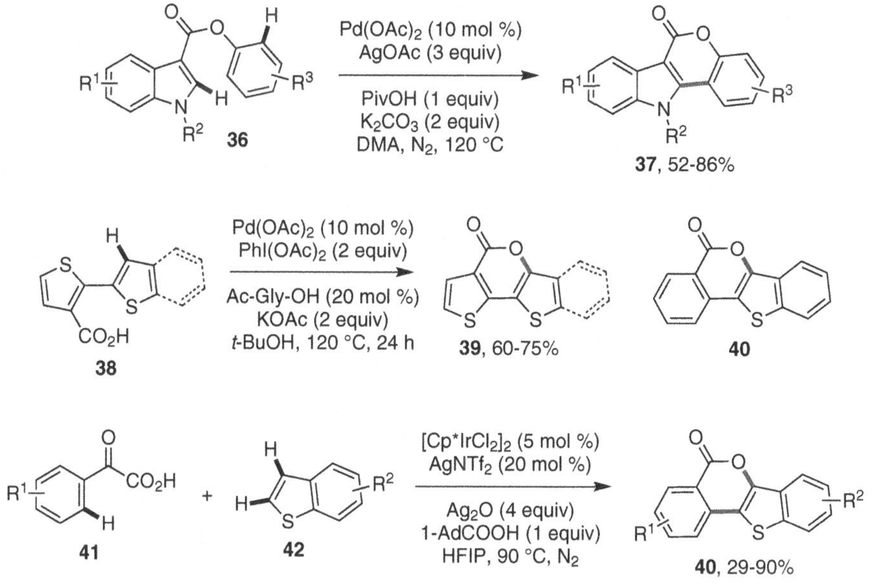

SCHEME 5.13

5.7 BENZO[C]CHROMENE

Pd-catalyzed intramolecular dehydrogenative coupling reaction of 3-(benzyloxy) phenyl pyridine-2-sulfonates **43** to generate 6*H*-benzo[*c*]chromenes **44** was reported by Guo et al. (2017). Reactions were performed using $Pd(OAc)_2$ (10 mol%) as the catalyst in HFIP at 55°C under O_2 pressure, giving 6*H*-benzo[*c*]chromen-1-yl pyridine-2-sulfonate derivatives **44** in 52–95% yields. The *O*-(2-pyridyl)sulfonyl group-directed palladation of 3-(benzyloxy)phenyl pyridine-2-sulfonate **43**, followed by concerted metalation/deprotonation (CMD) and then reductive elimination, gave the desired chromene product **44** along with Pd(0), which was reoxidized by O_2 to regenerate the Pd(II) species to complete the catalytic cycle (Scheme 5.14).

Reddy Chidipudi et al. (2013) developed Pd(II)- or Ru(II)-catalyzed dehydrogenative coupling of 2-aryl-3-hydroxy-2-cyclohexenones **45** with activated alkenes **46**. Reactions were conducted using either $Pd(OAc)_2$ (5 mol%) and $Cu(OAc)_2$ (2.1 equiv) in DMF at 120°C, or $[RuCl_2(p\text{-cymene})]_2$ (2.5 mol%), $Cu(OAc)_2$ (2.1 equiv), and K_2CO_3 (2 equiv) in *t*-AmOH at 90°C, giving tetrahydro-1*H*-benzo[*c*]chromene derivatives **47** in 45–73% or 19–76% yields, respectively. Methyl acrylates, acrylamide, and acrylonitrile were quite compatible under the Ru(II)-catalyzed reaction, while methyl vinyl ketone and phenyl vinyl sulfone were the best substrates under the Pd(II)-catalyzed coupling reaction. Dihydropyrano[4,3-*c*]isochromene derivatives **49** were synthesized by Rh(III)-catalyzed dehydrogenative coupling of 4-hydroxy-3-aryl-2*H*-pyran-2-ones **48** (X=O) with activated alkenes **46** (1.5 equiv) using $[Cp^*RhCl_2]_2$ (2.5 mol%) and $Cu(OAc)_2$ (2.1 equiv) in DMF at 70°C. Various activated alkenes, including acrylates, acrylonitrile, acrylamides, vinyl ketones, and vinyl sulfones were tolerated under reaction conditions, and corresponding dihydropyrano[4,3-*c*]isochromenes **49** were obtained in 52–96% yields. The reaction of 4-hydroxy-3-arylpyridin-2(1*H*)-ones **48** (X=NH) led to the creation of dihydro-1*H*-isochromeno[4,3-*c*]pyridines **49** in 52–83% yields. Coordination of the O atom with subsequent C–H activation of 4-hydroxy-3-aryl-2*H*-pyran-2-one **48** with $[Cp^*Rh(OAc)_2]$, generated in situ by ligand exchange between $[Cp^*RhCl_2]_2$ and $Cu(OAc)_2$, followed by coordination and migratory insertion of an alkene and then protonation by AcOH, gave alkenylated product **I** via *β*-hydride elimination, which was cyclized into the final desired product **49** (Scheme 5.15) (Bollikolla et al. 2017).

Pd-catalyzed dehydrogenative coupling of 2-aryl-3-hydroxy-2-cyclohexenones **45** with allyl acetate **51** was reported to produce 6-vinyltetrahydro-1*H*-benzo[*c*] chromen-1-one derivatives **52**. Reactions were carried out using 1.5 equiv of allyl acetate **51** in the presence of $Pd(OAc)_2$ (5 mol%) and $Cu(OAc)_2$ (4.2 equiv) in DMF

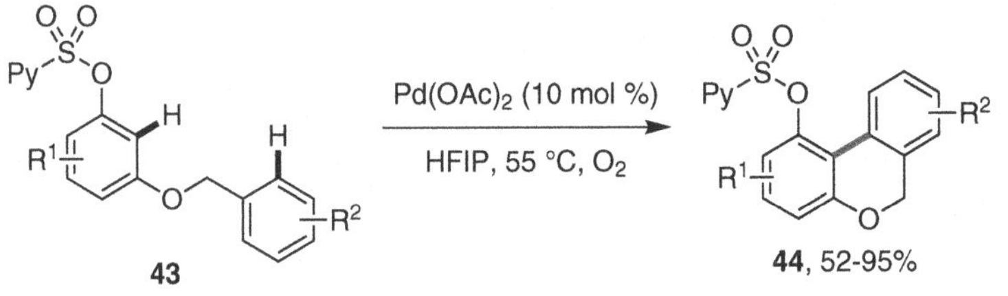

SCHEME 5.14

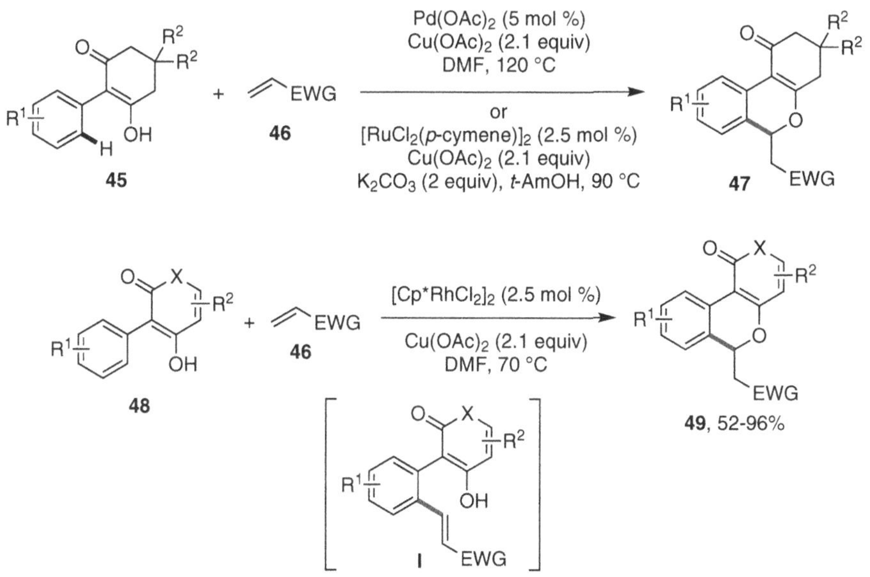

SCHEME 5.15

at 120°C, to afford 6-vinyltetrahydro-1*H*-benzo[*c*]chromen-1-ones **52** in 40–65% yields. Also, 6-vinyl-6*H*-isochromeno[4,3-*c*]quinolin-11(12*H*)-one derivatives **53** were synthesized in 57–67% yields by a similar approach starting from 4-hydroxy-3-phenylquinolin-2(1*H*)-ones **50**. In the proposed reaction mechanism, six-membered palladacycle **I** was generated by reaction of the enolate of **45** with $Pd(OAc)_2$, through C–H activation. By coordination of the allyl acetate **51** with subsequent migratory insertion, a palladacycle **II** was formed, which delivered allylated product **IV** upon protonation by AcOH to species **III**, followed by β-OAc elimination. Pd-catalyzed oxidative nucleophilic substitution of **IV** afforded a π-allyl palladium species **V**, which afforded the final desired product **52** by intramolecular nucleophilic substitution (Scheme 5.16) (Choppakatla, Dachepally, and Bollikolla 2016).

5.8 BENZO[C]CHROMEN-6-ONE

Electrochemical dehydrogenative lactonization of the C(sp^2)–H bond was developed for the construction of 6*H*-benzo[*c*]chromen-6-ones **55**. Reactions were conducted in an undivided cell, using a Pt anode and cathode ($J = 13.3$ mA/cm^2), in the presence of n-Bu_4NBF_4 in CH_3CN/MeOH at room temperature, resulting in the formation of 6*H*-benzo[*c*]chromen-6-ones **55** in 31–99% yields (Zhang et al. 2017). Similar Kolbe anodic cyclization via C(sp^2)–H functionalization/C–O formation was described in the presence of NaOH (10 mol%) in an MeOH/water mixture, and benzo[*c*]chromen-6-one derivatives **55** were obtained in 43–93% yields. 5*H*-Thieno[2,3-*c*]isochromen-5-one **55a** was synthesized in 55% yield when 2-(thiophen-3-yl)benzoic acid underwent dehydrogenative cyclization under anodic oxidation. By electrooxidation of the carboxylate anion, generated in situ by anodic dehydrogenation

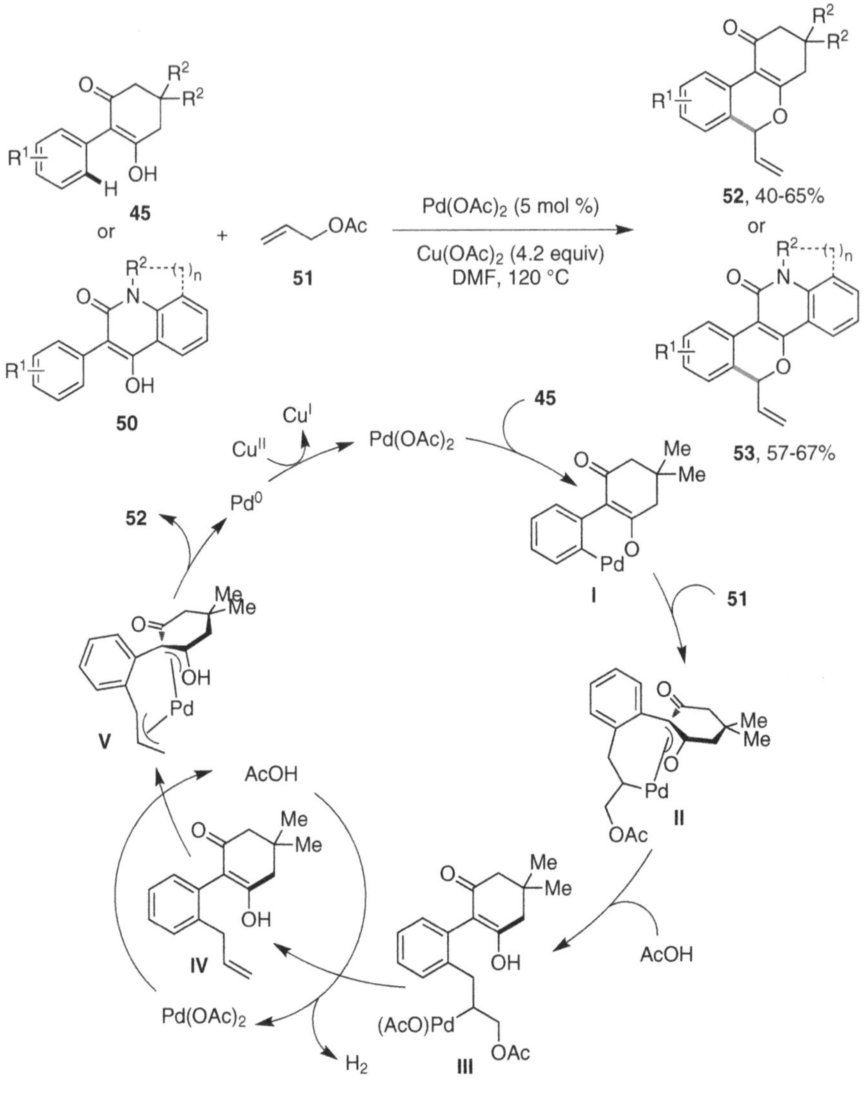

SCHEME 5.16

or deprotonation via electrogenerated MeO⁻, a carboxylate radical was generated, which underwent a 6-*endo*-trig cyclization to achieve products **55** by losing an electron and a proton (Scheme 5.17) (Zhang et al. 2018).

6*H*-Benzo[*c*]chromen-6-ones **55** were also accessed by Ru-catalyzed C–H arylation of benzoic acids **20** with 2-iodophenol **56**. Reactions were conducted by 1.5 equiv of 2-iodolphenol **56** in the presence of $[RuCl_2(p\text{-cymene})]_2$ (4 mol%), PCy_3 (8 mol%), and K_2CO_3 (1 equiv) in *N*-methyl-2-pyrrolidone (NMP) at 80°C, affording 6*H*-benzo[*c*]chromen-6-ones **55** in 39–96% yields. The reaction was proposed to proceed by carboxylic acid O-atom-directed Ru-catalyzed C–H arylation of benzoic

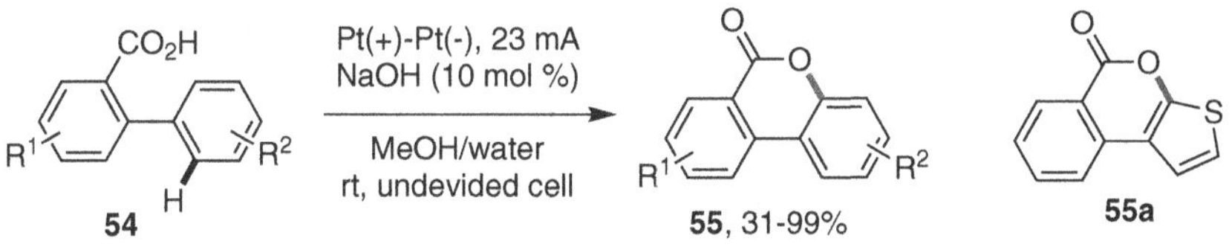

SCHEME 5.17

acid **20** with 2-iodophenol **56**, followed by lactonization (Huang and Weix 2016). Moreover, Pd-catalyzed direct dehydrogenative annulation of benzoic acids **20** with phenols **57** was developed to furnish 6*H*-benzo[*c*]chromen-6-ones **55** (Scheme 5.18) (Shiotani and Itatani 1974).

Rh(III)-catalyzed intermolecular C–H activation followed by an intramolecular esterification reaction between benzoic acids **20** and cyclic diazo-1,3-diketones **34a** was reported to synthesize 3,4-dihydro-1*H*-benzo[*c*]chromene-1,6(2*H*)-dione derivatives **58**. Treatment of an equimolar amount of cyclic 2-diazo-1,3-diketones **34a** and benzoic acids **20** with $[Cp^*RhCl_2]_2$ (2 mol%), in DCE at 100°C afforded 3,4-dihydro-1*H*-benzo[*c*]chromene-1,6(2*H*)-diones **58** in 80–96% yields. The reaction did not occur when heteroaromatic acids, including picolinic acid, thiophene-2-carboxylic acid, and furan-3-carboxylic acid were used. Acyclic 2-diazo-1,3-diketones **34a**, 3-diazopentane-2,4-dione, was also tolerated under reaction conditions, leading to corresponding 4-acetyl-3-methyl-1*H*-isochromen-1-ones **58a** in 60–65% yields. Acid *O*-atom-directed *ortho*-C–H activation of benzoic acid **20** with $[Cp^*RhCl_2]_2$, followed by coordination and release of N_2, generated a corresponding Rh(III)-carbene intermediate, which underwent migratory insertion of the carbene into the Rh–C bond and then protonation by HCl to give alkylated intermediate **I**, along with regeneration of the Rh(III) catalyst. Finally, by tautomerization with subsequent lactonization of **I**, the final product **58** was obtained (Yang et al. 2017). Similar 3,4-dihydro-1*H*-benzo[*c*]chromene-1,6(2*H*)-dione scaffolds **58b** were obtained in 41–76% yields by Rh-catalyzed C–H activation of phenacyl phosphoniums **59** in coupling with cyclic diazo-1,3-diketones **34a**, via C–H activation and subsequent intramolecular nucleophilic C–O formation (intermediate **II**) (Scheme 5.19) (Kim, Gressies, and Glorius 2016).

Synthesis of 5*H*-dibenzo[*c,f*]chromen-5-one derivatives **62** was achieved by Rh(III)-catalyzed C–H activation/cyclization of benzamides **60** and

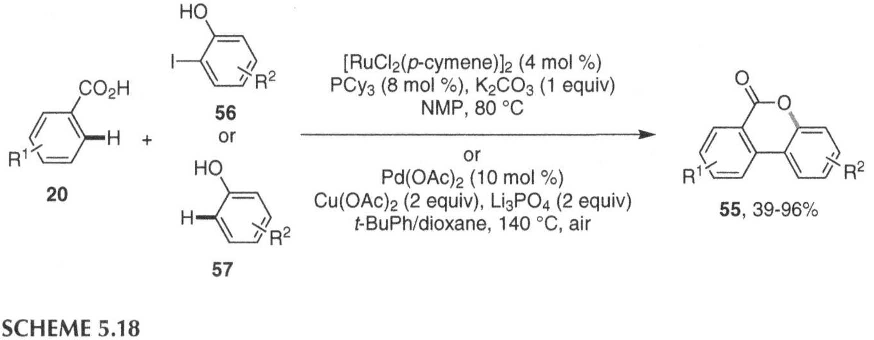

SCHEME 5.18

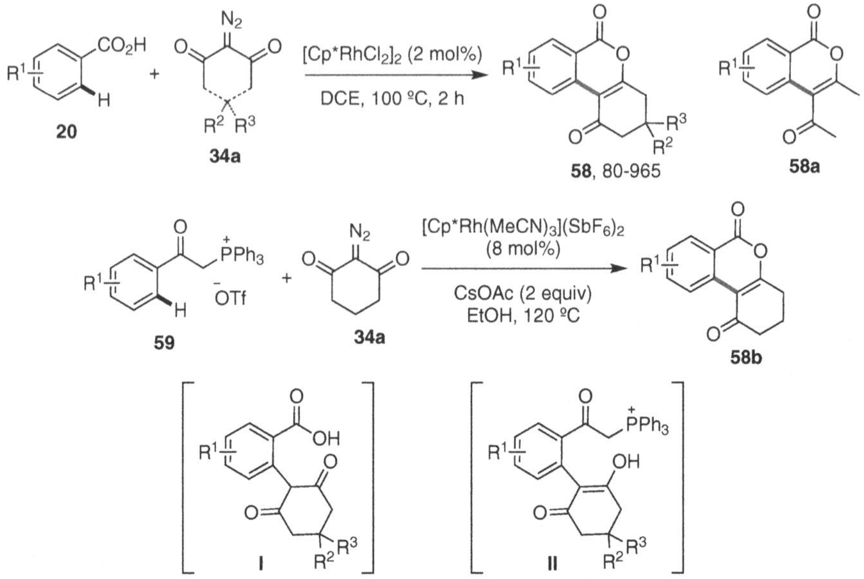

SCHEME 5.19

1-diazonaphthalen-2(1*H*)-ones **61**. Reactions were conducted by heating a solution of diazonaphthalen-2(1*H*)-ones **61** with 2 equiv of *N*-methoxybenzamides **60** in the presence of $[Cp^*RhCl_2]_2$ (5 mol%) and AgOAc (1 equiv) in toluene at 85°C, leading to the expected 5*H*-dibenzo[*c,f*]chromen-5-ones **62** in 49–72% yields. Heterocyclic carboxamides were also tolerated under reaction conditions, affording corresponding heterocyclic fused chromen-4-ones in 48–63% yields. 2-Diazonaphthalen-1(2*H*)-one was also subjected to the reaction, which led to 6*H*-dibenzo[*c,h*]chromen-6-one in 62% yield. The reaction was initiated by C–H activation of *N*-methoxybenzamides **60** with Rh(III) via the concerted metalation/deprotonation (CMD) pathway to give rhodacyclic intermediate **I**, followed by generation of carbene–Rh species **II** by N_2 release, which underwent migratory insertion of the carbene into the C–Rh bond to generate naphthol rhodium species **III**, with subsequent protonolysis by AcOH to give arylated product **IV**. Intramolecular cyclization of **IV** delivered the final desired product **62** (Scheme 5.20) (Chen and Cui 2017).

5.9 BENZO[*de*]CHROMENE

Thirunavukkarasu, Donati, and Ackermann (2012) developed a Ru(II)-catalyzed dehydrogenative coupling of 1-naphthols **63** with internal alkynes **13** for the construction of benzo[*de*]chromene derivatives **65**. Reactions were conducted in *m*-xylene using $[RuCl_2(p\text{-cymene})]_2$ (2 mol%) as the catalyst and $Cu(OAc)_2{\cdot}H_2O$ (1 equiv) as the oxidant at 80–110°C, leading to benzo[*de*]chromenes **65** in 45–71% yields. Unsymmetrical alkyl aryl alkynes furnished corresponding 3-alkyl-2-arylbenzo[*de*] chromenes **65** in 57–78% yields, regioselectively. Reacting 4-hydroxycoumarin **64** with internal alkynes **13** under the same reaction conditions resulted in the

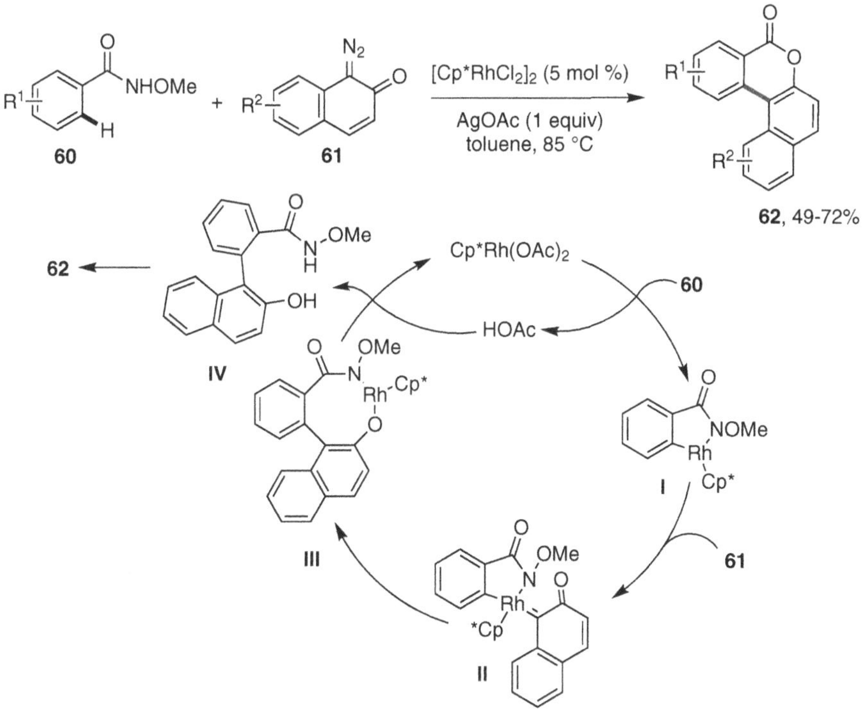

SCHEME 5.20

formation of 2*H*-pyrano[2,3,4-*de*]chromen-2-one derivatives **66** in 45–81% yields. 4-Hydroxyquinolin-2-one afforded pyrano[2,3,4-*de*]quinolin-2(1*H*)-one **66** in 92% yield. The reaction proceeded by naphthol *O*-atom coordination to the Ru center, followed by activation of the C8–H bond, with subsequent insertion of alkyne into the C–Ru bond and then reductive elimination to give the desired product (Scheme 5.21). A similar approach was developed by Mochida et al. (232010), using 5 mol% of $[Cp^*RhCl_2]_2$. Also, $[Cp^*Co(CO)I_2]$ (10 mol%) in the presence of CuO (2 equiv) as the oxidant was used for dehydrogenative coupling of 4-hydroxyquinolin-2-ones with alkynes. In the presence of a Co(III) catalyst, unsymmetrical alkyl aryl alkynes led to 5-alkyl-6-aryl-2*H*-pyrano[2,3,4-*de*]chromen-2-ones in 52–56% yields, regioselectively (Dutta, Ravva, and Sen 2019).

Rh(III)-catalyzed C–H activation of benzoylacetonitriles **67** with α-diazo ester and ketones **34b** was developed to synthesize benzo[*de*]chromenes **68**. Subjecting benzoylacetonitriles **67** to α-diazo carbonyls **34b** (2.5 equiv) in the presence of $[Cp^*RhCl_2]_2$ (4 mol%) and NaOAc (1 equiv) in DCE at room temperature overnight resulted in the formation of 9-cyano-2-hydroxy-2,3-dihydrobenzo[*de*]chromene-3,7-dicarboxylate derivatives **68** in 41–93% yields. The reaction did not occur when –CN in the substrate **67** was changed to other substituents, such as –COMe and –CO_2Et. A cyclic diazo-1,3-dione, 2-diazocycloheptane-1,3-dione, led to a fused polycyclic product **68b** in 83% yield. When reactions were carried out in the presence

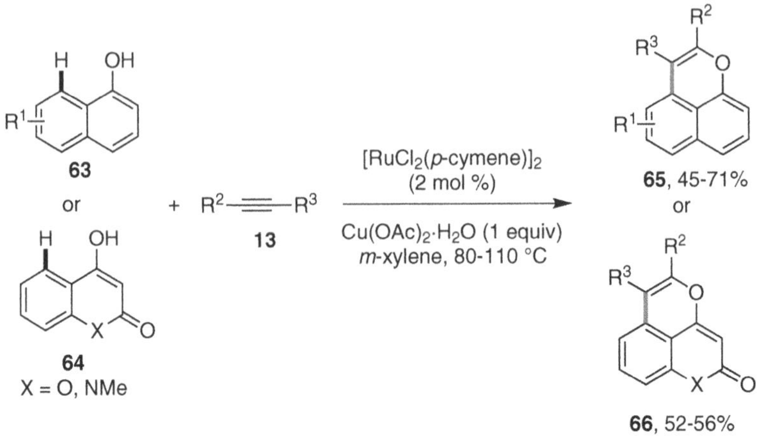

SCHEME 5.21

of CsOPiv at 85°C, corresponding decarboxylated products **68a** were obtained in 43–85% yields. The reaction was proposed to proceed in two steps: first, the dehydrogenative coupling of benzoylacetonitriles **67** with one molecule of α-diazo ester **34b** to give 3-cyano-4-hydroxynaphthalene-1-carboxylate **69**, followed by dehydrogenative coupling with a second α-diazo ester **34b** to give the final benzo[*de*]chromenes **68** (Fang et al. 2018). Also, benzo[*de*]chromene derivatives **70** were accessed by Rh(III)-catalyzed dehydrogenative coupling of benzoylacetonitrile **67** with internal alkynes **13**. Reactions were catalyzed with $[Cp^*RhCl_2]_2$ (5 mol%) and $Cu(OAc)_2{\cdot}H_2O$ (2 equiv) as the oxidant in DMF at 100°C, and benzo[*de*]chromene-9-carbonitriles **70** were obtained in 33–92% yields. Unsymmetrical alkyl aryl alkynes led to corresponding 3,7-dialkyl-2,8-diarylbenzo[*de*]chromene-9-carbonitriles **70** in 53–63% yields. Ethyl benzoylacetate and 2-nitro-1-phenylethanone reacted with diphenyl acetylene, affording low yields of expected products (47% and 23%, respectively). No reaction occurred when 1-phenylbutane-1,3-dione was used as the substrate (Scheme 5.22) (Tan et al. 2012).

Ru(II)-catalyzed methylphenyl sulfoximine-directed double annulation of both *ortho*-C–H bonds of arene carboxyamide with internal alkynes was described to construct pyrano[4,3,2-*ij*]isoquinoline skeletons **72**, via the formation of four (C–C)–(C–N) and (C–C)–(C–O)] bonds under single catalytic conditions. The reaction of *N*-[4-methylbenzoyl]methylphenyl sulfoximine **71** with 4 equiv of alkynes **13** in the presence of $[RuCl_2(p\text{-cymene})]_2$ (10 mol%), $AgSbF_6$ (40 mol%), and $Cu(OAc)_2{\cdot}H_2O$ (1.5 equiv) in 1,4-dioxane at 120°C led to pyrano[4,3,2-*ij*]isoquinoline derivatives **72** in 48–94% yields. Unsymmetrical double annulation of arenes with different alkynes **13** was also conducted by $[RuCl_2(p\text{-cymene})]_2$ (7.5 mol%)- and $AgSbF_6$ (30 mol%)-catalyzed reaction of **71** with 4-octyne (2 equiv) in the presence of AcOH (4 equiv) in DCE at 120°C, giving 3,4-dipropylisoquinolin-1(2*H*)-one **73** in 57–58% yields. This was followed by subjecting the obtained **73** to another alkyne (1.5 equiv) in the presence of $[RuCl_2(p\text{-cymene})]_2$ (5 mol%), $AgSbF_6$ (20 mol%), and $Cu(OAc)_2{\cdot}H_2O$ (1 equiv) in 1,4-dioxane at 120°C to furnish

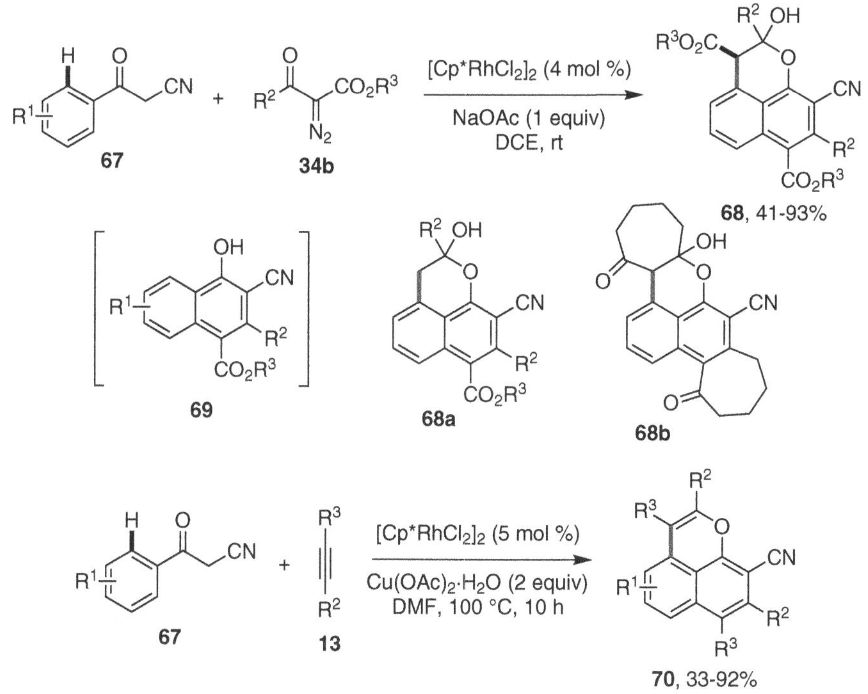

SCHEME 5.22

the final 2,3-disubstituted 7,8-dipropylpyrano[4,3,2-*ij*]isoquinolines **72** in 42–76% yields (Shankar et al. 2018). Similarly, Ru(II)-catalyzed dehydrogenative C–H/O–H annulations of 2-arylquinolin-4-ones **74** with internal alkynes **13** were developed for the preparation of pyrano[2,3,4-*de*]quinoline scaffolds **75** in 46–77% yields (Scheme 5.23) (Shaikh, Shinde, and Patil 2016).

5.10 XANTHENE AND XANTHENE-9-ONE

One example of diethyl 9*H*-xanthene-9,9-dicarboxylate **77** was synthesized by Cu(II)-catalyzed cyclization of diethyl 2-(2-phenoxyphenyl)malonate **76** via intramolecular dehydrogenative coupling of C(sp^2)–H/C(sp^3)–H bonds. Reactions were conducted using Cu(2-ethylhexanoate)$_2$ (2.5 equiv) and *N,N*-diisopropylethylamine (DIPEA) (2.5 equiv) in refluxing toluene under an atmosphere of argon (Scheme 5.24) (Hurst and Taylor 2017).

A metal-free intramolecular annulation of 2-aryloxybenzaldehydes **78** to xanthones **79** was described via direct dehydrogenative coupling of an aldehyde C–H bond and aromatic C(sp^2)–H bonds. Reactions were performed using tetrabutylammonium bromide (TBAB, 0.5 equiv) and TBHP in water at 120°C under air, and 2-aryloxybenzaldehydes **78** were cyclized to xanthones **79** in 52–86% yields. The reaction was initiated by the TBAB-promoted generation of a *tert*-butoxyl radical, which abstracted the H atom from the aldehyde. Cyclization of the generated acyl

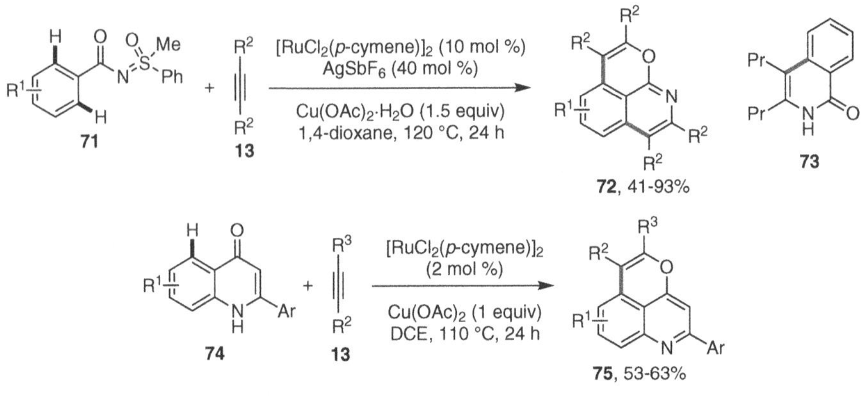

SCHEME 5.23

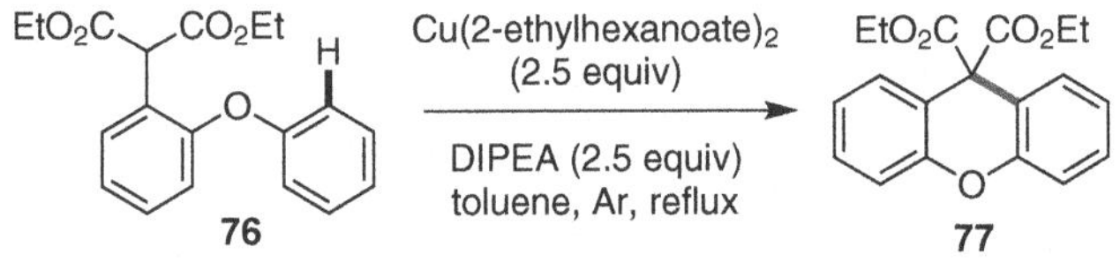

SCHEME 5.24

radical followed by a SET process and then a proton abstraction led to the xanthone product **79** (Rao et al. 2013). A similar cross-dehydrogenative-coupling reaction was developed using $Fe(Cp)_2$ (1 mol%) and *t*-BuOOH (2.2 equiv) in CH_3CN at 90°C, affording the expected xanthones in 30–78% yields (Wertz, Leifert, and Studer 2013). Also, intramolecular cross-dehydrogenative coupling of 2-aryloxybenzaldehydes **78** to xanthones **79** was carried out using $[RuCl_2(p\text{-cymene})_2]_2$ (10 mol%) and TBHP in CH_3CN at 120°C. Corresponding xanthones **79** were obtained in 41–78% yields. 4-Aryloxy-2*H*-chromene-3-carbaldehydes **80** afforded chromeno[4,3-*b*]chromen-7(6*H*)-one derivatives **81** in 61–75% yields, under the same reaction conditions. The plausible reaction mechanism involved the *O*-directed activation of the aldehyde C–H bond to generate an Ru(II)–H intermediate, which underwent oxidation with TBHP and *ortho*-C–H activation with subsequent reductive elimination to afford the desired scaffolds **79** (Manna, Manda, and Panda 2014). In addition, a dehydrogenative coupling process was conducted in order to construct chromeno[2,3-*b*]quinolin-12-ones, chromeno[2,3-*b*]pyridin-5-one **83** (Singh et al. 2017), and chromeno[2,3-*c*]pyrazol-4(1*H*)-one **85**, aza-like structures of xanthones (Scheme 5.25) (Li et al. 2015).

5.11 SEVEN-MEMBERED *O*-HETEROCYCLES

An intramolecular biaryl coupling reaction of dibenzyl ethers **86** to dihydrodibenzo[*c,e*] oxepines **87** was reported using phenyliodine(III) bis(trifluoroacetate) (PIFA) in the presence of $BF_3{\cdot}Et_2O$ in DCM at −40°C. Dibenzo[*c,e*]oxepines **87** were obtained in 38–85% yields. As strongly electron-donating substituted dibenzyl ethers led to the

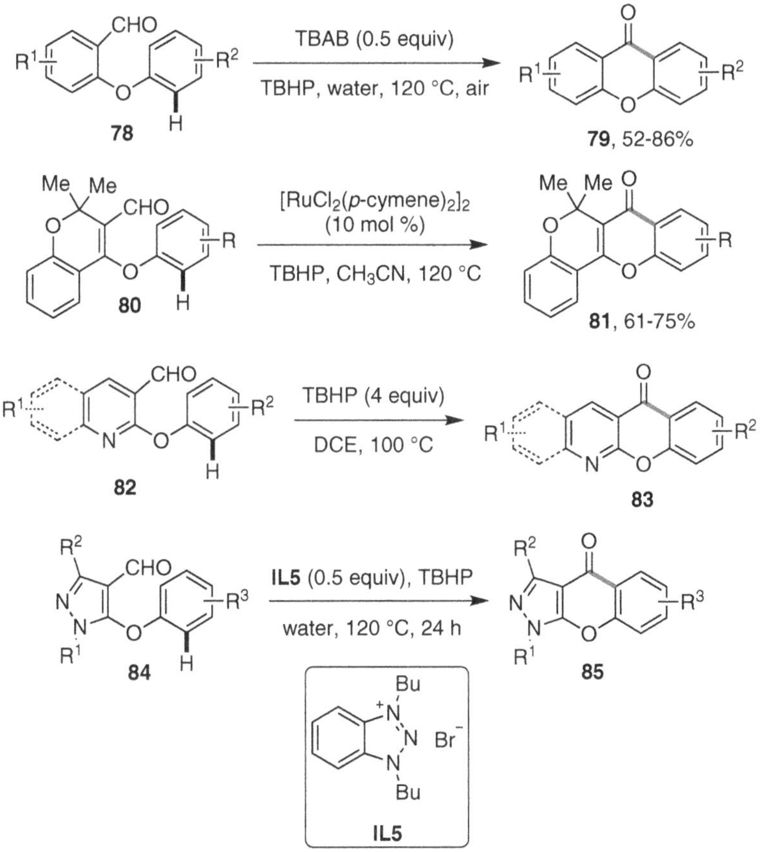

SCHEME 5.25

desired dibenzo[*c*,*e*]oxepine products, the cation radical of arenes was proposed as the intermediate (Scheme 5.26) (Takada et al. 1998).

Synthesis of dibenzo[*c*,*e*]oxepin-5(7*H*)-ones **89** was achieved by Rh-catalyzed double C–H activation of benzyl thioethers **88** and benzoic acids **20** through one-step cleavage of four bonds (two C–H, C–S, O–H) and the formation of two bonds (C–C, C–O). Reactions were performed using 3 equiv of benzyl thioethers **88** in the presence of $[Cp^*RhCl_2]_2$ (2.5 mol%), $AgSbF_6$ (40 mol%), and $AgNO_3$ (4 equiv) in toluene at 160°C, producing dibenzo[*c*,*e*]oxepin-5(7*H*)-ones **89** in 28–72% yields.

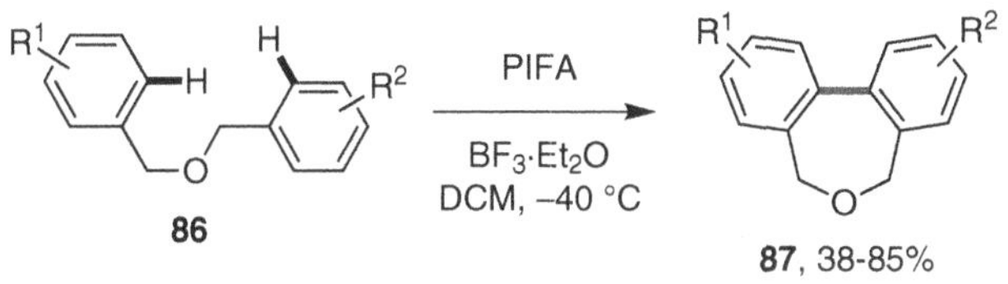

SCHEME 5.26

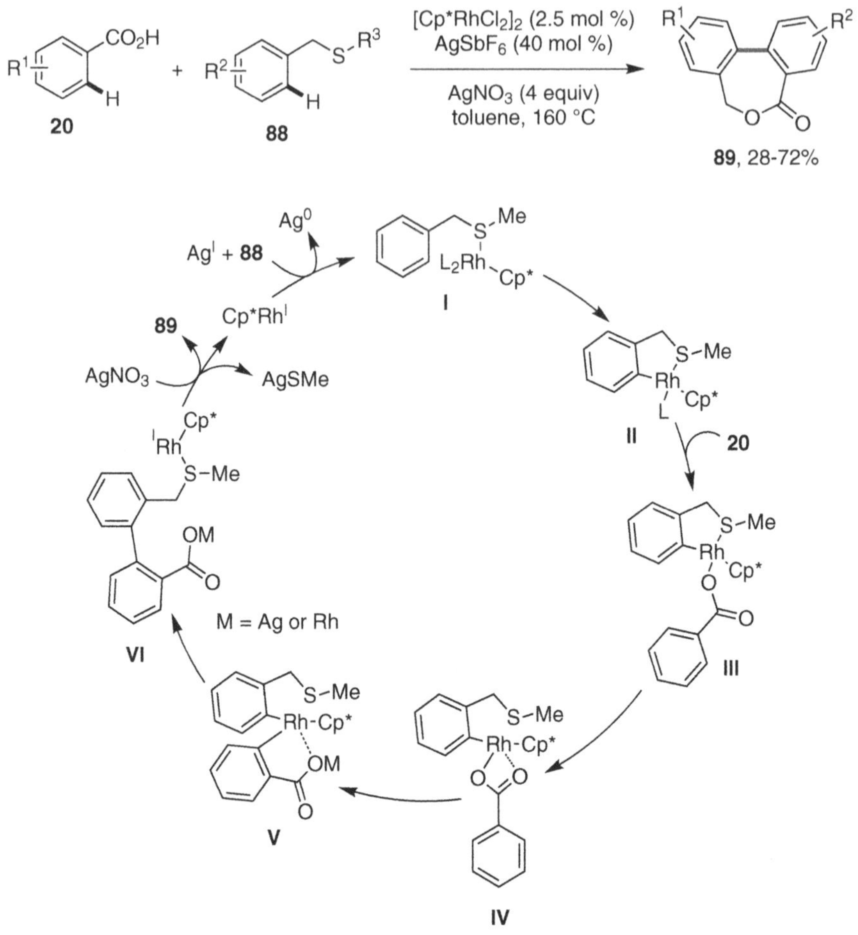

SCHEME 5.27

Reaction with 4-nitrobenzoic acid led to a lower yield of the product (7%). The first thioether-directed C–H bond activation generated the intermediate **II**, which underwent carboxylic acid coordination (**III**) and ligand exchange to form the intermediate **IV**. The second C–H bond activation occurred to give the intermediate **V**, which was converted to intermediate **VI** by reductive elimination along with generation of an Rh(I) species, which was oxidized by $AgNO_3$ to an Rh(III) species to complete the catalytic cycle. With the assistance of the Rh or Ag, **VI** was cyclized to the final product **89** (Scheme 5.27) (Zhang et al. 2015).

5.12 CONCLUSION

In summary, aromatic C(sp2)–H dehydrogenative coupling processes are widely used for the synthesis of six-membered oxygen-containing heterocyclic compounds, including chromenes, coumarins, isochromenes, isocoumarins, fused coumarins and isocoumarins, benzochromenes, and xanthenes. However, many different methods were developed for the construction of these types of heterocycles, due to the quantitative and one-step synthesis and a broad spectrum of substitution on the synthesized heterocycles, the aromatic C–H bond direct functionalization approach could be of interest in the synthesis of pharmaceutical, medicinal, and natural products.

REFERENCES

Ackermann, Lutz, Jola Pospech, Karolina Graczyk, and Karsten Rauch. 2012. "Versatile synthesis of isocoumarins and α-pyrones by ruthenium-catalyzed oxidative C–H/O–H bond cleavages." *Organic Letters* no. 14(3):930–933.

Bollikolla, Hari Babu, Subrahmanyam Choppakatla, Naresh Polam, Vijaya Durga Thripuram, and Suresh Reddy Chidipudi. 2017. "Synthesis of benzopyrans by enolate-directed rhodium-catalyzed oxidative C–H alkenylation of 1,3-dicarbonyl compounds." *Asian Journal of Organic Chemistry* no. 6(11):1598–1603.

Chen, Renjie, and Sunliang Cui. 2017. "Rh (III)-catalyzed C–H activation/cyclization of benzamides and diazonaphthalen-2 (1 H)-ones for synthesis of lactones." *Organic Letters* no. 19(15):4002–4005.

Chen, Xiaohong, Guangfan Zheng, Yunyun Li, Guoyong Song, and Xingwei Li. 2017. "Rhodium-catalyzed site-selective coupling of indoles with diazo esters: C4-alkylation versus C2-annulation." *Organic Letters* no. 19(22):6184–6187.

Choppakatla, Subrahmanyam, Aravind Kumar Dachepally, and Hari Babu Bollikolla. 2016. "Palladium-catalyzed double C–H functionalization of 2-aryl-1, 3-dicarbonyl compounds: A facile access to alkenylated benzopyrans." *Tetrahedron Letters* no. 57(23):2488–2491.

Dutta, Pratip K, Mahesh Kumar Ravva, and Subhabrata Sen. 2019. "Cobalt-catalyzed, hydroxyl-assisted C–H Bond functionalization: Access to diversely substituted polycyclic pyrans." *The Journal of Organic Chemistry* no. 84(3):1176–1184.

Eftekhari-Sis, Bagher, Masoumeh Sarvari Karajabad, and Shiva Haqverdi. 2017. "Pyridylmethylaminoacetic acid functionalized Fe3O4 magnetic nanorods as an efficient catalyst for the synthesis of 2-aminochromene and 2-aminopyran derivatives." *Scientia Iranica* no. 24(6):3022–3031.

Eftekhari-Sis, Bagher, and Maryam Zirak. 2014. "Chemistry of α-oxoesters: A powerful tool for the synthesis of heterocycles." *Chemical Reviews* no. 115(1):151–264.

Eftekhari-Sis, Bagher, Maryam Zirak, and Ali Akbari. 2013. "Arylglyoxals in synthesis of heterocyclic compounds." *Chemical Reviews* no. 113(5):2958–3043.

Fang, Feifei, Chunmei Zhang, Chaofan Zhou, Yazhou Li, Yu Zhou, and Hong Liu. 2018. "Rh (III)-catalyzed C–H activation of benzoylacetonitriles and tandem cyclization with diazo compounds to substituted benzo [de] chromenes." *Organic Letters* no. 20(7):1720–1724.

Feng, Shangbiao, Xingang Xie, Weiwei Zhang, Lin Liu, Zhuliang Zhong, Dengyu Xu, and Xuegong She. 2016. "Visible-light-promoted dual C–C bond formations of alkynoates via a domino radical addition/cyclization reaction: A synthesis of coumarins." *Organic Letters* no. 18(15):3846–3849.

Frasco, Daniel A, Cassandra P Lilly, Paul D Boyle, and Elon A Ison. 2013. "Cp* IrIII-catalyzed oxidative coupling of benzoic acids with alkynes." *ACS Catalysis* no. 3(10):2421–2429.

Gandeepan, Parthasarathy, Thomas Müller, Daniel Zell, Gianpiero Cera, Svenja Warratz, and Lutz Ackermann. 2018. "3d transition metals for C–H activation." *Chemical Reviews* no. 119(4):2192–2452.

Guo, Dong-Dong, Bin Li, Da-Yu Wang, Ya-Ru Gao, Shi-Huan Guo, Gao-Fei Pan, and Yong-Qiang Wang. 2017. "Synthesis of 6 H-benzo [c] chromenes via palladium-catalyzed intramolecular dehydrogenative coupling of two aryl C–H bonds." *Organic Letters* no. 19(4):798–801.

Huang, Liangbin, Agostino Biafora, Guodong Zhang, Valentina Bragoni, and Lukas J Gooßen. 2016. "Regioselective C–H hydroarylation of internal alkynes with arenecarboxylates: Carboxylates as deciduous directing groups." *Angewandte Chemie International Edition* no. 55(24):6933–6937.

Huang, Liangbin, and Daniel J Weix. 2016. "Ruthenium-catalyzed C–H arylation of diverse aryl carboxylic acids with aryl and heteroaryl halides." *Organic Letters* no. 18(20):5432–5435.

Hurst, Timothy E, and Richard JK Taylor. 2017. "A Cu-catalysed radical cross-dehydrogenative coupling approach to acridanes and related heterocycles." *European Journal of Organic Chemistry* no. 2017(1):203–207.

Jiang, Guangbin, JianXiao Li, Chuanle Zhu, Wanqing Wu, and Huanfeng Jiang. 2017. "Palladium-catalyzed sequential nucleophilic addition/oxidative annulation of bromoalkynes with benzoic acids to construct functionalized isocoumarins." *Organic Letters* no. 19(17):4440–4443.

Kim, Ju Hyun, Steffen Gressies, and Frank Glorius. 2016. "Cooperative lewis acid/Cp* CoIII catalyzed C–H bond activation for the synthesis of isoquinolin-3-ones." *Angewandte Chemie International Edition* no. 55(18):5577–5581.

Li, He, Chenjiang Liu, Yonghong Zhang, Yadong Sun, Bin Wang, and Wenbo Liu. 2015. "Green method for the synthesis of chromeno [2, 3-c] pyrazol-4 (1 H)-ones through ionic liquid promoted directed annulation of 5-(aryloxy)-1 H-pyrazole-4-carbaldehydes in aqueous media." *Organic Letters* no. 17(4):932–935.

Lu, Yi, Dong-Hui Wang, Keary M Engle, and Jin-Quan Yu. 2010. "Pd (II)-catalyzed hydroxyl-directed C–H olefination enabled by monoprotected amino acid ligands." *Journal of the American Chemical Society* no. 132(16):5916–5921.

Manna, Sudipta Kumar, Srinivas Lavanya Kumar Manda, and Gautam Panda. 2014. "[RuCl2 (p-cymene) 2] 2 catalyzed cross dehydrogenative coupling (CDC) toward xanthone and fluorenone analogs through intramolecular C–H bond functionalization reaction." *Tetrahedron Letters* no. 55(42):5759–5763.

Mochida, Satoshi, Masaki Shimizu, Koji Hirano, Tetsuya Satoh, and Masahiro Miura. 2010. "Synthesis of naphtho [1, 8-bc] pyran derivatives and related compounds through hydroxy group directed C–H bond cleavage under rhodium catalysis." *Chemistry – An Asian Journal* no. 5(4):847–851.

Mohammadi, Reza, Somayeh Esmati, Mahdi Gholamhosseini-Nazari, and Reza Teimuri-Mofrad. 2019. "Novel ferrocene substituted benzimidazolium based ionic liquid immobilized on magnetite as an efficient nano-catalyst for the synthesis of pyran derivatives." *Journal of Molecular Liquids* no. 275:523–534.

Morimoto, Keisuke, Koji Hirano, Tetsuya Satoh, and Masahiro Miura. 2011. "Synthesis of isochromene and related derivatives by rhodium-catalyzed oxidative coupling of benzyl and allyl alcohols with alkynes." *The Journal of Organic Chemistry* no. 76(22):9548–9551.

Nandi, Debkumar, Debalina Ghosh, Shih-Ji Chen, Bing-Chiuan Kuo, Nancy M Wang, and Hon Man Lee. 2013. "One-step synthesis of isocoumarins and 3-benzylidenephthalides via ligandless pd-catalyzed oxidative coupling of benzoic acids and vinylarenes." *The Journal of Organic Chemistry* no. 78(7):3445–3451.

Qin, Xurong, Xiaoyu Li, Quan Huang, Hu Liu, Di Wu, Qiang Guo, Jingbo Lan, Ruilin Wang, and Jingsong You. 2015. "Rhodium (III)-catalyzed ortho C–H heteroarylation of (hetero) aromatic carboxylic acids: A rapid and concise access to π-conjugated poly-heterocycles." *Angewandte Chemie International Edition* no. 54(24):7167–7170.

Rao, Honghua, Xinyi Ma, Qianzi Liu, Zhongfeng Li, Shengli Cao, and Chao-Jun Li. 2013. "Metal-free oxidative coupling: Xanthone formation via direct annulation of 2-aryloxybenzaldehyde using tetrabutylammonium bromide as a promoter in aqueous medium." *Advanced Synthesis & Catalysis* no. 355(11–12):2191–2196.

Reddy Chidipudi, Suresh, Martin D Wieczysty, Imtiaz Khan, and Hon Wai Lam. 2013. "Synthesis of benzopyrans by Pd (II)-or Ru (II)-catalyzed C–H alkenylation of 2-Aryl-3-hydroxy-2-cyclohexenones." *Organic Letters* no. 15(3):570–573.

Romanov-Michailidis, Fedor, Benjamin D Ravetz, Daniel W Paley, and Tomislav Rovis. 2018. "Ir (III)-catalyzed carbocarbation of alkynes through undirected double C–H bond activation of anisoles." *Journal of the American Chemical Society* no. 140(16):5370–5374.

Satoh, Tetsuya, Kenji Ueura, and Masahiro Miura. 2008. "Rhodium-and iridium-catalyzed oxidative coupling of benzoic acids with alkynes and alkenes." *Pure & Applied Chemistry* no. 80(5):1127–1134.

Sen, Malay, Pardeep Dahiya, J Richard Premkumar, and Basker Sundararaju. 2017. "Dehydrative Cp* Co (III)-catalyzed C–H bond allenylation." *Organic Letters* no. 19(14):3699–3702.

Shaikh, Aslam C, Dinesh R Shinde, and Nitin T Patil. 2016. "Gold vs rhodium catalysis: Tuning reactivity through catalyst control in the C–H alkynylation of isoquinolones." *Organic Letters* no. 18(5):1056–1059.

Shaikh, Tanveer Mahamadali, and Fung-E Hong. 2016. "Recent developments in the preparation of N-heterocycles using Pd-catalysed C–H activation." *Journal of Organometallic Chemistry* no. 801:139–156.

Shankar, Majji, Koushik Ghosh, Kallol Mukherjee, Raja K Rit, and Akhila K Sahoo. 2018. "One-pot unsymmetrical {[4+2] and [4+2]} double annulations of o/o′-C–H bonds of arenes: Access to unusual pyranoisoquinolines." *Organic Letters* no. 20(17):5144–5148.

Shiotani, Akinori, and Hiroshi Itatani. 1974. "Dibenzofurans by intramolecular ring closure reactions." *Angewandte Chemie International Edition in English* no. 13(7):471–472.

Singh, Jay B, Kalpana Mishra, Tanu Gupta, and Radhey M Singh. 2017. "TBHP promoted cross-dehydrogenative coupling (CDC) reaction: Metal/additive-free synthesis of chromone-fused quinolines." *ChemistrySelect* no. 2(3):1207–1210.

Takada, Takeshi, Mitsuhiro Arisawa, Michiyo Gyoten, Ryuji Hamada, Hirofumi Tohma, and Yasuyuki Kita. 1998. "Oxidative biaryl coupling reaction of phenol ether derivatives using a hypervalent iodine (III) reagent." *The Journal of Organic Chemistry* no. 63(22):7698–7706.

Tan, Xing, Bingxian Liu, Xiangyu Li, Bin Li, Shansheng Xu, Haibin Song, and Baiquan Wang. 2012. "Rhodium-catalyzed cascade oxidative annulation leading to substituted naphtho [1, 8-bc] pyrans by Sequential Cleavage of C (sp^2)–H/C (sp^3)–H and C (sp^2)–H/O–H Bonds." *Journal of the American Chemical Society* no. 134(39):16163–16166.

Thirunavukkarasu, Vedhagiri S, Margherita Donati, and Lutz Ackermann. 2012. "Hydroxyl-directed ruthenium-catalyzed C–H bond functionalization: Versatile access to fluorescent pyrans." *Organic Letters* no. 14(13):3416–3419.

Ueura, Kenji, Tetsuya Satoh, and Masahiro Miura. 2007a. "An efficient waste-free oxidative coupling via regioselective C–H bond cleavage: Rh/Cu-catalyzed reaction of benzoic acids with alkynes and acrylates under air." *Organic Letters* no. 9(7):1407–1409.

Ueura, Kenji, Tetsuya Satoh, and Masahiro Miura. 2007b. "Rhodium-and iridium-catalyzed oxidative coupling of benzoic acids with alkynes via regioselective C–H bond cleavage." *The Journal of Organic Chemistry* no. 72(14):5362–5367.

Wang, Fen, Zisong Qi, Jiaqiong Sun, Xuelin Zhang, and Xingwei Li. 2013. "Rh (III)-catalyzed coupling of benzamides with propargyl alcohols via hydroarylation–lactonization." *Organic Letters* no. 15(24):6290–6293.

Wang, Zhigang, Mufan Yang, and Yudong Yang. 2018. "Ir (III)-catalyzed oxidative annulation of phenylglyoxylic acids with benzo [b] thiophenes." *Organic Letters* no. 20(10):3001–3005.

Wertz, Sebastian, Dirk Leifert, and Armido Studer. 2013. "Cross dehydrogenative coupling via base-promoted homolytic aromatic substitution (BHAS): Synthesis of fluorenones and xanthones." *Organic Letters* no. 15(4):928–931.

Wu, Jie, Jingbo Lan, Siyuan Guo, and Jingsong You. 2014. "Pd-Catalyzed C–H carbonylation of (hetero) arenes with formates and intramolecular dehydrogenative coupling: A shortcut to indolo [3,2-c] coumarins." *Organic Letters* no. 16(22):5862–5865.

Yang, Cheng, Xinwei He, Lanlan Zhang, Guang Han, Youpeng Zuo, and Yongjia Shang. 2017. "Synthesis of isocoumarins from cyclic 2-diazo-1, 3-diketones and benzoic acids via Rh (III)-catalyzed C–H activation and esterification." *The Journal of Organic Chemistry* no. 82(4):2081–2088.

Zhang, Hang, Kang Wang, Bo Wang, Heng Yi, Fangdong Hu, Changkun Li, Yan Zhang, and Jianbo Wang. 2014. "Rhodium (III)-catalyzed transannulation of cyclopropenes with N-phenoxyacetamides through C–H activation." *Angewandte Chemie International Edition* no. 53(48):13234–13238.

Zhang, Lei, Zhenxing Zhang, Junting Hong, Jian Yu, Jianning Zhang, and Fanyang Mo. 2018. "Oxidant-free C (sp^2)–H functionalization/C–O bond formation: A Kolbe oxidative cyclization process." *The Journal of Organic Chemistry* no. 83(6):3200–3207.

Zhang, Mingliang, Hui-Jun Zhang, Tiantian Han, Wenqing Ruan, and Ting-Bin Wen. 2014. "Rh (III)-catalyzed oxidative coupling of benzoic acids with geminal-substituted vinyl acetates: Synthesis of 3-substituted isocoumarins." *The Journal of Organic Chemistry* no. 80(1):620–627.

Zhang, Sheng, Lijun Li, Huiqiao Wang, Qian Li, Wenmin Liu, Kun Xu, and Chengchu Zeng. 2017. "Scalable electrochemical dehydrogenative lactonization of C (sp^2/sp^3)–H bonds." *Organic Letters* no. 20(1):252–255.

Zhang, Xi-Sha, Yun-Fei Zhang, Zhao-Wei Li, Fei-Xian Luo, and Zhang-Jie Shi. 2015. "Synthesis of dibenzo [c, e] oxepin-5 (7H)-ones from benzyl thioethers and carboxylic acids: Rhodium-catalyzed double C–H activation controlled by different directing groups." *Angewandte Chemie* no. 127(18):5568–5572.

Zhao, Yingwei, Feng Han, Lei Yang, and Chungu Xia. 2015. "Access to coumarins by rhodium-catalyzed oxidative annulation of aryl thiocarbamates with internal alkynes." *Organic Letters* no. 17(6):1477–1480.

Zhou, Zhi, Mengyao Bian, Lixin Zhao, Hui Gao, Junjun Huang, Xiawen Liu, Xiyong Yu, Xingwei Li, and Wei Yi. 2018. "2H-chromene-3-carboxylic acid synthesis via solvent-controlled and rhodium (III)-catalyzed redox-neutral C–H activation/[3+3] annulation cascade." *Organic Letters*.

6 N,O-Heterocycles

6.1 INTRODUCTION

Heterocycles containing nitrogen and oxygen atoms in their structures have gained a lot of importance because of their use in intermediates for the preparation of biologically active compounds, natural products, and pharmaceuticals. They also exhibit a wide spectrum of pharmacological activities such as antibacterial, anticancer, anti-inflammatory, antihistamine, antiparkinson, inhibition of hepatitis C virus, 5-HT3 antagonistic effect, melatonin receptor antagonism, amyloidogenesis inhibition, and Rho-kinase inhibition (Kakkar et al. 2018). There are many reports on the construction of *N,O*-heterocyclic compounds by multi-step synthesis in the literature; the development of new approaches to synthesis of these types of *N,O*-heterocycles bearing various functional groups are of interest to synthetic and medicinal scientists.

Cross-dehydrogenative-coupling (CDC) (Gandeepan et al. 2018, Shaikh and Hong 2016) processes are widely used in the synthesis of a variety of types of heterocyclic compounds. In this chapter, intramolecular and intermolecular CDC of aromatic $C(sp^2)$–H bonds are discussed to construct *N,O*-heterocycles, including benzoxazole, benzoxazines, and benzoxazepines.

6.2 BENZOXAZOLE

Synthesis of 2-arylbenzoxazoles **2** was reported by Cu-catalyzed intramolecular dehydrogenative coupling of benzanilides **1**. Ueda and Nagasawa (Ueda and Nagasawa 2008) used $Cu(OTf)_2$ (20 mol%) in *o*-xylene at 140°C under an O_2 atmosphere for dehydrogenative C–O coupling of benzanilides **1** to give 2-arylbenzoxazoles **2** in 40–92% yields. Regioselectivity in the case of *meta*-substituted benzanilides **1** depends on the structure of the substituent, as C–O bond formation occurred exclusively at the less sterically hindered position in the reaction of *meta*-methyl- or methoxy-substituted anilides to give 2,5-disubstituted benzoxazoles. In the case of *meta*-pyrrolidin-2-one and *meta*-methoxycarbonyl substituents, dehydrogenative cyclization took place at the more sterically hindered position to give 2,7-disubstituted benzoxazoles, because these substituents acted as an additional directing group to promote the formation of a doubly coordinated intermediate such as **I**, leading to selective formation of 7-substituted benzoxazoles (Scheme 6.1). 2-Alkylbenzoxazoles were also prepared by a similar Cu(II)-catalyzed dehydrogenative coupling of corresponding benzanilides through an additional directing-group-induced reaction at the more sterically hindered position. The reaction proceeded by coordination of benzanilide to $Cu(OTf)_2$, followed by *ortho* metalation by electrophilic aromatic substitution at the Cu center, with subsequent reductive elimination to the final benzoxazole product (Ueda and Nagasawa 2009). $Pd(OAc)_2$ (10 mol%), in the presence of $K_2S_2O_8$ (1.5 equiv), and TfOH (1 equiv), in an AcOH/*N,N*-dimethylformamide (DMF) mixture at 100°C in air, catalyzed the intramolecular dehydrogenative C–O

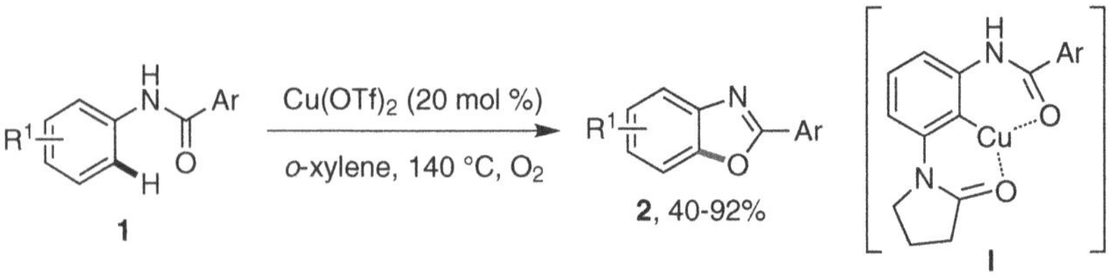

SCHEME 6.1

coupling of acetanilides to 2-methylbenzoxazoles in 28–93% yields (Wang et al. 2018). Similar approaches for the creation of benzoxazole derivatives were developed using phenyliodine bis(trifluoroacetate) as the oxidant (Yu, Ma, and Yu 2012, Alla, Sadhu, and Punniyamurthy 2014).

Visible-light-induced intramolecular C(sp^2)–H amination of aryl azidoformates **3** was conducted to afford benzoxazol-2-ones **4** under an Ir catalytic system. Reactions were carried out by irradiation of a solution of aryl azidoformates **3**, [Ir(dtbbpy)(ppy)$_2$]PF$_6$ (3 mol%) in dichloromethane (DCM) with a blue LED lamp (24 W) in the presence of a 4 Å molecular sieve (MS), leading to benzoxazol-2-one derivatives **4** in 73–96% yields. *Meta*-substituted aryl azidoformates **3** exhibited low regioselectivity. The proposed reaction mechanism involves the generation of triplet azide **I** by the triplet energy transfer from the excited photocatalyst to the azide **3**, followed by release of N_2 to form the triplet nitrene **II**. Subsequently, intramolecular radical addition and then 1,2-hydrogen atom transfer with concomitant rearomatization led to the desired product **4** (Scheme 6.2) (Zhang et al. 2018).

6.3 BENZO[1,3]OXAZINE

The electrochemical dehydrogenative cyclization of *N*-benzyl arylamides **5** was independently described by Yu et al. (2019) and Xu et al. (2017). In Yu et al.'s

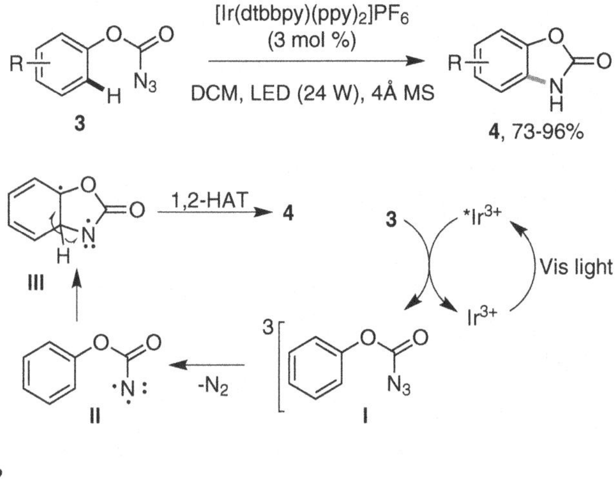

SCHEME 6.2

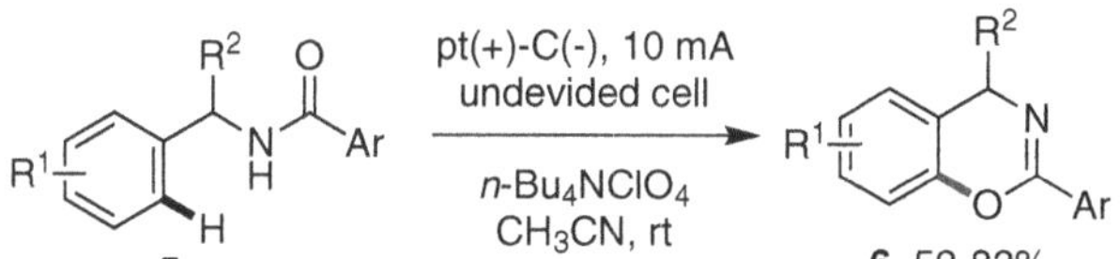

pt(+)-C(-), 10 mA undevided cell; *n*-Bu_4NClO_4, CH_3CN, rt. **5** → **6**, 52-83%

SCHEME 6.3

work, reactions were carried out in an undivided cell, with a Pt plate anode and graphite rod cathode with 10 mA of constant current at room temperature, leading to 4*H*-1,3-benzothiazines **6** in 52–83% yields. Also, *N*-benzyl acrylamide was compatible with the reaction to give the expected 2-vinyl substituted 4*H*-1,3-benzothiazine in 70% yield. However, no reaction occurred in the case of *N*-benzyl aliphatic amides (Scheme 6.3). In Xu et al.'s work, electrochemical dehydrogenative coupling was conducted using a reticulated vitreous carbon (RVC) anode and a Pt plate cathode with $j = 0.15$ mA cm^{-2} in the presence of Et_4NPF_6 (1 equiv) in CH_3CN/THF mixture under reflux conditions, affording corresponding 4*H*-1,3-benzoxazines **6** in 31–81% yields.

4*H*-3,1-Benzoxazines **8** were accessed by Cu-catalyzed dehydrogenative cross-coupling reactions of *N-para*-tolylamides **7**, by subjecting them to $Cu(OTf)_2$ (10 mol%), H_2O (10 mol%), Selectfluor (2 equiv), and $HNTf_2$ (1 equiv) in anhydrous 1,2-dichloroethane (DCE) at 120°C. 4*H*-3,1-Benzoxazines **8** were obtained in 54–91% yields. The transformation was initiated by successive activation of benzylic methyl C(sp^3)–H and aromatic C(sp^2)–H bonds through C–H activation by the Cu(III)FOH complex catalyst generated in situ, providing the intermolecular C–C bond-coupling intermediate **I**, followed by the intramolecular C–O coupling reactions to give the final annulation products **8** (Scheme 6.4) (Xiong et al. 2011).

Pd-catalyzed regioselective C–H bond carbonylation of *N*-alkyl anilines **9** for the synthesis of isatoic anhydrides, 1*H*-3,1-benzooxazine-2,4-diones **10**, was developed. Reactions were catalyzed with $Pd(OAc)_2$ (5 mol%), $Cu(OAc)_2$ (2.2 equiv), and KI (0.2 equiv) in CH_3CN under CO pressure (1 atm) at 60°C, resulting in the formation of 1*H*-3,1-benzooxazine-2,4-diones **10** in 62–85% yields. The carbonylation of tetrahydroquinoline under reaction conditions afforded the tricyclic isatoic anhydride **10a** in 75% yield, while indoline did not give the expected product. The

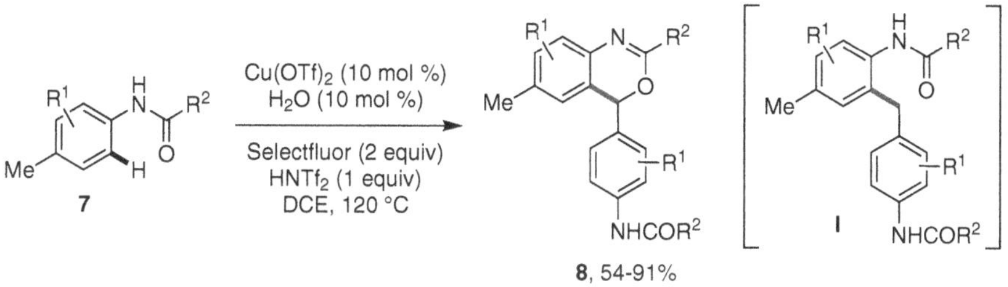

$Cu(OTf)_2$ (10 mol %), H_2O (10 mol %); Selectfluor (2 equiv), $HNTf_2$ (1 equiv), DCE, 120 °C. **7** → **8**, 54-91% (via intermediate **I**)

SCHEME 6.4

plausible reaction mechanism involves the electrophilic palladation of the C–H bond of *N*-methyl aniline **9** by $Pd(OAc)_2$ under a CO atmosphere to generate a dimeric palladium intermediate **I**, which underwent insertion of CO, giving intermediate **II**, followed by a reductive elimination to give *N*-methylanthranilic acid **III** and Pd(0). Upon reacting with $Pd(OAc)_2$, *N*-methylanthranilic acid **III** produced the intermediate **IV**, which underwent coordination and a second insertion of CO to afford the intermediate **V**. By a nucleophilic attack of the amino group on the acylpalladium moiety, the isatoic anhydride **10** was generated along with a Pd(0) species, which was reoxidized by $Cu(OAc)_2$ to regenerate the $Pd(OAc)_2$ active catalyst (Scheme 6.5) (Guan, Chen, and Ren 2012).

Rh(III)-catalyzed C–H activation/cyclization of *N*-carboxamide indoles and pyrroles **11** with diazonaphthalen-2(1*H*)-ones **12** was developed for the synthesis of naphtho[1,3]oxazino[3,4-*a*]indol-8-one and naphtho[1,2-*e*]pyrrolo[1,2-*c*][1,3]oxazin-5-one scaffolds **13**. By heating a solution of diazonaphthalen-2(1*H*)-ones **12** with 2 equiv of *N*-carboxamide indoles or pyrroles **11** in the presence of $[Cp^*RhCl_2]_2$ (5 mol%) and AgOAc (1 equiv) in toluene at 85°C, 8*H*-naphtho[1',2':5,6][1,3]oxazino[3,4-*a*]indol-8-one or 5*H*-naphtho[1,2-*e*]pyrrolo[1,2-*c*][1,3]oxazin-5-one derivatives **13** were obtained in 50–68% or 47–68% yields, respectively. The reaction was initiated by C–H activation of *N*-carboxamide indoles **11** with Rh(III) via

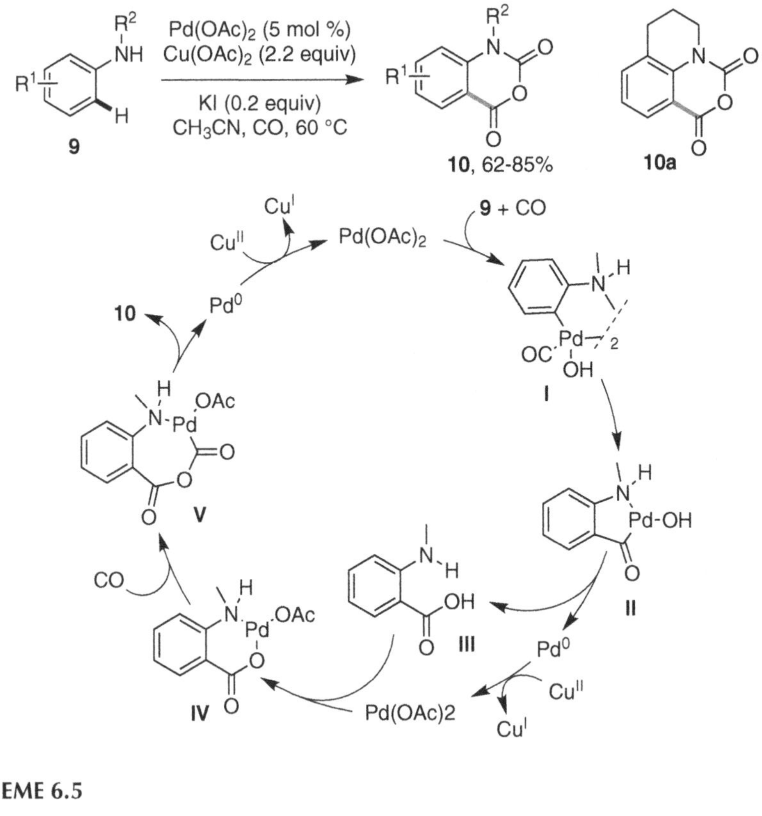

SCHEME 6.5

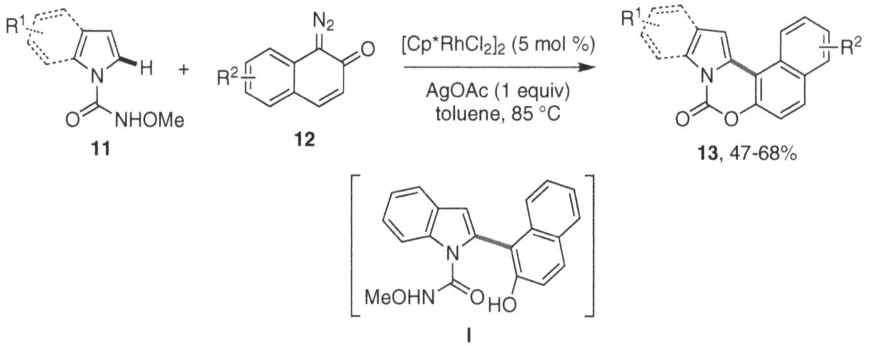

SCHEME 6.6

a concerted metalation/deprotonation (CMD) process, followed by the generation of carbene-Rh species by N_2 release, which underwent migratory insertion of carbene into a C–Rh bond, with subsequent protonolysis by AcOH to give arylated product **I**. Intramolecular cyclization of **I** delivered the final desired product **13** (Scheme 6.6) (Chen and Cui 2017).

6.4 BENZO[1,2]OXAZEPINE

Rh(III)-catalyzed dehydrogenative coupling of nitrones **14** with alkylidenecyclopropanes (ACPs) **15** was accomplished in order to synthesize 5-benzylidene-2-(*tert*-butyl)tetrahydro-1,4-methanobenzo[*d*][1,2]oxazepines **16**. Treating nitrone **14** with 2.5 equiv of alkylidenecyclopropanes **15** in the presence of [Cp*Rh(OAc)$_2$] (8 mol%) and AgOAc (2.5 equiv) in TFE at 40°C furnished benzo[*d*][1,2]oxazepine derivatives **16** in 33–91% yields. A reaction with 1,2-disubstituted ACPs and 1-alkyl substituted ACPs **15** failed to access corresponding benzo[*d*][1,2]oxazepines. The reaction proceeded by C–H activation of **14**, giving a rhodacyclic intermediate **I**, which underwent coordination with subsequent migratory insertion of ACP **15** to the Rh-aryl bond to provide an Rh(III)-alkyl intermediate **II**. β–C elimination and ring scission of **II** generated an Rh(III) alkyl species **III**, which was transformed into a diene **IV** together with an Rh(III) hydride upon β–H elimination. The diene **IV** underwent intramolecular [3+2] cycloaddition to afford the desired product **16**. The Rh(III) active catalyst was regenerated upon oxidation by AgOAc (Scheme 6.7) (Bai et al. 2018).

1,2-Dihydro-1,4-methanobenzo[*d*][1,2]oxazepin-5(4*H*)-one derivatives **18** were synthesized by Rh(III)-catalyzed dehydrogenative cross-coupling of nitrones **14** with diaryl cyclopropenones **17**, via a C–H acylation/[3+2] dipolar addition. The reaction of *N-tert*-butyl aryl nitrones **14** bearing *ortho* substituent were conducted in the presence of [Cp*RhCl$_2$]$_2$ (3 mol%), AgSbF$_6$ (40 mol%), and a 4 Å MS in DCE at 120°C, leading to 2-(*tert*-butyl)-4,10-diaryl-1,2-dihydro-1,4-methanobenzo[*d*][1,2]oxazepin-5(4*H*)-ones **18** in 56–87% yields. The reaction of *ortho*-unsubstituted nitrones **14** gave rise to formation of α-naphthol derivatives. However, reaction of *N*-arylnitrones **19** led to corresponding 3,4-dihydro-1,4-epoxybenzo[*b*]

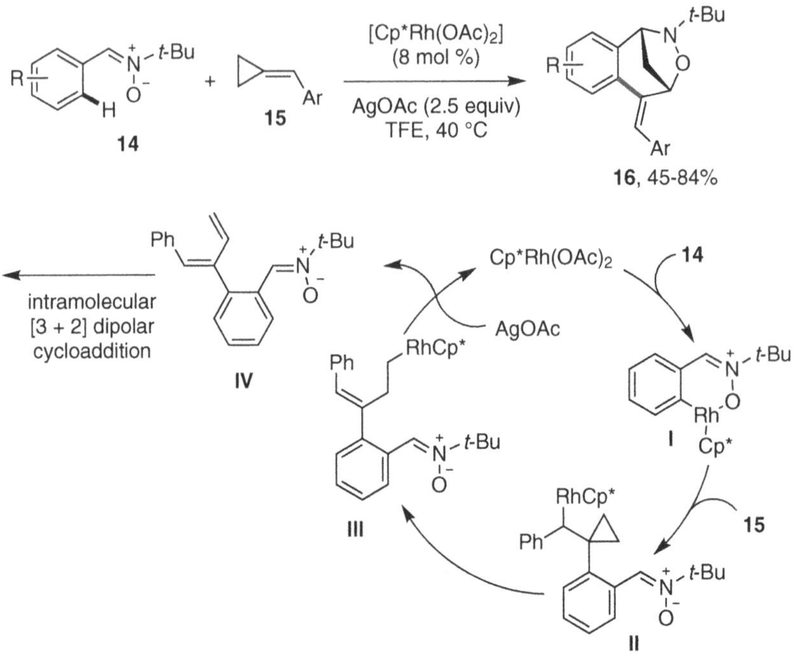

SCHEME 6.7

azepin-5(2*H*)-one derivatives **20** in 32–91% yields. A proposed reaction mechanism involves the cyclometalation of nitrone **14** to generate a six-membered rhodacycle **I**, which underwent coordination and migratory insertion of the carbonyl group of **17** into the Rh–C bond to give an eight-membered rhodacycle **II**. β–C elimination afforded the Rh(III) alkenyl **III**, which gave the chalcone intermediate **IV** upon protonolysis of the Rh–C(alkenyl) bond. Intramolecular dipolar addition led to the bicyclic product **18** (Scheme 6.8) (Xie et al. 2016).

Synthesis of 1,4-methanobenzo[*d*][1,2]oxazepine-4-carboxylate derivatives **22** was achieved by Rh(III)-catalyzed dehydrogenative cross-coupling of (hetero)aryl nitrones **14** with Morita–Baylis–Hillman (MBH) adducts **21**. Reactions were performed using 2 equiv of MBH adducts **21** in the presence of $[Cp^*RhCl_2]_2$ (2.5 mol%) and $AgSbF_6$ (10 mol%) in DCE at 60°C under an O_2 atmosphere, and 2-(*tert*-butyl)-1,2,4,5-tetrahydro-1,4-methanobenzo[*d*][1,2]oxazepine-4-carboxylates **22** were delivered in 72–99% yields. In addition, the reaction of polycyclic aromatic nitrones, including 1-naphthyl, 2-naphthyl, fluoren-2-yl, and pyren-1-yl nitrones with MBH adducts **21** proceeded smoothly to afford corresponding oxazepine-4-carboxylates in 44–79% yields. 3-Furyl nitrone led to the creation of the expected 4,7-methanofuro[3,2-*d*][1,2]oxazepine-7-carboxylate **22a** in 23% yield. Reaction was initiated by *O*-atom-directed *ortho*-C–H activation of nitrone **14** with an Rh(III) catalyst to generate intermediate **I**, which underwent coordination and migratory insertion of MBH adducts **21** to afford a rhodacyclic intermediate **II**. β–O elimination of **II** gave allylated intermediate **III**, which

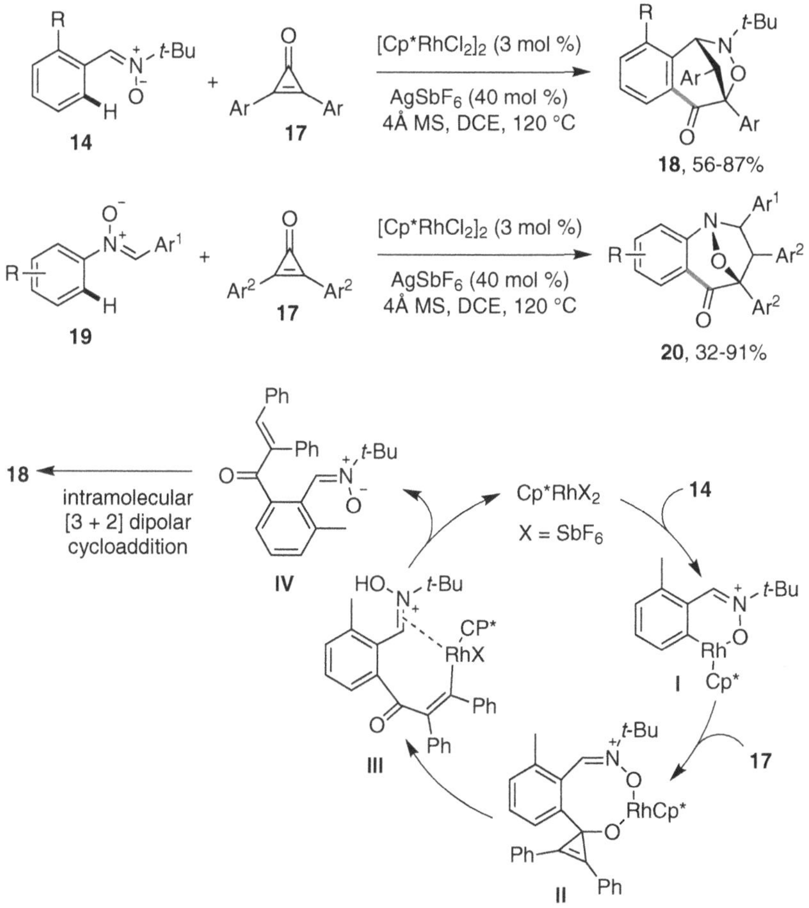

SCHEME 6.8

furnished bridged benzoxazepine product **22** via exotype [3 + 2] dipolar cycloaddition, with regeneration of catalytically active Rh(III) species (Scheme 6.9) (Pandey et al. 2018).

6.5 DIBENZOOXAZEPINE

Synthesis of different isomeric dibenzoxazepinones **25** was developed through Cu-catalyzed dehydrogenative etherification and C–N bond formation. Coupling reactions of an equimolar amount of *N*-(quinolin-8-yl)benzamide **23** and 2-bromophenols **24** were carried out using CuI (20 mol%) as catalyst and K_3PO_4 (3 equiv) as a base at reflux under air condition, delivering 10-(quinolin-8-yl)dibenzo[*b,f*][1,4]oxazepin-11(10*H*)-ones **25a** in 40–84% yields. When reactions were catalyzed with $Cu(OAc)_2$ (20 mol%) in the presence of K_2CO_3 as a base (2 equiv) in DMF at 74°C, followed by treatment with *t*-BuOK (1 equiv) at room temperature, rearranged 10-(quinolin-8-yl)

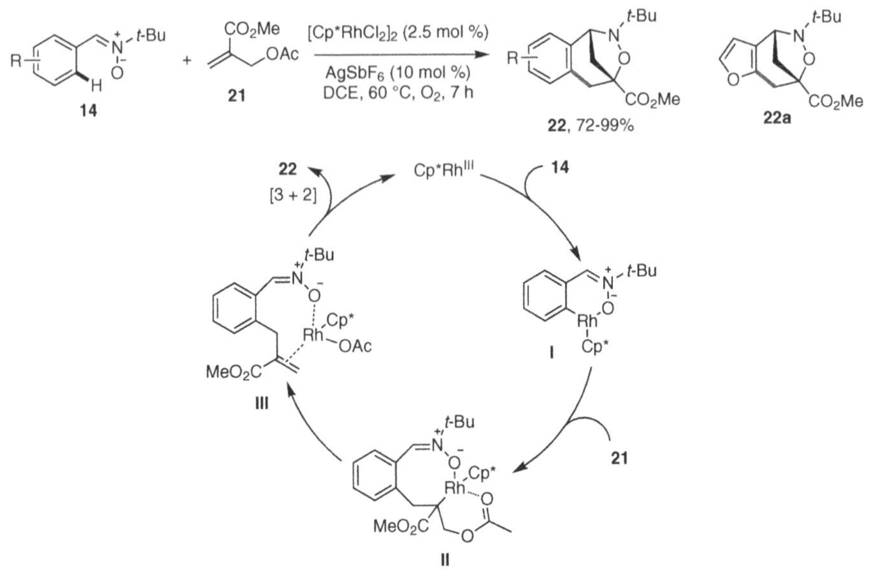

SCHEME 6.9

dibenzo[*b,f*][1,4]oxazepin-11(10*H*)-ones products **25b** were obtained in 58–80% yields. In the proposed reaction mechanism, aerobic dehydrogenative C–H etherification gave intermediate **I**, which was converted into nonrearranged product **25a** via a Goldberg reaction assisted by 8-aminoquinoline chelating Cu(I) (**II**) at a higher temperature (82°C). Alternatively, at moderate temperature (74°C) and in the presence of Cu(II), **I** did not undergo the cascade transformation into **25a**, and was transformed to **III** through a base-mediated Smiles rearrangement reaction followed by cyclization to produce the rearranged product **25b** (Scheme 6.10) (Zhou et al. 2016). Also, similar Cu(I)-catalyzed dehydrogenative coupling of C(sp^2)–H and O–H was described to synthesize 10-(quinolin-8-yl)dibenzo[*b,f*][1,4]oxazepin-11(10*H*)-one **25** derivatives using Cu_2O (10 mol%) and Na_2CO_3 (3.5 equiv) in DMF at 150°C under air atmosphere. Reaction with 2-bromo-3-hydroxypyridine resulted in formation of 11-(quinolin-8-yl)benzo[*f*]pyrido[3,2-*b*][1,4]oxazepin-10(11*H*)-ones **25c** in 62–85% yields (Zhang et al. 2016).

6.6 CONCLUSION

In summary, aromatic C(sp^2)–H dehydrogenative coupling processes are described for the synthesis of *N,O*-containing heterocyclic compounds, such as benzoxazole, benzoxazines, and benzoxazepines. However, many different methods were developed for the creation of these types of heterocycles, due to the quantitative and one-step synthesis and a broad spectrum of substitution on the synthesized heterocycles, the aromatic C–H bond direct functionalization approach could be of interest in the synthesis of pharmaceutical, medicinal, and natural products. On the other hand, there is still room for further studies in the field of synthesis of *N,O*-heterocycles via aromatic C–H dehydrogenative coupling reactions.

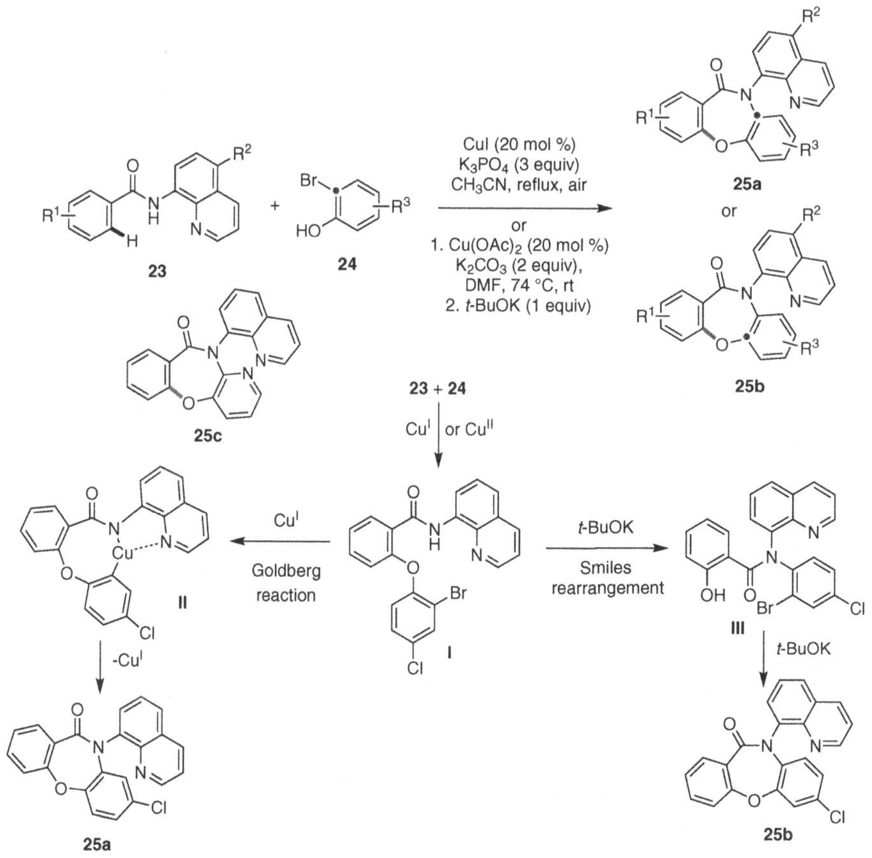

SCHEME 6.10

REFERENCES

Alla, Santhosh Kumar, Pradeep Sadhu, and Tharmalingam Punniyamurthy. 2014. "Organocatalytic syntheses of benzoxazoles and benzothiazoles using aryl iodide and oxone via C–H functionalization and C–O/S bond formation." *The Journal of Organic Chemistry* no. 79(16):7502–7511.

Bai, Dachang, Teng Xu, Chaorui Ma, Xin Zheng, Bingxian Liu, Fang Xie, and Xingwei Li. 2018. "Rh (III)-catalyzed mild coupling of nitrones and azomethine imines with alkylidenecyclopropanes via C–H activation: Facile access to bridged cycles." *ACS Catalysis* no. 8(5):4194–4200.

Chen, Renjie, and Sunliang Cui. 2017. "Rh (III)-catalyzed C–H activation/cyclization of benzamides and diazonaphthalen-2 (1 H)-ones for synthesis of lactones." *Organic Letters* no. 19(15):4002–4005.

Gandeepan, Parthasarathy, Thomas Müller, Daniel Zell, Gianpiero Cera, Svenja Warratz, and Lutz Ackermann. 2018. "3d transition metals for C–H activation." *Chemical Reviews* no. 119(4):2192–2452.

Guan, Zheng-Hui, Ming Chen, and Zhi-Hui Ren. 2012. "Palladium-catalyzed regioselective carbonylation of C–H bonds of N-alkyl anilines for synthesis of isatoic anhydrides." *Journal of the American Chemical Society* no. 134(42):17490–17493.

Kakkar, Saloni, Sumit Tahlan, Siong Meng Lim, Kalavathy Ramasamy, Vasudevan Mani, Syed Adnan Ali Shah, and Balasubramanian Narasimhan. 2018. "Benzoxazole derivatives: Design, synthesis and biological evaluation." *Chemistry Central Journal* no. 12(1):92.

Pandey, Ashok Kumar, Dahye Kang, Sang Hoon Han, Heeyoung Lee, Neeraj Kumar Mishra, Hyung Sik Kim, Young Hoon Jung, Sungwoo Hong, and In Su Kim. 2018. "Reactivity of Morita–Baylis–Hillman adducts in C–H functionalization of (hetero) aryl nitrones: Access to bridged cycles and carbazoles." *Organic Letters* no. 20(15):4632–4636.

Shaikh, Tanveer Mahamadali, and Fung-E Hong. 2016. "Recent developments in the preparation of N-heterocycles using Pd-catalysed C–H activation." *Journal of Organometallic Chemistry* no. 801:139–156.

Ueda, Satoshi, and Hideko Nagasawa. 2008. "Synthesis of 2-arylbenzoxazoles by copper-catalyzed intramolecular oxidative C–O coupling of benzanilides." *Angewandte Chemie* no. 120(34):6511–6513.

Ueda, Satoshi, and Hideko Nagasawa. 2009. "Copper-catalyzed synthesis of benzoxazoles via a regioselective C–H functionalization/C–O bond formation under an air atmosphere." *The Journal of Organic Chemistry* no. 74(11):4272–4277.

Wang, Biying, Chengfei Jiang, Jiasheng Qian, Shuwei Zhang, Xiaodong Jia, and Yu Yuan. 2018. "Synthesis of 2-methylbenzoxazoles directly from N-phenylacetamides catalyzed by palladium acetate." *Organic & Biomolecular Chemistry* no. 16(1):101–107.

Xie, Fang, Songjie Yu, Zisong Qi, and Xingwei Li. 2016. "Nitrone Directing Groups in rhodium (III) catalyzed C–H Activation of arenes: 1,3-dipoles versus traceless directing groups." *Angewandte Chemie International Edition* no. 55(49):15351–15355.

Xiong, Tao, Yan Li, Xihe Bi, Yunhe Lv, and Qian Zhang. 2011. "Copper-catalyzed dehydrogenative cross-coupling reactions of N-para-tolylamides through successive C–H activation: Synthesis of 4H-3, 1-benzoxazines." *Angewandte Chemie* no. 123(31):7278–7281.

Xu, Fan, Xiang-Yang Qian, Yan-Jie Li, and Hai-Chao Xu. 2017. "Synthesis of 4 H-1, 3-benzoxazines via metal-and oxidizing reagent-free aromatic C–H oxygenation." *Organic Letters* no. 19(23):6332–6335.

Yu, Hui, Mingdong Jiao, Ruohe Huang, and Xiaowei Fang. 2019"Electrochemical intramolecular dehydrogenative coupling of N-benzyl (thio) amides: A direct and facile synthesis of 4H-1, 3-benzoxazines and 4H-1, 3-benzothiazines." *European Journal of Organic Chemistry* no.11.

Yu, Zhengsen, Lijuan Ma, and Wei Yu. 2012. "Phenyliodine bis (trifluoroacetate) mediated intramolecular oxidative coupling of electron-rich N-phenyl benzamides." *Synlett* no. 23(10):1534–1540.

Zhang, Yipin, Xunqing Dong, Yanan Wu, Guigen Li, and Hongjian Lu. 2018. "Visible-light-induced intramolecular C (sp^2)–H amination and aziridination of azidoformates via a triplet nitrene pathway." *Organic Letters* no. 20(16):4838–4842.

Zhang, Zeyuan, Zhen Dai, Xinkun Ma, Yihan Liu, Xiaojun Ma, Wanli Li, and Chen Ma. 2016. "Cu-catalyzed one-pot synthesis of fused oxazepinone derivatives via sp^2 C–H and O–H cross-dehydrogenative coupling." *Organic Chemistry Frontiers* no. 3(7):799–803.

Zhou, Yunfei, Jianming Zhu, Bo Li, Yong Zhang, Jia Feng, Adrian Hall, Jiye Shi, and Weiliang Zhu. 2016. "Access to different isomeric dibenzoxazepinones through copper-catalyzed C–H etherification and C–N bond construction with controllable smiles rearrangement." *Organic Letters* no. 18(3):380–383.

7 *N,S*-Heterocycles

7.1 INTRODUCTION

Nitrogen- and sulfur-containing heterocycles occur in a variety of natural and non-natural products, with anticonvulsant, sedative, antidepressant, anti-inflammatory, antihypertensive, antihistaminic, and antiarthritic activities (Negwer 1994). In addition, these heterocycles are useful in synthetic organic chemistry as chiral auxiliaries (Baiget et al. 2008) and in organic functional materials such as fluorescent dyes. There are different methods available for the synthesis of thiazine derivatives in the literature (Eftekhari-Sis, Zirak, and Akbari 2013, Badshah and Naeem 2016).

Cross-dehydrogenative-coupling (CDC) (Gandeepan et al. 2018, Shaikh and Hong 2016) processes are widely used in the synthesis of a variety of types of heterocyclic compounds. In this chapter, intramolecular and intermolecular CDC of aromatic C(sp^2)–H bonds are discussed to create *N,O*-heterocycles, including benzoisothiazole, benzothiazine, benzothiazepine, benzothiazole, dibenzothiazine, and dibenzothiazepine derivatives.

7.2 BENZOISOTHIAZOLES

Rh(III)-catalyzed *ortho*-C–H olefination of aryl sulfonamide **I** directed by the SO_2NHAc group was developed in order to create benzoisothiazole derivatives **3**. The reaction of *N*-(phenylsulfonyl)acetamides **1** with 3 equiv of acrylates **2** was performed in the presence of $[Cp^*RhCl_2]_2$ (3 mol%) and $Cu(OAc)_2{\cdot}H_2O$ (2 equiv) in toluene at 100°C, giving (2,3-dihydrobenzo[*d*]isothiazol-7-yl)acrylates **3** in 41–97% yields. *N*-(phenylsulfonyl)butyramide was also compatible with the reaction to afford the corresponding benzoisothiazolylacrylate in 97% yield, while the reaction failed in the case of *N*-(phenylsulfonyl)pivalamide, *N*-(phenylsulfonyl)acrylamide, and *N*-methoxybenzenesulfonamide. The reaction occurred by bis-olefination of both *ortho* positions of aryl sulfonamide, followed by intramolecular conjugate addition of an *N* atom to one acrylate moiety. In the case of *ortho*-substituted aryl sulfonamides and *meta*-CF_3 aryl sulfonamide, only mono-olefination took place, and therefore corresponding (2,3-dihydrobenzo[*d*]isothiazol-3-yl)acetates **4** were produced in 65–97% and 43–59% yields, respectively. The reaction of *N*-(*p*-tolylsulfonyl)acetamide with *N,N*-dimethyl acrylamide gave a mono-alkenylation and cyclization product in 63% yield (Scheme 7.1) (Xie et al. 2014).

7.3 BENZOTHIAZOLES

Intramolecular dehydrogenative coupling for aromatic C–H thiolation by visible-light photoredox cobalt catalysis was developed for the creation of benzothiazoles **6**. Irradiation of *N*-arylbenzothioamides **5** solution in CH_3CN in the presence of

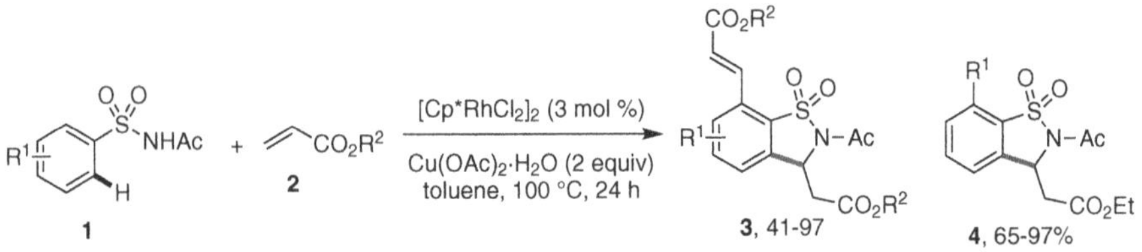

SCHEME 7.1

$Ru(bpy)_3(PF_6)_2$ (3 mol%), $Co(dmgH)_2(4\text{-}NMe_2Py)Cl$ (8 mol%), sodium-Gly (1 equiv), and 4-dimethylaminopyridine (DMAP) (0.4 equiv) under an Ar atmosphere with 3 W blue LEDs for 12 h at room temperature furnished 2-arylbenzothiazoles **6** in 45–99% yields. Also, substituted 2-alkylbenzothiazoles **6** were accessed by a similar approach, starting from *N*-arylalkanethioamides **5** bearing more bulky alkyl groups, including 2,2-dimethyl-*N*-phenylpropanethioamide and *N*-phenylcyclohexanecarbothioamide, and using tetrabutylammonium hydroxide (10 mol%) instead of sodium-Gly and DMAP, in 59–99% yields. In the proposed reaction mechanism, initially, excitation of $Ru(bpy)_3^{2+}$ by visible light produced an excited state $^*Ru(bpy)_3^{2+}$, which was reduced by single electron transfer (SET) with an anion intermediate of thioamide, derived from *N*-phenylbenzothioamide **5**, generating a sulfur-centered radical. The produced $Ru(bpy)_3^{+}$ was reoxidized to $Ru(bpy)_3^{2+}$ via SET to the catalyst Co(III), generating Co(II) to complete the photoredox cycle. On the other hand, the addition of the sulfur radical to the benzene ring generated the aryl radical, which released an electron to Co(II), followed by release of a proton and rearomatization to produce benzothiazole **6** (Scheme 7.2) (Zhang et al. 2015). Similar methodology was also developed using 1,8-diazabicyclo[5.4.0]undec-7-ene (DBU) (1 equiv) and $Ru(bpy)_3(PF_6)_2$ (1 mol%) under irradiation with 14 W compact fluorescent lamp (CFL) light in *N,N*-dimethylformamide (DMF) at room temperature under a 5% O_2 balloon, leading to 2-arylbenzothiazoles **6** in 34–88% yields (Cheng et al. 2011). Moreover, TEMPO-catalyzed electrochemical C–H thiolation of *N*-arylthioamides was developed. A variety of alkyl and aryl thioamides **5** were tolerated under reaction conditions, furnishing 2-alkyl and 2-arylbenzothiazole derivatives **6** in 48–95% yields. Thiazolo-pyridine derivatives **6a** were also prepared, starting from the corresponding *N*-pyridylbutanethioamides in 67–97% yields. No benzothiazole was formed in the case of a highly electron-withdrawing nitro group under all catalytic systems, and the substrate was decomposed (Qian et al. 2017). 1-Iodo-4-nitrobenzene-catalyzed synthesis of benzothiazoles was also reported via intramolecular dehydrogenative coupling of *N*-arylthioamides with oxone as an oxidant (Alla, Sadhu, and Punniyamurthy 2014).

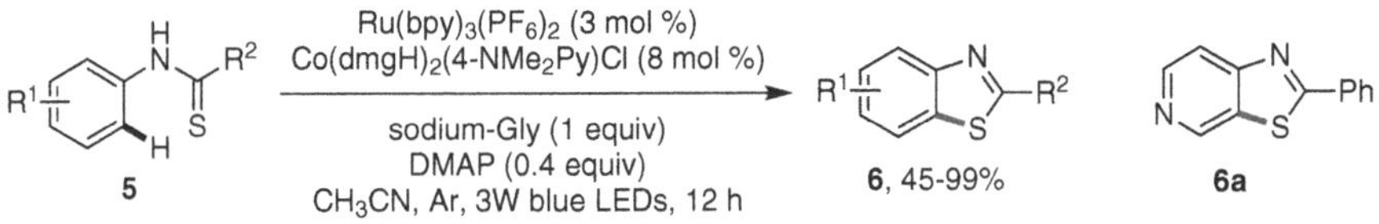

SCHEME 7.2

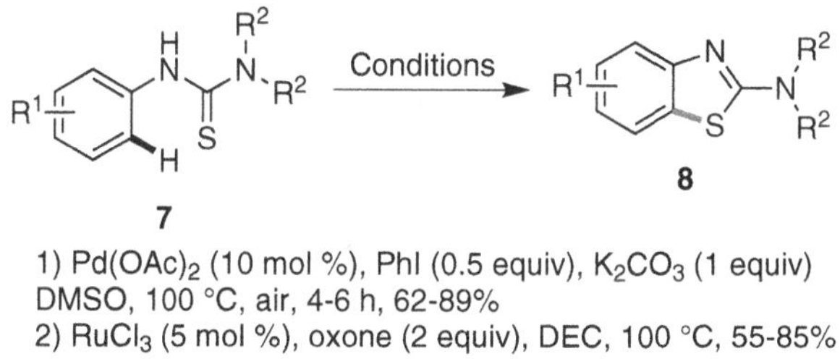

Conditions

1) $Pd(OAc)_2$ (10 mol %), PhI (0.5 equiv), K_2CO_3 (1 equiv) DMSO, 100 °C, air, 4-6 h, 62-89%
2) $RuCl_3$ (5 mol %), oxone (2 equiv), DEC, 100 °C, 55-85%

SCHEME 7.3

Iodobenzene-promoted Pd-catalyzed *ortho*-directed C–H activation of *N*-arylthioureas **7** was developed for the synthesis of 2-aminobenzothiazoles **8**. Reactions were performed in dimethyl sulfoxide (DMSO) using 10 mol% of $Pd(OAc)_2$ as catalyst, PhI (0.5 equiv) as an additive, and K_2CO_3 (1 equiv), as a base under air conditions, delivering 2-aminobenzothiazoles **8** in 62–89% yields (Zeng et al. 2017). In addition, an Ru-catalyzed intramolecular dehydrogenative C–S coupling reaction of *N*-arylthioureas **7** was developed by Sharma et al. (2016). $RuCl_3$ (5 mol%) in the presence of oxone (2 equiv) in 1,2-dichloroethane (DEC) catalyzed the conversion of *N*-arylthioureas **7** into corresponding 2-aminobenzothiazoles **8** in 55–91% yields. *N,N'*-diarylthioureas were also compatible with the reaction to afford corresponding 2-aminobenzothiazoles in 55–85% yields, in which cyclization occurred exclusively on the electron-rich arene ring (Scheme 7.3).

7.4 BENZO[4,5]ISOTHIAZOLO[2,3-*a*]INDOLES

Benzo[4,5]isothiazolo[2,3-*a*]indole 5,5-dioxides **10** were synthesized by a Pd-catalyzed intramolecular dehydrogenative coupling process. Treatment of *N*-(arylsulfonyl)indoles **9** with $Pd(OAc)_2$ (10 mol%), CsOPiv (20 mol%), and AgOAc (3 equiv) in PivOH at 130°C led to benzoisothiazolo[2,3-*a*]indole derivatives **10** in 30–85% yields. Similarly, benzo[*d*]pyrrolo[1,2-*b*]isothiazole 5,5-dioxides **10** were obtained from *N*-(arylsulfonyl)pyrroles **9** in 60–68% yields. The proposed reaction mechanism involves regioselective palladation at the C-2 position of the indole, followed by concerted metalation/deprotonation (CMD) at the *ortho* position of the sulfonyl group in the benzene ring, with subsequent reductive elimination (Scheme 7.4) (Laha et al. 2015). One example of *N*-phenylbenzo[4,5]isothiazolo[2,3-*a*] indole-11-carboxamide 5,5-dioxide was obtained in 70% yield via intermolecular

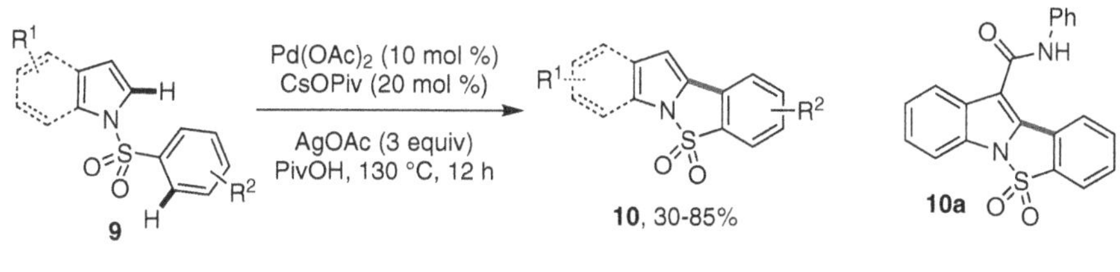

$Pd(OAc)_2$ (10 mol %)
CsOPiv (20 mol %)
AgOAc (3 equiv)
PivOH, 130 °C, 12 h

10, 30-85%

SCHEME 7.4

dehydrogenative coupling of *N*-phenyl-1-(phenylsulfonyl)-1*H*-indole-3-carboxamide under a $Pd(OAc)_2$ and $Cu(OAc)_2$ catalytic system in PivOH (Mahajan et al. 2016).

7.5 BENZOTHIAZINES

3-Aryl-2-(quinolin-8-yl)-2*H*-benzo[*e*][1,2]thiazine 1,1-dioxide derivatives **14a** were synthesized by Co-catalyzed C–H and N–H annulation of aryl sulfonamide with aryl acetylenes **12**. Treatment of *N*-(quinolin-8-yl)sulfonamides **11** with 2 equiv of aryl acetylenes **12** in the presence of $Co(OAc)_2{\cdot}4H_2O$ (10 mol%), $Mn(OAc)_3{\cdot}2H_2O$ (2 equiv), and NaOPiv (2 equiv) in 2,2,2-trifluoroethanol (TFE) at 100°C under O_2 atmosphere gave corresponding 3-arylbenzothiazine 1,1-dioxides **14a** in 45–90% yields. The reaction of aliphatic terminal alkynes led to the expected 3-alkylbenzothiazine 1,1-dioxides **14a** in 50–87% yields. Ethyl propiolate and trimethylsilyl (TMS)-acetylene were tolerated under reaction conditions, leading to the corresponding benzothiazine-3-carboxylate 1,1-dioxide **14c** and 3-TMS-benzothiazine 1,1-dioxides in 60% and 30% yields, respectively. Internal alkynes were also compatible with the reaction, delivering 3,4-disubstituted benzothiazine 1,1-dioxides in 77% yield. The reaction of unsymmetrical alkyl aryl alkyne 1-phenylpropyne afforded 4-methyl-3-phenylbenzothiazine 1,1-dioxide in 61% yield, regioselectively. The reaction was initiated by 8-aminoquinoline directed C–H bond activation, followed by coordination and insertion with alkyne, and then reductive elimination (Kalsi and Sundararaju 2015). A similar catalytic system was also reported for the creation of (1,1-dioxidobenzothiazin-3-yl)acetates **14b** ($R = CO_2R$) via dehydrogenative coupling of *N*-(quinolin-8-yl)sulfonamides **11** with allene carboxylates **13**. Annulated products **14b** were obtained in 55–75% yields. The reaction of internal disubstituted allenes afforded corresponding 4-substituted (1,1-dioxidobenzothiazin-3-yl)acetates in 71–88% yields. A nine-membered cyclic allene was also compatible with the reaction condition to furnish the three ring-fused heterocycle products, 6-(quinolin-8-yl) hexahydrobenzo[*e*]cyclohepta[*c*][1,2]thiazine 5,5-dioxides **14d** in 47–64% yields. 3-Vinyl substituted benzothiazine 1,1-dioxides **14e** were obtained when alkoxymethyl-substituted allenes were subjected to *N*-(quinolin-8-yl)sulfonamides **11** under reaction conditions. Thiophene- and pyrrol-2-sulfonamides provided corresponding thieno- and pyrrolo[3,2-*e*][1,2]thiazine 1,1-dioxides **14f** in 57–81% and 42% yields, respectively. The plausible reaction mechanism involves *ortho*-C–H activation followed by allene coordination and insertion into the C–Co bond, with subsequent reductive elimination, and then a 1,3-*H* shift (Scheme 7.5) (Lan, Wang, and Rao 2017). $[Cp^*Rh(OAc)_2]$ (5 mol%) in the presence of $Cu(OAc)_2$ (20 mol%) in *t*-AmOH at 100°C under an atmosphere of O_2 catalyzed the dehydrogenative cross-coupling of *N*-acetyl arylsulfonamides with internal alkynes, to afford 3,4-disubstituted benzothiazine 1,1-dioxide derivatives in 47–99% yields. Unsymmetrical alkyl aryl alkyne led to the formation of the corresponding annulated products in 99% yields, with 2:1–8:1 regioselectivity, favoring 3-alkyl-4-arylbenzothiazine 1,1-dioxide isomers (Pham, Ye, and Cramer 2012).

4-Unsubstituted 1,2-benzothiazines **17a** were prepared from sulfoximines **15** and allyl methyl carbonate **16** through an Rh(III)-catalyzed dehydrogenative cross-coupling reaction. Reactions were carried out by 3 equiv of allyl carbonate **16** in

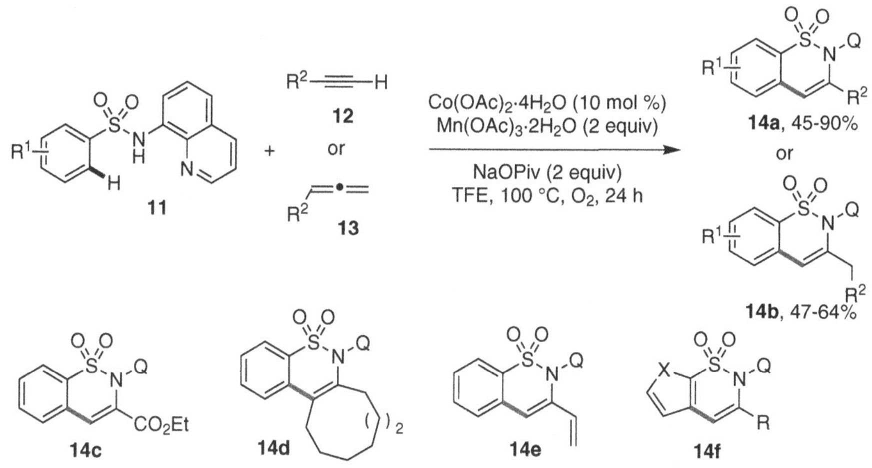

SCHEME 7.5

the presence of $[Cp^*RhCl_2]_2$ (4 mol%), $AgSbF_6$ (16 mol%), and $Cu(OAc)_2{\cdot}H_2O$ (2.1 equiv) in DME at 100°C under Ar atmosphere to provide 3-methyl-1,2-benzothiazine derivatives **17a** in 36–72% yields. Presumably due to steric reasons, no product was formed in the case of *ortho*-chloro-substituted sulfoximine. The reaction occurred by sulfoximine-directed *ortho*-C–H activation to generate a five-membered rhodacycle **I**. Through allyl coordination and alkene insertion, intermediate **II** was formed, which was converted into olefin-coordinated Rh(III) complex **III** upon β–O elimination. Subsequent Cu-mediated ring closure and β–H elimination followed by double-bond isomerization, producing the final desired product **17a** (Wen, Tiwari, and Bolm 2017). Also, the synthesis of 4-unsubstituted 1,2-benzothiazines **17b** was achieved by Rh(III)-catalyzed dehydrogenative alkylation of sulfoximines **15** with α-MsO/TsO ketones **18** followed by cyclocondensation (Scheme 7.6) (Yu, de Azambuja, and Glorius 2014).

Rh-catalyzed dehydrogenative coupling between *S*-aryl sulfoximines **15** and internal alkynes **19** was developed in order to create 1,2-benzothiazine derivatives **20**. Reactions were conducted using $[Cp^*Rh(MeCN)_3][BF_4]_2$ (5 mol%) in the presence of $Fe(OAc)_2$ (20 mol%) in toluene at 100°C for 48 h, affording 1,2-benzothiazine derivatives in 67–93% yields (Dong et al. 2013). Also, Rh-catalyzed dehydrogenative coupling of the aromatic C–H bond of *S*-aryl sulfoximines **15** with carbene precursors was developed to create benzo[*e*][1,2]thiazine-4-carboxylate 1-oxide derivatives **22**. Reactions were performed by heating a solution of *S*-aryl sulfoximines **15** and α-diazo-β-ketoesters **21** (1.1 equiv) in 1,2-dichloroethane (DCE) in the presence of $[Cp^*Rh(MeCN)_3][SbF_6]_2$ (2 mol%), and NaOAc (1 equiv) at 100°C under argon atmosphere, leading to the corresponding benzo[*e*][1,2]thiazine-4-carboxylates **22** in 71–99% yields. Dehydrogenative annulation of **15** with the formyl diazo compound gave 3-unsubstituted benzo[*e*][1,2]thiazine-4-carboxylate in 77% yield. Diazosulfone and diazophosphonate were also compatible with the reaction, affording the corresponding products in 98% and 88% yields, respectively. The reaction

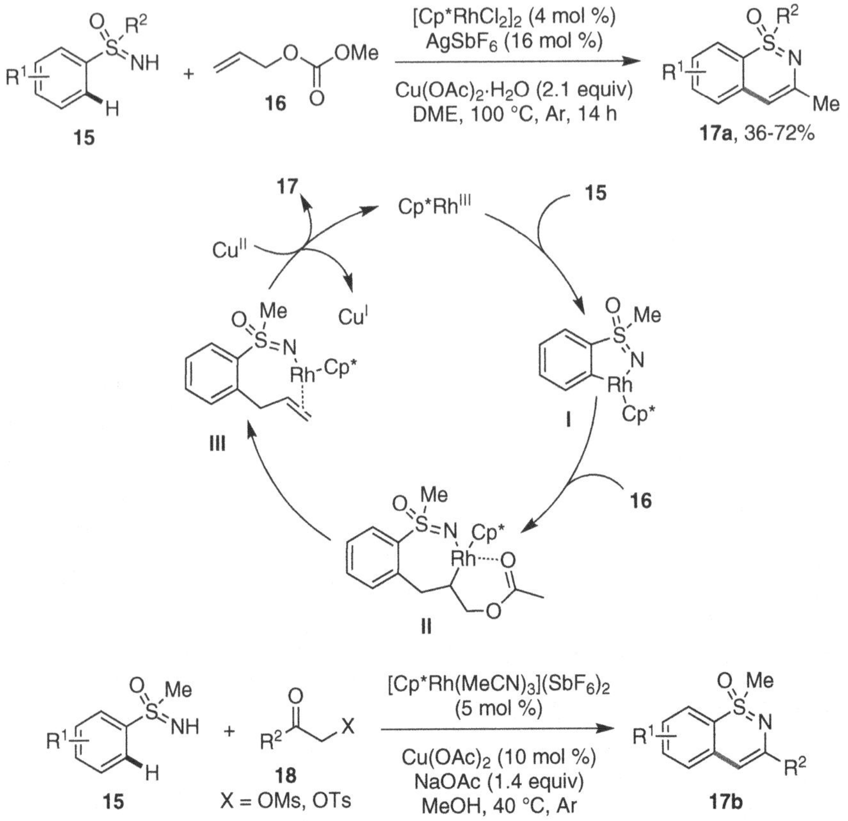

SCHEME 7.6

with α-diazo-β-diketones produced the expected 4-acyl substituted 1,2-benzothiazines in moderate yields (32–42%). The reaction was proposed to proceed by the sulfoximine *N*-atom-directed *ortho*-C–H activation followed by coordination and insertion of the diazo moiety through release of N_2 giving carbene-Rh species, which underwent migratory insertion of the carbene moiety into the C–Rh bond and then was transformed into the final annulated product by protonation with AcOH, followed by an intramolecular condensation reaction (Cheng and Bolm 2015). Likewise, the synthesis of 4-(pyridin-2-yl)benzo[*e*][1,2]thiazin-3(4*H*)-one 1-oxides **24** was accomplished by Rh-catalyzed N–H/C–H activation of simple alkyl aryl sulfoximines **15** in coupling with pyridotriazoles **23** using similar catalytic systems in toluene. Benzo[*e*][1,2]thiazin-3(4*H*)-one derivatives **24** were obtained in 60–99% yields. The proposed reaction mechanism involves the in situ generation of the diazoimine **I** from the pyridotriazole **23** via liberation of N_2 gas, with subsequent coordination and insertion into the C–Rh bond of a five-membered rhodacycle (Scheme 7.7) (Jeon et al. 2016).

The electrochemical dehydrogenative cyclization of *N*-benzyl arylthioamides **25** was accomplished to access 4*H*-1,3-benzothiazines **26**, by carrying the reaction

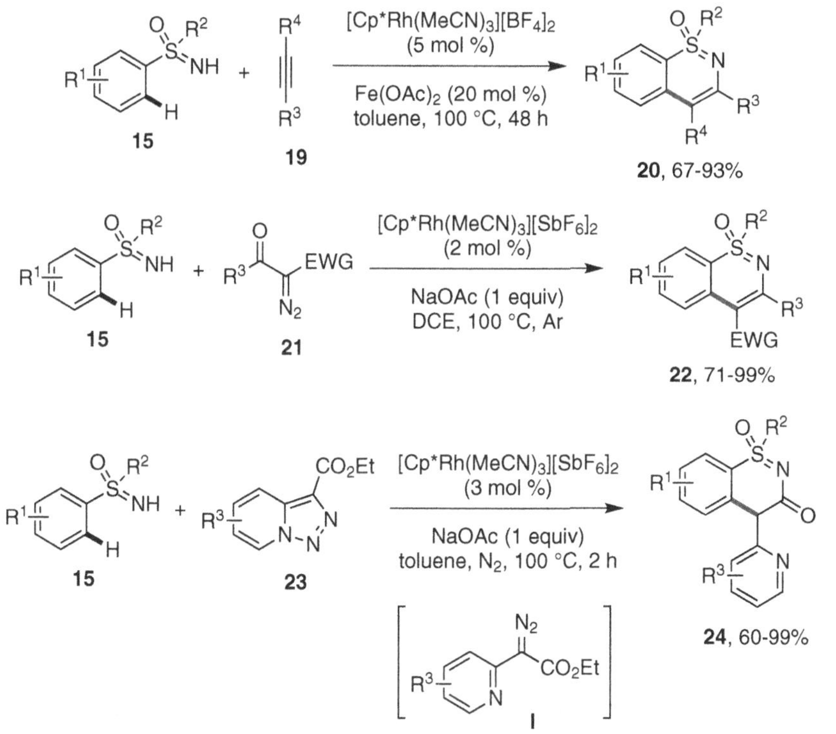

SCHEME 7.7

in an undivided cell with Pt plate anode and graphite rod cathode with 10 mA of constant current at room temperature, in 66–83% yields. The radical cation **I** was first generated by the oxidation of the benzylic moiety under electrolysis conditions. Cyclization and deprotonation of **I** gave the intermediate radical **II**, which was oxidized and then rearomatized to give rise to the desired product **26** (Scheme 7.8) (Yu et al.).

7.6 DIBENZO[*c,e*][1,2]THIAZINES

Pd(II) promoted the intramolecular dehydrogenative C(sp^2)–H arylation of 2-bromo-*N*-alkyl-*N*-phenylbenzenesulfonamides **27**, affording dibenzo[*c,e*][1,2]thiazine 5,5-dioxide derivatives **28a**. Reactions were performed by subjecting 2-bromo-*N*-phenylbenzenesulfonamide derivatives **27** with $Pd(OAc)_2$ (1 mol%), KOAc (2 equiv) in *N,N*-dimethylacetamide (DMA) at 150°C under argon, and related coupling products **28a** were obtained in 88–90% yield. No reaction occurred in the case of *NH*-free sulfonamide, 2-bromo-*N*-phenylbenzenesulfonamide. Benzothiazino[4,3,2-*hi*]indole 7,7-dioxide **28b** was produced in 91% yield when *N*-(phenylsulfonyl)indoline was subjected to a reaction (Bheeter, Bera, and Doucet 2012). Also, Pd-catalyzed intramolecular dehydrogenative coupling involving two C(sp^2)–H b bonds for the creation of dibenzo[*c,e*][1,2]thiazine 5,5-dioxide

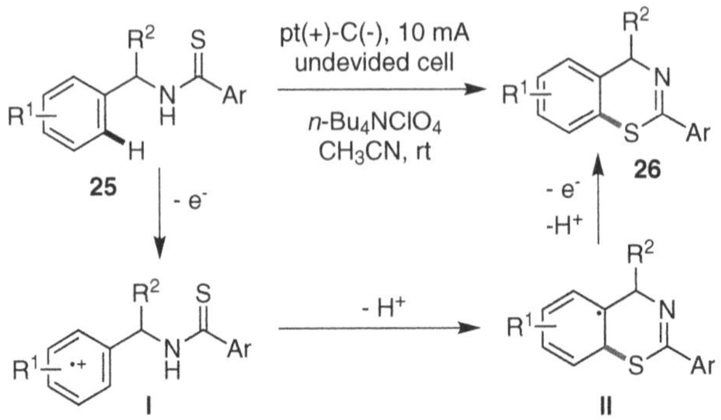

SCHEME 7.8

derivatives **28c** was developed by Laha et al. (Laha, Jethava, and Dayal 2014). Reactions were conducted by heating a solution of *N*-arylbenzenesulfonamides **29** in PivOH/AcOH in the presence of $Pd(OAc)_2$ (10 mol%), KO*t*-Bu (20 mol%) and AgOAc (3 equiv) at 130°C, leading to annulated biaryl sultams **28c** in 51–76% yields. No product was obtained in the case of a strong electron-donating OMe group at the C-2 or C-4 position. Also, the synthesis of *N*-substituted biaryl sultams was unsuccessful (Scheme 7.9).

Jeganmohan et al. (Chinnagolla, Vijeta, and Jeganmohan 2015) developed a Pd-catalyzed dehydrogenative cyclization of *S*-alkyl-*S*-(1,6-diarylphenyl)sulfoximines **30**, leading to dibenzo[*c,e*][1,2]thiazine derivatives **31**. Reactions were performed using $Pd(OAc)_2$ (10 mol%) in the presence of $PhI(OAc)_2$ (2 equiv) in toluene at 120°C, and 5-alkyl-4-aryldibenzo[*c,e*][1,2]thiazine 5-oxides **31** were obtained in 41–85% yields. Chiral sulfoximines afforded corresponding dibenzothiazine 5-oxides with excellent enantioselectivity (>99% ee). The starting *S*-alkyl-*S*-(1,6-diarylphenyl)sulfoximine substrates **30** were obtained by Ru-catalyzed *ortho* arylation of aromatic sulfoximines with aromatic boronic acids. The reaction occurred by sulfoximine *N*-atom-directed C–H activation, followed by reductive elimination

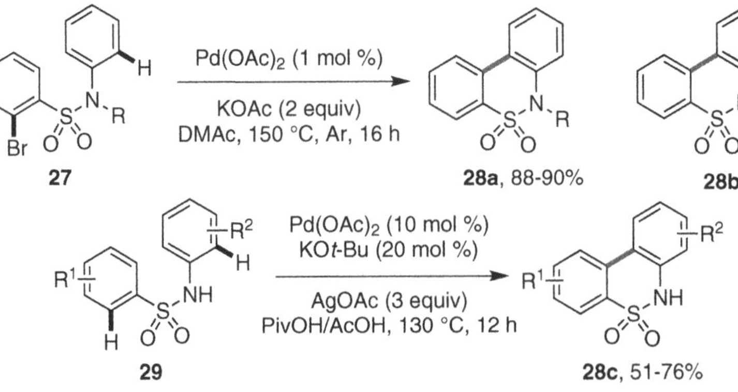

SCHEME 7.9

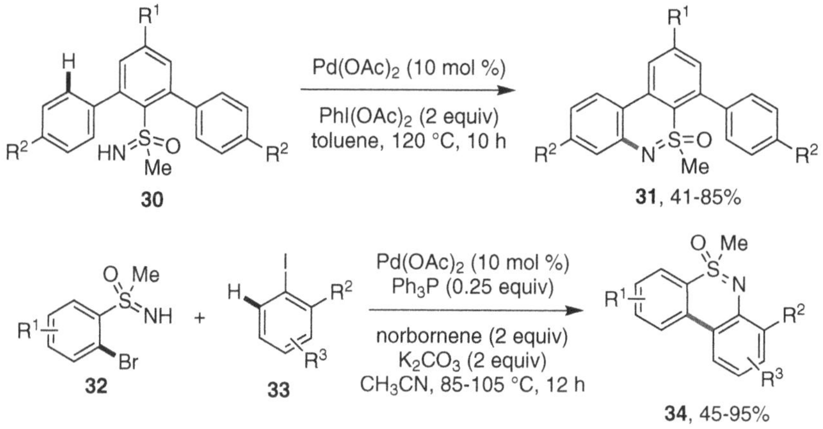

$Pd(OAc)_2$ (10 mol %)
$PhI(OAc)_2$ (2 equiv)
toluene, 120 °C, 10 h

30 → **31**, 41-85%

$Pd(OAc)_2$ (10 mol %)
Ph_3P (0.25 equiv)
norbornene (2 equiv)
K_2CO_3 (2 equiv)
CH_3CN, 85-105 °C, 12 h

32 + **33** → **34**, 45-95%

SCHEME 7.10

in the presence of $PhI(OAc)_2$. Synthesis of similar dibenzothiazine 5-oxide scaffolds was also reported by Cheng et al. (2017). Pd/norbornene cocatalyzed tandem dehydrogenative annulation reactions of free *NH*-sulfoximines **32** with aryl iodides **33** was accomplished to produce dibenzothiazine 5-oxide derivatives **34**. Subjecting *NH*-sulfoximine to 1.2 equiv of aryl iodide **33** in the presence of $Pd(OAc)_2$ (10 mol%), Ph_3P (0.25 equiv), norbornene (2 equiv), and K_2CO_3 (2 equiv) in CH_3CN at 85–105°C gave 10-substituted dibenzo[*c,e*][1,2]thiazine 5-oxides **34** in 45–95% yields (Scheme 7.10) (Zhou, Chen, and Chen 2018).

7.7 BENZOTHIADIAZINES

Benzo[*e*][1,2,4]thiadiazine derivatives **36** were synthesized by Cp*Co(III)-catalyzed direct C–H activation/dual C–N bond formation reaction of *S,S*-diaryl *NH*-sulfoximines **15** with 1,4,2-dioxazol-5-ones **35**. By reacting *NH*-sulfoximine **15** with 2 equiv of 3-aryl-1,4,2-dioxazol-5-one **35** in the presence of $[Cp^*Co(CO)I_2]$ (5 mol%), $AgNTf_2$ (10 mol%), and PivOH (2 equiv) in *t*-AmOH/$CHCl_3$ at 110°C under microwave heating conditions, 3-substituted 1-arylbenzo[*e*][1,2,4]thiadiazine 1-oxides **36** were obtained in 36–92% yields. 3-alkyl and vinyl-substituted 1,4,2-dioxazol-5-ones were also tolerated under reaction conditions, leading to corresponding benzo[*e*][1,2,4]thiadiazines in 59–67% and 72% yields, respectively. *S,S*-di(thiophen-2-yl) *NH*-sulfoximine was compatible with the reaction to give the expected 1-(thiophen-2-yl)thieno[3,2-*e*][1,2,4]thiadiazine 1-oxides **36a** in 47–68% yields. No selectivity was observed when two different aryl group-substituted *S,S*-diaryl *NH*-sulfoximines **15** were subjected to 1,4,2-dioxazol-5-one **35** under reaction conditions. The plausible reaction mechanism involves the generation of a five-membered cobaltacycle **II** by C–H activation of *NH*-sulfoximines **15** with reactive cationic species **I**, produced in situ from the reaction of neutral complex $[Cp^*Co(CO)I_2]$ with $AgNTf_2$ in the presence of PivOH. Coordination of 1,4,2-dioxazol-5-one **35** with **II** gave the cobaltacycle **III**, which underwent subsequent amido insertion to give amido complex

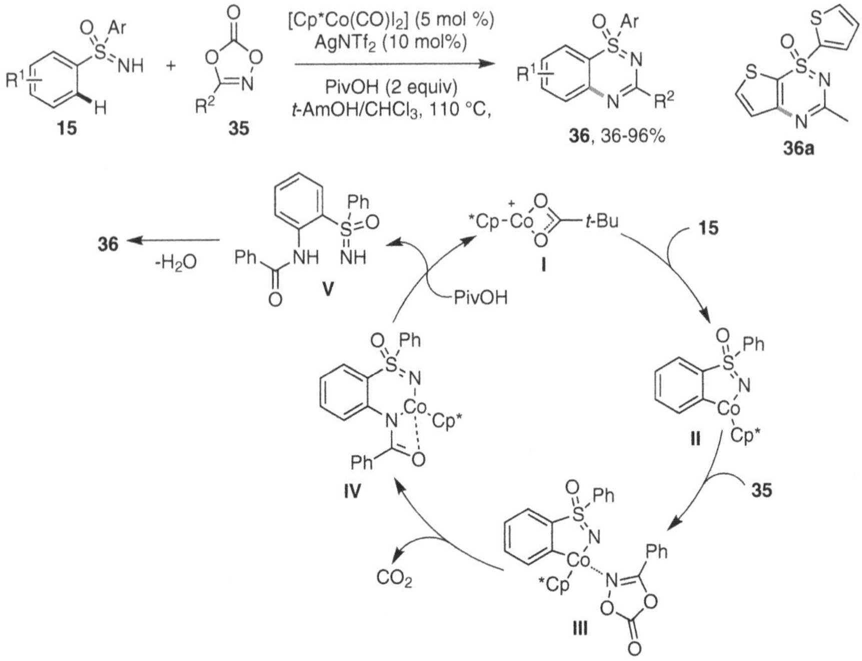

SCHEME 7.11

species **IV** with release of CO_2. Protodecobaltation of **IV** delivered the key intermediate **V**, along with the regeneration of the reactive metal catalyst **I** to complete the catalytic cycle. Finally, cyclodehydration of compound **V** furnished the product **36** (Scheme 7.11) (Huang et al. 2017).

7.8 BENZOTHIAZEPINES

In 2017, Wen et al. (2017) described Rh-catalyzed dehydrogenative annulations of sulfoximines **15** with *α,β*-unsaturated ketones **37** to construct 1,2-benzothiazepine 1-oxides **38**. By conducting the reaction of sulfoximine **15** (3 equiv) with ketone **37** in the presence of $[Cp^*RhCl_2]_2$ (2.5 mol%), $AgSbF_6$ (10 mol%), and 1-$AdCO_2H$ (3 equiv) in toluene at 100°C, 3-arylbenzo[*f*][1,2]thiazepine 1-oxides **38** were obtained in 45–76% yields. Reaction with 2-methylene-dihydroindenone led to an indene-fused product **38a** in 41% yield. No reaction occurred with alkyl vinyl ketones to give the corresponding 1,2-benzothiazepine 1-oxide. When reactions were carried out using 3 equiv of ketone **37** in the presence of $[Cp^*RhCl_2]_2$ (2.5 mol %), $AgSbF_6$ (10 mol %), and $Cu(OAc)_2{\cdot}H_2O$ (2 equiv) in DCE at 70°C, the expected 1,2-benzothiazepine 1-oxides **39** were obtained along with alkylation at another *ortho* position by a second *α,β*-unsaturated ketone molecule, in 73–92% yields. The reaction proceeded by sulfoximine *N*-atom-directed *ortho*-C–H activation, followed by coordination and insertion of a C=C double bond, with subsequent protonation to alkylated

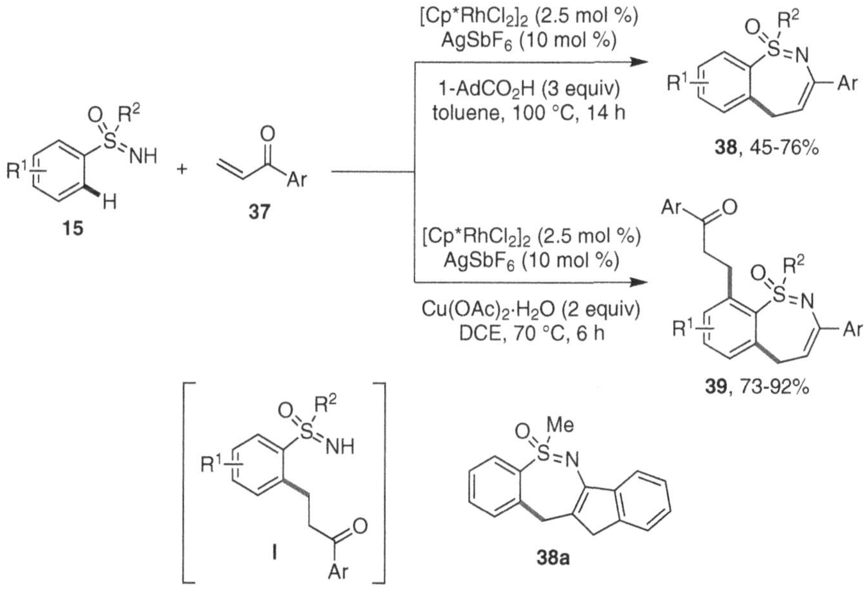

SCHEME 7.12

intermediate **I**, which underwent dehydrative cyclization to achieve the desired product (Scheme 7.12).

7.9 DIBENZOTHIAZEPINES

6,7-Dihydrodibenzo[*d,f*][1,2]thiazepine 5,5-dioxides **42a** were synthesized by Pd-catalyzed domino *N*-benzylation/intramolecular dehydrogenative direct arylation of sulfonanilides **40** and 2-bromobenzyl bromides **41**. Reactions were performed by heating a solution of an equimolar amount of sulfonanilides **40** and 2-bromobenzyl bromides **41** in dioxane at 110°C, in the presence of $Pd(OAc)_2$ (10 mol%), PPh_3 (20 mol%), and Cs_2CO_3 (2.5 equiv) under an atmosphere of N_2, affording 6,7-dihydrodibenzo[*d,f*][1,2]thiazepine 5,5-dioxides **42a** in 40–78% yields. However, arylation could also occur at the *N*-benzyl moiety, to give *N*-tosyl dibenzoazepines (Laha et al. 2014). Also, dehydrogenative coupling of *N*-alkylbenzene sulfonamides with 2-bromobenzyl bromides was described under the same reaction conditions, leading to corresponding dihydrodibenzo[*d,f*][1,2]thiazepines in 40–87% yields (Laha, Sharma, and Dayal 2015). In addition, Pd(II)-catalyzed intramolecular dehydrogenative $C(sp^2)$-H arylation of *N*-benzyl-2-bromo-*N*-methylbenzenesulfonamides **43** (X = Br) was developed in order to construct dibenzothiazepine structural motifs **42b** in 60–86% yields (Bheeter, Bera, and Doucet 2012). 7,8-Dihydro-6*H*-benzo[6,7][1,2]thiazepino[4,5-*b*]indole 5,5-dioxide derivatives **44** were prepared by a similar intramolecular arylation of *N*-((indol-2-yl)methyl)benzenesulfonamide, generated in situ by Pd-catalyzed Sonogashira coupling/hydroamination of

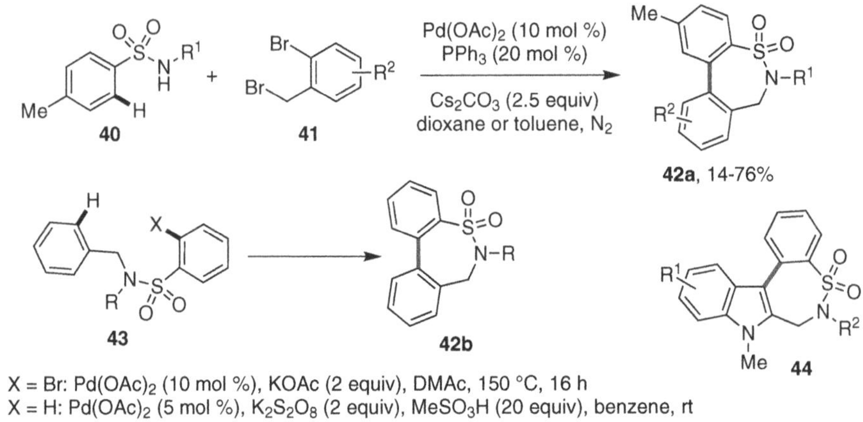

X = Br: $Pd(OAc)_2$ (10 mol %), KOAc (2 equiv), DMAc, 150 °C, 16 h
X = H: $Pd(OAc)_2$ (5 mol %), $K_2S_2O_8$ (2 equiv), $MeSO_3H$ (20 equiv), benzene, rt

SCHEME 7.13

propargylsulfonamides with *o*-iodoaniline derivatives (Debnath and Mondal 2018). However, dihydrodibenzothiazepine 5,5-dioxides **42b** were also produced by intramolecular arylation of *N*-benzylbenzenesulfonamides **43** (X=H) in the presence of $Pd(OAc)_2$ (5 mol%), $K_2S_2O_8$ (2 equiv), and $MeSO_3H$ (20 equiv) in benzene at room temperature, in 22–53% yields (Scheme 7.13) (Jeong et al. 2015).

7.10 BENZOTHIAZOCINES

Benzothiazocine 9-oxide derivatives **48** were accessed by Pd/norbornene cocatalyzed tandem dehydrogenative annulation reactions of free *NH*-sulfoximines **45** with aryl iodides **46** bearing an electron-withdrawing substituent at the *meta*-position of the phenyl ring. Reaction with *ortho*-substituted aryl iodides led to the corresponding dibenzothiazine 5-oxide. The reactions were catalyzed with $Pd(OAc)_2$ (10 mol%) and Ph_3P (0.25 equiv), in the presence of norbornene **47** (2 equiv) and K_2CO_3 (2 equiv) in CH_3CN at 105°C, to afford hexahydro-1,4-methanotribenzo[*c,e,g*][1,2]thiazocine 9-oxides **48** in 61–94% yields. *Meta*-methoxy phenyl iodide gave a complex mixture under reaction conditions. When increasing the amount of 2-bromo-*NH*-sulfoximine **45** to 2.4 equiv, benzothiazocine 9-oxides **49** bearing another sulfoximine moiety at the *ortho* position of a 13-phenyl ring were formed in 42–64% yields. The proposed reaction mechanism involves the generation of Pd(II) intermediate **III** by C–I bond activation (**I**), followed by norbornene C=C double-bond insertion (**II**) and then *ortho*-C–H activation. Oxidative addition of *NH*-sulfoximine **45** gave the Pd(IV) intermediate **IV**. An intramolecular attack by the *NH* moiety on Pd-X gave intermediate **V**, which underwent reductive elimination to afford intermediate **VI**. Finally, reductive elimination produced benzothiazocine structural motif **48** (Scheme 7.14) (Zhou, Chen, and Chen 2018).

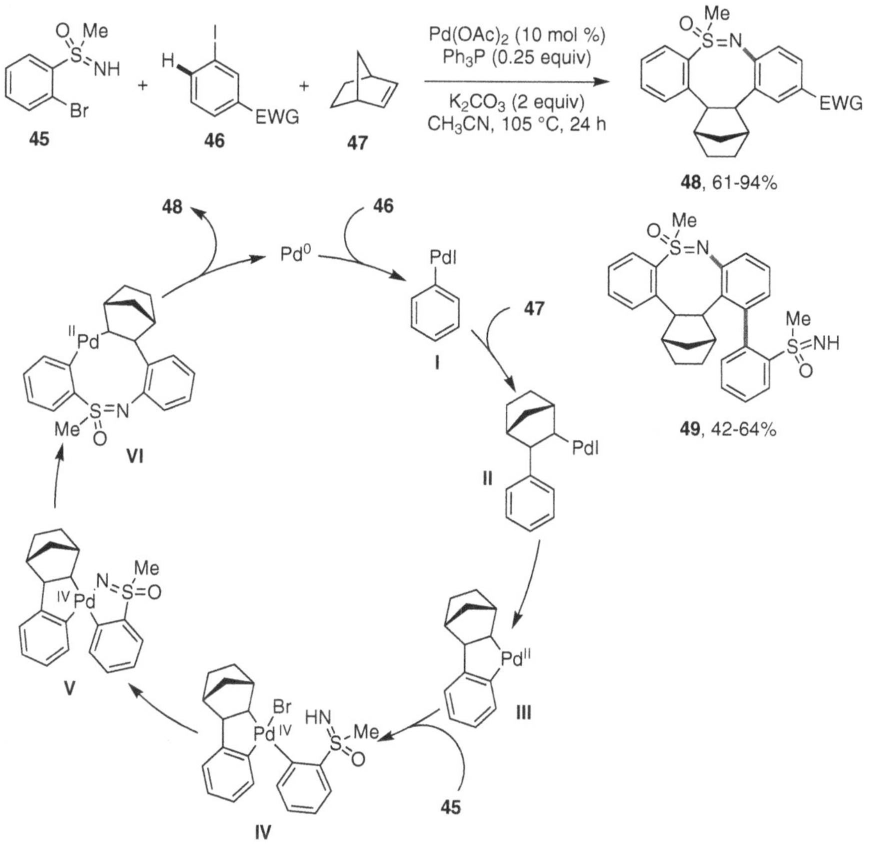

SCHEME 7.14

7.11 CONCLUSION

In summary, benzo-*N,O*-heterocycles, including benzothiazole, benzoisothiazole, benzothiazine, dibenzothiazine, benzothiazepine, and dibenzothiazepine derivatives were obtained by the aromatic C(sp^2)–H dehydrogenative coupling process. There are many different methods for the synthesis of these types of heterocycles in the literature, but because of the quantitative and one-step synthesis and a broad spectrum of substitution on the synthesized heterocycles, the direct functionalization of aromatic C–H bonds could be of interest in the synthesis of pharmaceutical, medicinal, and natural products.

REFERENCES

Alla, Santhosh Kumar, Pradeep Sadhu, and Tharmalingam Punniyamurthy. 2014. "Organocatalytic syntheses of benzoxazoles and benzothiazoles using aryl iodide and oxone via C–H functionalization and C–O/S bond formation." *The Journal of Organic Chemistry* no. 79(16):7502–7511.

Badshah, Syed, and Abdul Naeem. 2016. "Bioactive thiazine and benzothiazine derivatives: Green synthesis methods and their medicinal importance." *Molecules* no. 21(8):1054.

Baiget, Jessica, Annabel Cosp, Erik Gálvez, Loreto Gomez-Pinal, Pedro Romea, and Fèlix Urpí. 2008. "On the influence of chiral auxiliaries in the stereoselective cross-coupling reactions of titanium enolates and acetals." *Tetrahedron* no. 64(24):5637–5644.

Bheeter, Charles Beromeo, Jitendra K Bera, and Henri Doucet. 2012. "Palladium-catalysed intramolecular direct arylation of 2-bromobenzenesulfonic acid derivatives." *Advanced Synthesis & Catalysis* no. 354(18):3533–3538.

Cheng, Yannan, Jun Yang, Yue Qu, and Pixu Li. 2011. "Aerobic visible-light photoredox radical C–H functionalization: Catalytic synthesis of 2-substituted benzothiazoles." *Organic Letters* no. 14(1):98–101.

Cheng, Ying, and Carsten Bolm. 2015. "Regioselective syntheses of 1, 2-benzothiazines by rhodium-catalyzed annulation reactions." *Angewandte Chemie International Edition* no. 54(42):12349–12352.

Cheng, Ying, Wanrong Dong, Kanniyappan Parthasarathy, and Carsten Bolm. 2017. "Rhodium (III)-catalyzed ortho halogenations of N-acylsulfoximines and synthetic applications toward functionalized sulfoximine derivatives." *Organic Letters* no. 19(3):726–729.

Chinnagolla, Ravi Kiran, Arjun Vijeta, and Masilamani Jeganmohan. 2015. "Ruthenium- and palladium-catalyzed consecutive coupling and cyclization of aromatic sulfoximines with phenylboronic acids: An efficient route to dibenzothiazines." *Chemical Communications* no. 51(65):12992–12995.

Debnath, Sudarshan, and Shovan Mondal. 2018. "One-pot Sonogashira coupling, hydroamination of alkyne and intramolecular CH arylation reactions toward the synthesis of indole-fused benzosultams." *Tetrahedron Letters* no. 59(23):2260–2263.

Dong, Wanrong, Long Wang, Kanniyappan Parthasarathy, Fangfang Pan, and Carsten Bolm. 2013. "Rhodium-catalyzed oxidative annulation of sulfoximines and alkynes as an approach to 1,2-benzothiazines." *Angewandte Chemie International Edition* no. 52(44):11573–11576.

Eftekhari-Sis, Bagher, Maryam Zirak, and Ali Akbari. 2013. "Arylglyoxals in synthesis of heterocyclic compounds." *Chemical Reviews* no. 113(5):2958–3043.

Gandeepan, Parthasarathy, Thomas Müller, Daniel Zell, Gianpiero Cera, Svenja Warratz, and Lutz Ackermann. 2018. "3d transition metals for C–H activation." *Chemical Reviews* no. 119(4):2192–2452.

Huang, Jiapian, Yangfei Huang, Tao Wang, Qin Huang, Zhihua Wang, and Zhiyuan Chen. 2017. "Microwave-assisted Cp* CoIII-catalyzed C–H activation/double C–N bond formation reactions to thiadiazine 1-oxides." *Organic Letters* no. 19(5):1128–1131.

Jeon, Woo Hyung, Jeong-Yu Son, Ji Eun Kim, and Phil Ho Lee. 2016. "Synthesis of 1, 2-benzothiazines by a rhodium-catalyzed domino C–H activation/cyclization/elimination process from S-aryl sulfoximines and pyridotriazoles." *Organic Letters* no. 18(14):3498–3501.

Jeong, Eun Joo, Yoon Hyung Jo, Min Jung Jang, and So Won Youn. 2015. "Pd (II)-catalyzed ortho-hydroxylation and intramolecular oxidative C–C coupling of N-benzylbenzenesulfonamides." *Bulletin of the Korean Chemical Society* no. 36(2):453–454.

Kalsi, Deepti, and Basker Sundararaju. 2015. "Cobalt catalyzed C–H and N–H bond annulation of sulfonamide with terminal and internal alkynes." *Organic Letters* no. 17(24):6118–6121.

Laha, Joydev K, Neetu Dayal, Roli Jain, and Ketul Patel. 2014. "Palladium-catalyzed regiocontrolled domino synthesis of N-sulfonyl dihydrophenanthridines and dihydrodibenzo [c, e] azepines: Control over the formation of biaryl sultams in the intramolecular direct arylation." *The Journal of Organic Chemistry* no. 79(22):10899–10907.

Laha, Joydev K, Neetu Dayal, Krupal P Jethava, and Dilip V Prajapati. 2015. "Access to biaryl sulfonamides by palladium-catalyzed intramolecular oxidative coupling and subsequent nucleophilic ring opening of Heterobiaryl sultams with amines." *Organic Letters* no. 17(5):1296–1299.

Laha, Joydev K, Krupal P Jethava, and Neetu Dayal. 2014. "Palladium-catalyzed intramolecular oxidative coupling involving double C (sp^2)–H bonds for the synthesis of annulated biaryl Sultams." *The Journal of Organic Chemistry* no. 79(17):8010–8019.

Laha, Joydev K, Shubhra Sharma, and Neetu Dayal. 2015. "Palladium-catalyzed regio and chemoselective reactions of 2-bromobenzyl bromides: Expanding the scope for the synthesis of biaryls fused to a seven-membered sultam." *European Journal of Organic Chemistry* no. 2015(36):7885–7891.

Lan, Tianlong, Liguo Wang, and Yu Rao. 2017. "Regioselective annulation of aryl sulfonamides with allenes through cobalt-promoted C–H functionalization." *Organic Letters* no. 19(5):972–975.

Mahajan, Pankaj S, Vivek T Humne, Subhash D Tanpure, and Santosh B Mhaske. 2016. "Radical Beckmann rearrangement and its application in the formal total synthesis of antimalarial natural product isocryptolepine via C–H activation." *Organic Letters* no. 18(14):3450–3453.

Negwer, Martin. 1994. *Organic-Chemical Drugs and Their Synonym: (An International Syrvey).*

Pham, Manh V, Baihua Ye, and Nicolai Cramer. 2012. "Access to sultams by rhodium (III)-catalyzed directed C–H activation." *Angewandte Chemie* no. 124(42):10762–10766.

Qian, Xiang-Yang, Shu-Qi Li, Jinshuai Song, and Hai-Chao Xu. 2017. "TEMPO-catalyzed electrochemical C–H thiolation: Synthesis of benzothiazoles and thiazolopyridines from thioamides." *ACS Catalysis* no. 7(4):2730–2734.

Shaikh, Tanveer Mahamadali, and Fung-E Hong. 2016. "Recent developments in the preparation of N-heterocycles using Pd-catalysed C–H activation." *Journal of Organometallic Chemistry* no. 801:139–156.

Sharma, Shivani, Ramdas S Pathare, Antim K Maurya, Kandasamy Gopal, Tapta Kanchan Roy, Devesh M Sawant, and Ram T Pardasani. 2016. "Ruthenium catalyzed intramolecular C–S coupling reactions: Synthetic scope and mechanistic insight." *Organic Letters* no. 18(3):356–359.

Wen, Jian, Hanchao Cheng, Gerhard Raabe, and Carsten Bolm. 2017. "Rhodium-catalyzed [4+3] annulations of sulfoximines with α, β-unsaturated ketones leading to 1,2-benzothiazepine 1-oxides." *Organic Letters* no. 19(21):6020–6023.

Wen, Jian, Deo Prakash Tiwari, and Carsten Bolm. 2017. "1,2-benzothiazines from sulfoximines and allyl methyl carbonate by rhodium-catalyzed cross-coupling and oxidative cyclization." *Organic Letters* no. 19(7):1706–1709.

Xie, Weijia, Jie Yang, Baiquan Wang, and Bin Li. 2014. "Regioselective ortho olefination of aryl sulfonamide via rhodium-catalyzed direct C–H bond activation." *The Journal of Organic Chemistry* no. 79(17):8278–8287.

Yu, Da-Gang, Francisco de Azambuja, and frank Glorius. 2014. "α-MsO/TsO/Cl ketones as oxidized alkyne equivalents: Redox-neutral rhodium (III)-catalyzed C–H activation for the synthesis of N-heterocycles." *Angewandte Chemie International Edition* no. 53(10):2754–2758.

Yu, Hui, Mingdong Jiao, Ruohe Huang, and Xiaowei Fang. 2019 "Electrochemical intramolecular dehydrogenative coupling of N-benzyl (thio) amides: A direct and facile synthesis of 4H-1, 3-benzoxazines and 4H-1, 3-benzothiazines." *European Journal of Organic Chemistry* no. 2019 (10): 2004–2009.

Zeng, Meng-Tian, Wan Xu, Min Liu, Xing Liu, Cai-Zhu Chang, Hui Zhu, and Zhi-Bing Dong. 2017. "Iodobenzene-promoted Pd-catalysed ortho-directed C–H activation: The synthesis of benzothiazoles via intramolecular coupling." *SynOpen* no. 1(01):0001–0007.

Zhang, Guoting, Chao Liu, Hong Yi, Qingyuan Meng, Changliang Bian, Hong Chen, Jing-Xin Jian, Li-Zhu Wu, and Aiwen Lei. 2015. "External oxidant-free oxidative cross-coupling: A photoredox cobalt-catalyzed aromatic C–H thiolation for constructing C–S bonds." *Journal of the American Chemical Society* no. 137(29):9273–9280.

Zhou, Hao, Weihao Chen, and Zhiyuan Chen. 2018. "Pd/norbornene collaborative catalysis on the divergent preparation of heterocyclic sulfoximine frameworks." *Organic Letters* no. 20(9):2590–2594.

8 S- and S,O-Heterocycles

8.1 INTRODUCTION

Sulfur-containing heterocycles and their polycyclic derivatives comprise an important heterocyclic class that exhibit remarkable electrochemical and optical properties (Roncali 1992, 1997). Also, benzothiophenes and their conjugates have a wide range of applications in advanced materials such as conjugated polymers, semiconductors, and light-emitting devices (Huynh, Dittmer, and Alivisatos 2002). In addition, *S*-heterocyclic cores are found in some pharmaceuticals (Holmes et al. 1994) and display a wide range of biological activities (Al Nakib et al. 1990). Therefore, the development of new approaches for the preparation of *S*-heterocycles is of interest to synthetic and material scientists. Cross-dehydrogenative coupling of the aromatic C–H bond is one of the most powerful methods for the preparation of these types of heterocycles in a one-step and quantitative synthesis.

8.2 FIVE-MEMBERED *S*-HETEROCYCLES

8.2.1 Benzo[*b*]thiophene

Pd(II)-catalyzed intramolecular dehydrogenative coupling of phenethyl thioacetates **1** was developed. Reactions were catalyzed with $Pd(OAc)_2$ (10 mol%) in the presence of $P(t\text{-}Bu)_3 \cdot HBF_4$ (2 equiv), Ag_2CO_3 (2 equiv), and water in toluene at 120°C, leading to 2,3-dihydrobenzo[*b*]thiophene derivatives **2** in 32–80% yields. *Meta*-substituted arylethyl thioacetates afforded two regioisomers, favoring C–S bond formation at the less hindered *para* position. C3-substituted 2,3-dihydrobenzo[*b*]thiophenes **2** were also prepared in 61–77% yields under reaction conditions. *S*-(2-(thiophen-3-yl)ethyl) thioacetate afforded 2,3-dihydrothieno[2,3-*b*]thiophene **2a** in 40% yield. Biphenyl thioacetates provided dibenzothiophene derivatives in 79–90% yields. The proposed reaction mechanism involves the coordination of acetate-masked thiol **1** with the palladium, followed by metalation-deprotonation to activate the *ortho*-C–H bond. Reductive elimination with hydrolysis of the thioacetate group afforded the desired product **2** and generated a Pd(0) species, which was oxidized by silver to an active Pd(II) species to complete the catalytic cycle (Scheme 8.1) (Chen, Wang, and Jiang 2018).

8.2.2 Dibenzo[*b,d*]thiophene

Pd(II)-catalyzed intramolecular dehydrogenative sulfuration of 2-biphenylthiols **3** was described for the creation of dibenzo[*b,d*]thiophene derivatives **4**. The reactions were catalyzed with $PdCl_2$ (5 mol%) in dimethyl sulfoxide (DMSO) at 140°C under N_2 atmosphere, which gave rise to the formation of dibenzo[*b,d*]thiophenes **4** in 73–96% yields. Benzo[4,5]thieno[2,3-*c*]pyridine **4a** was obtained in 82% yield, starting from 2-(pyridin-4-yl)benzenethiol. Also, a highly twisted helical thienoacene, dinaphtho[2,1-*b*:1',2'-*d*]

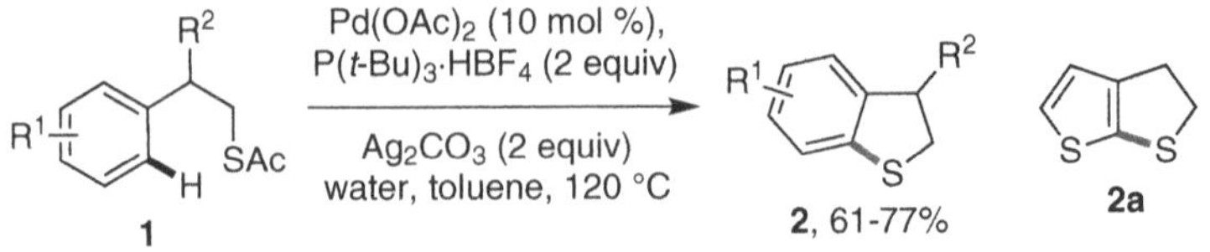

1

$Pd(OAc)_2$ (10 mol %), $P(t\text{-}Bu)_3 \cdot HBF_4$ (2 equiv)

Ag_2CO_3 (2 equiv) water, toluene, 120 °C

2, 61-77%

2a

SCHEME 8.1

thiophene **4b**, was prepared by a similar approach in 83% yield. 2,2'-(Thiophene-2,5-diyl)dibenzenethiol afforded dibenzo[*d,d'*]thieno[3,2-*b*;4,5-*b'*]dithiophene (DBTDT) **4c** in 41% yield under reaction conditions. An attempt to create trithiasumanene **4e** was undertaken using triphenylene-1,5,9-trithiol in the presence of $PdCl_2$ (90 mol%), giving trithiasumanene **4e** in only 2% yield, along with triphenyleno[1,12-*bcd*:4,5-*b'c'd'*]dithiophene **4d** in 5% yield. The proposed reaction mechanism involves ligand exchange between $PdCl_2$ and ArSH **3**, followed by *ortho*-C–H activation, with subsequent reductive elimination to afford the desired product **4** with generation of Pd(0) species, which were oxidized to the catalytically active Pd(II) species by in situ generated disulfide through the action of DMSO (Scheme 8.2) (Zhang et al. 2018).

Synthesis of dibenzothiophene **4** was achieved by Pd(II)-catalyzed cleavage of C–H and C–S bonds. Subjecting [1,1'-biphenyl]-2-yl(phenyl)sulfides **5** to $Pd(OAc)_2$ (15 mol%) and a $2,6\text{-}Me_2C_6H_3CO_2H$ ligand (45 mol%) in toluene at 130°C gave rise to the creation of dibenzothiophene derivatives **4** in 33–98% yields. However, in the presence of an oxidant (AgOAc), dual C–H bond activation and intramolecular arylation occurred to give 4-phenyldibenzo[*b,d*]thiophene, regioselectively. In the plausible reaction mechanism, ligand exchange of $Pd(OAc)_2$ with $2,6\text{-}Me_2C_6H_3CO_2H$ generated active $Pd(OCOAr)_2$ species, which reacted with [1,1'-biphenyl]-2-yl(phenyl)sulfide **5** to form palladacycle **I** via a sulfur-directed cyclometalation process. Reductive elimination of **I** gave ion pair **II** consisting of a dibenzosulfonium cation and an anionic Pd(0) fragment. Then, the oxidative addition of the Ph–S bond to the Pd(0) center led to the cleavage of the C–S bond to give complex **III**, which transformed into complex **IV** by release of benzene upon protonolysis with $2,6\text{-}Me_2C_6H_3CO_2H$, followed by delivery of the desired product **4** along with regeneration of active species $Pd(OCOAr)_2$ to complete the next catalytic cycles (Scheme 8.3) (Tobisu et al. 2016).

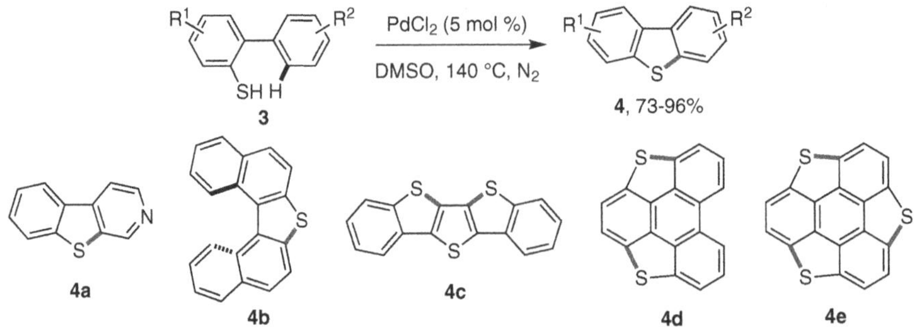

3

$PdCl_2$ (5 mol %)

DMSO, 140 °C, N_2

4, 73-96%

4a 4b 4c 4d 4e

SCHEME 8.2

Che et al. (2014) developed a Pd(II)-catalyzed intramolecular arylation of diaryl sulfides **6**. Reactions were conducted using $Pd(TFA)_2$ (10 mol%), K_2CO_3 (1 equiv), and AgOAc (4 equiv) in PivOH at 130°C to produce dibenzo[*b,d*]thiophene derivatives **4** in 48–88% yields. Reaction did not occur in the case of strongly electron-withdrawing group substituents. Synthesis of benzo[1,2-*b*:4,5-*b'*]bis[*b*]benzothiophene **4f** was successful from 1,4-bis(phenylthio)benzene substrate under reaction conditions (42% yield). Also, dibenzo[*b,d*]thiophene **4** scaffolds were prepared by Pd-catalyzed arylthiolation of electron-rich arenes with *N*-(2-bromophenylthio) succinimide, leading to (2-bromophenyl)(phenyl)sulfide, followed by intramolecular dehydrogenative arylation via C(sp^2)–H and C(sp^2)–Br cleavage, in 59–79% overall yields (Saravanan and Anbarasan 2014). In addition, Rh(III)-catalyzed sequential dehydrogenative coupling/deoxygenation of diaryl sulfoxides **7** was developed for the synthesis of dibenzothiophene derivatives **4**. The reactions were catalyzed with $RhCl_3{\cdot}3H_2O$ (5 mol%) and Ag_2CO_3 (2 equiv) in trifluoroacetic acid (TFA) under N_2 atmosphere, to afford dibenzothiophenes **4** in 24–88% yields. No reaction occurred in the case of electron-withdrawing substituted diaryl sulfoxides. The reaction was initiated by *ortho*-C–H activation with a subsequent second C–H activation followed by reductive elimination to give dibenzo[*b,d*]thiophene 5-oxides, which transformed into the desired dibenzo[*b,d*]thiophene **4** upon Rh-TFA-assisted deoxygenation (Scheme 8.4) (Huang et al. 2015).

An unexpected synthesis of dibenzothiophenes **9** was described by Pd-catalyzed dual C–H activation of benzyl aryl sulfoxides **8**, by treatment with $PdCl_2$ (15

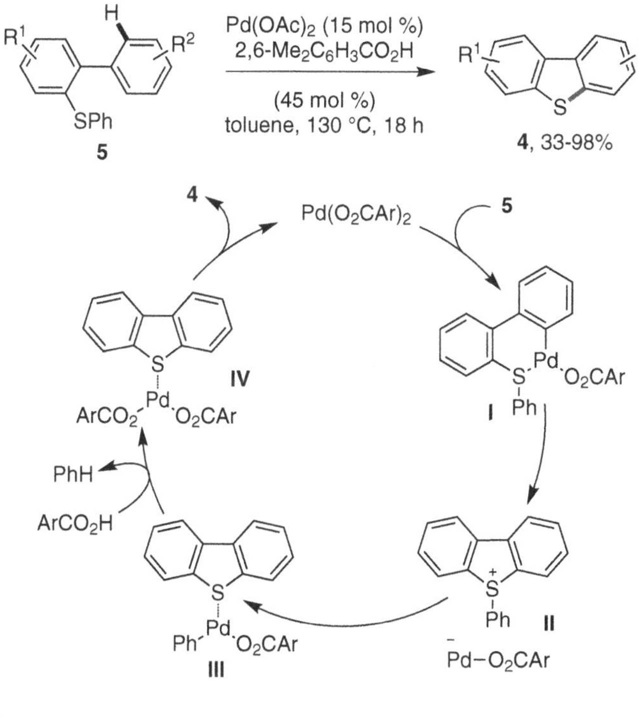

SCHEME 8.3

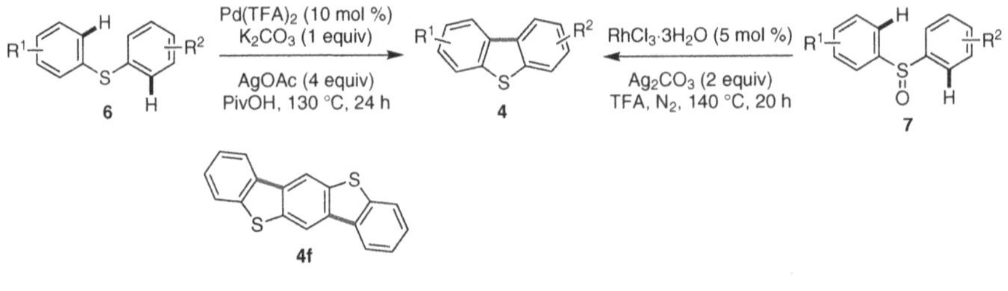

SCHEME 8.4

mol%), AgOAc (2 equiv), and *para*-fluoroiodobenzene (2 equiv) in AcOH at 110°C. Dibenzo[*b,d*]thiophene-1-carbaldehyde derivatives **9** were obtained in 45–78% yields. The reaction took place by sulfoxide-directed electrophilic attack of Pd(II) to give bimetallic species **I**, which underwent subsequent oxidative addition of aryl iodide to provide complex **II**. Reductive elimination and C–H activation generated palladacycle **III** and aryl acetate, which underwent reductive elimination to cyclic sulfoxide **IV** and Pd(0), which was reoxidized to Pd(II) by AgOAc. In the next cycle, sulfoxide **IV** was converted to mercaptoaldehyde **VI** through AgOAc and acetic-acid-promoted Pummerer rearrangement. Coordination of Pd(II) at the sulfur atom (**VII**) and subsequent C–H activation led to palladacycle **IX**. Reductive elimination of **IX** led to C–S bond formation to produce the desired dibenzothiophene **9** and Pd(0) species (Scheme 8.5) (Samanta and Antonchick 2011).

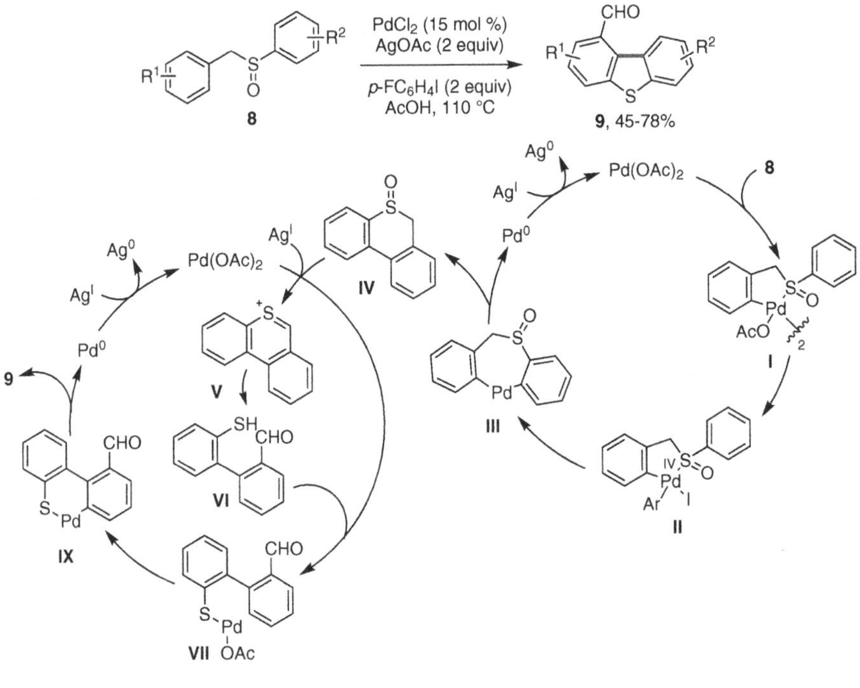

SCHEME 8.5

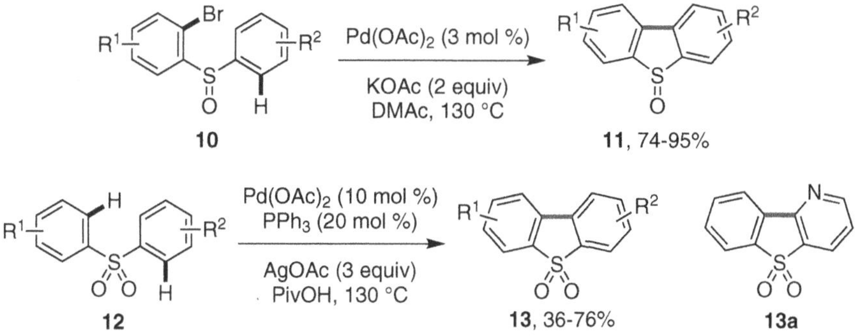

SCHEME 8.6

Wesch et al. (2013) developed a Pd-catalyzed intramolecular direct arylation of 2-bromo-diaryl sulfoxides **10** via C–H bond activation. Reactions were carried out using $Pd(OAc)_2$ (3 mol%) and KOAc (2 equiv) in *N,N*-dimethylacetamide (DMA) at 130°C, affording dibenzo[*b,d*]thiophene 5-oxides **11** in 74–95% yields. Also, dibenzo[*b,d*]thiophene 5-oxides **10** were accessed by a similar coupling route using $PdCl_2$ in the presence of AgOAc (Wang et al. 2014). Intramolecular dehydrogenative cyclization of biaryl sulfones **12** was catalyzed with $Pd(OAc)_2$ (10 mol%) in the presence of PPh_3 (20 mol%) and AgOAc (3 equiv) in PivOH at 130°C, resulting in the formation of dibenzo[*b,d*]thiophene 5,5-dioxide derivatives **13** in 36–76% yields. No reaction occurred when 2- and 4-(phenylsulfonyl)pyridines were subjected to $Pd(OAc)_2$ under reaction conditions, while 3-(phenylsulfonyl)pyridine afforded benzo[4,5]thieno[3,2-*b*]pyridine 5,5-dioxide **13a** in 76% yield, via exclusive C-2 dehydrogenative arylation of pyridine. The reaction of 3-(phenylsulfonyl)benzo[*b*] thiophene and 3-tosylindoles resulted in intramolecular dehydrogenative coupling to achieve the desired products in 89% and 81–83% yields, respectively (Scheme 8.6) (Laha and Sharma 2018).

8.2.3 Fused Thiophenes

2-Arylbenzo[4,5]thieno[3,2-*d*]thiazoles **15** were prepared by Cu(I)-catalyzed coupling of aryl ketoxime acetates **14** with aldehydes in the presence of elemental sulfur. Reactions were conducted in DMSO in the presence of CuBr (10 mol%) and Li_2CO_3 (50 mol%) at 120°C, leading to fused benzo[4,5]thieno[3,2-*d*]thiazoles **15** in 24–91% yields. Heteroaryl ketoxime acetates were also tolerated under reaction conditions, leading to heterocyclic-fused thieno[3,2-*d*]thiazoles **15a–d** in 35–96% yields. The reaction was initiated by Cu(I)-assisted formation of imino radical, which was transformed into intermediate **I**, with subsequent cyclocondensation with aldehyde and then *ortho*-C–H activation and sulfuration to achieve the final desired product (Scheme 8.7) (Huang et al. 2018).

Intramolecular enantioselective C–H arylation of *ortho*-bromo phenyl ferrocenyl sulfide **16** was performed with $Pd(OAc)_2$ (5 mol%) in the presence of Cs_2CO_3 (1.5 equiv), ligand **L2** (15 mol%), and PivOH (30 mol%) in toluene at 120°C for

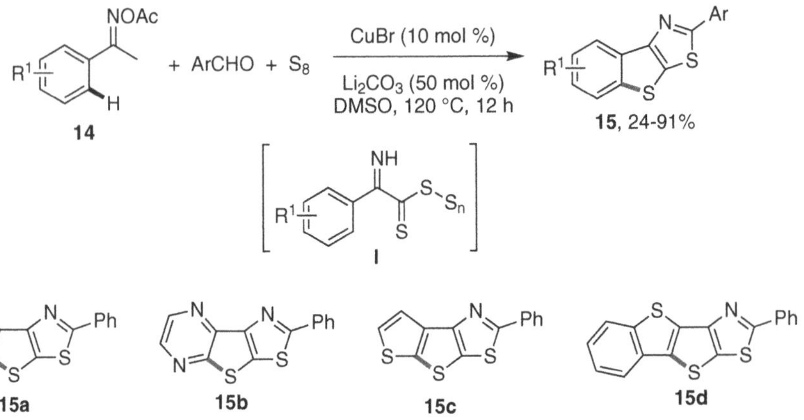

SCHEME 8.7

12 h. Corresponding benzothiophene-fused ferrocenes **17** were obtained in 74–99% yields, with 95–98% ee. Also, double arylation of ferrocene substrates containing a 2-bromoaryl sulfide moiety on each Cp ring was conducted under similar reaction conditions, leading to corresponding chiral bis-benzothiophene-fused ferrocenes **18** in 71–95% yields, with up to 99.5% ee and 15:1 diastereoselectivity (Scheme 8.8) (Xu et al. 2018).

8.3 SIX-MEMBERED *S*- AND *Se*-HETEROCYCLES

8.3.1 Thiochroman

The Pd(II)-catalyzed intramolecular dehydrogenative coupling of 2-((acetylthio) methyl)-2-benzylmalonates **19** was developed under $Pd(OAc)_2$ (10 mol%), P(*t*-Bu)$_3$·HBF_4 (2 equiv), and Ag_2CO_3 (2 equiv) catalytic system in AcOH at 120°C, resulting in the formation of thiochroman-3,3-dicarboxylate derivatives **20** in 49–90% yields. *Meta*-substituted aryl substrate gave two regioisomers, favoring C–S

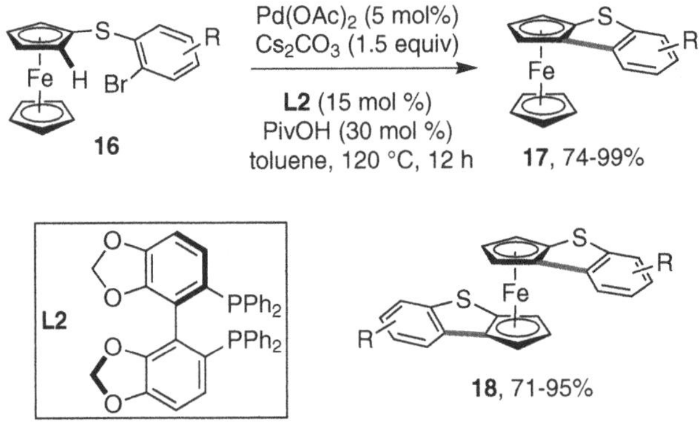

SCHEME 8.8

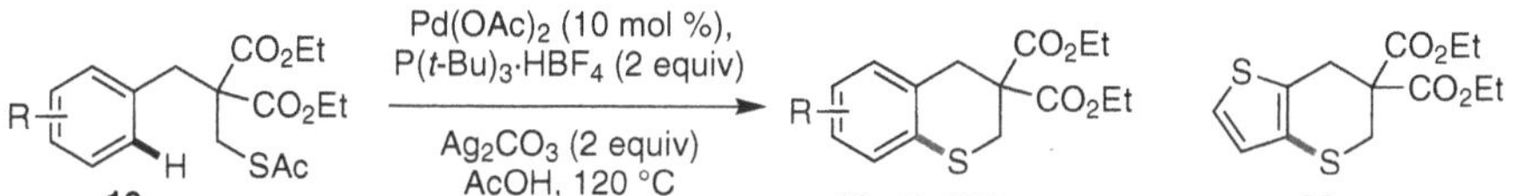

$Pd(OAc)_2$ (10 mol %), $P(t\text{-}Bu)_3{\cdot}HBF_4$ (2 equiv)
Ag_2CO_3 (2 equiv)
AcOH, 120 °C

19

20, 49-90%

20a

SCHEME 8.9

bond formation at the less hindered *para* position. 2-((Acetylthio)methyl)-2-(thio phen-2-ylmethyl)malonate delivered 5*H*-thieno[3,2-*b*]thiopyran-6,6(7*H*)-dicarboxylate **20a** in 80% yield, under reaction conditions (Scheme 8.9) (Chen, Wang, and Jiang 2018).

8.3.2 Benzo[c]thiochromene

Synthesis of 6*H*-benzo[*c*]thiochromene 5-oxide derivatives **21** was achieved by Pd(II)-catalyzed dual C(sp^2)–H bond dehydrogenative cyclization. Treatment of benzyl aryl sulfoxides **8** with $PdCl_2$ (10 mol%), AgOAc (4 equiv), PhI (2 equiv), and AcOH (4 equiv) in 1,1,2,2-tetrachloroethane at 100°C delivered 6*H*-benzo[*c*]thiochromene 5-oxides **21** in 30–82% yields. 1-(*p*-Tolylsulfinyl)naphthalene, 9-(*p*-tolylsulfinyl)phenanthrene and 4-(*p*-tolylsulfinyl)pyrene were also compatible with the reaction, leading to corresponding sulfur-based polycyclic compounds in 51%, 45%, and 61% yields, respectively (Wang et al. 2014). One example of 6*H*-benzo[*c*]thiochromene 5,5-dioxide **23** was synthesized by $PdCl(C_3H_5)$(dppb)- (1 mol%) catalyzed intramolecular C(sp^2)–H/C(sp^2)–Br coupling of 1-(benzylsulfonyl)-2-bromobenzene **22** in only 20% yield (Scheme 8.10) (Bheeter, Bera, and Doucet 2012).

8.3.3 Thioxanthene and Thioxanthene-9-one

Cu(II)-catalyzed intramolecular dehydrogenative C(sp^2)–H/C(sp^3)–H coupling of diethyl 2-(2-(phenylthio)phenyl)malonate **24** was developed to deliver diethyl 9*H*-thioxanthene-9,9-dicarboxylate **25**. Reactions were conducted using

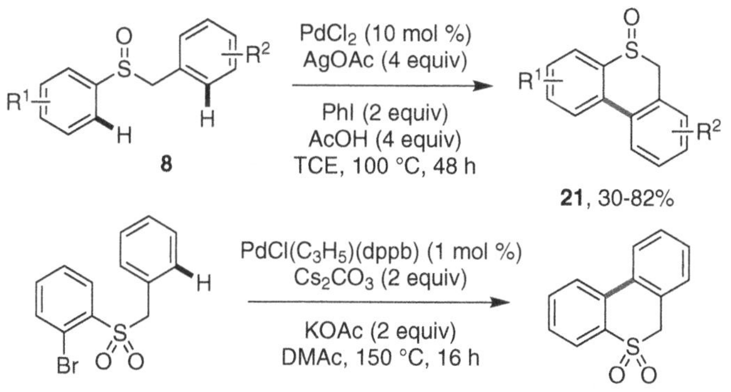

$PdCl_2$ (10 mol %)
AgOAc (4 equiv)
PhI (2 equiv)
AcOH (4 equiv)
TCE, 100 °C, 48 h

8

21, 30-82%

$PdCl(C_3H_5)$(dppb) (1 mol %)
Cs_2CO_3 (2 equiv)
KOAc (2 equiv)
DMAc, 150 °C, 16 h

22

23, 20%

SCHEME 8.10

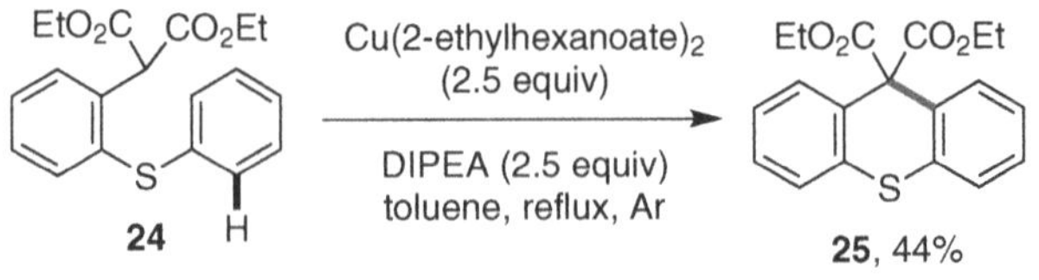

EtO_2C CO_2Et S H **24** — $Cu(2\text{-ethylhexanoate})_2$ (2.5 equiv), DIPEA (2.5 equiv), toluene, reflux, Ar → EtO_2C CO_2Et S **25**, 44%

SCHEME 8.11

$Cu(2\text{-ethylhexanoate})_2$ (2.5 equiv) and *N,N*-diisopropylethylamine (DIPEA) (2.5 equiv) in refluxing toluene under an atmosphere of argon, leading to the desired product **25** in 44% yield (Scheme 8.11) (Hurst and Taylor 2017).

Ru-catalyzed intramolecular cross-dehydrogenative coupling of 2-(phenylthio) benzaldehyde **26** to the corresponding 9*H*-thioxanthen-9-one **27** was carried out using $[RuCl_2(p\text{-cymene})_2]_2$ (10 mol%) in the presence of *tert*-butyl hydroperoxide (TBHP) in CH_3CN at 120°C, in 58% yield. 5*H*-Benzo[*c*]thioxanthen-7(6*H*)-one was obtained in 67% yield, when 1-(phenylthio)-3,4-dihydronaphthalene-2-carbaldehyde was treated via reaction. The reaction of 4-(phenylthio)-2*H*-chromene-3-carbaldehyde **28** afforded thiochromeno[3,2-*c*]chromen-7(6*H*)-one **29** in 67% yield, under the same reaction conditions. The plausible reaction mechanism involves the *S*-directed activation of the aldehyde C–H bond to generate an Ru(II)-H intermediate, which underwent oxidation with TBHP and *ortho*-C–H activation. Reductive elimination afforded the desired product (Manna, Manda, and Panda 2014). Also, the synthesis of pyrazole-fused thioxanthenone, 1-phenylthiochromeno[2,3-*c*]pyrazol-4(1*H*)-one **31**, was accomplished by an intramolecular dehydrogenative coupling process (Scheme 8.12) (Li et al. 2015).

8.3.4 Selenoxanthones

Ru(II)-catalyzed *ortho*-C–H chalcogenation of benzoic acids **32** with consecutive intramolecular cyclization was developed for the creation of selenoxanthones **33**. Reactions were performed in two steps, by treating benzoic acid **32** with diphenyl diselenide (2 equiv) in the presence of $[RuCl_2(p\text{-cymene})]_2$ (4 mol%), PCy_3 (8 mol%), and $NaHCO_3$ (1 equiv) in *N,N*-dimethylformamide (DMF) at 100°C, followed by subjection to TfOH at 100°C, giving related selenoxanthones **33** in 58–86% yields. 9*H*-Selenochromeno[3,2-*b*]thiophen-9-one and 9*H*-thieno[3,2-*b*]thiochromen-9-one **33a** were also obtained from the reaction of thiophene-2-carboxylic acid with diphenyl diselenide and diphenyl disulfide in 81% and 70% yields, respectively (Scheme 8.13) (Mandal et al. 2017).

8.4 SEVEN-MEMBERED *S*-HETEROCYCLES

The phenyliodine(III) bis(trifluoroacetate) (PIFA)-induced intramolecular biaryl coupling reaction of dibenzyl sulfoxides **34** to 5,7-dihydrodibenzo[*c,e*]thiepine 6-oxide **35** was reported in the presence of $BF_3{\cdot}Et_2O$ in dichloromethane (DCM) at −40°C. 5,7-Dihydrodibenzo[*c,e*]thiepine 6-oxides **35** were obtained in 42–73% yields. Also, 5,7-dihydrodibenzo[*c,e*]thiepine 6,6-dioxides were accessed by intramolecular

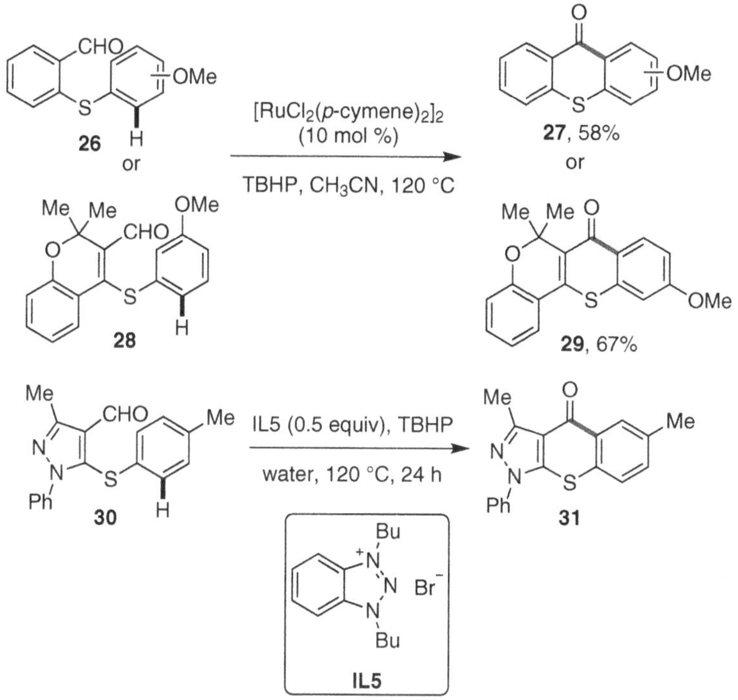

SCHEME 8.12

biaryl coupling of dibenzyl sulfones under the same reaction conditions, in 72–78% yields (Takada et al. 1998). Synthesis of 6,7-dihydrodibenzo[*b,d*]thiepine 5-oxide derivatives **37** was achieved by Pd(II)-catalyzed biaryl coupling via two C(sp^2)–H bond activation. By heating aryl phenethyl sulfoxides **36** in 1,1,2,2-tetrachloroethane at 100°C in the presence of $PdCl_2$ (10 mol%), AgOAc (4 equiv), PhI (2 equiv), and AcOH (4 equiv), 6,7-dihydrodibenzo[*b,d*]thiepine 5-oxides **37** were obtained in 37–67% yields (Scheme 8.14) (Wang et al. 2014).

8.5 *S,O*-HETEROCYCLES

Pd(II)-catalyzed dual *ortho*-C–H bond activation of aryl thiocarbamates **38** was developed to create 3-(2-oxobenzo[*d*][1,3]oxathiol-7-yl)acrylates **40**. Reactions were carried out by subjecting *O*-aryl thiocarbamates **38** with acrylates **39** (5 equiv)

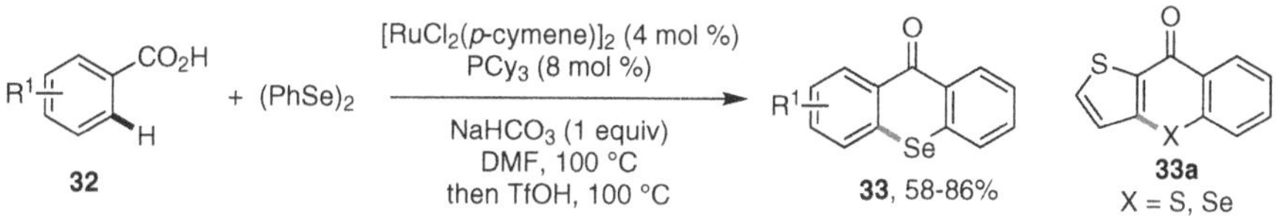

SCHEME 8.13

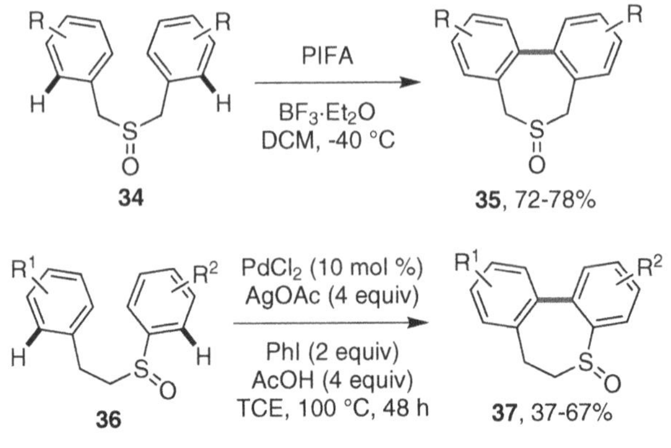

SCHEME 8.14

in the presence of $Pd(OAc)_2$ (10 mol%) and 1,4-benzoquinone (BQ) (2 equiv) as an oxidant, in AcOH/hexafluoroisopropanol (HFIP) under air atmosphere, to give 3-(2-oxobenzo[*d*][1,3]oxathiol-7-yl)acrylate derivatives **40** in 32–75% yields. The reaction of *O*-(2-naphthyl)thiocarbamate with *n*-butyl acrylate led to the formation of 3-(2-oxonaphtho[1,2-*d*][1,3]oxathiol-4-yl)acrylate in 53% yield, regioselectively, favoring sulfuration at the α-position. 7-Styrylbenzo[*d*][1,3]oxathiol-2-one was obtained in 28% yield, when phenyl thiocarbamate **38** was reacted with styrene under reaction conditions. The reaction proceeded by thiocarbamate-directed Pd-catalyzed *ortho*-C–H alkenylation to generate intermediate **I**, followed by a further S-atom-directed *ortho*-C–H activation and sulfuration by reductive elimination of the six-membered palladacycle **II**, followed by hydrolysis (Scheme 8.15) (Li et al. 2018).

Dibenzo[*c*,*e*][1,2]oxathiine 6,6-dioxide derivatives **42** were synthesized by Pd(II)-catalyzed intramolecular direct arylation of phenyl 2-bromobenzenesulfonate derivatives **41**. Reactions were carried out by subjecting phenyl 2-bromobenzenesulfonates **41** to $Pd(OAc)_2$ (1 mol%) and KOAc (2 equiv) in DMA at 150°C under argon, to furnish dibenzo[*c*,*e*][1,2]oxathiine 6,6-dioxides **42** in 62–96% yield. No

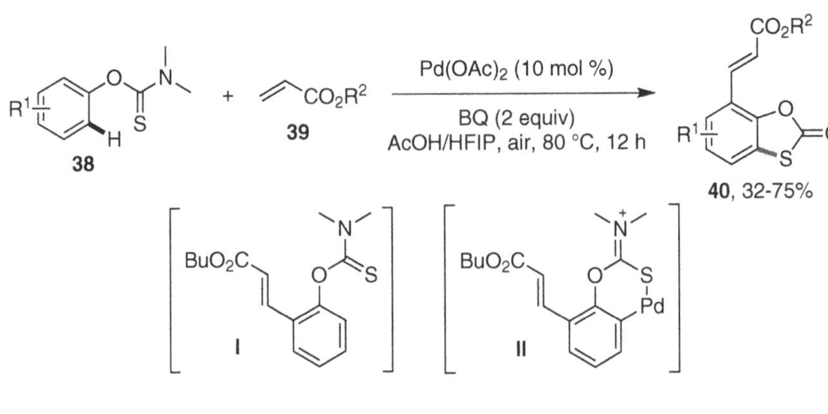

SCHEME 8.15

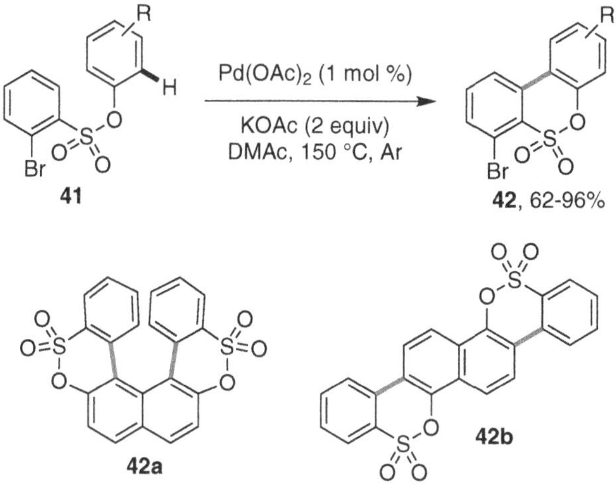

SCHEME 8.16

reaction occurred in the case of electron-withdrawing substituted aryl 2-bromobenzenesulfonates (Bheeter, Bera, and Doucet 2012). Similar scaffolds were prepared by $Pd(PPh_3)_4$-catalyzed intramolecular dehydrogenative coupling of 2-bromoaryl and bromonaphthyl benzenesulfonates. 1,8-Dibromonaphthalene-2,7-diyl dibenzenesulfonate and 2,6-dibromonaphthalene-1,5-diyl dibenzenesulfonate provided corresponding bis-annulated products **42a** and **42b** in 79% and 78% yields, respectively (Scheme 8.16) (Majumdar, Mondal, and Ghosh 2009).

8.6 CONCLUSION

In summary, the aromatic Csp^2–H dehydrogenative coupling process may be used for the synthesis of sulfur-containing heterocyclic compounds, including benzothiophene, dibenzothiophene, thiochromane, benzothiochromene, thioxanthone, and dibenzothiepine derivatives. Seleno-heterocycles were also produced by the aromatic dehydrogenative coupling process. In addition, a few *S,O*-heterocycles were synthesized. However, there are different approaches to the construction of these types of heterocycles, due to the quantitative and one-step synthesis, the aromatic C–H bond direct functionalization approach could be of interest in the synthesis of pharmaceutical, medicinal, and natural products.

REFERENCES

Al Nakib, T, V Bezjak, MJ Meegan, and R Chandy. 1990. "Synthesis and antifungal activity of some 3-benzylidenechroman-4-ones, 3-benzylidenethiochroman-4-ones and 2-benzylidene-1-tetralones." *European Journal of Medicinal Chemistry* no. 25(5):455–462.

Bheeter, Charles Beromeo, Jitendra K Bera, and Henri Doucet. 2012. "Palladium-catalysed intramolecular direct arylation of 2-bromobenzenesulfonic acid derivatives." *Advanced Synthesis & Catalysis* no. 354(18):3533–3538.

Che, Rui, Zhiqing Wu, Zhengkai Li, Haifeng Xiang, and Xiangge Zhou. 2014. "Synthesis of dibenzothiophenes by Pd-catalyzed dual C-H activation from diaryl sulfides." *Chemistry – A European Journal* no. 20(24):7258–7261.

Chen, Shihao, Ming Wang, and Xuefeng Jiang. 2018. "Pd-Catalyzed C–S Cyclization via C–H functionalization strategy: Access to sulfur-containing benzoheterocyclics." *Chinese Journal of Chemistry* no. 36(10):921–924.

Holmes, Judy M, Gary CM Lee, Mercy Wijono, Robert Weinkam, Larry A Wheeler, and Michael E Garst. 1994. "Synthesis and carbonic anhydrase Inhibitory activity of 4-substituted 2-thiophenesulfonamides." *Journal of Medicinal Chemistry* no. 37(11):1646–1651.

Huang, Huawen, Zhenhua Xu, Xiaochen Ji, Bin Li, and Guo-Jun Deng. 2018. "Thiophene-fused heteroaromatic systems enabled by internal oxidant-induced cascade bis-heteroannulation." *Organic Letters* no. 20(16):4917–4920.

Huang, Qiufeng, Shurong Fu, Shaojia Ke, Hanbing Xiao, Xiaofeng Zhang, and Shen Lin. 2015. "Rhodium-catalyzed sequential dehydrogenation/deoxygenation in one-pot: Efficient synthesis of dibenzothiophene derivatives from diaryl sulfoxides." *European Journal of Organic Chemistry* no. 2015(30):6602–6605.

Hurst, Timothy E, and Richard JK Taylor. 2017. "A Cu-catalysed radical cross-dehydrogenative coupling approach to acridanes and related heterocycles." *European Journal of Organic Chemistry* no. 2017(1):203–207.

Huynh, Wendy U, Janke J Dittmer, and A Paul Alivisatos. 2002. "Hybrid nanorod-polymer solar cells." *Science* no. 295(5564):2425–2427.

Laha, Joydev K, and Shubhra Sharma. 2018. "Palladium-catalyzed intramolecular oxidative arylations for the synthesis of fused biaryl sulfones." *ACS Omega* no. 3(5):4860–4870.

Li, He, Chenjiang Liu, Yonghong Zhang, Yadong Sun, Bin Wang, and Wenbo Liu. 2015. "Green method for the synthesis of chromeno [2, 3-c] pyrazol-4 (1 H)-ones through ionic liquid promoted directed annulation of 5-(aryloxy)-1 H-pyrazole-4-carbaldehydes in aqueous media." *Organic Letters* no. 17(4):932–935.

Li, Wendong, Yingwei Zhao, Shaoyu Mai, and Qiuling Song. 2018. "Thiocarbamate-directed tandem olefination–intramolecular sulfuration of two ortho C–H bonds: Application to synthesis of a COX-2 inhibitor." *Organic Letters* no. 20(4):1162–1166.

Majumdar, KC, Shovan Mondal, and Debankan Ghosh. 2009. "Synthesis of tricyclic and tetracyclic sultones by Pd-catalyzed intramolecular cyclization." *Tetrahedron Letters* no. 50(33):4781–4784.

Mandal, Anup, Suman Dana, Harekrishna Sahoo, Gowri Sankar Grandhi, and Mahiuddin Baidya. 2017. "Ruthenium (II)-catalyzed ortho-C–H chalcogenation of benzoic acids via weak O-coordination: Synthesis of chalcogenoxanthones." *Organic Letters* no. 19(9):2430–2433.

Manna, Sudipta Kumar, Srinivas Lavanya Kumar Manda, and Gautam Panda. 2014. "[RuCl2 (p-cymene) 2] 2 catalyzed cross dehydrogenative coupling (CDC) toward xanthone and fluorenone analogs through intramolecular C–H bond functionalization reaction." *Tetrahedron Letters* no. 55(42):5759–5763.

Roncali, Jean. 1992. "Conjugated poly (thiophenes): Synthesis, functionalization, and applications." *Chemical Reviews* no. 92(4):711–738.

Roncali, Jean. 1997. "Synthetic principles for bandgap control in linear π-conjugated systems." *Chemical Reviews* no. 97(1):173–206.

Samanta, Rajarshi, and Andrey P Antonchick. 2011. "Palladium-catalyzed double C–H activation directed by sulfoxides in the synthesis of dibenzothiophenes." *Angewandte Chemie International Edition* no. 50(22):5217–5220.

Saravanan, Perumal, and Pazhamalai Anbarasan. 2014. "Palladium catalyzed aryl (alkyl) thiolation of unactivated arenes." *Organic Letters* no. 16(3):848–851.

Takada, Takeshi, Mitsuhiro Arisawa, Michiyo Gyoten, Ryuji Hamada, Hirofumi Tohma, and Yasuyuki Kita. 1998. "Oxidative biaryl coupling reaction of phenol ether derivatives using a hypervalent iodine (III) reagent." *The Journal of Organic Chemistry* no. 63(22):7698–7706.

Tobisu, Mamoru, Yoshihiro Masuya, Katsuaki Baba, and Naoto Chatani. 2016. "Palladium (II)-catalyzed synthesis of dibenzothiophene derivatives via the cleavage of carbon–sulfur and carbon–hydrogen bonds." *Chemical Science* no. 7(4):2587–2591.

Wang, Binjie, Yue Liu, Cong Lin, Yiming Xu, Zhanxiang Liu, and Yuhong Zhang. 2014. "Synthesis of sulfur-bridged polycycles via Pd-catalyzed dehydrogenative cyclization." *Organic Letters* no. 16(17):4574–4577.

Wesch, Thomas, Anaïs Berthelot-Bréhier, Frederic R Leroux, and Francoise Colobert. 2013. "Palladium-catalyzed intramolecular direct arylation of 2-bromo-diaryl sulfoxides via C–H bond activation." *Organic Letters* no. 15(10):2490–2493.

Xu, Bing-Bin, Jie Ye, Yu Yuan, and Wei-Liang Duan. 2018. "Palladium-catalyzed asymmetric C–H arylation for the synthesis of planar chiral benzothiophene-fused ferrocenes." *ACS Catalysis* no. 8(12):11735–11740.

Zhang, Tao, Guigang Deng, Hanjie Li, Bingxin Liu, Qitao Tan, and Bin Xu. 2018. "Cyclization of 2-biphenylthiols to dibenzothiophenes under PdCl2/DMSO catalysis." *Organic Letters* no. 20(17):5439–5443.

9 *P*- and *Si*-Heterocycles

9.1 INTRODUCTION

Phosphorus heterocycles are promising building blocks existing in numerous functional materials (Baumgartner and Réau 2006, Hirai et al. 2019), such as luminogenic molecules (Zhen et al. 2018), optoelectronics (Wu et al. 2019), semiconductors, and bioimaging probes (Tsuji et al. 2009). Also, they are used as synthetic intermediates in organic synthesis and medicinal chemistry (Börner 2008, Mathey 2004). In addition, *Si*-heterocycles have gained much interest in recent years because of their applications in light-emitting materials, thin-film transistors, host materials for electroluminescent devices, and solar cells (Lu et al. 2008).

Several reports exist in the literature about the synthesis of *P*- and *Si*-heterocyclic compounds; among them, the dehydrogenative coupling process has also been developed. However, there is still room for further studies in the field of synthesis of *P*- and *Si*-containing heterocyclic compounds via aromatic C–H dehydrogenative coupling reactions.

9.2 *P*-HETEROCYCLES

9.2.1 Phosphindol

Cu-mediated C–H functionalization and annulation of trisubstituted phosphines **1** with alkynes **2** was developed to access phosphindolium salts **3**. Reactions were carried out using 1.2 equiv of an alkyne **2** in the presence of $Cu(BF_4)_2 \cdot 6H_2O$ (2 equiv), in CH_3CN at 100°C, under Ar atmosphere, delivering phosphindolium salts **3** in 54–85% yields. Aliphatic internal alkynes **2** were also compatible with the reaction, leading to corresponding annulated products **3** in 68–71% yields. The reaction with unsymmetrical alkyl aryl alkynes afforded 2-alkyl-3-phenyl-1*H*-phosphindol-1-iums **3** in 64–69% yields, regioselectively. Naphthalen-1-yldiphenylphosphane and anthracen-9-yldiphenylphosphane **4** resulted in formation of the six-membered products benzo- and dibenzophosphinolin-1-iums **5** in 62–92% and 91% yields, respectively. The proposed reaction mechanism involves the generation of a phosphorus-centered radical cation through a single electron transfer (SET) process between the Cu(II) salt and the trisubstituted phosphine. Subsequently, the addition to the triple bond of an alkyne **2** gave the vinyl radical **I**, which underwent intramolecular cyclization to furnish the cyclic radial intermediate **II**. By SET from **II** to Cu(II), followed by deprotonation, the desired product **3** was produced. Also, biphenyl-substituted phosphanes **6** were subjected to a Cu(II) catalyst in the absence of alkynes, which furnished the corresponding 5*H*-benzo[*b*]phosphindol-5-iums **7** in 83–96% yields, via intramolecular P–C cross-coupling through the cleavage of the C–H bond (Scheme 9.1) (Ge et al. 2017).

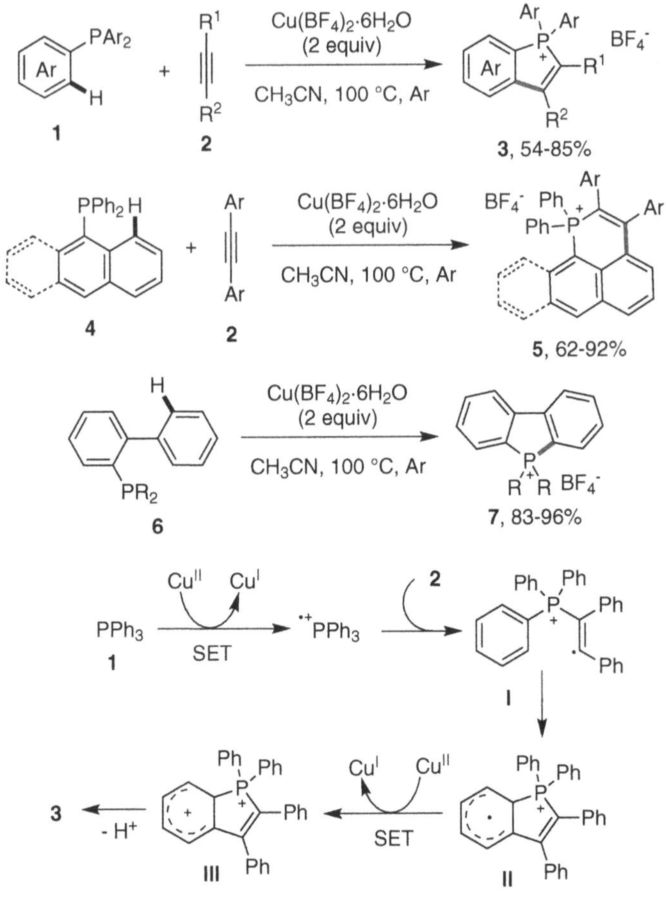

SCHEME 9.1

Dehydrogenative cross-coupling of diarylphosphine oxides **8** with alkynes **2** was studied using different catalytic systems. Unoh et al. (2013) reported AgOAc- (2 equiv) mediated annulation of diphenylphosphine oxides **8** with internal alkynes **2** in *N,N*-dimethylformamide (DMF) at 100°C under an atmosphere of N_2, giving 2,3-disubstituted 1-phenyl-1*H*-phosphindole 1-oxides **9** in 35–89% yields. Reaction with 4-octyne furnished the corresponding 2,3-dipropyl phosphindole 1-oxide **9** in 63% yield. Reaction with unsymmetrical phenyl acetylenes afforded 2-substituted 3-phenylphosphindole 1-oxides in 41–86% yields. 1,4-Di(prop-1-yn-1-yl)benzene afforded 3,3'-(1,4-phenylene)bis(2-methyl-1-phenyl-1*H*-phosphindole 1-oxide) **9b** in 38% yield. Alkyl phenylphosphine oxides were also tolerated under reaction conditions, delivering corresponding phosphindole 1-oxides in 53–66% yields. When substituted diarylphosphine oxides were subjected to diphenyl acetylene under reaction conditions, in addition to the expected annulation products **9a**, other unexpected regioisomers **9b** were generated. The reaction was initiated by Ag(I)-salt-promoted generation of a *P*-centered radical from **I** through homolytic P–H bond cleavage, followed by addition of the radical across the triple bond of an alkyne **2**, with subsequent cyclization of the phenyl moiety (**V**), then SET to Ag(I) and deprotonation to form benzophosphole oxide **9a**. The formation of the unexpected regioisomer **9b** could be rationalized by the attack of alkenyl radical **III** on the *ipso*-carbon atom of the ring to generate the spirocyclohexadienyl radical **VI**, which underwent ring expansion (**VII**), with subsequent annulation (**IX**). Phosphindole 1-sulfide **9c** was also prepared when diphenylphosphine sulfide was reacted with $Mn(OAc)_3$ instead of AgOAc (Scheme 9.2). Similar results were reported with Ag(I)- (Chen and Duan 2013) and Cu(II)/*tert*-butyl hydroperoxide (TBHP)- (Zhang et al. 2016) mediated dehydrogenative coupling of diarylphosphine oxides with alkynes. Also, $K_2S_2O_8$ as radical initiator was used for direct P–H/C–H functionalization to provide benzo[*b*]phosphole 1-oxides from annulation of diarylphosphine oxides with alkynes (Ma et al. 2016).

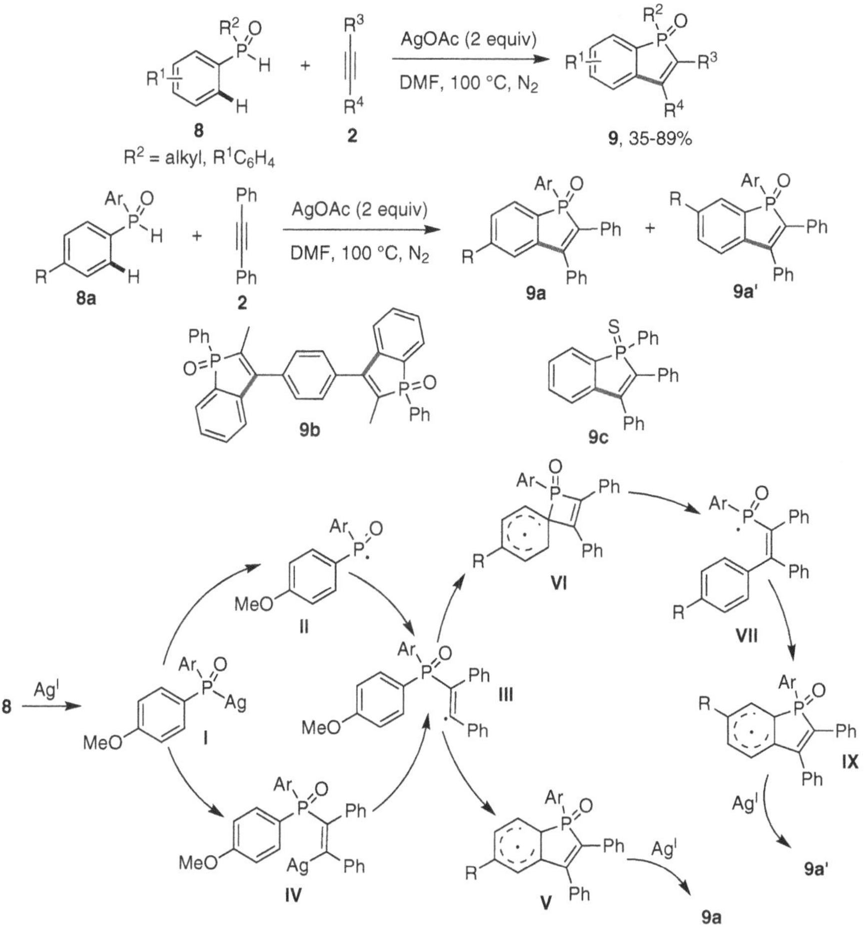

R² = alkyl, R¹C₆H₄

AgOAc (2 equiv)

DMF, 100 °C, N₂

8 **2** **9**, 35-89%

8a **2** **9a** **9a'** **9b** **9c**

I **II** **III** **IV** **V** **VI** **VII** **IX**

SCHEME 9.2

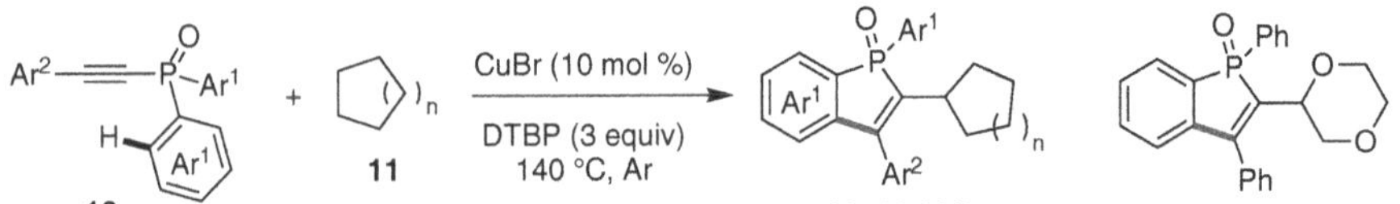

SCHEME 9.3

Synthesis of 2-cycloalkyl substituted 3-aryl-1*H*-phosphindole 1-oxide derivatives **12** was achieved by sequential C–H functionalization along with two new C–C bond formations through Cu-catalyzed radical addition/dehydrogenative cyclization of unactivated cycloalkanes **11** with diaryl(arylethynyl)-phosphine oxides **10**. Reactions were carried out by heating a solution of diaryl(arylethynyl)-phosphine oxides **10** in cycloalkane **11** in the presence of CuBr (10 mol%) and di-*tert*-butyl peroxide (DTBP) (3 equiv) at 140°C under Ar atmosphere, to give phosphindole 1-oxides **12** in 42–90% yields. An acyclic alkane, *n*-hexane, was also investigated, leading to corresponding three regioisomers 2-(1-hexyl)-, 2-(2-hexyl)-, and 2-(3-hexyl)phosphindole 1-oxides in 70% yield, in a ratio of 1:2:2. Dioxane as an alkane substrate gave the corresponding 2-(1,4-dioxan-2-yl)phosphindole 1-oxide **13** in 52% yield. The proposed reaction mechanism involves the selective addition of an alkyl radical, produced via H abstraction by a Cu(I)-promoted in-situ-generated *tert*-butoxyl radical from DTBP, to the α-position of the triple bond in **10** to afford the corresponding alkenyl radical, which underwent two possible pathways, direct cyclization or rearrangement with subsequent cyclization pathways, to offer the desired product **12** (Scheme 9.3) (Ma et al. 2018).

9.2.2 Benzophosphindole

Pd(II)-catalyzed synthesis of dibenzophosphole oxides **15** was developed by intramolecular dehydrogenative coupling reaction of secondary hydrophosphine oxide **14** bearing a biphenyl group. Reactions were performed using 5 mol% $Pd(OAc)_2$ in tetrahydrofuran (THF) at 65°C, affording dibenzophosphole oxide derivatives **15** in 61–94% yields. Phenyl[2-(thiophene-3-yl)phenyl]phosphine oxide gave a mixture of two regioisomers, phosphindolo[2,3-*b*]thiophene 8-oxide **15a** and phosphindolo[2,3-*c*]thiophene 4-oxide **15b,** in a ratio of 57:25. The reaction proposed proceeds by P–H bond activation via elimination of AcOH followed by sequential C–H bond activation, with subsequent reductive elimination to give dibenzophosphole oxide **15** and Pd(0) species. Alternatively, oxidative addition of generated Pd(0) species into the P–H bond followed by sequential C–H bond activation by σ-bond metathesis, followed by reductive elimination, also gave dibenzophosphole oxide **15** (Kuninobu, Yoshida, and Takai 2011). Also, dibenzophosphole oxides **15** were accessed by $Pd(OAc)_2$-catalyzed intramolecular dehydrogenative cyclization of biarylphosphine **16** to 5-phenyl-5*H*-benzo[*b*]phosphindoles **III**, which underwent oxidation to benzophosphindole oxides **15** under air or in the presence of H_2O_2 solution at room temperature in 54–96% overall yields. Heterocycle-fused phosphole oxides, phosphindolo[3,2-*b*]furan 4-oxide **15c**, benzo[*d*]pyrrolo[1,2-*a*][1,3]azaphosphole 9-oxide **15d**, and phosphindolo[2,3-*c*]pyridine 9-oxide **15e**, were also synthesized in 54%, 58%, and 85% yields, respectively.

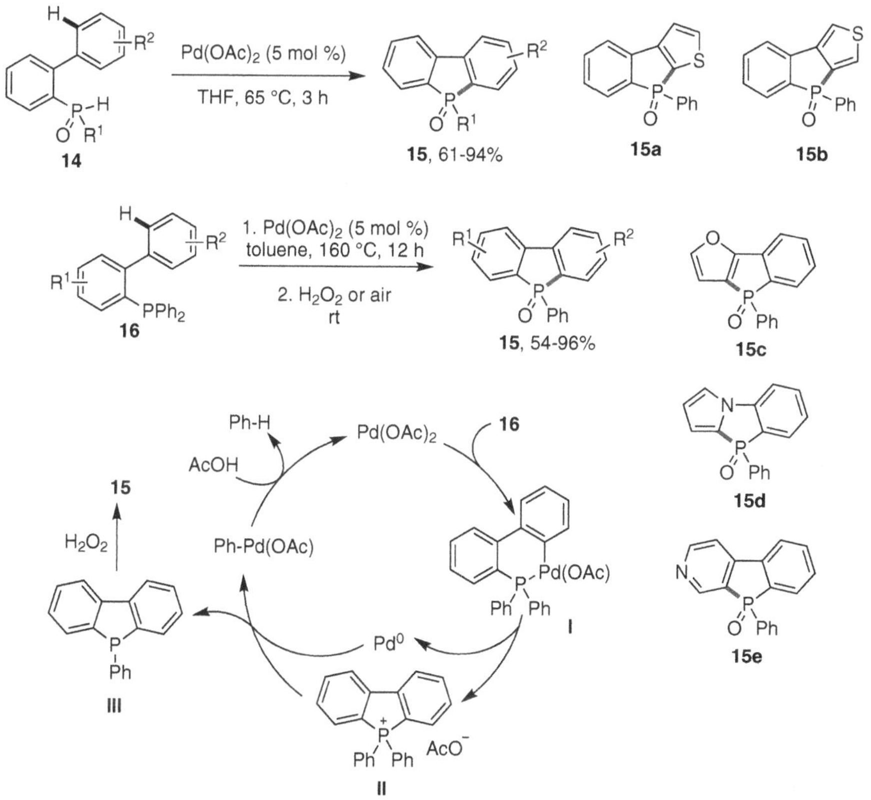

SCHEME 9.4

In the proposed reaction mechanism, palladacycle complex **I** was generated by the reaction of Pd(II) with biarylphosphine **16**, with subsequent reductive elimination to form phosphonium **II** along with Pd(0). The phosphonium **II** immediately underwent oxidative addition to Pd(0) to provide dibenzophosphole **III** and PhPd(OAc) through cleavage of a C–P bond. The obtained dibenzophosphole **III** is sensitive to O_2 and was oxidized to benzophosphindole oxides **15** in the presence of $Pd(OAc)_2$ (Scheme 9.4) (Baba, Tobisu, and Chatani 2013).

9.3 *P,O*-HETEROCYCLES

9.3.1 Benzo[1,2]oxaphosphole

Rh-catalyzed tandem dehydrogenative alkenylation and an intramolecular oxy-Michael reaction of arylphosphonic acid monoethyl esters **17** with activated alkenes **18** was developed to produce benzoxaphosphole 1-oxides **19**. Reaction of *ortho*-substituted arylphosphonic acid monoethyl esters **17** was conducted using 2 equiv of an alkene **18** in the presence of $[Cp^*RhCl_2]_2$ (4 mol%), AgOAc (2 equiv), and Na_2HPO_4 (1 equiv) in CH_3CN at 110°C, leading to 2-(1-ethoxy-1-oxido-1,3-dihydrobenzo[*c*]

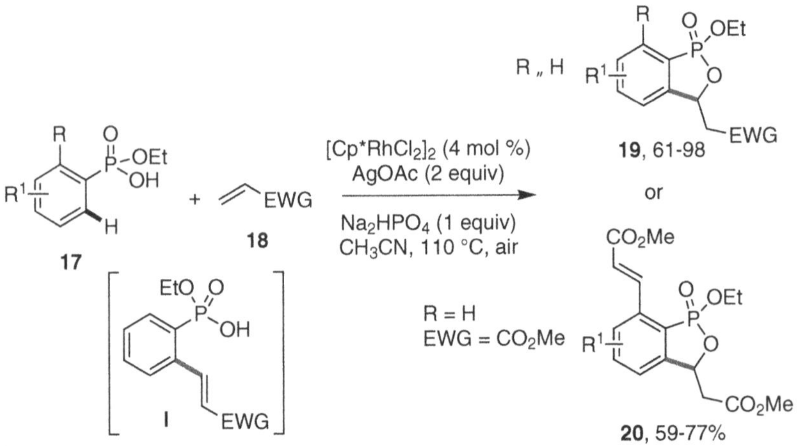

SCHEME 9.5

[1,2]oxaphosphol-3-yl)acetates **19** in 61–98% yields. In addition to acrylates, alkyl vinyl ketones, acrylonitrile, acrylamide, vinyl phosphonate, and vinyl sulfone are compatible with the reaction, affording corresponding benzoxaphosphole 1-oxides **19** in high yields. Using *ortho*-unsubstituted arylphosphonic acid monoethyl esters **17** (R = H) led to formation of (1-oxido-1,3-dihydrobenzo[*c*][1,2]oxaphosphol-7-yl)acrylates **20** in 59–77% yields. A proposed catalytic cycle was initiated by coordination and *ortho*-C–H activation of arylphosphonic acid monoethyl esters **17** with Cp^*RhX_2, followed by alkene coordination and insertion then *β*-hydride elimination, to afford *ortho*-alkenylated arylphosphonic acid monoethyl esters **I**, producing benzoxaphosphole 1-oxides **19** by oxy-Michael reaction (Scheme 9.5) (Ryu et al. 2013).

9.3.2 Benzo[1,2]oxaphosphinine

Phosphaisocoumarins **22** were accessed by the Rh-catalyzed dehydrogenative coupling of diarylphosphinic acid derivatives **21** with alkynes **2**. Under $[Cp^*Rh(MeCN)_3][SbF_6]_2$ (4 mol%) and AgOAc (3 equiv) catalytic systems in diglyme at 120°C under N_2, reactions afforded 1-aryl-1*H*-benzo[*c*][1,2]oxaphosphinine 1-oxides **22** in 45–95% yields. Unsymmetrical alkyl aryl alkyne led to corresponding 4-alkylbenzo[*c*][1,2]oxaphosphinine 1-oxide **22**, regioselectively (Unoh et al. 2013). Also, arylphosphonic acid **17** was reacted with diaryl alkynes **2** in the presence of $[Cp^*RhCl_2]_2$ (2 mol%), Ag_2CO_3 (1 equiv), and AgOAc (1 equiv) in *t*-BuOH at 90°C, resulting in the formation of 1-ethoxy-1*H*-benzo[*c*][1,2]oxaphosphinine 1-oxides **23** in 70–92% yields. Aliphatic internal alkynes were also tolerated under reaction conditions to afford corresponding phosphaisocoumarins **23** in 75–92% yields. Reaction with unsymmetrical alkyl aryl alkynes gave 4-alkylbenzo[*c*][1,2]oxaphosphinine 1-oxide in 80–91% yields, regioselectively. Heterocyclic-fused oxaphosphinine 1-oxides, thieno[2,3-*c*][1,2]oxaphosphinine 1-oxide **23b** and oxaphosphinino[3,4-*b*]indole 1-oxide **23a**, were obtained from ethyl hydrogen thiophen-2-ylphosphonate and ethyl hydrogen 1*H*-indol-2-ylphosphonate substrate in 76% and 60% yields, respectively (Scheme 9.6) (Seo et al. 2013).

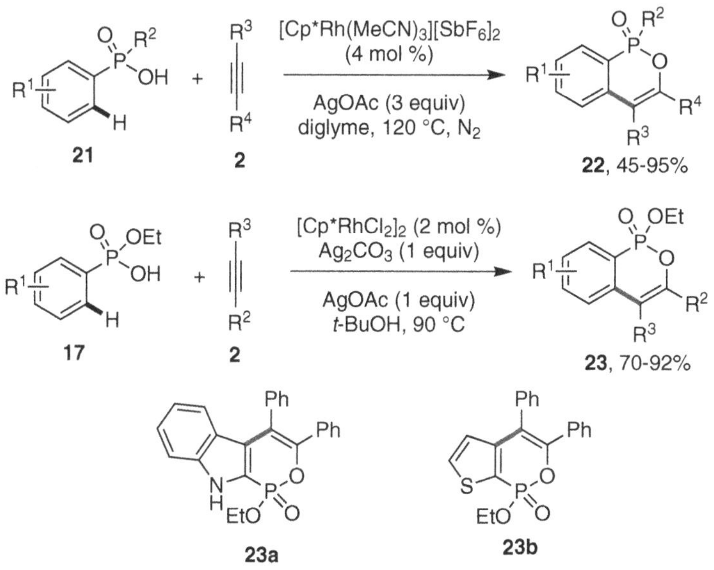

SCHEME 9.6

9.3.3 Dibenzo[1,2]oxaphosphinine

Synthesis of dibenzo[*c,e*][1,2]oxaphosphinine 6-oxide derivatives **25** were developed by Pd-catalyzed C(sp^2)–H activation/C–O formation. Intramolecular dehydrogenative coupling reactions of ethyl hydrogen [1,1'-biphenyl]-2-ylphosphonates **24** were catalyzed with $Pd(OAc)_2$ (10 mol%) in the presence of $PhI(OAc)_2$ (2 equiv), KOAc (2 equiv), and an *N*-acetyl-L-leucine ligand (30 mol%) in *t*-BuOH at 100°C under air, leading to dibenzo[*c,e*][1,2]oxaphosphinine 6-oxides **25** in 52–72% yields. Fused benzo[*c*]thieno[2,3-*e*][1,2]oxaphosphinine 5-oxide was prepared in 52% yield starting from ethyl hydrogen (2-(thiophen-2-yl)phenyl)phosphonate. The reaction took place by *N*-acetyl-L-leucine-ligand-assisted O(sp^2)-atom-directed C–H palladation, followed by oxidation to Pd(IV) species by $PhI(OAc)_2$. Finally, reductive elimination gave the desired product along with regeneration of the Pd(II) catalyst to proceed to the next catalytic cycle (Scheme 9.7) (Shin et al. 2014).

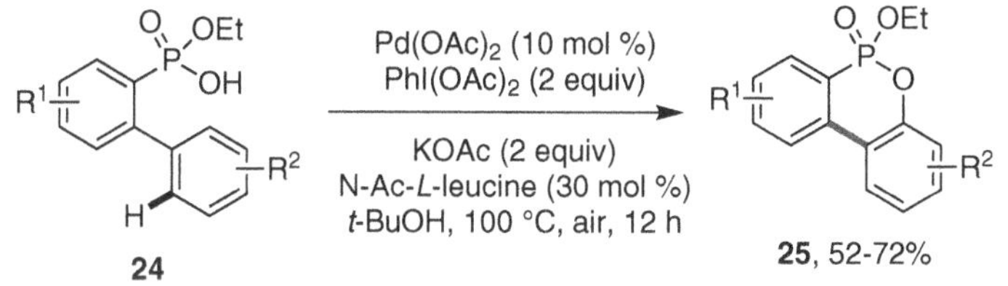

SCHEME 9.7

9.4 *P,N*-HETEROCYCLES

9.4.1 Benzo[1,2]azaphosphinine

Co(II)-catalyzed synthesis of dihydrobenzo[*c*][1,2]azaphosphinine 1-oxide derivatives **27** was reported via dehydrogenative coupling of arylphosphinic acid aminoquinoline amides **26** with alkynes **2**. Treatment of phosphinic amide **26** with an internal alkyne **2** (1.1 equiv) in the presence of $Co(NO_3)_2{\cdot}6H_2O$ (20 mol%), NaOPiv (2 equiv), $Mn(OAc)_3{\cdot}2H_2O$ (1 equiv), and dipivaloylmethane (0.5 equiv) in EtOH at air led to the formation of 3,4-disubstituted 1,2-dihydrobenzo[*c*][1,2]azaphosphinine 1-oxides **27** in 61–87% yields. Reaction with an unsymmetrical alkyne, 1-phenylpropyne, afforded 4-methyl-3-phenyl-1,2-dihydrobenzo[*c*][1,2]azaphosphinine 1-oxide **27** in 72% yield, regioselectively. Also, terminal alkynes including alkyl and aryl acethylenes were tolerated under reaction conditions, delivering 4-unsubstituted benzoazaphosphinine 1-oxides in 24–81% yields. However, reaction of allyl pivalate did not afford a corresponding cyclic product; vinyl pivalate produced related 3,4-unsubstituted benzoazaphosphinine 1-oxide in 60% yield when treated with phosphinic amide (Scheme 9.8) (Nguyen, Grigorjeva, and Daugulis 2015).

9.4.2 Dibenzo[1,2]azaphosphinine

Pd-catalyzed C–H arylation of *N*-(*ortho*-bromoaryl)-diarylphosphinic amides **28** was developed by Lin et al. (2015) to furnish 5,6-dihydrodibenzo[*c,e*][1,2]azaphosphinine 6-oxides **29**. Reactions were performed using $Pd(OAc)_2$ (10 mol%) in the presence of TADDOL-based *N*-Mephosphoramidite ligand **L** (10 mol%), K_3PO_4 (1.5 equiv), and PivOH (30 mol%) in toluene at 80°C, affording 6-aryl-5,6-dihydrodibenzo[*c,e*][1,2]azaphosphinine 6-oxide derivatives **29** in 58–94% yields. Using a chiral ligand gave rise to the formation of dibenzo[*c,e*][1,2]azaphosphinine 6-oxides **29** in enantioselective manner with 83–93% ee (Scheme 9.9). A similar approach was conducted using Pd(dba)2 (8 mol%), Cs_2CO_3 (1.5 equiv), and PivOH (40 mol%) in hexane at 60°C. 6-aryl-5,6-dihydrodibenzo[*c,e*][1,2]azaphosphinine 6-oxides **29** were obtained in 61–99% yields, with 8–97% ee. The reaction was initiated by oxidative addition of the C(sp²)–Br bond onto Pd(0), followed by Br anion exchange with pivalate to afford Pd pivalate. C–H bond activation followed by reductive elimination furnished a *P*-heterocycle **29** along with regeneration of the Pd(0) species, for a further catalytic cycle (Liu et al. 2015).

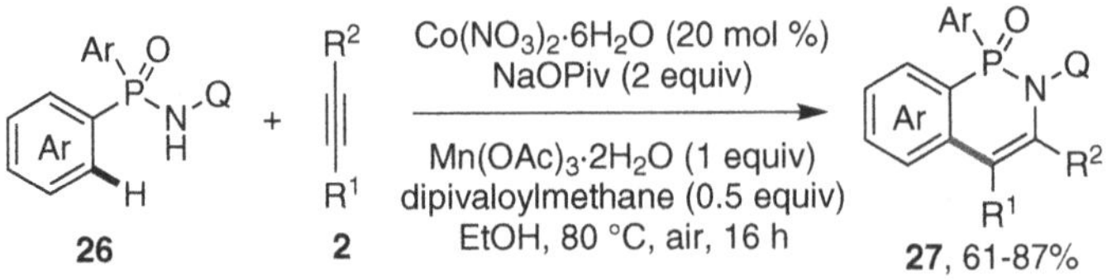

SCHEME 9.8

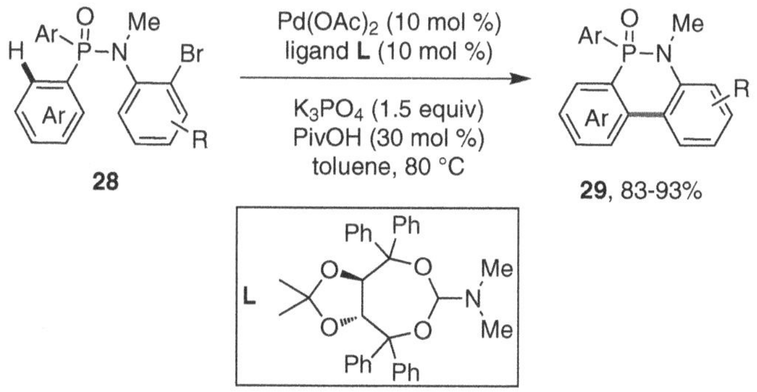

SCHEME 9.9

9.5 *Si*-HETEROCYCLES

9.5.1 BENZOSILOLE AND BENZOSILINE

Intramolecular dehydrogenative coupling of phenethylsilane **30a** and 3-phenylpropylsilane **30b** was developed to afford dihydro-1*H*-benzo[*b*]silole **31a** and tetrahydrobenzo[*b*]saline **31b**. Reactions were conducted using 5 mol% of $Tp^{Me2}PtMe_2H$ at 200°C to give corresponding cyclic products in 64–70% yields (Scheme 9.10) (Tsukada and Hartwig 2005).

Synthesis of 3-(Dippoxy)-1*H*-benzo[*b*]silole derivatives **33** was achieved by Pd/acid-catalyzed intramolecular *anti*-hydroarylation of aryloxyethynyl(aryl)silanes **32** via *ortho*-C–H bond activation. Reactions were catalyzed by $Pd(dba)_2$ (5 mol%), in the presence of PEt_3 (5 mol%) and PivOH (10 mol%) in toluene at 120°C, leading to benzo[*b*]siloles **33** in 73–99% yields. Heterocycle-fused siloles, including silolo[2,3-*b*]thiophenes, silolo[3,2-*b*]thiophene, benzo[*b*]silolo[3,2-*d*]thiophene and silolo[2,3-*b*]indole, were also prepared in 45–98% yields under reaction conditions starting from corresponding substrates. 1-Naphthyl(ethynyl)silane afforded a C2–H bond activation product, naphthosilole **33a**, along with a C8–H bond activation product, 1-silaphenalene **34**, in 43% and 20% yields, respectively. The proposed reaction mechanism involves the generation of η^2-alkyne-palladium complex **I**, followed by *syn*-hydropalladation to provide vinyl palladium pivalate **II**. Oxygen

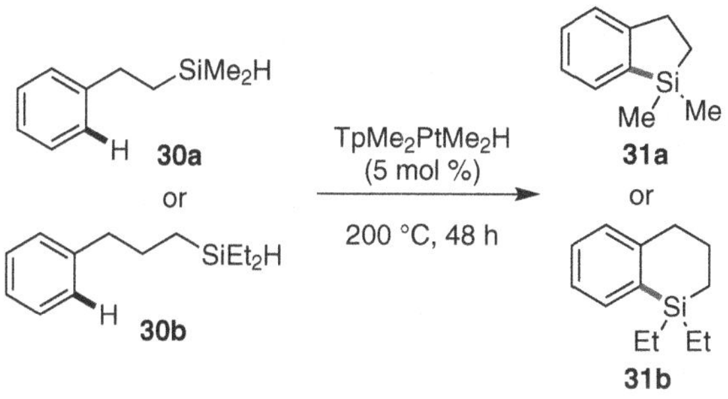

SCHEME 9.10

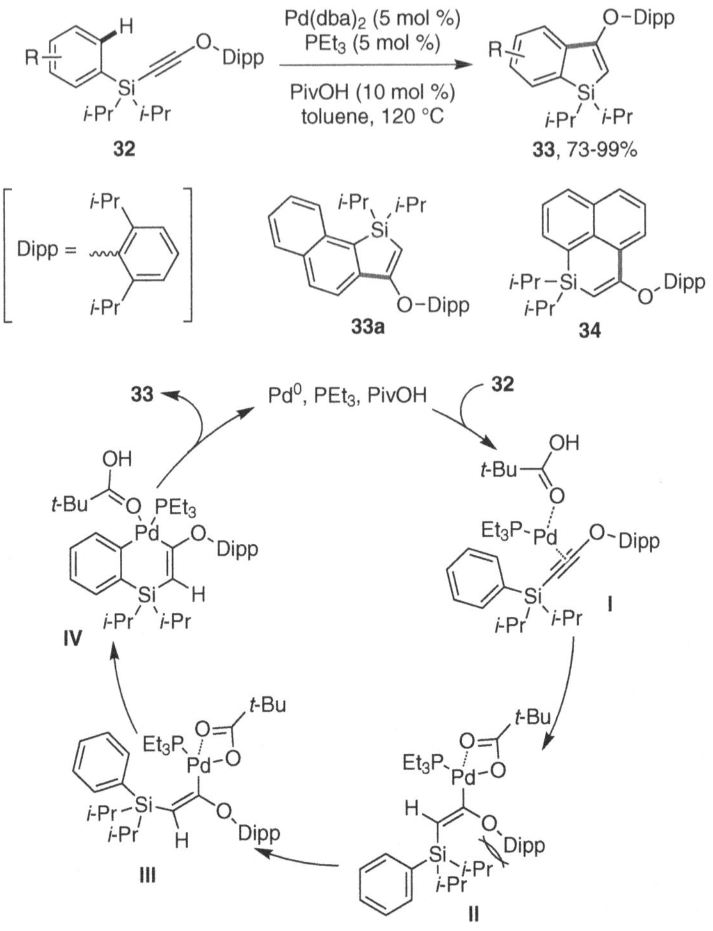

SCHEME 9.11

lone-pair electron-assisted stereoisomerization to *Z*-complex **III** was accelerated due to the steric repulsion between Dipp and two isopropyl groups on silicon. Subsequent C–H activation via the concerted metalation deprotonation (CMD) pathway gave palladacycle **IV**, which underwent reductive elimination to furnish the desired product **33** with regeneration of Pd(0) species (Scheme 9.11) (Minami, Noguchi, and Hiyama 2017).

9.5.2 Dibenzosilole

Pd-catalyzed intramolecular dehydrogenative coupling of 2-(arylsilyl)aryl triflates **35** was described to furnish dibenzo[*b,d*]silole derivatives **36**. Subjecting 2-(arylsilyl)aryl triflates **35** to $Pd(OAc)_2$ (5 mol%), PCy_3 (10 mol%), and Et_2NH (2 equiv) in *N,N*-dimethylacetamide (DMA) at 100°C led to dibenzo[*b,d*]siloles **36** in 39–98% yields. Also, heterocycle-fused benzosiloles **36a,b** were obtained in 56–95% yields, starting from thiophene-, benzothiophene-, benzofuran- and indole-triflates under reaction conditions. Furthermore, the helicene-type molecule, benzo[4,5]silolo[2,3-*b*]benzo[4,5]silolo[3,2-*d*]thiophene **36c**, was also synthesized in 85% yield by two-fold intramolecular dehydrogenative coupling of 2,5-bis(silyl)thiophene (Shimizu, Mochida, and Hiyama 2008). Moreover, similar Pd-catalyzed intramolecular coupling of 2-[(2-indolyl)silyl]aryl triflates **37** was reported by Mochida, Shimizu, and Hiyama (2009) in which unexpected benzo[4,5]silolo[3,2-*b*]indoles **38a** were obtained instead of benzo[4,5]silolo[2,3-*b*]indole **38b** through 1,2-silicon migration. Unexpected benzo[4,5]silolo[3,2-*b*]indoles **38a** were produced in 59–91% yields. 2-[(2-Pyrrolyl)silyl]aryl triflates **37** were also tolerated under reaction conditions, affording corresponding benzo[4,5]silolo[3,2-*b*]pyrroles **38a** in 31–78% yields, along with expected products **38b** in 0–42% yields. In the proposed reaction mechanism, arylpalladium intermediate **I** was generated by the oxidative addition of 2-[(2-pyrrolyl)silyl]aryl triflate **37** to a Pd(0) complex, which transformed into spirocyclic cationic intermediate **III** either by intramolecular electrophilic substitution (ES) at the 3-position of the pyrrole to give **II**, followed by migration of the Pd atom to the 2-position (route a), or by direct palladation at the 2-position (route b). Subsequently, 1,2-*Si* migration to intermediate **IV** followed by deprotonation by a base gave six-membered palladacycle **V**, which underwent reductive elimination (RE) to produce **38a** (Scheme 9.12).

Rh-catalyzed synthesis of dibenzo[*b,d*]siloles **40** was developed by intramolecular dehydrogenative coupling of biphenylhydrosilanes **39** via both Si–H and C–H bond activation. Reactions were performed using $RhCl(PPh_3)_3$ (0.5 mol%) in dioxane at 135°C, delivering dibenzo[*b,d*]silole derivatives **40** in 60–99% yields. Also, ladder-type bis-silicon-bridged *p*-terphenyl was obtained in 87% yield. The reaction was initiated by Si–H bond activation via oxidative addition of a hydrosilane to the Rh center to generate an Rh–H intermediate, followed by aromatic C–H bond activation with subsequent reductive elimination (Ureshino et al. 2010). In addition, trisilasumanene and trigermasumanene **40a** were produced by a similar approach in 45–52% and 80% yields, respectively (Zhou et al. 2017). Similar dehydrogenative silylation of the C–H bond was developed to construction of sila[*n*]helicenes **40b** using $[RhCl(cod)]_2$ (2.5 mol%) as catalyst. Sila[*n*]helicenes **40b** were obtained in 48–98% yields (Murai et al. 2016). Benzosilolothiophene derivatives were also accessed in 26–81% yields by Rh-catalyzed Si–H/C–H bonds cleavage using $[RhCl(cod)]_2$ (1 mol%), in the presence of a dppe-F_{20} ligand (3 mol%) and $(EtO)_3SiH$ (15 mol%) in toluene at 145°C (Mitsudo et al. 2017). Moreover, an Rh-catalyzed intramolecular dehydrogenative coupling reaction was accomplished to prepare benzosiloloferrocene derivatives **40c**. Reactions were catalyzed with $[RhCl(cod)]_2$ (10 mol%) in the presence of chiral diene ligand (24 mol%) and 3,3-dimethyl-1-butene (10 equiv) in toluene at 135°C, which gave rise to the formation of benzosiloloferrocenes **40c** in 39–75% yields, with 5–86% ee. Also, (dimethylhydrogermyl)phenylferrocene was compatible with the reaction, affording benzogermoloferrocene **40c** (X = Ge) in 40% yield, with 33% ee (Scheme 9.13) (Shibata, Shizuno, and Sasaki 2015).

Construction of chiral dibenzo[*b,d*]silole derivatives **42** was achieved by tandem desymmetrization of silacyclobutanes/intermolecular dehydrogenative silylation. Reacting biaryl silacyclobutanes **41** with arenes (2 equiv) in the presence of [Rh(cod)

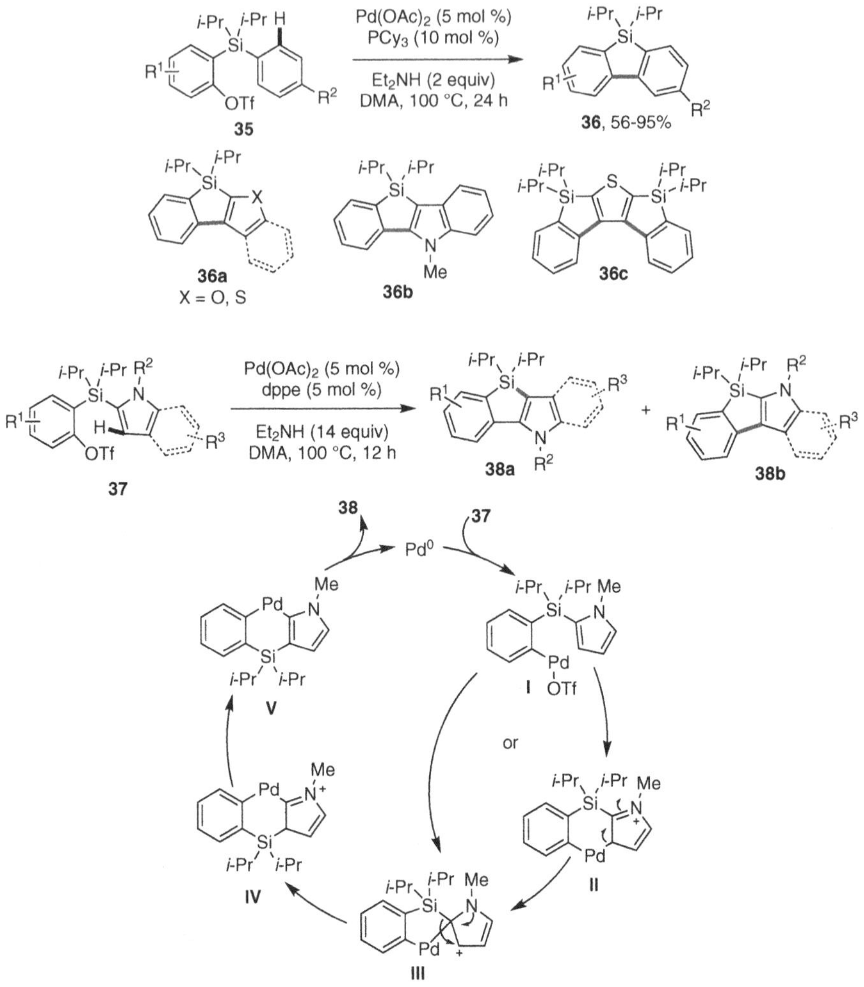

SCHEME 9.12

OH] (10 mol%) and a TMS-segphos ligand (10 mol%) at 40°C led to corresponding chiral 5-aryl-5-*n*-propyl-5*H*-dibenzo[*b,d*]siloles **42** in 54–91% yields, with 65–93% ee. The reaction occurred by intramolecular desymmetric Si–C activation of C(sp^2)–H silylation to 5-*n*-propyl-5*H*-dibenzo[*b,d*]silole **I**, followed by intermolecular dehydrogenative silylation with an arene (Scheme 9.14) (Zhang et al. 2017).

9.5.3 Tribenzo[*b,d,f*]silepine

Ir-catalyzed intramolecular dehydrogenative C–H/Si–H coupling of 2′,6′-diaryl-2-(hydrosilyl)biphenyls **43** afforded tribenzosilepine scaffolds **44**. Reactions were conducted using [IrCl(cod)]$_2$ (10 mol%) in the presence of a DPPBen ligand (20 mol%) and 3,3-dimethyl-1-butene (1 equiv) in toluene at 120°C, resulting in the formation

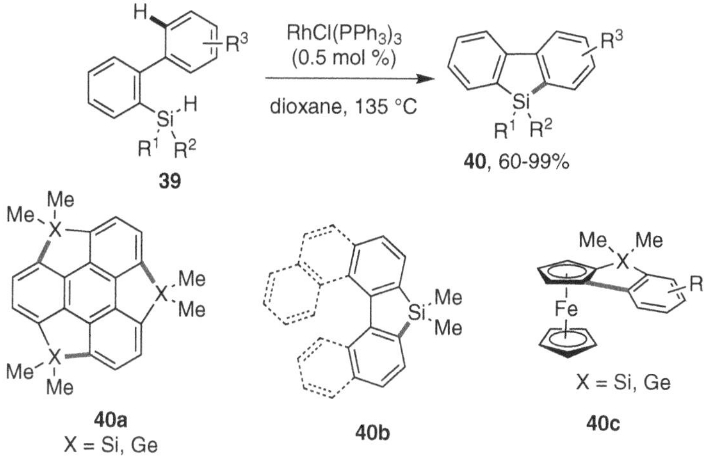

SCHEME 9.13

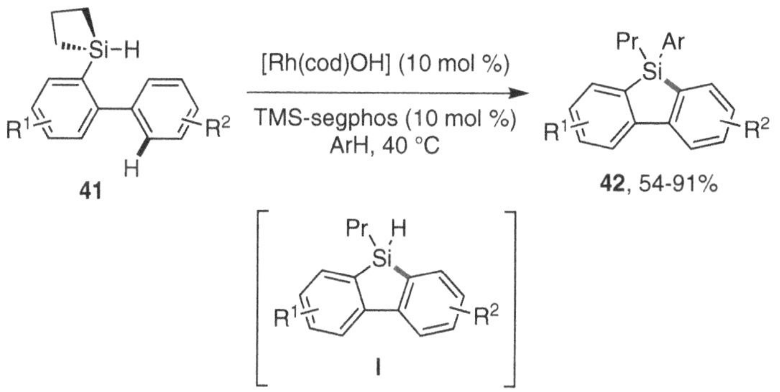

SCHEME 9.14

of tribenzosilepines **44** in 45–90% yields. The reaction of 2'-aryl-2-(hydrosilyl) biphenyls led to a corresponding five-membered dibenzosilole. The naphthalene-containing substrate gave the corresponding tribenzosilepine in 57% yield. However, a six-membered benzosiline was not obtained (Scheme 9.15) (Shibata et al. 2018).

9.6 *Si,O*-HETEROCYCLES

One example of (*R*)-1,1-diethyl-3-phenyl-1,3-dihydrobenzo[*c*][1,2]oxasilole **46** was prepared via Rh-catalyzed dehydrogenative coupling of (benzhydryloxy)diethylsilane **45**. The reaction was catalyzed with $[RhCl(cod)]_2$ or $[RhCl(C_2H_4)_2]_2$ (0.5 mol%) in the presence of a chiral ligand and norbornene in THF at 50°C, delivering the intramolecular Si–H/C–H activation product benzooxasilole **46** in 90–94% yields, with 99% ee (Scheme 9.16) (Lee and Hartwig 2017).

Pd(II)-catalyzed intramolecular dehydrogenative arylation of (2-bromophenoxy)(phenyl)silanes or (2-bromophenyl)(phenoxy)silanes **47** was developed

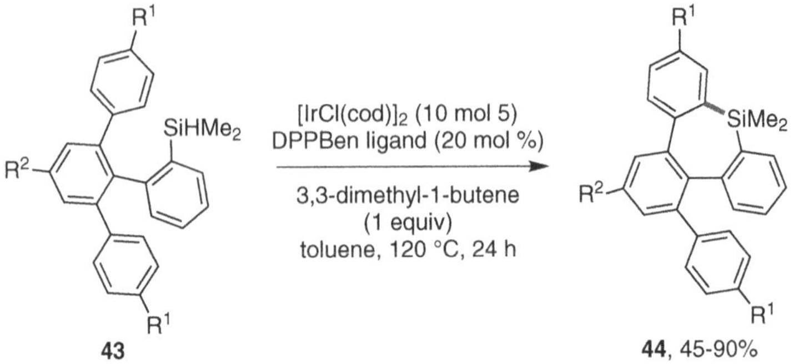

SCHEME 9.15

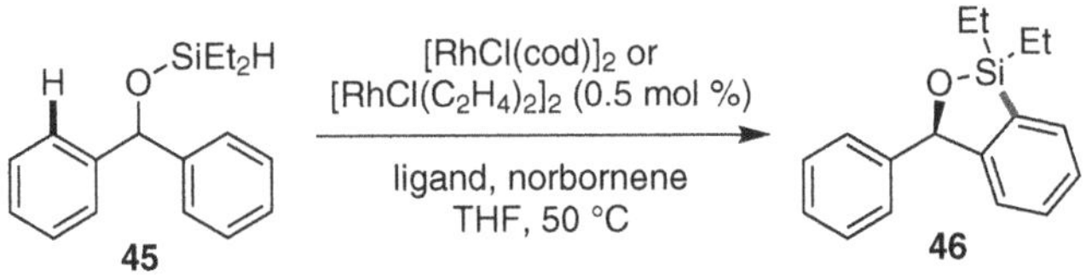

SCHEME 9.16

in order to create dibenzooxasiline derivatives **48**. Treating (2-bromophenoxy)(phenyl)silanes **47** with $Pd(OAc)_2$ (5 mol%), PCy_3-HBF_4 (10 mol%), PivOH, and Cs_2CO_3 in *p*-xylene at 140°C led to the formation of dibenzo[*c,e*][1,2]oxasilines **48** in 30–100% yields. 5,6-Dihydrodibenzo[*c,e*][1,2]azasiline **49** was also obtained in 77% yield, starting from *N*-(2-Br-TBDPS)-protected aniline (Huang and Gevorgyan 2009). In another work, dibenzooxasilines **48a** were accessed by Rh-catalyzed dehydrogenative C–H/Si–H cleavage of ([1,1'-biphenyl]-2-yloxy) silanes **50**. $[RhCl(cod)]_2$ (2 mol%) in the presence of the XantPhos ligand (4 mol%) in toluene at 120°C catalyzed the conversion of ([1,1'-biphenyl]-2-yloxy)silanes **50**, which produced in one-pot manner from 2-arylphenoles the corresponding dibenzo[*c,e*][1,2]oxasilines **48a** in 52–90% yields. Fused benzo[*e*]thieno[3,2-*c*][1,2]oxasiline was obtained in 65% yield, starting from (2-(thiophen-2-yl)phenoxy)silane substrate. 2-Ferrocenyl phenols were transformed into benzosiline ferrocene derivatives **51** under similar reaction conditions in 51–83% yields, with 90–95% ee (Scheme 9.17) (Zhao et al. 2018).

9.7 CONCLUSION

In summary, aromatic C(sp²)–H dehydrogenative coupling processes are used for the synthesis of *P*-, *P,O*-, *P,N*-, *Si*-, and *Si,O*-heterocycles, such as phosphindole, benzophosphindole, benzooxaphosphole, benzooxaphosphinine, benzoazaphosphinine, benzosilole, benzosiline, tribenzosilepine, benzooxasilole, and dibenzooxasilines. Because of the quantitative and one-step synthesis of these types of heterocycles, the aromatic C–H bond activation approach could be of interest in the synthesis of

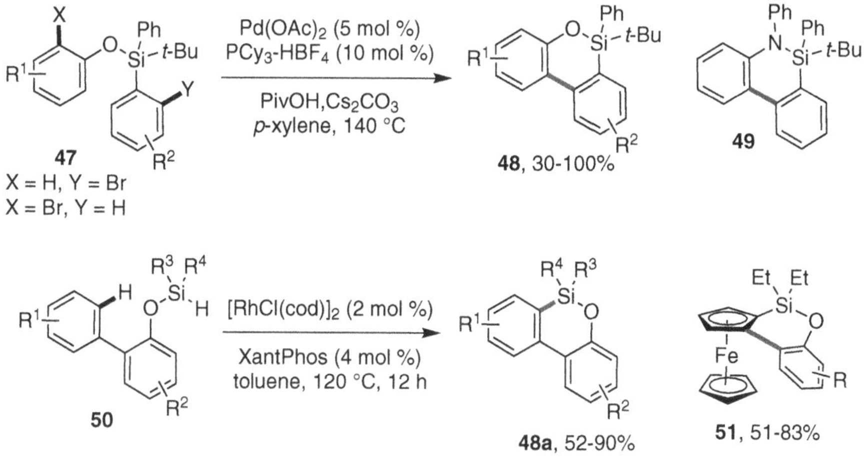

SCHEME 9.17

pharmaceutical, medicinal, and natural products. On the other hand, there is still room for further studies in the field of synthesis of *P*-, *P,O*-, *P,N*-, *Si*-, and *Si,O*-heterocycles via aromatic C–H bond dehydrogenative coupling reactions.

REFERENCES

Baba, Katsuaki, Mamoru Tobisu, and Naoto Chatani. 2013. "Palladium-catalyzed direct synthesis of phosphole derivatives from triarylphosphines through cleavage of carbon–hydrogen and carbon–phosphorus bonds." *Angewandte Chemie* no. 125(45):12108–12111.

Baumgartner, Thomas, and Régis Réau. 2006. "Organophosphorus π-conjugated materials." *Chemical Reviews* no. 106(11):4681–4727.

Börner, Armin. 2008. *Phosphorus Ligands in Asymmetric Catalysis: Synthesis and Applications*. Vol. 1: Wiley-VCH.

Chen, Yun-Rong, and Wei-Liang Duan. 2013. "Silver-mediated oxidative C–H/P–H functionalization: An efficient route for the synthesis of benzo [b] phosphole oxides." *Journal of the American Chemical Society* no. 135(45):16754–16757.

Ge, Qingmei, Jiarui Zong, Bin Li, and Baiquan Wang. 2017. "Copper-mediated annulation of phosphorus-containing arenes with alkynes: An approach to phosphindolium salts." *Organic Letters* no. 19(24):6670–6673.

Hirai, Masato, Naoki Tanaka, Mika Sakai, and Shigehiro Yamaguchi. 2019. "Structurally constrained boron-, nitrogen-, silicon-, and phosphorus-centered polycyclic π-conjugated systems." *Chemical Reviews*.

Huang, Chunhui, and Vladimir Gevorgyan. 2009. "TBDPS and Br–TBDPS protecting groups as efficient aryl group donors in Pd-catalyzed arylation of phenols and anilines." *Journal of the American Chemical Society* no. 131(31):10844–10845.

Kuninobu, Yoichiro, Takuya Yoshida, and Kazuhiko Takai. 2011. "Palladium-catalyzed synthesis of dibenzophosphole oxides via intramolecular dehydrogenative cyclization." *The Journal of Organic Chemistry* no. 76(18):7370–7376.

Lee, Taegyo, and John F Hartwig. 2017. "Mechanistic studies on rhodium-catalyzed enantioselective silylation of aryl C–H bonds." *Journal of the American Chemical Society* no. 139(13):4879–4886.

Lin, Zi-Qi, Wei-Zhen Wang, Shao-Bai Yan, and Wei-Liang Duan. 2015. "Palladium-catalyzed enantioselective C–H arylation for the synthesis of P-stereogenic compounds." *Angewandte Chemie International Edition* no. 54(21):6265–6269.

Liu, Lantao, An-An Zhang, Yanfang Wang, Fuqiang Zhang, Zhenzhen Zuo, Wen-Xian Zhao, Cui-Lan Feng, and Wenjin Ma. 2015. "Asymmetric synthesis of P-stereogenic phosphinic amides via Pd (0)-catalyzed enantioselective intramolecular C–H arylation." *Organic Letters* no. 17(9):2046–2049.

Lu, Gang, Hakan Usta, Chad Risko, Lian Wang, Antonio Facchetti, Mark A Ratner, and Tobin J Marks. 2008. "Synthesis, characterization, and transistor response of semiconducting silole polymers with substantial hole mobility and air stability: Experiment and theory." *Journal of the American Chemical Society* no. 130(24):7670–7685.

Ma, Dumei, Weizhu Chen, Gaobo Hu, Yun Zhang, Yuxing Gao, Yingwu Yin, and Yufen Zhao. 2016. "K2S2O8-mediated metal-free direct P–H/C–H functionalization: A convenient route to benzo [b] phosphole oxides from unactivated alkynes." *Green Chemistry* no. 18(12):3522–3526.

Ma, Dumei, Jiaoting Pan, Lu Yin, Pengxiang Xu, Yuxing Gao, Yingwu Yin, and Yufen Zhao. 2018. "Copper-catalyzed Direct Oxidative C–H functionalization of unactivated cycloalkanes into cycloalkyl benzo [b] phosphole oxides." *Organic Letters.*

Mathey, Francois. 2004. "Transient 2 H-Phospholes as powerful synthetic intermediates in organophosphorus chemistry." *Accounts of Chemical Research* no. 37(12):954–960.

Minami, Yasunori, Yuta Noguchi, and Tamejiro Hiyama. 2017. "Synthesis of benzosiloles by intramolecular anti-hydroarylation via ortho-C–H activation of aryloxyethynyl silanes." *Journal of the American Chemical Society* no. 139(40):14013–14016.

Mitsudo, Koichi, Seiichi Tanaka, Ryota Isobuchi, Tomohiro Inada, Hiroki Mandai, Toshinobu Korenaga, Atsushi Wakamiya, Yasujiro Murata, and Seiji Suga. 2017. "Rh-catalyzed dehydrogenative cyclization leading to benzosilolothiophene derivatives via Si–H/C–H bond cleavage." *Organic Letters* no. 19(10):2564–2567.

Mochida, Kenji, Masaki Shimizu, and Tamejiro Hiyama. 2009. "Palladium-catalyzed intramolecular coupling of 2-[(2-pyrrolyl) silyl] aryl triflates through 1, 2-silicon migration." *Journal of the American Chemical Society* no. 131(24):8350–8351.

Murai, Masahito, Ryo Okada, Atsushi Nishiyama, and Kazuhiko Takai. 2016. "Synthesis and properties of sila [n] helicenes via dehydrogenative silylation of C–H bonds under rhodium catalysis." *Organic Letters* no. 18(17):4380–4383.

Nguyen, Tung Thanh, Liene Grigorjeva, and Olafs Daugulis. 2015. "Cobalt-catalyzed, aminoquinoline-directed functionalization of phosphinic amide sp² C–H bonds." *ACS Catalysis* no. 6(2):551–554.

Ryu, Taekyu, Jaeeun Kim, Youngchul Park, Sanghyuck Kim, and Phil Ho Lee. 2013. "Rhodium-catalyzed oxidative cyclization of arylphosphonic acid monoethyl esters with alkenes: Efficient synthesis of benzoxaphosphole 1-oxides." *Organic Letters* no. 15(15):3986–3989.

Seo, Jungmin, Youngchul Park, Incheol Jeon, Taekyu Ryu, Sangjune Park, and Phil Ho Lee. 2013. "Synthesis of phosphaisocoumarins through rhodium-catalyzed cyclization using alkynes and arylphosphonic acid monoesters." *Organic Letters* no. 15(13):3358–3361.

Shibata, Takanori, Tsubasa Shizuno, and Tomoya Sasaki. 2015. "Enantioselective synthesis of planar-chiral benzosiloloferrocenes by Rh-catalyzed intramolecular C–H silylation." *Chemical Communications* no. 51(37):7802–7804.

Shibata, Takanori, Ninna Uno, Tomoya Sasaki, Hideaki Takano, Tatsuki Sato, and Kyalo Stephen Kanyiva. 2018. "Ir-catalyzed synthesis of substituted tribenzosilepins by dehydrogenative C–H/Si–H coupling." *The Journal of Organic Chemistry* no. 83(7):3426–3432.

Shimizu, Masaki, Kenji Mochida, and Tamejiro Hiyama. 2008. "Modular approach to silicon-bridged biaryls: Palladium-catalyzed intramolecular coupling of 2-(arylsilyl) aryl triflates." *Angewandte Chemie International Edition* no. 47(50):9760–9764.

Shin, Seohyun, Dongjin Kang, Woo Hyung Jeon, and Phil Ho Lee. 2014. "Synthesis of ethoxy dibenzooxaphosphorin oxides through palladium-catalyzed C (sp^2)–H activation/C–O formation." *Beilstein Journal of Organic Chemistry* no. 10:1220.

Tsuji, Hayato, Kosuke Sato, Yoshiharu Sato, and Eiichi Nakamura. 2009. "Benzo [b] phosphole sulfides. Highly electron-transporting and thermally stable molecular materials for organic semiconductor devices." *Journal of Materials Chemistry* no. 19(21):3364–3366.

Tsukada, Naofumi, and John F Hartwig. 2005. "Intermolecular and intramolecular, platinum-catalyzed, acceptorless dehydrogenative coupling of hydrosilanes with aryl and aliphatic methyl C–H bonds." *Journal of the American Chemical Society* no. 127(14):5022–5023.

Unoh, Yuto, Yuto Hashimoto, Daisuke Takeda, Koji Hirano, Tetsuya Satoh, and Masahiro Miura. 2013. "Rhodium (III)-catalyzed oxidative coupling through C–H bond cleavage directed by phosphinoxy groups." *Organic Letters* no. 15(13):3258–3261.

Unoh, Yuto, Koji Hirano, Tetsuya Satoh, and Masahiro Miura. 2013. "An approach to benzophosphole oxides through silver-or manganese-mediated dehydrogenative annulation involving C–C and C–P bond formation." *Angewandte Chemie International Edition* no. 52(49):12975–12979.

Ureshino, Tomonari, Takuya Yoshida, Yoichiro Kuninobu, and Kazuhiko Takai. 2010. "Rhodium-catalyzed synthesis of silafluorene derivatives via cleavage of silicon–hydrogen and carbon– hydrogen bonds." *Journal of the American Chemical Society* no. 132(41):14324–14326.

Wu, Di, Jueting Zheng, Chenyong Xu, Dawei Kang, Wenjing Hong, Zheng Duan, and François Mathey. 2019. "Phosphindole fused pyrrolo [3,2-b] pyrroles: A new single-molecule junction for charge transport." *Dalton Transactions* no. 48(19):6347–6352.

Zhang, Pengbo, Yuzhen Gao, Liangliang Zhang, Zhiqiang Li, Yan Liu, Guo Tang, and Yufen Zhao. 2016. "Copper-catalyzed cycloaddition between secondary phosphine oxides and alkynes: Synthesis of benzophosphole oxides." *Advanced Synthesis and Catalysis* no. 358(1):138–142.

Zhang, Qing-Wei, Kun An, Li-Chuan Liu, Qi Zhang, Huifang Guo, and Wei He. 2017. "Construction of chiral Tetraorganosilicons by tandem desymmetrization of silacyclobutanes/intermolecular dehydrogenative silylation." *Angewandte Chemie* no. 129(4):1145–1149.

Zhao, Wen-Tao, Zhuo-Qun Lu, Hanliang Zheng, Xiao-Song Xue, and Dongbing Zhao. 2018. "Rhodium-catalyzed 2-arylphenol-derived six-membered silacyclization: Straightforward access toward dibenzooxasilines and silicon-containing planar chiral metallocenes." *ACS Catalysis* no. 8(9):7997–8005.

Zhen, Shijie, Jingjing Guo, Wenwen Luo, Anjun Qin, Zujin Zhao, and Ben Zhong Tang. 2018. "Synthesis, structure, photoluminescence and photochromism of phosphindole oxide and benzo [b] thiophene S, S-dioxide derivatives." *Journal of Photochemistry and Photobiology A: Chemistry* no. 355:274–282.

Zhou, Dandan, Ya Gao, Bingxin Liu, Qitao Tan, and Bin Xu. 2017. "Synthesis of silicon and germanium-containing heterosumanenes via rhodium-catalyzed cyclodehydrogenation of silicon/germanium–hydrogen and carbon–hydrogen bonds." *Organic Letters* no. 19(17):4628–4631.

10 Natural Products Synthesis

10.1 INTRODUCTION

Many natural products, isolated from plants, animals, microorganisms, etc. have a heterocyclic core and exhibit many biological activities, including anti-inflammatory, antioxidant, cytoprotective, antimicrobial, proteasome inhibitor in breast cancer cells, hepatoprotective, antiproliferative, antibacterial, antifungal, anticancer, genotoxic, and cytotoxic properties. Therefore, the synthesis and development of new approaches to the creation of natural products, especially natural products with heterocyclic cores, are of interest to synthetic and medicinal scientists. The dehydrogenative coupling process of the aromatic $C(sp^2)$–H bond is an interesting method for the preparation of heterocyclic systems with different types of substitutions, which could be transformed into desired natural products.

10.2 NATURAL PRODUCTS CONTAINING NITROGEN HETEROCYCLES

The aristolactam BII alkaloid, isolated from the stem bark of *Goniothalamus velutinus* (Airy Shaw) (Iqbal et al. 2018), exhibits inhibitory activity against T and B lymphocyte proliferation and also exhibits cytotoxic activity (Zhang et al. 2007). Ji et al. (2017) developed a total synthesis of aristolactam BII, starting by formation of 3-alkylidene isoindolinone **3** cores via dehydrogenative coupling of benzamide **1** with 2,2-difluorovinyl tosylate **2** in the presence of $[Cp^*Rh(CH_3CN)_3][SbF_6]_2$ (5 mol%), CsOPiv, and $AgSbF_6$ in hexafluoroisopropanol (HFIP) at 60°C for 16 h, in 63% yield. Aristolactam BII **4** was obtained from **3** in four further steps (Scheme 10.1).

The rosettacin class of topoisomerase I (topl) inhibitors possess an indenoisoquinoline core, exhibiting antiproliferative activity and anti-topl activity (Fox et al. 2003). Some substituted rosettacin derivatives are isolated from *Camptotheca acuminate* (Xiao et al. 2006, Lin and Cordell 1989). An efficient Rh(III)-catalyzed intramolecular annulation of benzamide- **5** bearing tethered alkyne was developed as a key step in the synthesis of the indenoisoquinoline cores of rosettacin. The reaction was performed using $[Cp^*RhCl_2]_2$ (5 mol%) in the presence of $Cu(OAc)_2$ (2 equiv) and CsOAc (50 mol%) in *t*-AmOH at 110°C for 8 h, to afford compound **6** in 71% yield, which was converted to rosettacin **7** in treatment with tetra-*n*-butylammonium fluoride (TBAF) in tetrahydrofuran (THF), in 88% yield. A similar approach was also useful for the creation of an oxypalmatine natural product **9** in 66% yield, starting from benzamide **8** (Scheme 10.2) (Song et al. 2017). Rosettacin **7** was also synthesized in six steps, starting with Rh-catalyzed intramolecular dehydrogenative coupling of *N*-(4-pentynyl)oxybenzamide **10** using $[Cp^*RhCl_2]_2$ (1 mol%) in

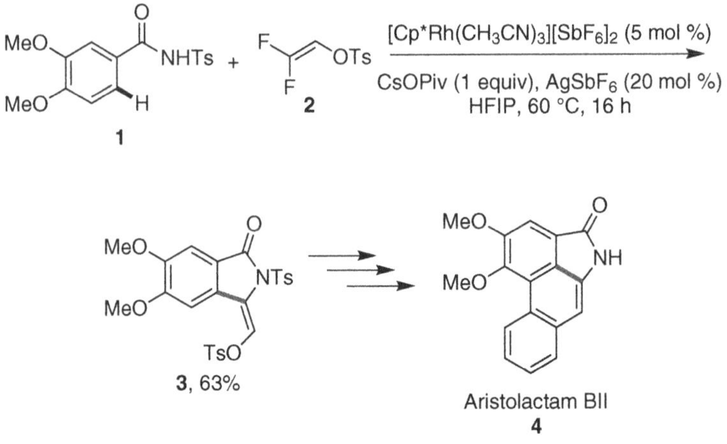
MeO
MeO
NHTs
H
1
F
F
OTs
2
[Cp*Rh(CH3CN)3][SbF6]2 (5 mol %)
CsOPiv (1 equiv), AgSbF6 (20 mol %)
HFIP, 60 °C, 16 h
NTs
TsO
3, 63%
NH
Aristolactam BII
4

SCHEME 10.1

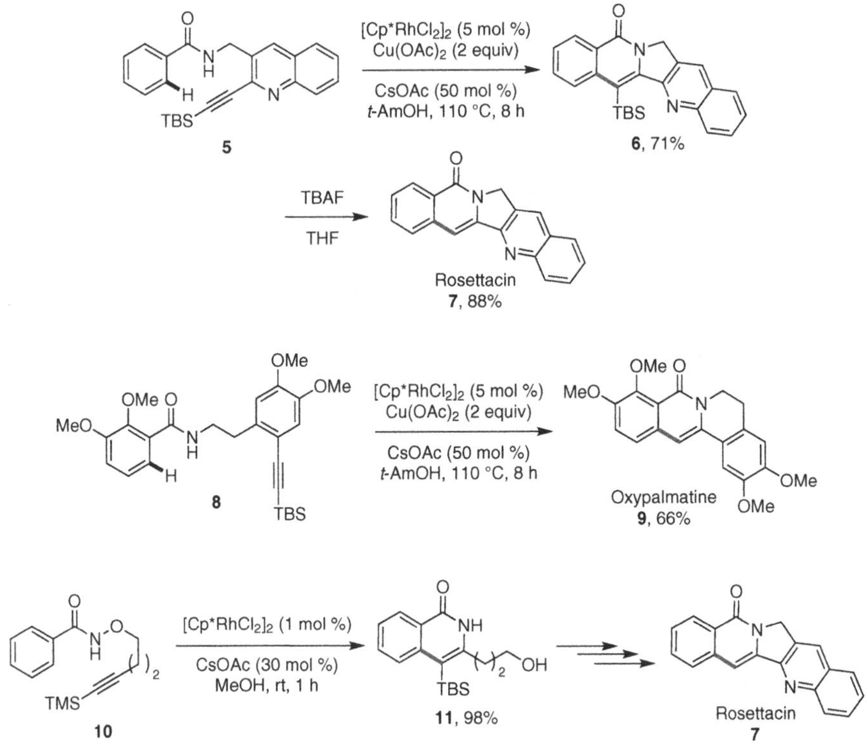
[Cp*RhCl2]2 (5 mol %)
Cu(OAc)2 (2 equiv)
CsOAc (50 mol %)
t-AmOH, 110 °C, 8 h
TBS
5
6, 71%
TBAF
THF
Rosettacin
7, 88%
OMe
MeO
8
Oxypalmatine
9, 66%
[Cp*RhCl2]2 (1 mol %)
CsOAc (30 mol %)
MeOH, rt, 1 h
TMS
10
NH
OH
11, 98%
Rosettacin
7

SCHEME 10.2

the presence of CsOAc (30 mol%) in MeOH at room temperature for 1 h, to give 3-hydroxypropylisoquinolin-1-one **11** in 98% yield. 3-Hydroxypropylisoquinolin-1-one **11** was transformed into rosettacin **7** in a further five steps in 59% overall yield (Scheme 10.2) (Xu, Liu, and Park 2012).

Calothrixin A (**14**) and B (**15**), indolo[3,2-*j*]phenanthridine alkaloids isolated from *Calothrix* cyanobacteria (Rickards et al. 1999), show potent antimalarial activity, inhibition of RNA polymerase, and DNA topoisomerase I poisoning activity (Bernardo et al. 2004, Khan, Lu, and Hecht 2009). In the Ramkumar and Nagarajan synthesis of calothrixin A and B, Pd(II)-catalyzed intramolecular cross-dehydrogenative coupling of *N*-(*ortho*-iodophenyl)carbazole carboxamides **12** is a key step of the synthesis. The dehydrogenative coupling process was conducted using $Pd(OAc)_2$ (5 mol%), KOAc (3 equiv), and Ag_2O (10 mol%) in *N,N*-dimethylformamide (DMF) at 110°C for 8 h, leading to indolophenanthridinone **13**, which was converted to calothrixin A and B in further steps (Scheme 10.3) (Ramkumar and Nagarajan 2013).

Synthesis of isoquinolinone alkaloid oxychelerythrine, exhibiting antitumor properties and the ability to stimulate glutathione (GSH) transport and inhibit the BclXL function, was described by Jayakumar, Parthasarathy, and Cheng (2012) through Rh(III)-catalyzed dehydrogenative coupling of benzaldehyde **16**, $MeNH_2$, and alkyne **17**. Reactions were performed using $[Cp^*RhCl_2]_2$ (2 mol%), $AgBF_4$ (1 equiv), and $Cu(OAc)_2$ (1 equiv) in *t*-AmOH at 110°C for 3 h, affording isoquinolinium salt **18** in 82% yield. Oxychelerythrine **19** was obtained from **18** in three steps by hydrolysis of the iminium moiety, followed by oxidation of alcohol with subsequent acid-catalyzed ring closing and a dehydration reaction (Scheme 10.4).

The pyrrolophenanthridine alkaloid oxoassoanine, isolated from various species of *Amarilidacea* (Meyers and Hutchings 1993, Grundon 1989), possessing reversible inhibitors to fertility in male rats and antitumor activity, was synthesized by cross-dehydrogenative coupling of *N*-homoallyl benzamide **20** with α,β-unsaturated ketoester **21**. Treating *N*-homoallyl benzamide **20** with α,β-unsaturated ketoester **21** in the presence of $[Cp^*RhCl_2]_2$ (5 mol%), $AgSbF_6$ (20 mol%), and AgOAc (20 mol%)

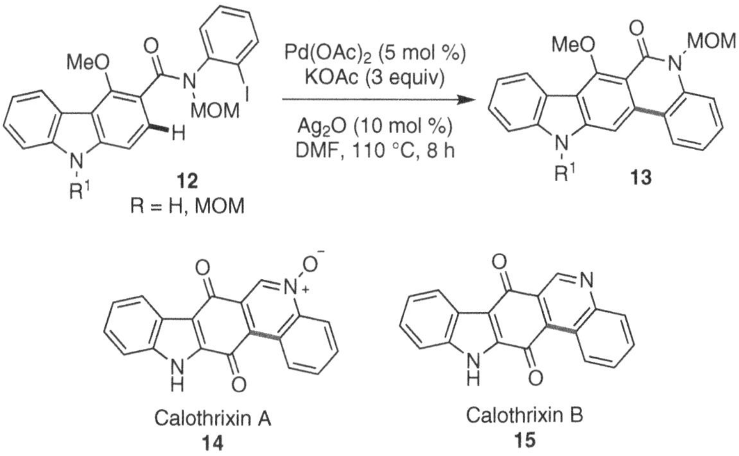

SCHEME 10.3

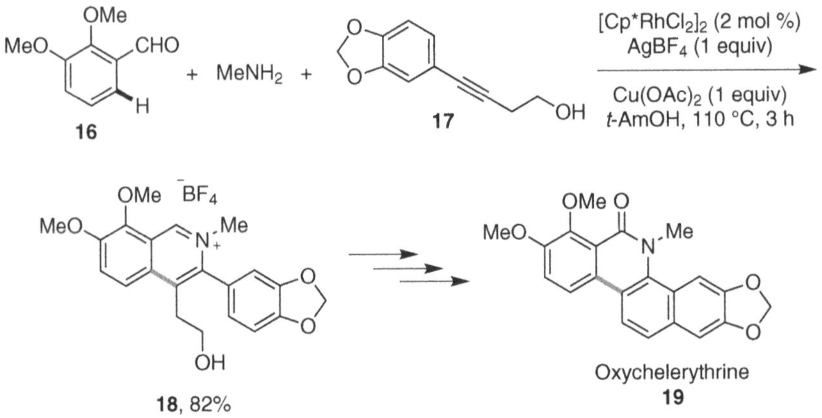

SCHEME 10.4

in 1,2-dichloroethane (DCE) at 20°C for 2.5 h and then with TFA (3 equiv) in toluene at 90°C for 15 h, followed by adding TFAA (6 equiv), led to isoquinolin-1,3-dion **22**, which was transformed to oxoassoanine **23** in three further steps (Scheme 10.5) (Weinstein and Ellman 2016).

A Pd(II)-catalyzed dehydrogenative coupling process was developed in order to synthesize the natural alkaloid trisphaeridine, isolated from *Hymenocallis* × festalis, exhibiting excellent antiproliferative effects (Zupko et al. 2009). Subjecting *N*-phenylmethanesulfonamide **24** to 5-bromo-6-(bromomethyl)benzo[*d*][1,3]dioxole **25** in the presence of $Pd(TFA)_2$ (5 mol%), PPh_3 (20 mol%), and Cs_2CO_3 (4 equiv) in DMF under N_2 at 160°C for 24 h led to trisphaeridine **26** in 55% yield (Han et al. 2015), which could be converted to the phenanthrene-type alkaloid dihydrobicolorine **27** (Scheme 10.6) (Chen et al. 2015).

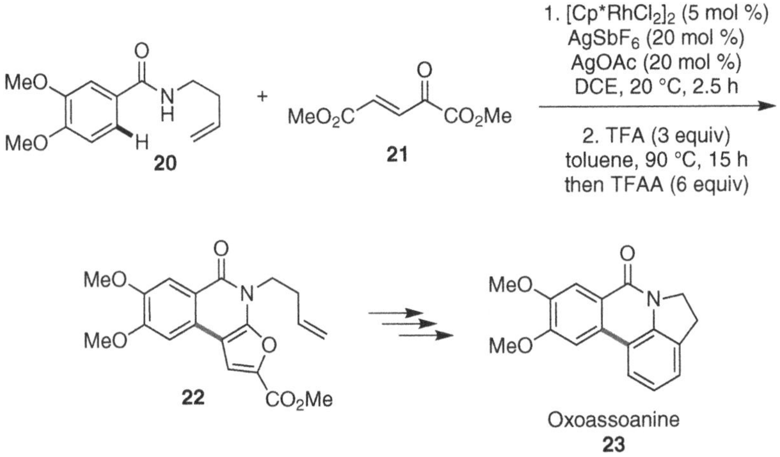

SCHEME 10.5

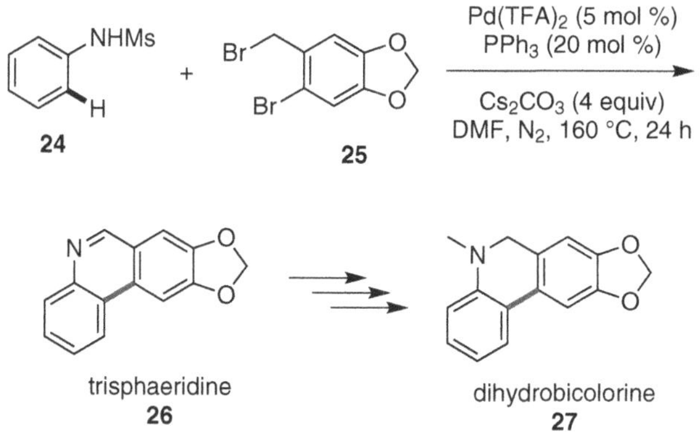

SCHEME 10.6

Cu-catalyzed *ortho*-C–H/N–H annulation of *N*-(5-methoxyquinoline-8-yl)benzamides **28** with benzyne precursor **29** was developed to synthesize the phenaglydon **31** and crinasiadine **32** alkaloids. Reactions were conducted using $Cu(OAc)_2$ (35 mol%), CsF (1.2 equiv), and tetrabutylammonium iodide (TBAI) (0.5 equiv) in DMF/MeCN at 80°C under O_2 for 12 h, leading to phenanthridinones **30**, which were converted to phenaglydon **31** and crinasiadine **32**, isolated from the lipophilic leaf extract of *Glycosmis cyanocarpa* (Rutaceae) (Wurz, Hofer, and Greger 1993) through the next two steps (Scheme 10.7) (Zhang et al. 2017b).

PJ-34 **35**, demonstrated as a cell-permeable poly ADP ribose polymerase (PARP) inhibitor, was prepared via Co-catalyzed C(sp^2)–H carbonylation of *ortho*-arylaniline **33** using diisopropyl azodicarboxylate as the CO source and oxygen as the sole oxidant. The reaction was performed in dioxane in the presence of $CoCl_2$ (30 mol%), diisopropyl azodicarboxylate (DIAD) (2 equiv), and NaOPiv (2 equiv) under O_2 atmosphere at 130°C, affording (*NH*)-phenanthridinone **34** in 95% yield, which was converted to PJ-34 **35** in a further three steps in 66% overall yield (Scheme 10.8) (Ling et al. 2018). Pd(II)-catalyzed C(sp^2)–H carbonylation of *ortho*-arylaniline **36** using CO and then conversion to PJ-34 **35** was also described (Liang et al. 2013).

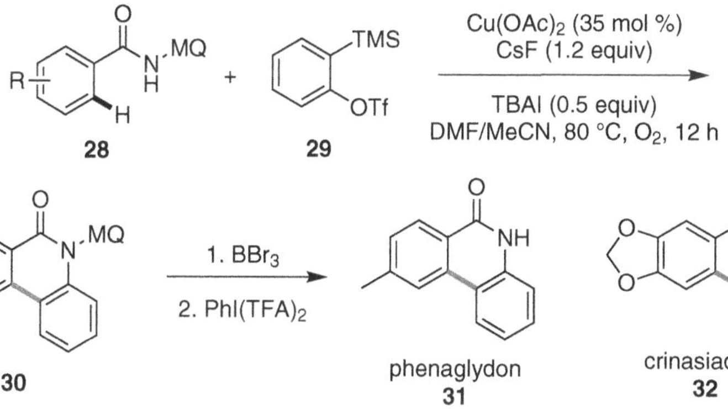

SCHEME 10.7

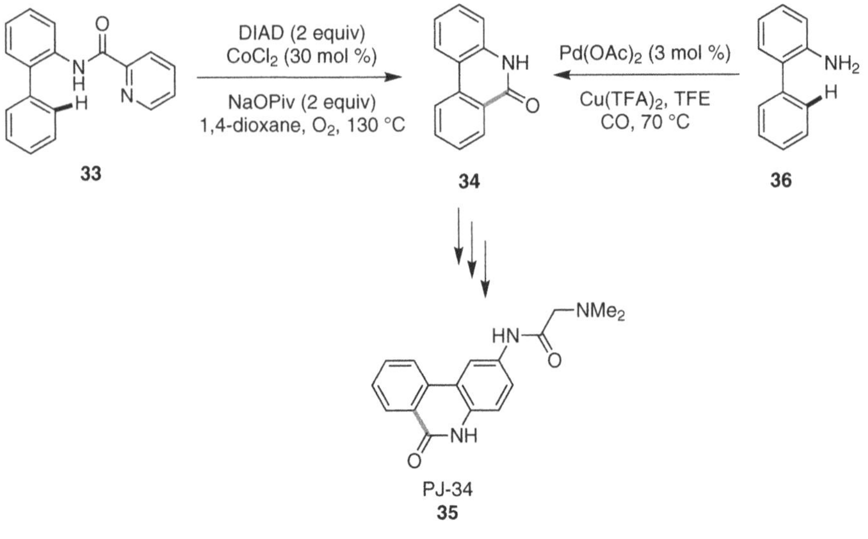

SCHEME 10.8

Isocryptolepine, isolated from the West African climbing shrub *Cryptolepis sanguinolenta* (Lindl), is used in Central and West Africa as traditional medicine for the treatment of malaria. Zhang et al. (2017a) reported the synthesis of isocryptolepine by Cu-mediated cascade of C–H/N–H annulation of indolocarboxamides with arynes. The reaction of 8-amino-5-methoxyquinoline (MQ)-protected indol-3-carboxamide **37** with benzyne precursor **29** was catalyzed with $Cu(OAc)_2$ (0.5 equiv) in the presence of CsF (1.2 equiv) and TBAI (1 equiv) in DMF/MeCN at 80°C under O_2 for 12 h, leading to indolo[3,2-*c*]quinoline **38**, which was transformed into the final isocryptolepine **39** in further steps. Synthetic analog isoneocryptolepine **41** could be also prepared by a similar approach starting from indol-2-carboxamide **40** (Scheme 10.9). Isocryptolepine **39** was also synthesized through intramolecular dehydrogenative coupling of *N*-phenyl indol-3-carboxamide **42** using $Pd(OAc)_2$ (20 mol%) and $Cu(OAc)_2$ (1 equiv) in AcOH at 120°C for 18 h, to furnish indoloquinoline **43** in 66% yield, which was converted to isocryptolepine **35** by further steps (Scheme 10.10) (Mahajan et al. 2016).

Tylophorine, a phenanthraindolizidine alkaloid isolated from *Tylophora indica* possessing antiangiogenic and antitumor activity (Saraswati et al. 2013), was prepared by intramolecular dehydrogenative radical cyclization of phenantroline carboxamide **44**. The reaction was carried out using $Pd(PPh_3)_4$ (10 mol%) in the presence of K_3PO_4 (2 equiv) in dioxane at 100°C for 24 h, affording phenanthraindolizidinone **45**, which then gave tylophorine **46** through reduction of the amide moiety using $LiAlH_4$ in THF at room temperature (Scheme 10.11) (Liu, Reimann, and Opatz 2016).

A homoprotoberberine alkaloid isolated from the roots of *Berberis actinucantha* (*Berberidaceae*) (Rahimizadeh 1996), was synthesized in a one-pot manner, involving C–H activation and cyclization, with subsequent Pictet–Spengler cyclization. The reaction between benzamide **47** and acrolein **48** in the presence of $[Cp^*RhCl_2]_2$

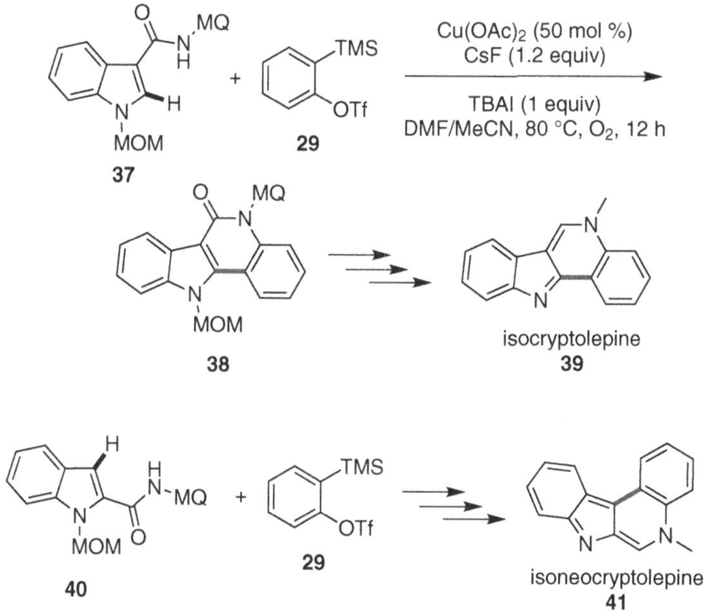

SCHEME 10.9

(2.5 mol%), $AgSbF_6$ (10 mol%), and PivOH (2 equiv) in dioxane at 60°C, under Ar atmosphere for 12 h, led to homoprotoberberine **49** in 65% yield (Scheme 10.12) (Shi, Grohmann, and Glorius 2013).

10.3 NATURAL PRODUCTS CONTAINING OXYGEN HETEROCYCLES

Guo et al. (2016) developed a synthesis of lithospermic acid, isolated from *Salvia miltiorrhiza* Bunge, starting by formation of a dihydrobenzofuran core via a Rh(III)-catalyzed intramolecular dehydrogenative coupling reaction. Treatment of imine of

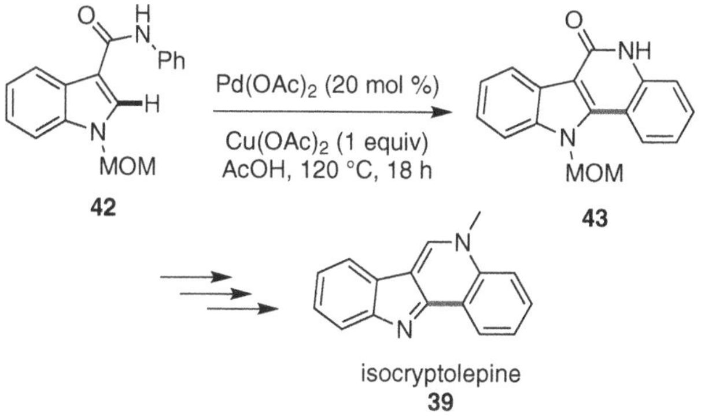

SCHEME 10.10

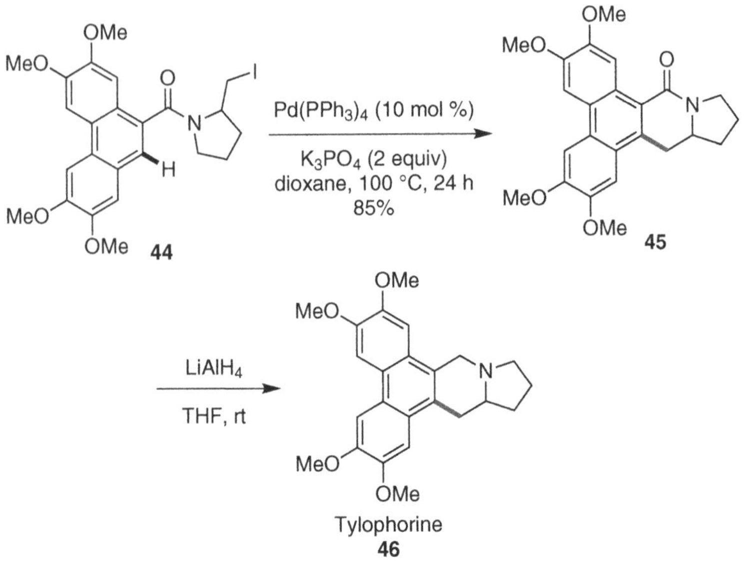

SCHEME 10.11

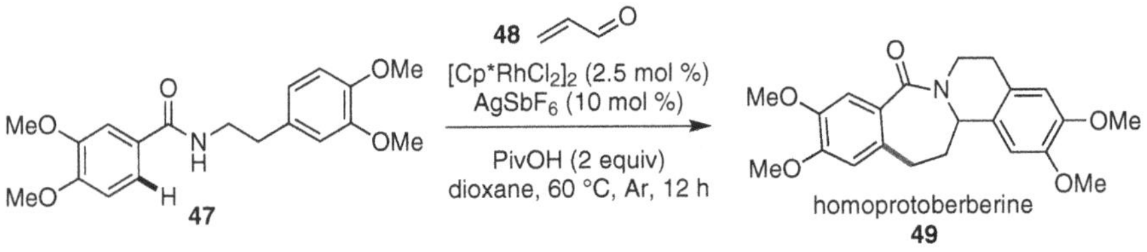

SCHEME 10.12

(*Z*)-methyl 3-(3,4-dimethoxyphenyl)-3-(5-formyl-2-methoxyphenoxy)acrylate **50** with $[RhCl(cod)_2]_2$ (10 mol%) and $FcPCy_2$ (30 mol%) in toluene at 75°C for 20 h, followed by hydrolysis of imine moiety, led to dihydrobenzofuran-3-carboxylate **51** in 56% yield with 99% ee after recrystallization, which was transformed into the final lithospermic acid **52** in a further two steps, in 68% yield (Scheme 10.13) (O'Malley et al. 2005).

Flemichapparin C, a coumestan, isolated from the roots of *Flemingia chappar* (Leguminosae; Lotoideae) (Adityachaudhury and Gupta 1973) was prepared by Pd(II)-catalyzed dehydrogenative coupling of 4-(benzo[*d*][1,3]dioxol-5-yloxy)-7-methoxy-2*H*-chromen-2-one **53**. The reaction was performed using $Pd(OAc)_2$ (10 mol%) in the presence of AgOAc (2 equiv), and CsOAc (2 equiv) in PivOH at 100°C for 15 h, which resulted in the formation of flemichapparin C **54** in 92% yield. Also, coumestrol **56** was synthesized by a similar approach, followed by hydrolysis of methyl ether moieties in **55** (Scheme 10.14) (Cheng et al. 2016). Preparation of flemichapparin C was developed by a similar methodology by another research group (Mackey et al. 2016).

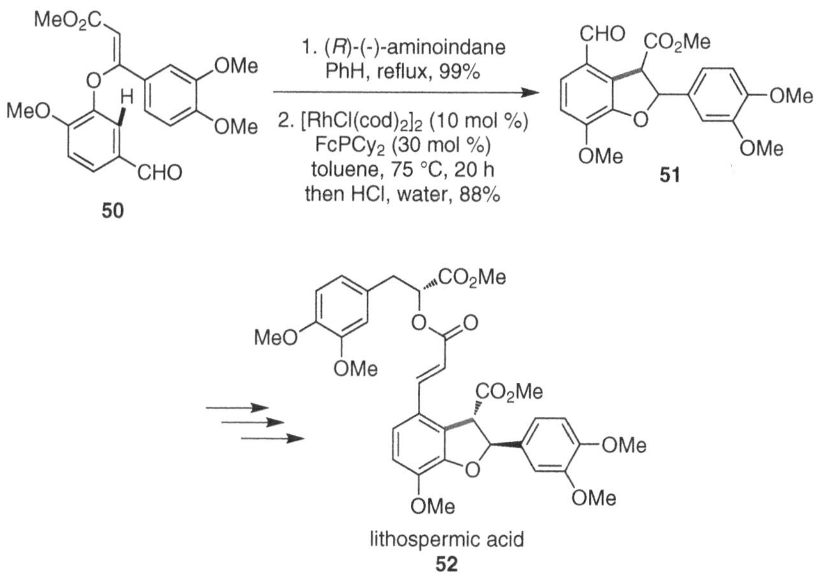

50

1. (*R*)-(-)-aminoindane
PhH, reflux, 99%

2. $[RhCl(cod)_2]_2$ (10 mol %)
$FcPCy_2$ (30 mol %)
toluene, 75 °C, 20 h
then HCl, water, 88%

51

lithospermic acid
52

SCHEME 10.13

The key step of the synthesis of cannabinol, obtained from the red oil of Minnesota wild hemp (Adams, Pease, and Clark 1940), involves the intramolecular dehydrogenative coupling of 3-((4-methylbenzyl)oxy)-5-pentylphenyl pyridine sulfonate **57**. The reaction was performed using $Pd(OAc)_2$ (10 mol%) in HFIP under O_2 atmosphere at 55°C, affording a 6*H*-benzo[*c*]chromene core **58** in 85% yield, which was converted to cannabinol **59** in four further steps (Scheme 10.15) (Guo et al. 2017).

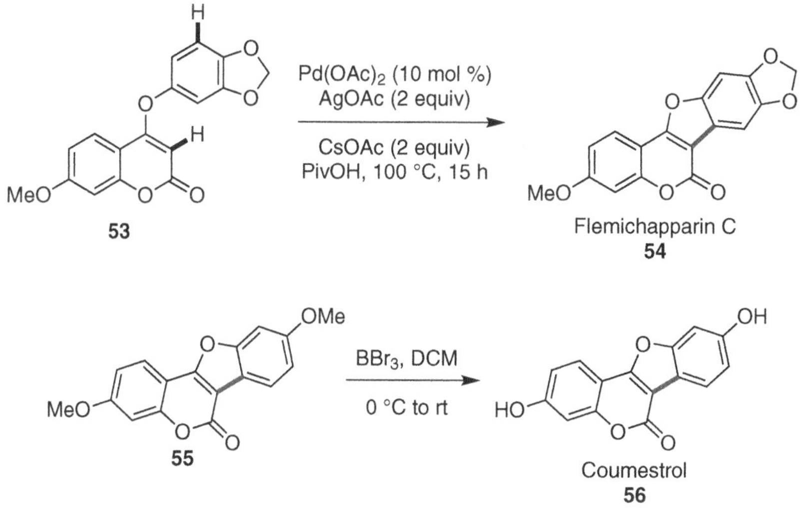

53

$Pd(OAc)_2$ (10 mol %)
AgOAc (2 equiv)

CsOAc (2 equiv)
PivOH, 100 °C, 15 h

Flemichapparin C
54

55

BBr_3, DCM

0 °C to rt

Coumestrol
56

SCHEME 10.14

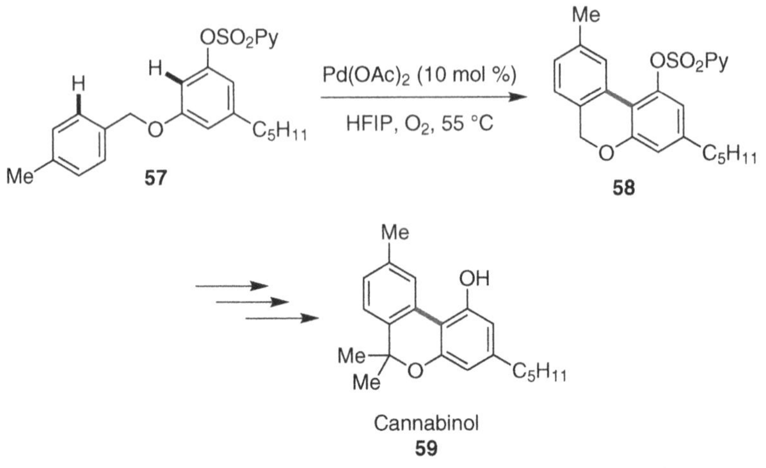

SCHEME 10.15

10.4 CONCLUSION

In summary, the aromatic C–H bond direct activation and functionalization are used for the synthesis of the heterocyclic core of natural products. The synthesis of the heterocyclic core in the structure of natural products, such as isoindole, indenoisoquinoline, indolophenanthridinone, isoquinoline, phenanthridinone, indoloquinoline, phenanthraindolizidinone, dihydrobenzofuran, and benzofurochromene, is described. These heterocyclic motifs were obtained in one step and high yields.

REFERENCES

Adams, Roger, DC Pease, and JH Clark. 1940. "Isolation of cannabinol, cannabidiol and quebrachitol from red oil of Minnesota wild hemp." *Journal of the American Chemical Society* no. 62(8):2194–2196.

Adityachaudhury, N, and PK Gupta. 1973. "A new pterocarpan and coumestan in the roots of Flemingia chappar." *Phytochemistry* no. 12(2):425–428.

Bernardo, Paul H, Christina LL Chai, Graham A Heath, Peter J Mahon, Geoffrey D Smith, Paul Waring, and Bronwyn A Wilkes. 2004. "Synthesis, electrochemistry, and bioactivity of the cyanobacterial calothrixins and related quinones." *Journal of Medicinal Chemistry* no. 47(20):4958–4963.

Chen, Wei-Lin, Chun-Yuan Chen, Yan-Fu Chen, and Jen-Chieh Hsieh. 2015. "Hydride-induced anionic cyclization: An efficient method for the synthesis of 6-H-phenanthridines via a transition-metal-free process." *Organic Letters* no. 17(6):1613–1616.

Cheng, Chao, Wen-Wen Chen, Bin Xu, and Ming-Hua Xu. 2016. "Intramolecular cross dehydrogenative coupling of 4-substituted coumarins: Rapid and efficient access to coumestans and indole [3,2-c] coumarins." *Organic Chemistry Frontiers* no. 3(9):1111–1115.

Fox, Brian M, Xiangshu Xiao, Smitha Antony, Glenda Kohlhagen, Yves Pommier, Bart L Staker, Lance Stewart, and Mark Cushman. 2003. "Design, synthesis, and biological evaluation of cytotoxic 11-alkenylindenoisoquinoline topoisomerase I inhibitors and indenoisoquinoline–camptothecin hybrids." *Journal of Medicinal Chemistry* no. 46(15):3275–3282.

Grundon, MF. 1989. "Indolizidine and quinolizidine alkaloids." *Natural Product Reports* no. 6(5):523–536.

Guo, Dong-Dong, Bin Li, Da-Yu Wang, Ya-Ru Gao, Shi-Huan Guo, Gao-Fei Pan, and Yong-Qiang Wang. 2017. "Synthesis of 6 H-benzo [c] chromenes via palladium-catalyzed intramolecular dehydrogenative coupling of two aryl C–H bonds." *Organic Letters* no. 19(4):798–801.

Guo, Yong Xue, Chang Zhi Shi, Lei Zhang, Lin Lv, and Yue Yong Zhang. 2016. "Extraction and isolation of lithospermic acid B from salvia miltiorrhiza Bunge using aqueous two-phase extraction followed by high-performance liquid chromatography." *Journal of Separation Science* no. 39(18):3624–3630.

Han, Wenyong, Xiaojian Zhou, Siyi Yang, Guangyan Xiang, Baodong Cui, and Yongzheng Chen. 2015. "Palladium-catalyzed nucleophilic substitution/C–H activation/aromatization cascade reaction: One approach to construct 6-unsubstituted phenanthridines." *The Journal of Organic Chemistry* no. 80(22):11580–11587.

Iqbal, Erum, Linda BL Lim, Kamariah Abu Salim, Shaheen Faizi, Ayaz Ahmed, and Abddalla Jama Mohamed. 2018. "Isolation and characterization of aristolactam alkaloids from the stem bark of Goniothalamus velutinus (Airy Shaw) and their biological activities." *Journal of King Saud University-Science* no. 30(1):41–48.

Jayakumar, Jayachandran, Kanniyappan Parthasarathy, and Chien-Hong Cheng. 2012. "One-pot synthesis of isoquinolinium salts by rhodium-catalyzed." *Angewandte Chemie International Edition* no. 51(1):197–200.

Ji, Wei-Wei, E Lin, Qingjiang Li, and Honggen Wang. 2017. "Heteroannulation enabled by a bimetallic Rh (iii)/Ag (I) relay catalysis: Application in the total synthesis of aristolactam BII." *Chemical Communications* no. 53(41):5665–5668.

Khan, Qasim A, Jun Lu, and Sidney M Hecht. 2009. "Calothrixins, a new class of human DNA topoisomerase I poisons." *Journal of Natural Products* no. 72(3):438–442.

Liang, Zunjun, Jitan Zhang, Zhanxiang Liu, Kai Wang, and Yuhong Zhang. 2013. "Pd (II)-catalyzed C (sp^2)–H carbonylation of biaryl-2-amine: Synthesis of phenanthridinones." *Tetrahedron* no. 69(31):6519–6526.

Lin, Long-Ze, and Geoffrey A Cordell. 1989. "Quinoline alkaloids from Camptotheca acuminata." *Phytochemistry* no. 28(4):1295–1297.

Ling, Fei, Chaowei Zhang, Chongren Ai, Yaping Lv, and Weihui Zhong. 2018. "Metal oxidant free cobalt catalyzed C (sp^2)–H carbonylation of ortho-arylanilines: An approach towards free (NH)-phenanthridinones." *The Journal of Organic Chemistry.*

Liu, Gong-Qing, Marcel Reimann, and Till Opatz. 2016. "Total synthesis of phenanthroindolizidine alkaloids by combining iodoaminocyclization with free radical cyclization." *The Journal of Organic Chemistry* no. 81(14):6142–6148.

Mackey, Katrina, Leticia M Pardo, Aisling M Prendergast, Marie-T Nolan, Lorraine M Bateman, and Gerard P McGlacken. 2016. "Cyclization of 4-phenoxy-2-coumarins and 2-pyrones via a double C–H activation." *Organic Letters* no. 18(11):2540–2543.

Mahajan, Pankaj S, Vivek T Humne, Subhash D Tanpure, and Santosh B Mhaske. 2016. "Radical Beckmann rearrangement and its application in the formal total synthesis of antimalarial natural product isocryptolepine via C–H activation." *Organic Letters* no. 18(14):3450–3453.

Meyers, AI, and RH Hutchings. 1993. "A total synthesis of the pyrrolophenthridone alkaloid oxoassoanine." *Tetrahedron Letters* no. 34(39):6185–6188.

O'Malley, Steven J, Kian L Tan, Anja Watzke, Robert G Bergman, and Jonathan A Ellman. 2005. "Total synthesis of (+)-lithospermic acid by asymmetric intramolecular alkylation via catalytic C–H bond activation." *Journal of the American Chemical Society* no. 127(39):13496–13497.

Rahimizadeh, M. 1996. "Isolation and structural elucidation of the first known C-homoprotoberberine." *Journal of Sciences, Islamic Republic of Iran* no. 7:172–176.

Ramkumar, Nagarajan, and Rajagopal Nagarajan. 2013. “Total synthesis of calothrixin A and B via C–H activation.” *The Journal of Organic Chemistry* no. 78(6):2802–2807.

Rickards, Rodney W, Jennifer M Rothschild, Anthony C Willis, Nola M de Chazal, Julie Kirk, Kiaran Kirk, Kevin J Saliba, and Geoffrey D Smith. 1999. “Calothrixins A and B, novel pentacyclic metabolites from Calothrix cyanobacteria with potent activity against malaria parasites and human cancer cells.” *Tetrahedron* no. 55(47):13513–13520.

Saraswati, Sarita, Pawan K Kanaujia, Shakti Kumar, Ranjeet Kumar, and Abdulqader A Alhaider. 2013. “Tylophorine, a phenanthraindolizidine alkaloid isolated from Tylophora indica exerts antiangiogenic and antitumor activity by targeting vascular endothelial growth factor receptor 2–mediated angiogenesis.” *Molecular Cancer* no. 12(1):82.

Shi, Zhuangzhi, Christoph Grohmann, and Frank Glorius. 2013. “Mild rhodium (III)-catalyzed cyclization of amides with α, β-unsaturated aldehydes and ketones to azepinones: Application to the synthesis of the homoprotoberberine framework.” *Angewandte Chemie International Edition* no. 52(20):5393–5397.

Song, Liangliang, Guilong Tian, Yi He, and Erik V Van der Eycken. 2017. “Rhodium (iii)-catalyzed intramolecular annulation through C–H activation: Concise synthesis of rosettacin and oxypalmatime.” *Chemical Communications* no. 53(92):12394–12397.

Weinstein, Adam B, and Jonathan A Ellman. 2016. “Convergent synthesis of diverse nitrogen heterocycles via Rh (III)-catalyzed C–H conjugate addition/cyclization reactions.” *Organic Letters* no. 18(13):3294–3297.

Wurz, Gerald, Otmar Hofer, and Harald Greger. 1993. “Structure and synthesis of phenaglydon, a new quinolone derived phenanthridine alkaloid from Glycosmis cyanocarpa.” *Natural Product Letters* no. 3(3):177–182.

Xiao, Xiangshu, Smitha Antony, Yves Pommier, and Mark Cushman. 2006. “Total synthesis and biological evaluation of 22-hydroxyacuminatine.” *Journal of Medicinal Chemistry* no. 49(4):1408–1412.

Xu, Xianxiu, Yu Liu, and Cheol-Min Park. 2012. “Rhodium (III)-catalyzed intramolecular annulation through C–H activation: Total synthesis of (±)-antofine,(±)-septicin e,(±)-tylophorine, and rosettacin.” *Angewandte Chemie International Edition* no. 51(37):9372–9376.

Zhang, Ting-Yu, Chang Liu, Chao Chen, Jian-Xin Liu, Heng-Ye Xiang, Wei Jiang, Tong-Mei Ding, and Shu-Yu Zhang. 2017a. “Copper-mediated cascade C–H/N–H annulation of Indolocarboxamides with arynes: Construction of tetracyclic indoloquinoline alkaloids.” *Organic Letters* no. 20(1):220–223.

Zhang, Ting-Yu, Jun-Bing Lin, Quan-Zhe Li, Jun-Chen Kang, Jin-Long Pan, Si-Hua Hou, Chao Chen, and Shu-Yu Zhang. 2017b. “Copper-catalyzed Selective ortho-C–H/N–H annulation of benzamides with arynes: Synthesis of phenanthridinone alkaloids.” *Organic Letters* no. 19(7):1764–1767.

Zhang, Yi-Nan, Xiang-Gen Zhong, Zong-Ping Zheng, Xu-Dong Hu, Jian-Ping Zuo, and Li-Hong Hu. 2007. “Discovery and synthesis of new immunosuppressive alkaloids from the stem of Fissistigma oldhamii (Hemsl.) Merr.” *Bioorganic & Medicinal Chemistry* no. 15(2):988–996.

Zupko, Istvan, Borbala Rethy, Judit Hohmann, Joseph Molnar, Imre Ocsovszki, and George Falkay. 2009. “Antitumor activity of alkaloids derived from Amaryllidaceae species.” *In Vivo* no. 23(1):41–48.

Index

Printed in the United States
by Baker & Taylor Publisher Services